NONLINEAR INVERSE PROBLEMS IN IMAGING

NONLINEAR INVERSE PROBLEMS IN IMAGING

Jin Keun Seo
Yonsei University, Korea

Eung Je Woo
Kyung Hee University, Korea

WILEY
A John Wiley & Sons, Ltd., Publication

This edition first published 2013

Registered office
John Wiley & Sons Ltd, The Atrium, Southern Gate, Chichester, West Sussex, PO19 8SQ, United Kingdom

For details of our global editorial offices, for customer services and for information about how to apply for permission to reuse the copyright material in this book please see our website at www.wiley.com.

Library of Congress Cataloging-in-Publication Data

Seo, Jin Keun.
Nonlinear inverse problems in imaging / Jin Keun Seo and Eung Je Woo.
pages cm
Includes bibliographical references and index.
ISBN 978-0-470-66942-6 (hardback)
1. Image processing–Mathematics. 2. Cross-sectional imaging–Mathematics. 3. Inverse problems (Differential equations) 4. Nonlinear theories. I. Woo, E. J. (Eung Je) II. Title.
TA1637.S375 2013
621.36′70151–dc23

2012031509

A catalogue record for this book is available from the British Library.

Print ISBN: 978-0-470-66942-6

Typeset in 10/12pt Times by Laserwords Private Limited, Chennai, India

Printed and bound in Singapore by Markono Print Media Pte Ltd

Contents

The methods mainly comprise mathematical and numerical tools to solve the problems. Instrumentation will be treated only in enough detail to describe practical limitations imposed by measurement methods.

Readers will acquire the diverse knowledge and skills needed to deal effectively with nonlinear inverse problems in imaging by following the steps below.

1. Understand the underlying physical phenomena and the constraints imposed on the problem, which may enable solutions of nonlinear inverse problems to be improved. Physics, chemistry and also biology play crucial roles here. No attempt is made to be comprehensive in terms of physics, chemistry and biology.
2. Understand forward problems, which usually are the processes of information loss. They provide strategic insights into seeking solutions of nonlinear inverse problems. The underlying principles are described here so that readers can understand their mathematical formulations.
3. Formulate forward problems in such a way that they can be dealt with systematically and quantitatively.
4. Understand how to probe the imaging object and what is measurable using available engineering techniques. Practical limitations associated with the measurement sensitivity and specificity, such as noise, artifacts, interface between target object and instrument, data acquisition time and so on, must be properly understood and analyzed.
5. Understand what is feasible in a specific nonlinear inverse problem.
6. Formulate proper nonlinear inverse problems by defining the image contrast associated with physical quantities. Mathematical formulations should include any interrelation between those qualities and measurable data.
7. Construct inversion methods to produce images of contrast information.
8. Develop computer programs and properly address critical issues of numerical analysis.
9. Customize the inversion process by including *a priori* information.
10. Validate results by simulations and experiments.

This book is for advanced graduate courses in applied mathematics and engineering. Prerequisites for students with a mathematical background are vector calculus, linear algebra, partial differential equations and numerical analysis. For students with an engineering background, we recommend taking linear algebra, numerical analysis, electromagnetism, signal and system and also preferably instrumentation.

Lecture notes, sample codes, experimental data and other teaching material are available at http://mimaging.yonsei.ac.kr/NIPI.

Preface

Imaging techniques in science, engineering and medicine have evolved to expand our ability to visualize the internal information in an object such as the human body. Examples may include X-ray computed tomography (CT), magnetic resonance imaging (MRI), ultrasound imaging and positron emission tomography (PET). They provide cross-sectional images of the human body, which are solutions of corresponding inverse problems. Information embedded in such an image depends on the underlying physical principle, which is described in its forward problem. Since each imaging modality has limited viewing capability, there have been numerous research efforts to develop new techniques producing additional contrast information not available from existing methods.

There are such imaging techniques of practical significance, which can be formulated as nonlinear inverse problems. Electrical impedance tomography (EIT), magnetic induction tomography (MIT), diffuse optical tomography (DOT), magnetic resonance electrical impedance tomography (MREIT), magnetic resonance electrical property tomography (MREPT), magnetic resonance elastography (MRE), electrical source imaging and others have been developed and adopted in application areas where new contrast information is in demand. Unlike X-ray CT, MRI and PET, they manifest some nonlinearity, which result in their image reconstruction processes being represented by nonlinear inverse problems.

Visualizing new contrast information on the electrical, optical and mechanical properties of materials inside an object will widen the applications of imaging methods in medicine, biotechnology, non-destructive testing, geophysical exploration, monitoring of industrial processes and other areas. Some are advantageous in terms of non-invasiveness, portability, convenience of use, high temporal resolution, choice of dimensional scale and total cost. Others may offer a higher spatial resolution, sacrificing some of these merits.

Owing primarily to nonlinearity and low sensitivity, in addition to the lack of sufficient information to solve an inverse problem in general, these nonlinear inverse problems share the technical difficulties of ill-posedness, which may result in images with a low spatial resolution. Deep understanding of the underlying physical phenomena as well as the implementation details of image reconstruction algorithms are prerequisites for finding solutions with practical significance and value.

Research outcomes during the past three decades have accumulated enough knowledge and experience that we can deal with these topics in graduate programs of applied mathematics and engineering. This book covers nonlinear inverse problems associated with some of these imaging modalities. It focuses on methods rather than applications.

List of Abbreviations

General Notation

$\mathbb{R}$: the set of real numbers
$\mathbb{R}^n$: n-dimensional Euclidean space
$\mathbb{C}$: the set of complex numbers
$\mathrm{i} = \sqrt{-1}$
$\mathbb{N}_0 = \{0, 1, 2, \ldots\}$: the set of non-negative integers
$\mathbb{N} = \{1, 2, \ldots\}$: the set of positive integers
$\mathbf{r} = (x, y, z)$: position
$B_r(\mathbf{a})$: the ball with radius r at the center $\mathbf{a}$ and $B_r = B_r(\mathbf{0})$
$\mathbf{e}_j$: jth unit vector in $\mathbb{R}^n$, for example, $\mathbf{e}_2 = (0, 1, 0, \ldots, 0)$
$\forall$:= for all
$\exists$:= there exist(s)
$\because$:= because
$\therefore$:= therefore

Electromagnetism

$\mathbf{E}$: electric field intensity
$\mathbf{B}$: magnetic flux density
$\mathbf{J}$: current density
$\mathbf{D}$: electric flux density
σ : electrical conductivity
ϵ : electrical permittivity
$\gamma = \sigma + \mathrm{i}\omega\epsilon$: admittivity
u : voltage (electrical potential)

Notations for Domains and Vector Spaces

Ω : a domain in $\mathbb{R}^n$

$\partial\Omega$: the boundary of the domain Ω

$\mathbf{n}$: the unit outward normal vector to the boundary

$C(\Omega)$: the set of all continuous functions in Ω

$C^k(\Omega)$: the set of continuously kth differentiable functions defined in the domain Ω

1

Introduction

We consider a physical system where variables and parameters interact in a domain of interest. Variables are physical quantities that are observable or measurable, and their values change with position and time to form signals. We may express system structures and properties as parameters, which may also change with position and time. For a given system, we understand its dynamics based on underlying physical principles describing the interactions among the variables and parameters. We adopt mathematical tools to express the interactions in a manageable way.

A physical excitation to the system is an input and its response is an output. The response is always accompanied by some form of energy transfer. The input can be applied to the system externally or internally. For the internal input, we may also use the term "source". Observations or measurements can be done at the outside, on the boundary and also on the inside of the system. For the simplicity of descriptions, we will consider boundary measurements as external measurements. Using the concept of the generalized system, we will introduce the forward and inverse problems of a physical system.

1.1 Forward Problem

The generalized system H in Figure 1.1 has a system parameter p, input x and output y. We first need to understand how they are entangled in the system by understanding underlying physical principles. A mathematical representation of the system dynamics is the forward problem formulation. We formulate a forward problem of the system in Figure 1.1 as

$$y = H(p, x), \tag{1.1}$$

where H is a nonlinear or linear function of p and x. We should note that the expression in (1.1) may not be feasible in some cases where the relation between the input and output can only be described implicitly.

To treat the problem in a computationally manageable way, we should choose core variables and parameters of most useful and meaningful information to quantify their interrelations. The expression in (1.1), therefore, could be an approximation of complicated

Nonlinear Inverse Problems in Imaging, First Edition. Jin Keun Seo and Eung Je Woo.

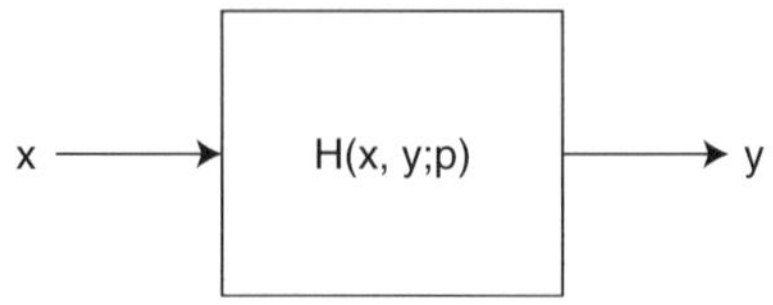

Figure 1.1 Forward problem for a system with parameter p, input x and output y

interactions among variables and parameters. In practice, we may not be able to control the input precisely because of technical limitations, and the measured output will always be contaminated by noise. Solving a forward problem is to find the output from a given input and system parameter. Evaluation of (1.1) suffices for its solution.

A simple example is a sound recording and reproduction system including a microphone, amplifier and speaker. An input sound wave enters the system through the microphone, goes through the amplifier and exits the system as an output sound wave through the speaker. The system characteristics are determined by the electrical and mechanical properties of the system components, including the gain or amplitude amplification factor, phase change, frequency bandwidth, power and so on.

Example 1.1.1 *The output signal $y(t)$ at time t of a system H is found to be $y(t) = Kx(t-\tau)$, where $x(t)$ is the input signal, K is a fixed gain and τ is a fixed time delay. Taking the Fourier transform of both sides, we can find the frequency transfer function $H(\mathrm{i}\omega)$ of the system to be*

$$H(\mathrm{i}\omega) = \frac{\mathcal{F}\{y(t)\}}{\mathcal{F}\{x(t)\}} = \frac{K\mathrm{e}^{-\mathrm{i}\omega\tau}X(\mathrm{i}\omega)}{X(\mathrm{i}\omega)} = K\mathrm{e}^{-\mathrm{i}\omega\tau}, \tag{1.2}$$

where $\mathrm{i} = \sqrt{-1}$ and ω is the angular frequency. This means that, for all ω,

$$|H(\mathrm{i}\omega)| = K \qquad (\textit{flat magnitude response}), \tag{1.3}$$

$$\theta(\mathrm{i}\omega) = \angle H(\mathrm{i}\omega) = -\tau\omega \qquad (\textit{linear phase response}). \tag{1.4}$$

When (1.3) and (1.4) are satisfied within a frequency range of the input signal, the output is a time-delayed amplified version of the input signal without any distortion. For a sinusoidal input signal

$$x(t) = A\cos(\omega t + \theta),$$

the corresponding output signal is

$$y(t) = KA\cos(\omega t + \theta - \tau\omega) = KA\cos\{\omega(t-\tau) + \theta\}.$$

When the input signal is a sum of many sinusoids with different frequencies, that is,

$$x(t) = \sum_{j=1}^{n} A_j\cos(\omega_j t + \theta_j),$$

the corresponding output signal is

$$y(t) = K \sum_{j=1}^{n} A_j \cos(\omega_j t + \theta_j - \tau\omega_j) = K \sum_{j=1}^{n} A_j \cos\{\omega_j (t - \tau) + \theta_j\}.$$

This system with parameters K and τ is a linear system with no distortion. Given the forward problem expressed as $y(t) = Kx(t - \tau)$ with known values of K and τ, we can find the output y for any given input x.

In most physical systems, inputs are mixed within the system to produce outputs. The mixing process is accompanied by smearing of information embedded in the inputs. Distinct features of the inputs may disappear in the outputs, and the effects of the system parameters may spread out in the observed outputs.

Exercise 1.1.2 *For a continuous-time linear time-invariant system with impulse response $h(t)$, where t is time, find the expression for the output $y(t)$ corresponding to the input $x(t)$.*

Exercise 1.1.3 *For a discrete-time linear time-invariant system with impulse response $h[n]$, where n is time, find the expression for the output $y[n]$ corresponding to the input $x[n]$.*

1.2 Inverse Problem

For a given forward problem, we may consider two types of related inverse problems as in Figure 1.2. The first type is to find the input from a measured output and identified system parameter. The second is to find the system parameter from a designed input and measured output. We symbolically express these two cases as follows:

$$x = H_1^+(p, y) \tag{1.5}$$

and

$$p = H_2^+(x, y), \tag{1.6}$$

where H_1^+ and H_2^+ are nonlinear or linear functions. We may need to design multiple inputs carefully to get multiple input–output pairs with enough information to solve the inverse problems.

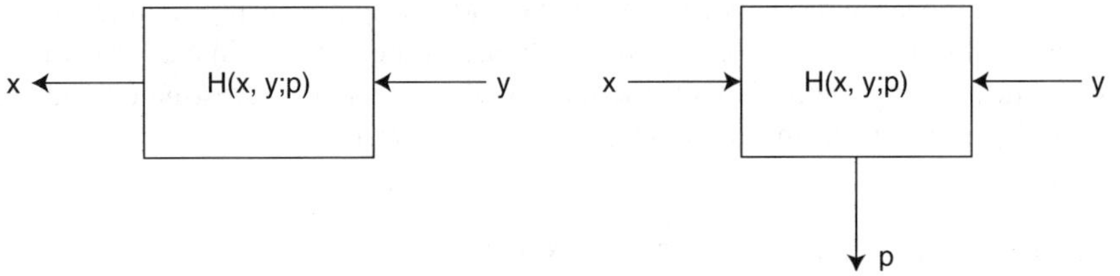

Figure 1.2 Two different inverse problems for a system with parameter p, input x and output y

Example 1.2.1 *We consider the linear system in Example 1.1.1 and assume that we have measured the output* $y(t)$ *subject to the sinusoidal input* $x(t) = A\cos(\omega t + \theta)$ *with unknown* A *and* θ. *We can find the amplitude and phase of the sinusoidal input,* A *and* θ, *respectively, by performing the following phase-sensitive demodulation process. For the in-phase channel,*

$$Y_{\mathrm{I}} = \frac{1}{T}\int_{t_0}^{t_0+T} y(t)\cos\omega t\,\mathrm{d}t = \frac{KA}{2}\cos(\theta - \tau\omega), \tag{1.7}$$

where $T = 2\pi/\omega$ *is the period of the sinusoid and* t_0 *is an arbitrary time. For the quadrature channel,*

$$Y_{\mathrm{Q}} = \frac{1}{T}\int_{t_0}^{t_0+T} y(t)\sin\omega t\,\mathrm{d}t = \frac{KA}{2}\sin(\theta - \tau\omega). \tag{1.8}$$

We recover A *and* θ *as*

$$A = \frac{\sqrt{4Y_{\mathrm{I}}^2 + 4Y_{\mathrm{Q}}^2}}{K} \quad \text{and} \quad \theta = \tan^{-1}\frac{Y_{\mathrm{Q}}}{Y_{\mathrm{I}}} + \tau\omega, \tag{1.9}$$

assuming that we know the system parameters K *and* τ.

Exercise 1.2.2 *Assume that we have measured the output* $y(t)$ *of the linear system in Example 1.1.1 for the known sinusoidal input* $x(t)$. *Find the system parameters: the gain* K *and the delay* τ.

Exercise 1.2.3 *Consider a discrete-time linear time-invariant system with impulse response* $h[n]$, *where* n *is time. We have measured its output* $y[n]$ *subject to the known input* $x[n]$. *Discuss how to find* $h[n]$.

In general, most inverse problems are complicated, since the dynamics among inputs, outputs and system parameters are attributed to complex, possibly nonlinear, physical phenomena. Within a given measurement condition, multiple inputs may result in the same output for given system parameters. Similarly, different system parameters may produce the same input–output relation. The inversion process, therefore, suffers from the uncertainty that originates from the mixing process of the corresponding forward problem.

To seek a solution of an inverse problem, we first need to understand how those factors are entangled in the system by understanding the underlying physical principles. Extracting core variables of most useful information, we should properly formulate a forward problem to quantify their interrelations. This is the reason why we should investigate the associated forward problem before trying to solve the inverse problem.

1.3 Issues in Inverse Problem Solving

In solving an inverse problem, we should consider several factors. First, we have to make sure that there exists at least one solution. This is the issue of the existence of a solution, which must be checked in the formulation of the inverse problem. In practice, it may not be a serious question, since the existence is obvious as long as the system deals with physically existing or observable quantities. Second is the uniqueness of a solution. This is a more serious issue in both theoretical and practical aspects, and finding a unique solution of an inverse problem requires careful analyses of the corresponding forward and inverse problems. If a solution is not unique, we must check its optimality in terms of its physical meaning and practical usefulness.

To formulate a manageable problem dealing with key information, we often go through a simplification process and sacrifice some physical details. Mathematical formulations of the forward and inverse problems, therefore, suffer from modeling errors. In practice, measured data always include noise and artifacts. To acquire a quantitative numerical solution of the inverse problem, we deal with discretized versions of the forward and inverse problems. The discretization process may add noise and artifacts. We must carefully investigate the effects of these practical restrictions in the context of the existence and uniqueness of a solution.

We introduce the concept of well-posedness as proposed by Hadamard (1902). When we construct a mathematical model of a system to transform the associated physical phenomena into a collection of mathematical expressions and data, we should consider the following three properties.

1. Existence: at least one solution exists.
2. Uniqueness: only one solution exists.
3. Continuity: a solution depends continuously on the data.

In the sense of Hadamard, a problem is well-posed when it meets the above requirements of existence, uniqueness and continuity. If these requirements are not met, the problem is ill-posed.

If we can properly formulate the forward problem of a physical system and also its inverse problem, we can safely assume that a solution exists. Non-uniqueness often becomes a practically important issue, since it is closely related with the inherent mixing process of the forward problem. Once the inputs are mixed, uniquely sorting out some inputs and system parameters may not be feasible. The mixing process may also cause sensitivity problems. When the sensitivity of a certain output to the inputs and/or system parameters is low, small changes in the inputs or system parameters may result in small and possibly discontinuous changes in the output, with measurement errors. The inversion process in general includes a step where the measured output values are divided by sensitivity factors. If we divide small measured values, including errors, by a small sensitivity factor, we may amplify the errors in the results. The effects of the amplified errors may easily dominate the inversion process and result in useless solutions, which do not comply with the continuity requirement.

Considering that mixing processes are embedded in most forward problems and that the related inverse problems are ill-posed in many cases, we need to devise effective methods to deal with such difficulties. One may incorporate as much *a priori* information as possible in the inversion process. Preprocessing methods such as denoising and feature extraction can be employed. One may also need to implement some regularization techniques to find a compromise between the robustness of an inversion method and the accuracy or sharpness of its solution.

1.4 Linear, Nonlinear and Linearized Problems

Linearity is one of the most desirable features in solving an inverse problem. We should carefully check whether the forward and inverse problems are linear or not. If not, we may try to approximately linearize the problems in some cases. We first define the linearity of the forward problem in Figure 1.1 as follows.

1. Homogeneity: if $y_1 = H(x)$, then $y_2 = H(Kx) = KH(x) = Ky_1$ for any constant K.
2. Additivity: if $y_1 = H(x_1)$ and $y_2 = H(x_2)$, then $y_3 = H(x_1 + x_2) = H(x_1) + H(x_2) = y_1 + y_2$.

For a linear system, we can, therefore, apply the following principle of superposition: if $y_1 = H(x_1)$ and $y_2 = H(x_2)$, then

$$y_3 = H(K_1x_1 + K_2x_2) = K_1y_1 + K_2y_2 \tag{1.10}$$

for any constants K_1 and K_2.

For the inverse problems in Figure 1.2, we may similarly define the linearity for two functions H_1^+ and H_2^+. Note that we should separately check the linearity of the three functions H, H_1^+ and H_2^+. Any problem, either forward or inverse, is nonlinear if it does not satisfy both of the homogeneity and additivity requirements.

Example 1.4.1 *Examples of nonlinear systems are*

1. $y = Kx^2$,
2. $y = K_1e^{K_2x}$.

For a nonlinear problem $y = H(x)$, we may fix $x = x_0$ and consider a small change Δx around x_0. As illustrated in Figure 1.3, we can approximate the corresponding change Δy as

$$\Delta y = H(x_0 + \Delta x) - H(x_0) \approx \partial_x H(x)|_{x0}\Delta x = S_{x0}\Delta x, \tag{1.11}$$

where S_{x_0} is the sensitivity of Δy to Δx at $x = x_0$, which can be found from the analysis of the problem. The approximation in (1.11) is called the linearization to find $y_1 = H(x_0 + \Delta x) \approx y_0 + \Delta y$ where $y_0 = H(x_0)$. The approximation is accurate only for a small Δx.

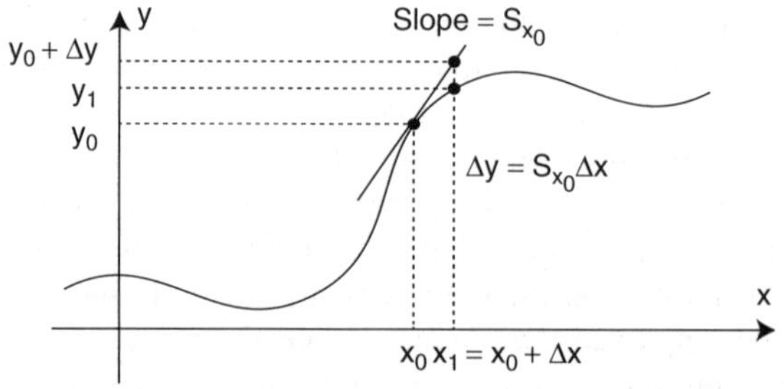

Figure 1.3 Illustration of a linearization process

Problems either forward or inverse can be linear or nonlinear depending on the underlying physical principles. Proper formulations of forward and inverse problems using mathematical tools are essential steps before any attempt to seek solution methods. In the early chapters of the book, we study mathematical backgrounds to deal with linear and nonlinear problems. In the later chapters, we will introduce several imaging modalities.

Reference

Hadamard J 1902 Sur les problèmes aux dérivées partielles et leur signification physique. *Bull. Univ. Princeton*, **13**, 49–52.

2

Signal and System as Vectors

To solve forward and inverse problems, we need mathematical tools to formulate them. Since it is convenient to use vectors to represent multiple system parameters, variables, inputs (or excitations) and outputs (or measurements), we first study vector spaces. Then, we introduce vector calculus to express interrelations among them based on the underlying physical principles. To solve a nonlinear inverse problem, we often linearize an associated forward problem to utilize the numerous mathematical tools of the linear system. After introducing such an approximation method, we review mathematical techniques to deal with a linear system of equations and linear transformations.

2.1 Vector Spaces

We denote a signal, variable or system parameter as $f(\mathbf{r}, t)$, which is a function of position $\mathbf{r} = (x, y, z)$ and time t. To deal with a number n of signals, we adopt the vector notation $(f_1(\mathbf{r}, t), \ldots, f_n(\mathbf{r}, t))$. We may also set vectors as $(f(\mathbf{r}_1, t), \ldots, f(\mathbf{r}_n, t))$, $(f(\mathbf{r}, t_1), \ldots, f(\mathbf{r}, t_n))$ and so on. We consider a set of all possible such vectors as a subset of a vector space. In the vector space framework, we can add and subtract vectors and multiply vectors by numbers. Establishing the concept of a subspace, we can project a vector into a subspace to extract core information or to eliminate unnecessary information. To analyze a vector, we may decompose it as a linear combination of basic elements, which we handle as a basis or coordinate of a subspace.

2.1.1 Vector Space and Subspace

Definition 2.1.1 *A non-empty set V is a vector space over a field $\mathbb{F} = \mathbb{R}$ or $\mathbb{C}$ if there are operations of vector addition and scalar multiplication with the following properties.*

Vector addition

1. $\mathbf{u} + \mathbf{v} \in V$ *for every* $\mathbf{u}, \mathbf{v} \in V$ *(closure).*
2. $\mathbf{u} + \mathbf{v} = \mathbf{v} + \mathbf{u}$ *for every* $\mathbf{u}, \mathbf{v} \in V$ *(commutative law).*

Nonlinear Inverse Problems in Imaging, First Edition. Jin Keun Seo and Eung Je Woo.

3. $(\mathbf{u} + \mathbf{v}) + \mathbf{w} = \mathbf{u} + (\mathbf{v} + \mathbf{w})$ *for every* $\mathbf{u}, \mathbf{v}, \mathbf{w} \in V$ *(associative law).*
4. *There exist* $\mathbf{0} \in V$ *such that* $\mathbf{u} + \mathbf{0} = \mathbf{u}$ *for every* $\mathbf{u} \in V$ *(additive identity).*
5. *For all* $\mathbf{u} \in V$ *there exists* $-\mathbf{u} \in V$ *such that* $\mathbf{u} + (-\mathbf{u}) = \mathbf{0}$ *and* $-\mathbf{u}$ *is unique (additive inverse).*

Scalar multiplication

1. *For* $a \in \mathbb{F}$ *and* $\mathbf{u} \in V$, $a\mathbf{u} \in V$ *(closure).*
2. *For* $a, b \in \mathbb{F}$ *and* $\mathbf{u}, \mathbf{v} \in V$, $a(b\mathbf{u}) = (ab)\mathbf{u}$ *(associative law).*
3. *For* $a \in \mathbb{F}$ *and* $\mathbf{u}, \mathbf{v} \in V$, $a(\mathbf{u} + \mathbf{v}) = a\mathbf{u} + a\mathbf{v}$ *(first distributive law).*
4. *For* $a, b \in \mathbb{F}$ *and* $\mathbf{u} \in V$, $(a + b)\mathbf{u} = a\mathbf{u} + b\mathbf{u}$ *(second distributive law).*
5. $1\mathbf{u} = \mathbf{u}$ *for every* $\mathbf{u} \in V$ *(multiplicative identity).*

A subset W of a vector space V over F is a *subspace* of V if and only if $a\mathbf{u} + \mathbf{v} \in W$ for all $a \in \mathbb{F}$ and for all $\mathbf{u}, \mathbf{v} \in W$. The subspace W itself is a vector space.

Example 2.1.2 *The following are examples of vector spaces.*

- $\mathbb{R}^n = \{(x_1, x_2, \ldots, x_n) : x_1, \ldots, x_n \in \mathbb{R}\}$, *n-dimensional Euclidean space.*
- $\mathbb{C}^n = \{(x_1, x_2, \ldots, x_n) : x_1, \ldots, x_n \in \mathbb{C}\}$.
- $C([a, b])$, *the set of all complex-valued functions that are continuous on the interval* $[a, b]$.
- $C^1([a, b]) := \{f \in C([a, b]) : f' \in C[a, b]\}$, *the set of all functions in* $C([a, b])$ *with continuous derivative on the interval* $[a, b]$.
- $L^2((a, b)) := \{f : (a, b) \to \mathbb{R} : \int_a^b |f(x)|^2 \, \mathrm{d}x < \infty\}$, *the set of all square-integrable functions on the open interval* (a, b).
- $H^1((a, b)) := \{f \in L^2(a, b) : \int_a^b |f'(x)|^2 \, \mathrm{d}x < \infty\}$.

Definition 2.1.3 *Let* $G = \{\mathbf{u}_1, \ldots, \mathbf{u}_n\}$ *be a subset of a vector space* V *over a field* $\mathbb{F} = \mathbb{R}$ *or* $\mathbb{C}$. *The set of all linear combinations of elements of* G *is denoted by* span G*:*

$$\operatorname{span} G := \left\{ \sum_{j=1}^{n} a_j \mathbf{u}_j : a_j \in \mathbb{F} \right\}.$$

The span G is the smallest subspace of V containing G. For example, if $V = \mathbb{R}^3$, then $\operatorname{span}\{(1, 2, 3), (1, -2, 0)\}$ is the plane $\{a(1, 2, 3) + b(1, -2, 0) : a, b \in \mathbb{R}\}$.

Definition 2.1.4 *The elements* $\mathbf{u}_1, \ldots, \mathbf{u}_n$ *of a vector space* V *are said to be linearly independent if*

$$\sum_{j=1}^{n} a_j \mathbf{u}_j = \mathbf{0} \quad \textit{holds only for } a_1 = a_2 = \cdots = a_n = 0.$$

Otherwise, $\mathbf{u}_1, \ldots, \mathbf{u}_n$ *are linearly dependent.*

If $\{\mathbf{u}_j\}_{j=1}^n$ is linearly independent, no vector $\mathbf{u}_j$ can be expressed as a linear combination of other vectors in the set. If $\mathbf{u}_1$ can be expressed as $\mathbf{u}_1 = a_2\mathbf{u}_2 + \cdots + a_n\mathbf{u}_n$, then $a_1\mathbf{u}_1 + a_2\mathbf{u}_2 + a_n\mathbf{u}_n = 0$ with $a_1 = -1$, so they are not linearly independent. For

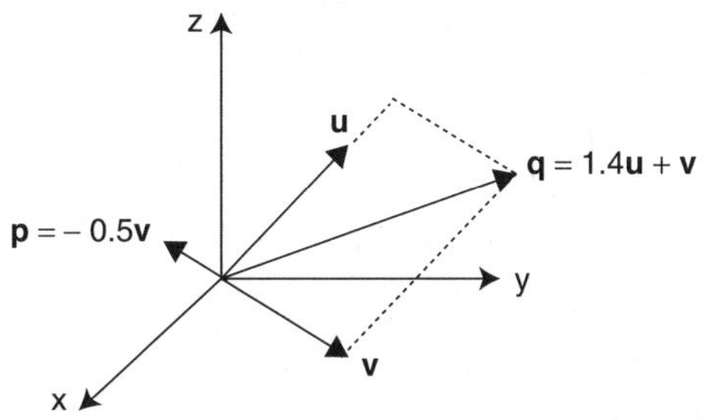

Figure 2.1 Linearly independent and dependent vectors

example, $\{(4, 1, 5), (2, 1, 3), (1, 0, 1)\}$ is linearly dependent since $-(4, 1, 5) + (2, 1, 3) + 2(1, 0, 1) = (0, 0, 0)$. The elements $(2, 1, 3)$ and $(1, 0, 1)$ are linearly independent because $a_1(2, 1, 3) + a_2(1, 0, 1) = (0, 0, 0)$ implies $a_1 = a_2 = 0$.

Example 2.1.5 *In Figure 2.1, the two vectors* $\{\mathbf{u}, \mathbf{v}\}$ *are linearly independent whereas the four vectors* $\{\mathbf{u}, \mathbf{v}, \mathbf{p}, \mathbf{q}\}$ *are linearly dependent. Note that* $\mathbf{p}$ *and* $\mathbf{q}$ *are linearly independent.*

2.1.2 Basis, Norm and Inner Product

Definition 2.1.6 *Let W be a subspace of a vector space V. If* $\text{span}\{\mathbf{u}_1, \ldots, \mathbf{u}_n\} = W$ *and* $\{\mathbf{u}_1, \ldots, \mathbf{u}_n\}$ *is linearly independent, then* $\{\mathbf{u}_1, \ldots, \mathbf{u}_n\}$ *is said to be a basis for W.*

If $\{\mathbf{u}_j\}_{j=1}^n$ is a basis for W, then any vector $\mathbf{v} \in W$ can be expressed uniquely as $\mathbf{v} = \sum_{j=1}^n a_j \mathbf{u}_j \in W$. If G' is another basis of W, then G' contains exactly the same number n of elements.

Definition 2.1.7 *Let W be a subspace of a vector space V. Then W is n-dimensional if the number of elements of the basis of W is n; W is finite-dimensional if* $\dim W < \infty$*; otherwise W is infinite-dimensional.*

To quantify a measure of similarity or dissimilarity among vectors, we need to define the magnitude of a vector and the distance between vectors. We use the norm $\|\mathbf{u}\|$ of a vector $\mathbf{u}$ to define such a magnitude. In the area of topology, the metric is also used for defining a distance. To distinguish different vectors in a vector space, we define a measure of distance or metric between two vectors $\mathbf{u}$ and $\mathbf{v}$ as the norm $\|\mathbf{u} - \mathbf{v}\|$. The norm must satisfy the following three rules.

Definition 2.1.8 *A normed vector space V is a vector space equipped with a norm* $\|\cdot\|$ *that satisfies the following:*

1. $0 \leq \|\mathbf{u}\| < \infty, \forall\, \mathbf{u} \in V and \|\mathbf{u}\| = 0$ *iff* $\mathbf{u} = \mathbf{0}$.
2. $\|a\mathbf{u}\| = |a|\, \|\mathbf{u}\|, \forall\, \mathbf{u} \in V$ *and* $\forall a \in \mathbb{F}$.
3. $\|\mathbf{u} + \mathbf{v}\| \leq \|\mathbf{u}\| + \|\mathbf{v}\|, \forall\, \mathbf{u}, \mathbf{v} \in V$ *(triangle inequality).*

Here, the notation $\forall$ stands for "for all" and iff stands for "if and only if".

Example 2.1.9 *Consider the vector space* $\mathbb{C}^n$. *For* $\mathbf{u} = (u_1, u_2, \ldots, u_n) \in \mathbb{C}^n$ *and* $1 \leq p < \infty$, *the p-norm of* $\mathbf{u}$ *is*

$$\|\mathbf{u}\|_p = \begin{cases} \left(\sum_{j=1}^n |u_j|^p\right)^{1/p} & \text{for } 1 \leq p < \infty, \\ \max_{1\leq j\leq n} |u_j| & \text{for } p = \infty. \end{cases} \tag{2.1}$$

In particular, $\|\mathbf{u} - \mathbf{v}\|_2 = \sqrt{(u_1 - v_1)^2 + \cdots + (u_n - v_n)^2}$ *is the standard distance between* $\mathbf{u}$ *and* $\mathbf{v}$. *We should note that, when* $0 < p < 1$, $\|\mathbf{u}\|_p$ *is not a norm because it does not satisfy the triangle inequality.*

Example 2.1.10 *Consider the vector space* $V = C([0, 1])$. *For* $f, g \in V$, *the distance between* f *and* g *can be defined by*

$$\|f - g\| = \sqrt{\int_0^1 |f(x) - g(x)|^2 \, dx}.$$

In addition to the distance between vectors, it is desirable to establish the concept of an angle between them. This requires the definition of an inner product.

Definition 2.1.11 *Let* V *be a vector space over* $\mathbb{F} = \mathbb{R}$ *or* $\mathbb{C}$. *We denote the complex conjugate of* $a \in \mathbb{C}$ *by* $\bar{a}$. *A vector space* V *with a function* $\langle \cdot, \cdot \rangle : V \times V \to \mathbb{C}$ *is an* inner product space *if:*

1. $0 \leq \langle \mathbf{u}, \mathbf{u} \rangle < \infty, \forall \mathbf{u} \in V$ *and* $\langle \mathbf{u}, \mathbf{u} \rangle = 0$ *if* $\mathbf{u} = \mathbf{0}$;
2. $\langle \mathbf{u}, \mathbf{v} \rangle = \overline{\langle \mathbf{v}, \mathbf{u} \rangle}, \forall \mathbf{u}, \mathbf{v} \in V$;
3. $\langle a\mathbf{u} + b\mathbf{v}, \mathbf{w} \rangle = a\langle \mathbf{u}, \mathbf{w} \rangle + b\langle \mathbf{v}, \mathbf{w} \rangle, \forall \mathbf{u}, \mathbf{v}, \mathbf{w} \in V$ *and* $\forall a, b \in \mathbb{F}$.

In general, $\langle \mathbf{u}, \mathbf{v} \rangle$ is a complex number, but $\langle \mathbf{u}, \mathbf{u} \rangle$ is real. Note that $\langle \mathbf{w}, a\mathbf{u} + b\mathbf{v} \rangle = \bar{a}\langle \mathbf{w}, \mathbf{u} \rangle + \bar{b}\langle \mathbf{w}, \mathbf{v} \rangle$ and $\langle \mathbf{u}, \mathbf{0} \rangle = 0$. If $\langle \mathbf{u}, \mathbf{v} \rangle = 0$ for all $\mathbf{v} \in V$, then $\mathbf{u} = \mathbf{0}$. Given any inner product, $\sqrt{\langle \mathbf{u}, \mathbf{u} \rangle} = \|\mathbf{u}\|$ is a norm on V.

For a real inner product space V, the inner product provides angle information between two vectors $\mathbf{u}$ and $\mathbf{v}$. We denote the angle θ between $\mathbf{u}$ and $\mathbf{v}$ as

$$\theta = \angle(\mathbf{u}, \mathbf{v}) = \cos^{-1} \frac{\langle \mathbf{u}, \mathbf{v} \rangle}{\|\mathbf{u}\| \, \|\mathbf{v}\|} \quad \text{and} \quad \langle \mathbf{u}, \mathbf{v} \rangle = \|\mathbf{u}\| \, \|\mathbf{v}\| \cos\theta.$$

We interpret the angle as follows.

1. If $\theta = 0$, then $\langle \mathbf{u}, \mathbf{v} \rangle = \|\mathbf{u}\| \, \|\mathbf{v}\|$ and $\mathbf{v} = a\mathbf{u}$ for some $a > 0$. The two vectors $\mathbf{u}, \mathbf{v}$ are in the same direction.
2. If $\theta = \pi$, then $\langle \mathbf{u}, \mathbf{v} \rangle = -\|\mathbf{u}\| \, \|\mathbf{v}\|$ and $\mathbf{v} = a\mathbf{u}$ for some $a < 0$. The two vectors $\mathbf{u}, \mathbf{v}$ are in opposite directions.
3. If $\theta = \pm\pi/2$, then $\langle \mathbf{u}, \mathbf{v} \rangle = 0$. The two vectors $\mathbf{u}, \mathbf{v}$ are orthogonal.

Definition 2.1.12 *The set* $\{\mathbf{u}_1, \ldots, \mathbf{u}_n\}$ *in an inner product space* V *is said to be an orthonormal set if* $\langle \mathbf{u}_j, \mathbf{u}_k \rangle = 0$ *for* $j \neq k$ *and* $\|\mathbf{u}_j\| = 1$. *A basis is an orthonormal basis if it is an orthonormal set.*

2.1.3 Hilbert Space

When we analyze a vector $\mathbf{f}$ in a vector space V having a basis $\{\mathbf{u}_j\}_{j=1}^{\infty}$, we wish to represent $\mathbf{f}$ as

$$\mathbf{f} = \sum_{j=1}^{\infty} a_j \mathbf{u}_j.$$

Computation of the coefficients a_j could be very laborious when the vector space V is not equipped with an inner product and $\{\mathbf{u}_j\}_{j=1}^{\infty}$ is not an orthonomal set. A Hilbert space is a closed vector space equipped with an inner product.

Definition 2.1.13 *A vector space H over $\mathbb{F} = \mathbb{R}$ or $\mathbb{C}$ is a Hilbert space if:*

1. *H is an inner product space;*
2. *$H = \bar{H}$ (H is a closed vector space), that is, whenever $\lim_{n\to\infty} \|\mathbf{u}_n - \mathbf{u}\| = 0$ for some sequence $\{\mathbf{u}_n\} \subset H$, $\mathbf{u}$ belongs to H.*

For $\mathbf{u}, \mathbf{v}, \mathbf{w} \in H$, a Hilbert space, we have the following properties.

- Cauchy–Schwarz inequality: $|\langle \mathbf{u}, \mathbf{v}\rangle| \le \|\mathbf{u}\|\,\|\mathbf{v}\|$.
- Triangle inequality: $\|\mathbf{u} + \mathbf{v}\| \le \|\mathbf{u}\| + \|\mathbf{v}\|$.
- Parallelogram law: $\|\mathbf{u} + \mathbf{v}\|^2 + \|\mathbf{u} - \mathbf{v}\|^2 = 2\|\mathbf{u}\|^2 + 2\|\mathbf{v}\|^2$.
- Polarization identity: $4\langle \mathbf{u}, \mathbf{v}\rangle = \|\mathbf{u} + \mathbf{v}\|^2 - \|\mathbf{u} - \mathbf{v}\|^2 + \mathrm{i}\|\mathbf{u} + \mathrm{i}\mathbf{v}\|^2 - \mathrm{i}\|\mathbf{u} - \mathrm{i}\mathbf{v}\|^2$.
- Pythagorean theorem: $\|\mathbf{u} + \mathbf{v}\|^2 = \|\mathbf{u}\|^2 + \|\mathbf{v}\|^2$ if $\langle \mathbf{u}, \mathbf{v}\rangle = 0$.

Exercise 2.1.14 (Gram–Schmidt process) *Let H be a Hilbert space with a basis $\{\mathbf{v}_1, \mathbf{v}_2, \ldots\}$. Assume that $\{\mathbf{u}_1, \mathbf{u}_2, \ldots\}$ is obtained from the following procedure depicted in Figure 2.2:*

1. *Set $\mathbf{w}_1 = \mathbf{v}_1$ and $\mathbf{u}_1 = \mathbf{w}_1/\|\mathbf{w}_1\|$;*
2. *Set $\mathbf{w}_2 = \mathbf{v}_2 - \langle \mathbf{v}_2, \mathbf{u}_1\rangle \mathbf{u}_1$ and $\mathbf{u}_2 = \mathbf{w}_2/\|\mathbf{w}_2\|$;*
3. *For $n = 2, 3, \ldots,$*

$$\mathbf{w}_{n+1} = \mathbf{v}_{n+1} - \sum_{j=1}^{n} \langle \mathbf{v}_{j+1} \mathbf{u}_j \rangle \mathbf{u}_j \quad \textit{and} \quad \mathbf{u}_{n+1} = \frac{\mathbf{w}_{n+1}}{\|\mathbf{w}_{n+1}\|}.$$

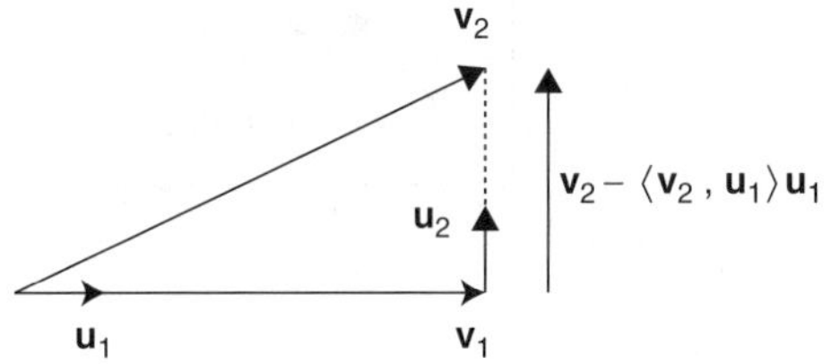

Figure 2.2 Illustration of the Gram–Schmidt process

Prove that $\{\mathbf{u}_1, \mathbf{u}_2, \ldots\}$ *is an orthonormal basis of* H*, that is,*

$$\text{span}\{\mathbf{u}_1, \ldots, \mathbf{u}_n\} = \text{span}\{\mathbf{v}_1, \ldots, \mathbf{v}_n\}.$$

Theorem 2.1.15 (Projection theorem) *Let* G *be a closed convex subset of a Hilbert space* H*. For every* $\mathbf{u} \in H$*, there exists a unique* $\mathbf{u}_* \in G$ *such that* $\|\mathbf{u} - \mathbf{u}_*\| \leq \|\mathbf{u} - \mathbf{v}\|$ *for all* $\mathbf{v} \in G$*.*

For the proof of the above theorem, see Rudin (1970). Let S be a closed subspace of a Hilbert space H. We define the orthogonal complement $S^\perp$ of S as

$$S^\perp := \{\mathbf{v} \in H \mid \langle \mathbf{u}, \mathbf{v} \rangle = 0 \text{ for all } \mathbf{u} \in S\}.$$

According to the projection theorem, we can define a projection map $P_S : H \to S$ such that the value $P_S(\mathbf{u})$ satisfies

$$\|\mathbf{u} - P_S(\mathbf{u})\| \leq \|\mathbf{u} - (P_S(\mathbf{u}) + t\mathbf{v})\| \quad \text{for all } \mathbf{v} \in S \text{ and } t \in \mathbb{R}.$$

This means that $f(t) = \|\mathbf{u} - P_S(\mathbf{u}) + t\mathbf{v}\|^2$ has its minimum at $t = 0$ for any $\mathbf{v} \in S$ and, therefore,

$$0 = f'(0) = \langle \mathbf{u} - P_S(\mathbf{u}), \mathbf{v} \rangle \quad \text{for all } \mathbf{v} \in S$$

or

$$\mathbf{u} - P_S(\mathbf{u}) \in S^\perp.$$

Hence, the projection theorem states that every $\mathbf{u} \in H$ can be uniquely decomposed as $\mathbf{u} = \mathbf{v} + \mathbf{w}$ with $\mathbf{v} \in S$ and $\mathbf{w} \in S^\perp$ and we can express the Hilbert space H as

$$H = S \oplus S^\perp.$$

From the Pythagorean theorem,

$$\|\mathbf{u}\|^2 = \|P_S(\mathbf{u})\|^2 + \|\mathbf{u} - P_S(\mathbf{u})\|^2.$$

Figure 2.3 illustrate the projection of a vector $\mathbf{u}$ onto a subspace S.

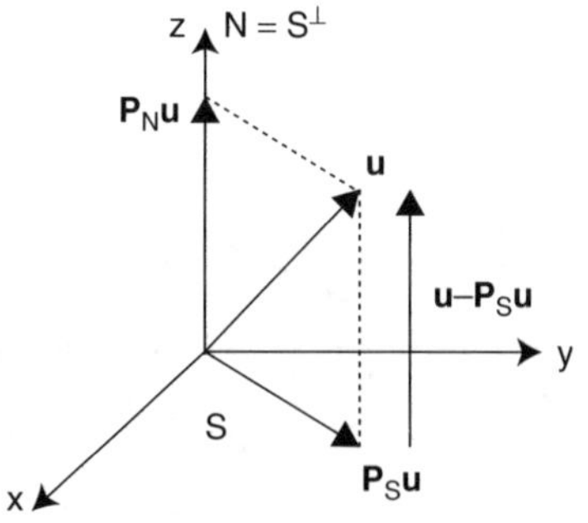

Figure 2.3 Illustration of a projection of a vector **u** onto a subspace S

Example 2.1.16 (Euclidean space $\mathbb{R}^n$ is a Hilbert space) *A Hilbert space is a generalization of the Euclidean space $\mathbb{R}^n$. For $\mathbf{x} = (x_1, x_2, \ldots, x_n)$, $\mathbf{y} = (y_1, y_2, \ldots, y_n) \in \mathbb{R}^n$, we define the inner product and norm as*

$$\langle \mathbf{x}, \mathbf{y} \rangle = \sum_{j=1}^{n} x_j y_j \quad \text{and} \quad \|\mathbf{x}\| = \sqrt{\langle \mathbf{x}, \mathbf{x} \rangle}.$$

The distance between $\mathbf{x}$ and $\mathbf{y}$ is defined by $\|\mathbf{x} - \mathbf{y}\|$, and $\|\mathbf{x} - \mathbf{y}\| = 0$ implies $\mathbf{x} = \mathbf{y}$. If $\langle \mathbf{x}, \mathbf{y} \rangle = 0$, then $\mathbf{x}$ and $\mathbf{y}$ are orthogonal and the following Pythagorean theorem holds:

$$\|\mathbf{x} + \mathbf{y}\|^2 = \|\mathbf{x}\|^2 + \|\mathbf{y}\|^2.$$

If $\{\mathbf{e}_1, \mathbf{e}_2, \ldots, \mathbf{e}_n\}$ is an orthonormal basis of $\mathbb{R}^n$, every $\mathbf{x} \in \mathbb{R}^n$ can be represented uniquely by

$$\mathbf{x} = \sum_{j=1}^{n} \langle \mathbf{x}, \mathbf{e}_j \rangle \, \mathbf{e}_j.$$

For example, we let $\mathbf{e}_1 = (1, 0, \ldots, 0)$, $\mathbf{e}_2 = (0, 1, 0, \ldots, 0)$, ... If $V_m = \operatorname{span}\{\mathbf{e}_1, \ldots, \mathbf{e}_m\}$ for $m < n$, the vector $P_{V_m}(\mathbf{x}) \in V_m$ is

$$P_{V_m}(\mathbf{x}) = \sum_{j=1}^{m} \langle \mathbf{x}, \mathbf{e}_j \rangle \, \mathbf{e}_j,$$

with the distance $\|\mathbf{x} - P_{V_m}(\mathbf{x})\| = \sqrt{\sum_{j=m+1}^{n} \langle \mathbf{x}, \mathbf{e}_j \rangle^2}$. We can generalize this dot product property in Euclidean space $\mathbb{R}^n$ to an infinite-dimensional Hilbert space.

Example 2.1.17 (L^2-space) *Let I be the interval $[0, 1]$. We denote the set of all square-integrable complex functions by $L^2(I)$, that is,*

$$L^2(I) = \left\{ f \;\middle|\; \int_I |f(x)|^2 \, \mathrm{d}x < \infty \right\}.$$

For $f, g \in L^2(I)$, we define the inner product as

$$\langle f, g \rangle = \int_I f(x) \overline{g(x)} \, \mathrm{d}x,$$

where $\overline{g(x)}$ denotes the complex conjugate of $g(x)$. Dividing the interval $[0, 1]$ into N subintervals with endpoints $x_0 = 0, x_1 = \Delta x, \ldots, x_N = N\Delta x = 1$ and equal gap $\Delta x = 1/N$, the inner product $\langle f, g \rangle$ in $L^2(I)$ can be viewed approximately as the inner product in Euclidean space $\mathbb{C}^n$:

$$\langle f, g \rangle \approx \begin{pmatrix} f(x_1) \\ \vdots \\ f(x_n) \end{pmatrix} \cdot \begin{pmatrix} \overline{g(x_1)} \\ \vdots \\ \overline{g(x_n)} \end{pmatrix} \Delta x = \sum_{j=1}^{N} f(x_j) \overline{g(x_j)} \Delta x.$$

The vector space $L^2(I)$ with the above inner product is a Hilbert space and retains features of Euclidean space. Indeed, in order for the vector space $L^2(I)$ to be a Hilbert space, we *need Lebesgue measure theory because there are infinitely many* f *with* $\|f\| = 0$ *but* $f \neq 0$ *in the pointwise sense. In Lebesgue measure theory,* $\|f\| = 0$ *implies* $f = 0$ *almost everywhere in the sense of the measure. In* L^2*-space,* $f = g$ *means that* $f = g$ *almost everywhere. For the details of Lebesgue measure theory, please refer to Rudin (1970).*

Exercise 2.1.18 *Prove that*

$$\{e^{2\pi inx} : n = 0, \pm 1, \pm 2, \ldots\}$$

is an orthonormal set in $L^2([0, 1])$.

2.2 Vector Calculus

Based on the underlying physical principles, we need to express interrelations among system parameters, variables, excitations and measurements. A scalar-valued function $f(\mathbf{x})$ defined in $\mathbb{R}^n$ represents a numerical quantity at a point $\mathbf{x} = (x_1, x_2, \ldots, x_n) \in \mathbb{R}^n$. Examples may include temperature, voltage, pressure, altitude and so on. A vector-valued function $\mathbf{F}(\mathbf{x})$ is a vector quantity at $\mathbf{x} \in \mathbb{R}^n$. It represents a vector field such as displacement, velocity, force, electric field intensity, magnetic flux density and so on. We now review the vector calculus of gradient, divergence and curl to handle basic dynamics among variables and parameters.

2.2.1 *Gradient*

Let $f \in C^1(\mathbb{R}^n)$. For a given unit vector $\mathbf{d} \in \mathbb{R}^n$, the directional derivative of f at $\mathbf{x}$ in the direction $\mathbf{d}$ is denoted by $\partial_{\mathbf{d}} f(\mathbf{x})$. It represents the rate of increase in f at $\mathbf{x}$ in the direction of $\mathbf{d}$:

$$\partial_{\mathbf{d}} f(\mathbf{x}) = \lim_{h \to 0} \frac{f(\mathbf{x} + h\mathbf{d}) - f(\mathbf{x})}{h}.$$

If $\mathbf{d} = \mathbf{e}_j$, the jth unit vector in the Cartesian coordinate system, we simply write $\partial_{\mathbf{e}_j} f(\mathbf{x}) = \partial_j f(\mathbf{x})$. The gradient of f, denoted as ∇f, is a vector-valued function that points in the direction of maximum increase of f:

$$\nabla f(\mathbf{x}) = \partial_{\mathbf{d}^*} f(\mathbf{x}) \mathbf{d}^* \quad \text{where} \quad \partial_{\mathbf{d}^*} f(\mathbf{x}) = \sup_{|\mathbf{d}|=1} \partial_{\mathbf{d}} f(\mathbf{r}).$$

Remark 2.2.1 *Let* $f \in C^1(\mathbb{R}^n)$. *Suppose that* $f(\mathbf{x}_0) = \lambda$ *and* $\nabla f(\mathbf{x}_0) \neq \mathbf{0}$. *The vector* $\nabla f(\mathbf{x}_0)$ *is perpendicular to the level set* $\mathcal{L}_\lambda = \{\mathbf{y} \in \mathbb{R}^n : f(\mathbf{y}) = \lambda\}$ *because there is no increase in* f *along the level set* $\mathcal{L}_\lambda$. *If* $\Omega_\lambda := \{\mathbf{y} \in \mathbb{R}^n : f(\mathbf{y}) < \lambda\}$ *is a domain enclosed by the level surface* $\mathcal{L}_\lambda$, *the unit outward normal vector* $\mathbf{n}(\mathbf{x}_0)$ *at* $\mathbf{x}_0$ *on the boundary* $\partial\Omega_\lambda$ *is*

$$\mathbf{n}(\mathbf{x}_0) = \frac{\nabla f(\mathbf{x}_0)}{|\nabla f(\mathbf{x}_0)|},$$

which points in the steepest ascending direction. The curvature along the level set $\mathcal{L}_\lambda$ *is given by*

$$\kappa(\mathbf{x}) := \nabla \cdot \frac{\nabla f(\mathbf{x})}{|\nabla f(\mathbf{x})|}, \quad \mathbf{x} \in \mathcal{L}_\lambda.$$

Proof. If $\gamma(t)$ is a smooth curve lying on a level surface

$$\mathcal{L}_\lambda = \{\mathbf{y} \in \mathbb{R}^n : f(\mathbf{y}) = f(\mathbf{x}_0) = \lambda\} \quad \text{with } \boldsymbol{\gamma}(0) = \mathbf{x}_0,$$

then

$$f(\gamma(t)) = \lambda \quad \text{and} \quad 0 = \frac{\mathrm{d}}{\mathrm{d}t} f(\gamma(t)) = (\partial_1 f(\gamma(t)), \ldots, \partial_n f(\gamma(t))) \cdot \gamma'(t).$$

The vector $(\partial_1 f(\gamma(t)), \ldots, \partial_n f(\gamma(t)))$ is perpendicular to the tangent direction $\gamma'(t)$ of the level set $\mathcal{L}_\lambda$. Since $(\partial_1 f(\gamma(t)), \ldots, \partial_n f(\gamma(t)))$ has the same direction as the gradient $\nabla f(\mathbf{x})$, we can write

$$\nabla f = (\partial_1 f, \ldots, \partial_n f).$$ □

Exercise 2.2.2 *Prove that* $\nabla f = (\partial_1 f, \ldots, \partial_n f)$.

2.2.2 Divergence

The divergence of $\mathbf{F}(\mathbf{r})$ at a point $\mathbf{r}$, written as $\operatorname{div}\mathbf{F}$, is the net outward flux of $\mathbf{F}$ per unit volume of a ball centered at $\mathbf{r}$ as the ball shrinks to zero:

$$\operatorname{div}\mathbf{F}(\mathbf{r}) := \lim_{r\to 0} \frac{3}{4\pi r^3} \int_{\partial B_r(\mathbf{r})} \mathbf{F}(\mathbf{r}') \cdot \mathrm{d}\mathbf{S}_{\mathbf{r}'},$$

where $\mathrm{d}\mathbf{S}$ is the surface element, $B_r(\mathbf{r})$ is the ball with radius r and center $\mathbf{r}$, and ∂B is the boundary of B, which is a sphere.

Theorem 2.2.3 (Divergence theorem) *Let Ω be a bounded smooth domain in $\mathbb{R}^3$. The volume integral of the divergence of a C^1-vector field $\mathbf{F} = (F_1, F_2, F_3)$ equals the total outward flux of the vector $\mathbf{F}$ through the boundary of Ω:*

$$\int_\Omega \operatorname{div} \boldsymbol{F}(\mathbf{y})\, \mathrm{d}\mathbf{y} = \int_{\partial\Omega} \boldsymbol{F}(\mathbf{y}) \cdot \mathrm{d}\boldsymbol{S}.$$

Exercise 2.2.4 *Prove* $\operatorname{div}\mathbf{F} = \partial_1 F_1 + \partial_2 F_2 + \partial_3 F_3$ *and the divergence theorem.*

2.2.3 Curl

The circulation of a vector field $\mathbf{F} = (F_1, F_2, F_3)$ around a closed path C in $\mathbb{R}^3$ is defined as a scalar line integral of the vector $\mathbf{F}$ over the path C:

$$\oint_C \mathbf{F} \cdot \mathrm{d}\mathbf{l} = \oint_C F_1\, \mathrm{d}x_1 + F_2\, \mathrm{d}x_2 + F_3\, \mathrm{d}x_3.$$

If $\mathbf{F}$ represents an electric field intensity, the circulation will be an electromotive force around the path C.

The curl of a vector field $\mathbf{F}$, denoted by $\operatorname{curl}\mathbf{F}$, is a vector whose magnitude is the maximum net circulation of $\mathbf{F}$ per unit area as the area shrinks to zero and whose direction is the normal direction of the area when the area is oriented to make the net circulation maximum. We can define the **d**-directional net circulation of **F** at **r** precisely by

$$\operatorname{curl}_{\mathbf{d}} \mathbf{F}(\mathbf{r}) = \lim_{r\to 0^+} \frac{1}{\pi r^2} \oint_{\partial D_{r,\mathbf{d}}(\mathbf{r})} \mathbf{F}(r') \cdot \mathrm{d}l_{\mathbf{r}'},$$

where $D_{r,\mathbf{d}}(\mathbf{r}) = \{\mathbf{r}' \in \mathbb{R}^3 : (\mathbf{r}' - \mathbf{r}) \cdot \mathbf{d} = 0, |\mathbf{r} - \mathbf{r}'| < r\}$ is the disk centered at **r** with radius r and normal to **d**. Then, curl **F** is its maximum net circulation:

$$\operatorname{curl} \mathbf{F}(\mathbf{r}) = \operatorname{curl}_{\mathbf{d}^*} \mathbf{F}(\mathbf{r}) \quad \text{where} \quad |\operatorname{curl}_{\mathbf{d}^*} \mathbf{F}(\mathbf{r})| = \max_{|\mathbf{d}|=1} |\operatorname{curl}_{\mathbf{d}} \mathbf{F}(\mathbf{r})|.$$

Theorem 2.2.5 (Stokes's theorem) *Let C_{area} be an open smooth surface with its boundary as a smooth contour C. The surface integral of the curl of a C^1-vector field* **F** *over the surface C_{area} is equal to the closed line integral of the vector* **F** *along the contour C:*

$$\int_{C_{\text{area}}} \nabla \times \mathbf{F}(\mathbf{y}) \cdot \mathrm{d}S = \oint_C \mathbf{F}(\mathbf{y}) \cdot \mathrm{d}\mathbf{l}.$$

Exercise 2.2.6 *Prove that the expression for* curl **F** *in Cartesian coordinates is*

$$\operatorname{curl} F = \nabla \times \mathbf{F} = \begin{vmatrix} \mathbf{e}_1 & \mathbf{e}_2 & \mathbf{e}_3 \\ \partial_1 & \partial_2 & \partial_3 \\ F_1 & F_2 & F_3 \end{vmatrix}.$$

Exercise 2.2.7 *Let $A, B, C, U \in [C^1(\mathbb{R}^3)]^3$ and $u \in C^2(\mathbb{R}^n)$. Prove the following vector identities.*

1. $A \times (B \times C) = (A \cdot C)B - (A \cdot B)C.$
2. $\nabla \times \nabla \times U = \nabla(\nabla \cdot U) - \Delta U.$
3. $\nabla \times \nabla u = 0.$
4. *If* $\nabla \times U = 0$, *then there exists* $v \in C^1(\mathbb{R}^3)$ *such that* $\nabla v = U$ *in* $\mathbb{R}^3$.
5. $\nabla \cdot (\nabla \times U) = 0.$
6. $\nabla \cdot (A \times B) = B \cdot (\nabla \times A) - A \cdot (\nabla \times B).$

Exercise 2.2.8 *Let $f, g \in C^2(\mathbb{R}^3)$. Let C be a closed curve and let C_{area} be a surface enclosed by C. Let Ω be a bounded domain in $\mathbb{R}^3$. Prove the following:*

1. $\displaystyle \oint_C f\nabla g \cdot \mathrm{d}\mathbf{l} = \int_{C_{\text{area}}} (\nabla f \times \nabla g) \cdot \mathbf{n}\, \mathrm{d}S.$
2. $\displaystyle \oint_C (f\nabla g + g\nabla f) \cdot \mathrm{d}\mathbf{l} = 0.$
3. $\displaystyle \int_{\partial\Omega} f\nabla g \cdot \mathbf{n}\, \mathrm{d}S = \int_\Omega (f\nabla^2 g + \nabla f \cdot \nabla g)\, \mathrm{d}\mathbf{y}.$

2.2.4 *Curve*

A curve $\mathcal{C}$ in $\mathbb{R}^2$ and $\mathbb{R}^3$ is represented, respectively, by

$$\mathbf{r}(t) = (x(t), y(t)) : I = [a, b] \to \mathbb{R}^2$$

and

$$\mathbf{r}(t) = (x(t), y(t), z(t)) : I = [a, b] \to \mathbb{R}^3.$$

The curve C is said to be *regular* if $\mathbf{r}'(t) \neq 0$ for all t. The *arc length* $s(t)$ of C between $\mathbf{r}(t_0)$ to $\mathbf{r}(t_1)$ is given by

$$s(t_1) - s(t_0) = \int_{t_0}^{t_1} \|\mathbf{r}'(t)\,\mathrm{d}t \quad \text{or} \quad s'(t) = \| \mathbf{r}_t \|.$$

The unit tangent vector T of the curve C is given by

$$T = \frac{\mathbf{r}_t}{\| \mathbf{r}_t \|} = \frac{\mathrm{d}\mathbf{r}}{\mathrm{d}s} \quad \text{or} \quad \mathbf{r}'(t) = s'(t)T.$$

The unit normal vector $\mathbf{n}$ of the curve C is determined by

$$\frac{\mathrm{d}^2\mathbf{r}}{\mathrm{d}s^2} = \frac{\mathrm{d}T}{\mathrm{d}s} = \kappa\,\mathbf{n} \quad \text{or} \quad \mathbf{r}'' = s''T + (s')^2\kappa\,\mathbf{n},$$

where κ is the curvature given by

$$|\kappa| = \frac{\| \mathbf{r}_t \times \mathbf{r}_{tt} \|}{\| \mathbf{r}_t \|^3}.$$

Here, we use $|\mathbf{r}'' \times \mathbf{r}'| = |(s')^2\kappa\, T \times (s'\mathbf{n})| = |(s')^3\kappa|$. Note that $\mathrm{d}\mathbf{n}/\mathrm{d}s = -\kappa\, T$ since $T \cdot \mathbf{n} = 0$ and $T_s \cdot \mathbf{n} + T \cdot \mathbf{n}_s = 0$.

2.2.5 *Curvature*

Consider a plane curve $\mathbf{r}(s) = (x(s), y(s))$ in $\mathbb{R}^2$, where s is the length parameter. If $\theta(s)$ stands for the angle between $T(s)$ and the x-axis, then $\kappa(s) = \mathrm{d}\theta/\mathrm{d}s$ because

$$\frac{\mathrm{d}T}{\mathrm{d}s} = \frac{\mathrm{d}(\cos\theta, \sin\theta)}{\mathrm{d}s} = \frac{\mathrm{d}\theta}{\mathrm{d}s}\mathbf{n}.$$

When the curve is represented as $\mathbf{r}(t) = (x(t), y(t))$, then

$$\kappa = \frac{|x'y'' - x''y'|}{\|\mathbf{r}_t\|^3}.$$

Now, consider a curve C given implicitly by the level set of $\phi(x, y) =: \mathbb{R}^2 \to \mathbb{R}$:

$$C := \{(x, y) : \phi(x, y) = 0\}.$$

Then, the normal and tangent vectors to the level curve are

$$\mathbf{n} = \pm\frac{\nabla\phi}{\|\nabla\phi\|} \quad \text{and} \quad T = \pm\frac{(-\phi_y, \phi_x)}{\|\nabla\phi\|}.$$

The curvature κ is

$$\kappa = \frac{\phi_{xx}u_y^2 - 2\phi_{xy}\phi_x\phi_y + \phi_{yy}\phi_x^2}{\|\nabla\phi\|^3} = \nabla \cdot \left(\frac{\nabla\phi}{\|\nabla\phi\|}\right).$$

To see this, assume that $\phi(\mathbf{r}(t)) = 0$ and set $y' = -\phi_x$ and $x' = \phi_y$ because $\nabla\phi(\mathbf{r}) \cdot \mathbf{r}'(t) = 0$. Then,

$$\frac{\mathrm{d}}{\mathrm{d}t}\nabla\phi(\mathbf{r}) \cdot \mathbf{r}'(t) = 0 \quad \text{and} \quad (x', y') = (\phi_y, -\phi_x)$$

imply

$$\phi_{xx}(x')^2 + 2\phi_{xy}x'y' + \phi_{yy}(y')^2 = -[\phi_x x'' + \phi_u y'']$$

Hence,

$$\phi_{xx}(\phi_y)^2 - 2\phi_{xy}\phi_x\phi_y + \phi_{yy}(\phi_x)^2 = y'x'' - x'y''.$$

The curvature κ at a given point is the inverse of the radius of a disk that best fits the curve $\mathcal{C}$ at that point.

Next, we consider a space curve $\mathcal{C}$ in $\mathbb{R}^3$ represented by

$$\mathbf{r}(t) = (x(t), y(t), z(t)) \quad : \quad I := (a, b) \to \mathbb{R}^3.$$

Then

$$T = \mathbf{r}_s, \quad \mathbf{r}_{ss} = T_s = \kappa N, \quad N_s = -\kappa T + \tau B, \quad B = T \times N, \quad s' = \|\mathbf{r}_t\|.$$

The curvature κ and torsion τ are computed by

$$\kappa = \frac{\| \mathbf{r}_t \times \mathbf{r}_{tt} \|}{\| \mathbf{r}_t \|^3} \quad \text{and} \quad \tau = \frac{[\mathbf{r}_t, \mathbf{r}_{tt}, \mathbf{r}_{ttt}]}{\| \mathbf{r}_t \times \mathbf{r}_{tt} \|^2},$$

where [ ] stands for the triple scalar product. If the relation between the moving coordinate system $\{T, N, B\}$ and the fixed coordinates $\{\mathbf{x}, \mathbf{y}, \mathbf{z}\}$ is

$$\{T, N, B\} = A(s)\{\mathbf{x}, \mathbf{y}, \mathbf{z}\} \quad \text{where } A(s) \text{ is a rotation matrix,}$$

then

$$\frac{\mathrm{d}}{\mathrm{d}s}\{T, N, B\} = C_A\{T, N, B\} \quad \text{and} \quad C_A = \begin{bmatrix} 0 & \kappa(s) & 0 \\ -\kappa(s) & 0 & \tau(s) \\ 0 & -\tau(s) & 0 \end{bmatrix}.$$

We consider a regular surface $\mathcal{S}$:

$$\mathbf{r}(u, v) : U \subset \mathbb{R}^2 \to \mathbb{R}^3, \quad \mathbf{r}(u, v) = (x(u, v), y(u, v), z(u, v)).$$

Note that $\mathcal{S} = \mathbf{r}(U) = \{r(u, v) \mid (u, v) \in U\}$. The condition corresponding to the regularity of a space curve is that

$$\forall\, u, v, \quad |\mathbf{r}_u \times \mathbf{r}_v| \neq 0$$

or

$$\text{Jacobian matrix} \begin{bmatrix} x_u & x_y \\ y_u & y_v \\ z_u & z_v \end{bmatrix} \text{ has rank 2.}$$

If $|\mathbf{r}_u \times \mathbf{r}_v| \neq 0$, the tangent vectors $\mathbf{r}_u$ and $\mathbf{r}_v$ generate a tangent plane that is spanned by $\mathbf{r}_u$ and $\mathbf{r}_v$. The surface normal vector is expressed as

$$\mathbf{n} = \frac{\mathbf{r}_u \times \mathbf{r}_v}{\|\mathbf{r}_u \times \mathbf{r}_v\|}.$$

For example, if $U = [0, 2\pi] \times [0, \pi]$ and $\mathbf{r}(u, v) = (\cos u \cos v,\ \cos u \sin v,\ \sin v)$, then $\mathcal{S} = \mathbf{r}(U)$ represents the unit sphere. If the surface $\mathcal{S}$ is written as $z = f(x, y)$, then

$$\mathbf{r}(x, y) = (x, y, f(x, y)), \quad \mathbf{r}_x = (1, 0, f_x), \quad \mathbf{r}_y = (0, 1, f_y),$$

and the unit normal vector $\mathbf{n}$ is

$$\mathbf{n} = \frac{\mathbf{r}_x \times \mathbf{r}_y}{|\mathbf{r}_x \times \mathbf{r}_y|} = \frac{(-f_x, -f_y, 1)}{\sqrt{f_x^2 + f_y^2 + 1}}.$$

If the surface $\mathcal{S}$ is given by $\phi(\mathbf{r}) = 0$, then

$$\mathbf{n} = \frac{\nabla \phi}{|\nabla \phi|}.$$

2.3 Taylor's Expansion

The dynamics among system parameters, variables, excitations and measurements are often expressed as complicated nonlinear functions. In a nonlinear inverse problem, we may adopt a linearization approach in its solution-seeking process. Depending on the given problem, one may need to adopt a specific linearization process. In this section, we review Taylor's expansion as a tool to perform such an approximation.

Taylor polynomials are often used to approximate a complicated function. Taylor's expansion for $f \in C^{m+1}(\mathbb{R})$ about x is

$$f(x + h) = f(x) + f'(x)h + \cdots + \frac{f^{(m)}(x)}{m!} h^m + O(|h|^{m+1}), \tag{2.2}$$

where $O(|h|^{m+1})$ is the remainder term containing the $(m + 1)$th order of h. Precisely, the remainder term is

$$R_m(x, h) = \int_x^{x+h} \frac{(x - y)^m}{m!} f^{m+1}(y)\, \mathrm{d}y = O(|h|^{m+1}).$$

This expansion leads to numerical differential formulas of f in various ways, for example:

1. $f'(x) = \dfrac{f(x + h) - f(x)}{h} + O(h)$ (forward difference);
2. $f'(x) = \dfrac{f(x) - f(x - h)}{h} + O(h)$ (backward difference);
3. $f'(x) = \dfrac{f(x + h) - f(x - h)}{2h} + O(h^2)$ (centered difference);
4. $f'(x) = \dfrac{1}{12h}[f(x - 2h) - 8f(x - h) + 8f(x + h) - f(x + 2h)] + O(h^4)$.

The Newton–Raphson method to find a root of $f(x) = 0$ can be explained from the first-order Taylor's approximation $f(x + h) = f(x) + hf'(x) + O(h^2)$ ignoring the term $O(h^2)$, which is negligible when h is small. The method is based on the approximation

$$0 \leftarrow f(x_{n+1}) \approx f(x_n) + h_n f'(x_n) \quad (\text{where } h_n := x_{n+1} - x_n).$$

It starts with an initial guess x_0 and generates a sequence $\{x_n\}$ by the formula

$$x_{n+1} \leftarrow x_n - \frac{f(x_n)}{f'(x_n)}.$$

It may not converge to a solution in general. The convergence issue will be discussed later.

We turn our attention to the second derivative $f''(x)$. By Taylor's expansion, we approximate $f''(x)$ by

$$f''(x) = \frac{f(x+h) + f(x-h) - 2f(x)}{h^2} + O(h^2).$$

The sign of $f''(x)$ gives local information about f for a sufficiently small positive h:

1. $f''(x) = 0 \implies f(x) \approx \dfrac{f(x+h) + f(x-h)}{2}$ (mean value property, MVP);
2. $f''(x) > 0 \implies f(x) < \dfrac{f(x+h) + f(x-h)}{2}$ (sub-MVP);
3. $f''(x) < 0 \implies f(x) > \dfrac{f(x+h) + f(x-h)}{2}$ (super-MVP).

We consider a multi-dimensional case $\mathbb{R}^n$. The following fourth-order Taylor's theorem in n variables will be used later.

Theorem 2.3.1 (Taylor's approximation) *For $f \in C^4(\mathbb{R}^n)$,*

$$f(\mathbf{x}+\mathbf{h}) = f(\mathbf{x}) + \nabla f(\mathbf{x}) \cdot \mathbf{h} + \frac{1}{2}(D^2 f(\mathbf{x})\mathbf{h}) \cdot \mathbf{h} + \frac{1}{3!}\sum_{i,j,k=1}^{n} h_i h_j h_k \frac{\partial^3 f}{\partial_i \partial_j \partial_k}(\mathbf{x}) + O(|\mathbf{h}|^4),$$

where $\mathbf{x} = (x_1, x_2, \ldots, x_n)$, $\mathbf{h} = (h_1, \ldots, h_n)$ and

$$D^2 f(\mathbf{x}) = \begin{pmatrix} \dfrac{\partial^2 f}{\partial x_1^2}(\mathbf{x}) & \cdots & \dfrac{\partial^2 f}{\partial x_1 \partial x_n}(\mathbf{x}) \\ \vdots & \ddots & \vdots \\ \dfrac{\partial^2 f}{\partial x_n \partial x_1}(\mathbf{x}) & \cdots & \dfrac{\partial^2 f}{\partial x_n^2}(\mathbf{x}) \end{pmatrix},$$

which is called the Hessian matrix.

This leads to

$$\frac{f(\mathbf{x}+\mathbf{h}) + f(\mathbf{x}-\mathbf{h})}{2} - f(x) = \frac{1}{2}(D^2 f(\mathbf{x})\mathbf{h}) \cdot \mathbf{h} + O(|\mathbf{h}|^4).$$

If $D^2 f(\mathbf{x})$ is a positive definite matrix, then, for a sufficiently small r,

$$f(\mathbf{x}) < \frac{f(\mathbf{x}+\mathbf{h}) + f(\mathbf{x}-\mathbf{h})}{2} \quad \text{for all } |\mathbf{h}| < r,$$

which leads to the sub-MVP

$$f(\mathbf{x}) < \frac{1}{|B_r(\mathbf{x})|} \int_{B_r(\mathbf{x})} f(\mathbf{y})\, \mathrm{d}\mathbf{y}.$$

Similarly, the super-MVP can be derived for a negative definite matrix $D^2 f(\mathbf{x})$.

Theorem 2.3.2 *Suppose $f : \Omega \subset \mathbb{R}^n \to \mathbb{R}$ is a C^3 function and $\nabla f(\mathbf{x}_0) = 0$.*

1. *If f has a local maximum (minimum) at $\mathbf{x}_0$, then the Hessian matrix $D^2 f(\mathbf{x}_0)$ is negative (positive) semi-definite.*
2. *If $D^2 f(\mathbf{x}_0)$ is negative (positive) definite, then f has a local maximum (minimum) at x_0.*

Example 2.3.3 *If $f \in C^2(\mathbb{R}^n)$ satisfies the Laplace equation $\nabla^2 f(\mathbf{x}) := \nabla \cdot \nabla f(\mathbf{x}) = \sum_{j=1}^n \partial_j^2 f(\mathbf{x}) = 0$, then, for a small $h > 0$, we have*

$$\frac{1}{2}\big(f(\mathbf{x} + h\mathbf{e}_j) + f(\mathbf{x} - h\mathbf{e}_j)\big) = f(\mathbf{x}) + O(h^4) \quad (j = 1, \ldots, n).$$

Hence, $f(\mathbf{x})$ can be viewed approximately as the average of neighboring points:

$$f(\mathbf{x}) \approx \frac{1}{2n} \sum_{j=1}^{n} (f(\mathbf{x} + h\mathbf{e}_j) + f(x - h\mathbf{e}_j)) \quad (h \approx 0).$$

This type of mean value property will be discussed later.

2.4 Linear System of Equations

We consider a linear system of equations including system parameters, variables, excitations and measurements. We may derive it directly from a linear physical system or through a linearization of a nonlinear system. Once we express a forward problem as a linear transform of inputs or system parameters to measured outputs, we can adopt numerous linear methods to seek solutions of the associated inverse problem. For the details of linear system of equations, please refer to Strang (2005, 2007).

2.4.1 Linear System and Transform

We consider a linear system in Figure 2.4. We express its outputs y_i for $i = 1, \ldots, m$ as a linear system of equations:

$$\begin{cases} y_1 &= a_{11}x_1 + a_{12}x_2 + \cdots + a_{1n}x_n, \\ y_2 &= a_{21}x_1 + a_{22}x_2 + \cdots + a_{2n}x_n, \\ \quad\vdots & \\ y_m &= a_{m1}x_1 + a_{m2}x_2 + \cdots + a_{mn}x_n. \end{cases} \tag{2.3}$$

Using two vectors $\mathbf{y} \in \mathbb{R}^m$, $\mathbf{x} \in \mathbb{R}^n$ and a matrix $\mathbf{A} \in \mathbb{R}^{m \times n}$, we can express the linear system as

$$\mathbf{y} = \mathbf{A}\mathbf{x} \tag{2.4}$$

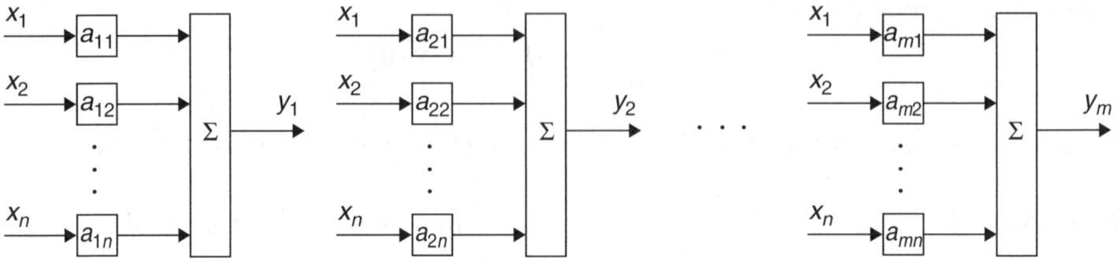

Figure 2.4 Linear system with multiple inputs and multiple outputs

or

$$\begin{bmatrix} y_1 \\ y_2 \\ \vdots \\ y_m \end{bmatrix} = \begin{bmatrix} a_{11} & a_{12} & \cdots & a_{1n} \\ a_{21} & a_{22} & \cdots & a_{2n} \\ \vdots & \vdots & \ddots & \vdots \\ a_{m1} & a_{m2} & \cdots & a_{mn} \end{bmatrix} \begin{bmatrix} x_1 \\ x_2 \\ \vdots \\ x_n \end{bmatrix}.$$

In most problems, the output vector $\mathbf{y}$ consists of measured data. If the input vector $\mathbf{x}$ includes external excitations or internal sources, the matrix $\mathbf{A}$ is derived from the system transfer function or gain determined by the system structure and parameters. When the vector $\mathbf{x}$ contains system parameters, the linear system of equations in (2.4) is formulated subject to certain external excitations or internal sources, information about which is embedded in the matrix $\mathbf{A}$.

For the simplest case of $m = n = 1$, it is trivial to find $x_1 = y_1/a_{11}$. If $m = 1$ and $n > 1$, that is, $y_1 = \sum_{j=1}^{n} a_{1j}x_j$, it is not possible in general to determine all x_j uniquely since they are summed to result in a single value y_1. This requires us to increase the number of measurements so that $m \geq n$. However, for certain cases, we will have to deal with the situation of $m < n$. Figure 2.5 illustrates three different cases of the linear system of equations. To obtain an optimal solution $\mathbf{x}$ by solving (2.4) for $\mathbf{x}$, it is desirable to understand the structure of the matrix $\mathbf{A}$ in the context of vector spaces.

2.4.2 Vector Space of Matrix

We denote by $\mathcal{L}(\mathbb{R}^n, \mathbb{R}^m)$ a set of linear transforms from $\mathbb{R}^n$ to $\mathbb{R}^m$, that is, $\mathcal{L}(\mathbb{R}^n, \mathbb{R}^m)$ is the vector space consisting of $m \times n$ matrices

$$\mathbf{A} = \begin{bmatrix} a_{11} & a_{12} & \cdots & a_{1n} \\ a_{21} & a_{22} & \cdots & a_{2n} \\ \vdots & \vdots & \ddots & \vdots \\ a_{m1} & a_{m2} & \cdots & a_{mn} \end{bmatrix} = [a_{ij}].$$

We call an $m \times 1$ matrix a column vector and an $1 \times n$ matrix a row vector. For the matrix $\mathbf{A}$, we denote its ith row vector as

$$\text{row}(A; i) = [a_{i1}\ a_{i2}\ \cdots\ a_{in}] \in \mathbb{R}^n$$

and its jth column vector as

$$\text{col}(A; j) = [a_{1j}\ a_{2j}\ \cdots\ a_{mj}]^{\mathrm{T}} \in \mathbb{R}^m.$$

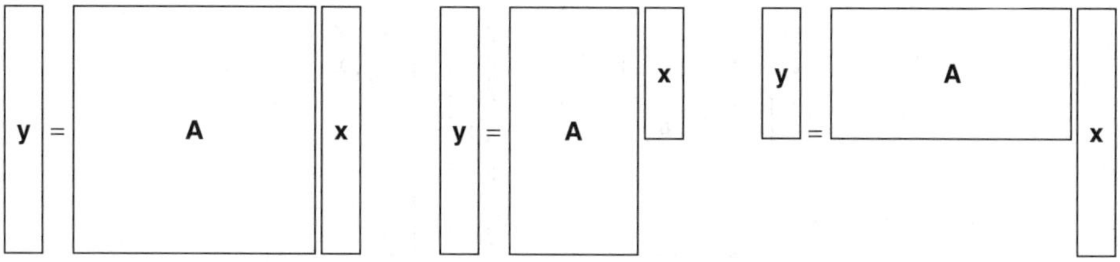

Figure 2.5 Three cases of linear systems of equations

For two vectors $\mathbf{u}, \mathbf{v} \in \mathbb{R}^n$, we define their inner product by

$$\mathbf{u}^{\mathrm{T}}\mathbf{v} = \langle \mathbf{u}, \mathbf{v} \rangle = \sum_{i=1}^{n} u_i v_i$$

and define the outer product as

$$\mathbf{u}\mathbf{v}^{\mathrm{T}} = \begin{bmatrix} u_1 v_1 & u_1 v_2 & \cdots & u_1 v_n \\ u_2 v_1 & u_2 v_2 & \cdots & u_2 v_n \\ \vdots & \vdots & \ddots & \vdots \\ u_n v_1 & u_n v_2 & \cdots & u_n v_n \end{bmatrix} \in \mathcal{L}(\mathbb{R}^n, \mathbb{R}^n).$$

For $\mathbf{y} \in \mathbb{R}^m$ and $\mathbf{x} \in \mathbb{R}^n$, we consider a linear transform or a linear system of equations:

$$\mathbf{y} = \mathbf{A}\mathbf{x}.$$

We can understand it as either

$$\mathbf{y} = \begin{bmatrix} y_1 \\ y_2 \\ \vdots \\ y_m \end{bmatrix} = \underbrace{\begin{bmatrix} \mathrm{row}(A;1) \\ \mathrm{row}(A;2) \\ \vdots \\ \mathrm{row}(A;m) \end{bmatrix}}_{=\mathbf{A}} \begin{bmatrix} x_1 \\ x_2 \\ \vdots \\ x_n \end{bmatrix} = \begin{bmatrix} \langle \mathrm{row}(A;1), \mathbf{x} \rangle \\ \langle \mathrm{row}(A;2), \mathbf{x} \rangle \\ \vdots \\ \langle \mathrm{row}(A;m), \mathbf{x} \rangle \end{bmatrix} \tag{2.5}$$

or

$$\mathbf{y} = \underbrace{\left[\mathrm{col}(A;1)\ \mathrm{col}(A;2)\ \ldots\ \mathrm{col}(A;n)\right]}_{=\mathbf{A}} \begin{bmatrix} x_1 \\ x_2 \\ \vdots \\ x_n \end{bmatrix} = \sum_{j=1}^{n} x_j \,\mathrm{col}(A;j). \tag{2.6}$$

Figure 2.6 shows these two different representations of a linear system of equations $\mathbf{y} = \mathbf{A}\mathbf{x}$ for the case of $m = n$. In (2.5), each output y_i is a weighted sum of all $\{x_j\}_{j=1}^n$, and the row vector, $\mathrm{row}(A;i)$, provides weights for y_i. In (2.6), the output vector $\mathbf{y}$ is expressed as a linear combination of n column vectors, $\{\mathrm{col}(A;j)\}_{j=1}^n$ with weights $\{x_j\}_{j=1}^n$. It is very useful to have these two different views about the linear transform in (2.4) to better understand a solution of an inverse problem as well as the forward problem itself.

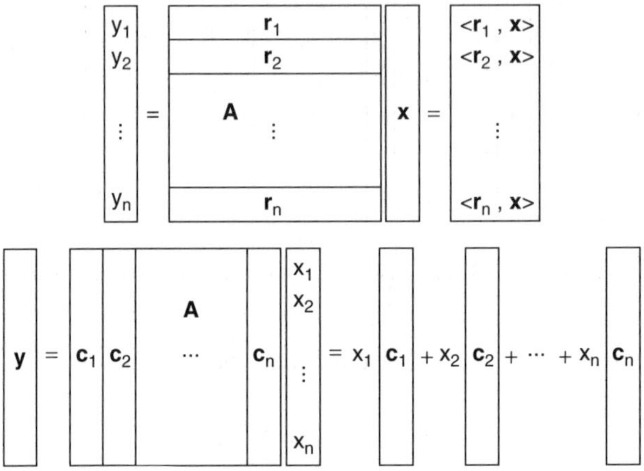

Figure 2.6 Two different representations of a linear system of equations $\mathbf{y} = \mathbf{A}\mathbf{x}$

We now summarize the null space, range and rank of a matrix $\mathbf{A}$ before we describe how to solve $\mathbf{y} = \mathbf{A}\mathbf{x}$ for $\mathbf{x}$.

Definition 2.4.1 *We have the following:*

- *The* **null space** *of a matrix* $\mathbf{A} \in \mathcal{L}(\mathbb{R}^n, \mathbb{R}^m)$ *is*

$$N(A) := \{\mathbf{x} \in \mathbb{R}^n : A\mathbf{x} = \mathbf{0}\}.$$

- *The* **range space** *of a matrix* $\mathbf{A} \in \mathcal{L}(\mathbb{R}^n, \mathbb{R}^m)$ *is*

$$R(A) := \{A\mathbf{x} : \mathbf{x} \in \mathbb{R}^n\}.$$

- *The* **rank** *of a matrix* $\mathbf{A} \in \mathbb{R}^{m\times n}$, *denoted by* rank($\mathbf{A}$), *is the maximum number of independent columns or rows.*

From (2.5) and (2.6), we have

$$N(\mathbf{A}) = \{\mathbf{x} : \text{row}(A; i) \cdot \mathbf{x} = 0, \ \forall\, i = 1, 2, \ldots, m\} \tag{2.7}$$

and

$$R(\mathbf{A}) = \text{span}\{\text{col}(A; j) : j = 1, 2, \ldots, n\}. \tag{2.8}$$

If $N(\mathbf{A}) = \{\mathbf{0}\}$, the following are true:

1. $\text{col}(A; 1), \ldots, \text{col}(A; n)$ are linearly independent.
2. $\mathbf{A} : \mathbb{R}^n \to R(\mathbf{A})$ is invertible, that is, $\mathbf{A}\mathbf{x} = \mathbf{y}$ uniquely determines $\mathbf{x}$.
3. Every $\mathbf{y} \in R(\mathbf{A})$ can be decomposed uniquely as $\mathbf{y} = \sum_{j=1}^{n} x_j \,\text{col}(A; j)$.
4. $\mathbf{A}$ has a left inverse, that is, there exists $\mathbf{B} \in \mathbb{R}^{n\times m}$ such that $\langle \text{row}(B; j), \text{col}(A; k)\rangle = 0$ for $j \neq k$ and $\langle \text{row}(B; j), \text{col}(A; j)\rangle = 1$.
5. $\det(\mathbf{A}^{\mathrm{T}}\mathbf{A}) \neq 0$.

If $R(\mathbf{A}) = \mathbb{R}^m$, the following hold:

1. $\text{span}\{\text{col}(A; j)\}_{j=1}^n = \mathbb{R}^m$.
2. $\{\text{row}(A; j)\}_{j=1}^m$ is linearly independent and $N(\mathbf{A}^T) = \{\mathbf{0}\}$.
3. $\mathbf{A}$ has a right inverse, that is, there exists $\mathbf{B} \in \mathbb{R}^{m\times n}$ such that $\langle \text{row}(A; j), \text{col}(B; k)\rangle = 0$ for $j \neq k$ and $\langle \text{row}(A; j), \text{col}(B; j)\rangle = 1$.
4. $\det(\mathbf{A}\mathbf{A}^T) \neq 0$.

We also note that

$$\text{rank}(\mathbf{A}) = \text{rank}(\mathbf{A}^T) = \dim[R(\mathbf{A})] = n - \dim[N(\mathbf{A})].$$

2.4.3 Least-Squares Solution

We consider a linear system of equations $\mathbf{y} = \mathbf{A}\mathbf{x}$. If there are more equations than unknowns, that is, $m > n$, the system is over-determined and may not have any solution. On the other hand, if there are fewer equations than unknowns, that is, $m < n$, the system is under-determined and has infinitely many solutions. In these cases, we need to seek a best solution of $\mathbf{y} = \mathbf{A}\mathbf{x}$ in an appropriate sense.

Definition 2.4.2 *Let* $\mathbf{A} \in \mathcal{L}(\mathbb{R}^n, \mathbb{R}^m)$. *Then*

- $\mathbf{x}_*$ *is called the* **least-squares solution** *of* $\mathbf{y} = \mathbf{A}\mathbf{x}$ *if*

$$\|\mathbf{A}\mathbf{x}_* - \mathbf{y}\| = \inf_{\mathbf{x}\in\mathbb{R}^n} \|\mathbf{A}\mathbf{x} - \mathbf{y}\|;$$

- $\mathbf{x}^\dagger$ *is called the* **minimum-norm solution** *of* $\mathbf{y} = \mathbf{A}\mathbf{x}$ *if* $\mathbf{x}^\dagger$ *is a least-squares solution of* $\mathbf{y} = \mathbf{A}\mathbf{x}$ *and*

$$\|\mathbf{x}^\dagger\| = \inf\{\|\mathbf{x}\| : \mathbf{x} \textit{ is the least-squares solution of } \mathbf{y} = \mathbf{A}\mathbf{x}\}.$$

If $\mathbf{x}_*$ is the least-squares solution of $\mathbf{y} = \mathbf{A}\mathbf{x}$, then $\mathbf{A}\mathbf{x}_*$ is the projection of $\mathbf{y}$ on $R(\mathbf{A})$, and the orthogonality principle yields

$$0 = \langle \mathbf{A}\mathbf{z}, \mathbf{A}\mathbf{x}_* - \mathbf{y}\rangle = \mathbf{z}^T(\mathbf{A}^T\mathbf{A}\mathbf{x}_* - \mathbf{A}^T\mathbf{y}) \quad \text{for all } \mathbf{z} \in \mathbb{R}^n.$$

If $\mathbf{A}^T\mathbf{A}$ is invertible, then

$$\mathbf{x}_* = (\mathbf{A}^T\mathbf{A})^{-1}\mathbf{A}^T\mathbf{y}$$

and the projection matrix on $R(\mathbf{A})$ can be expressed as

$$\mathbf{P}_\mathbf{A} = \mathbf{A}(\mathbf{A}^T\mathbf{A})^{-1}\mathbf{A}^T. \tag{2.9}$$

Since $\|\mathbf{x}\|^2 \geq \|\mathbf{x}_\dagger\|^2$ for all $\mathbf{x}$ such that $\mathbf{y} = \mathbf{A}\mathbf{x}$,

$$\left(\mathbf{x} - \mathbf{x}_\dagger\right) \perp \mathbf{x}_\dagger \quad \text{for all } \mathbf{x} \text{ satisfying } \mathbf{A}\mathbf{x} = \mathbf{y},$$

that is, $\mathbf{x}_\dagger$ is orthogonal to $N(\mathbf{A})$.

Exercise 2.4.3 *Considering Figure 2.7, explain the least-squares and minimum-norm solutions of* $\mathbf{y} = \mathbf{A}\mathbf{x}$.

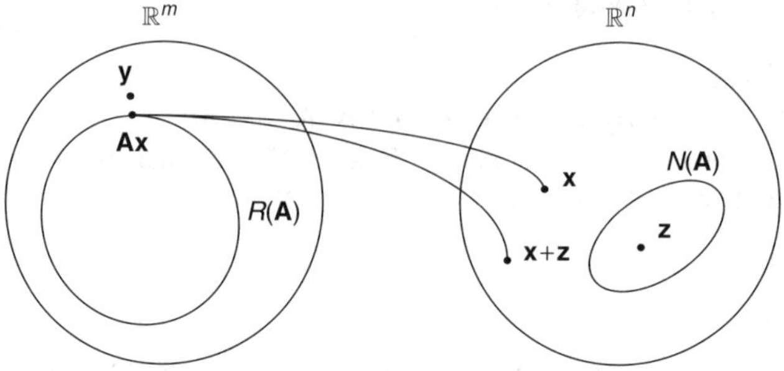

Figure 2.7 Least-squares and minimum-norm solutions of $\mathbf{y} = \mathbf{Ax}$

2.4.4 Singular Value Decomposition (SVD)

Let $\mathbf{A}$ be an $m \times n$ matrix. Then $\mathbf{A}^{\mathrm{T}}\mathbf{A}$ can be decomposed into

$$\mathbf{A}^{\mathrm{T}}\mathbf{A} = \underbrace{(\mathbf{v}_1 \ \cdots \ \mathbf{v}_n)}_{\mathbf{V}} \underbrace{\begin{pmatrix} \sigma_1^2 & & \\ & \ddots & \\ & & \sigma_n^2 \end{pmatrix}}_{\Lambda} \underbrace{\begin{pmatrix} \mathbf{v}_1^{\mathrm{T}} \\ \vdots \\ \mathbf{v}_n^{\mathrm{T}} \end{pmatrix}}_{\mathbf{V}^{\mathrm{T}}},$$

where $\mathbf{v}_1, \ldots, \mathbf{v}_n$ are orthonormal eigenvectors of $\mathbf{A}^{\mathrm{T}}\mathbf{A}$. Since $\mathbf{V}$ is an orthogonal matrix, $\mathbf{V}^{-1} = \mathbf{V}^{\mathrm{T}}$.

If we choose $\mathbf{u}_i = (1/\sigma_i)\mathbf{A}\mathbf{v}_i$, then

$$\|\mathbf{u}_i\|^2 = \frac{\mathbf{v}_i^{\mathrm{T}}\mathbf{A}^{\mathrm{T}}\mathbf{A}\mathbf{v}_i}{\sigma_i^2} = 1 \tag{2.10}$$

and

$$\mathbf{A}\mathbf{A}^{\mathrm{T}}\mathbf{u}_i = \mathbf{A}\mathbf{A}^{\mathrm{T}}\frac{1}{\sigma_i}\mathbf{A}\mathbf{v}_i = \sigma_i\mathbf{A}\mathbf{v}_i = \sigma_i^2\mathbf{u}_i. \tag{2.11}$$

Since $\mathbf{V}^{-1} = \mathbf{V}^{\mathrm{T}}$, we have the singular value decomposition (SVD) of $\mathbf{A}$:

$$\sigma_i\mathbf{u}_i = \mathbf{A}\mathbf{v}_i \quad \Longleftrightarrow \quad \mathbf{U}\Sigma = \mathbf{A}\mathbf{V} \quad \Longleftrightarrow \quad \mathbf{A} = \mathbf{U}\Sigma\mathbf{V}^{\mathrm{T}}, \tag{2.12}$$

where

$$\mathbf{U} = (\mathbf{u}_1 \ \ldots \ \mathbf{u}_n) \quad \text{and} \quad \Sigma = \begin{pmatrix} \sigma_1 & & \\ & \ddots & \\ & & \sigma_n \end{pmatrix}.$$

Suppose we number the $\mathbf{u}_i$, $\mathbf{v}_i$ and σ_i so that $\sigma_1 \geq \sigma_2 \geq \cdots \geq \sigma_r > 0 = \sigma_{r+1} = \cdots = \sigma_n$. Then, we can express the singular value decomposition of $\mathbf{A}$ as

$$\mathbf{A} = (\mathbf{u}_1 \ \ldots \ \mathbf{u}_r) \begin{pmatrix} \sigma_1 & & \\ & \ddots & \\ & & \sigma_r \end{pmatrix} \begin{pmatrix} \mathbf{v}_1^{\mathrm{T}} \\ \vdots \\ \mathbf{v}_r^{\mathrm{T}} \end{pmatrix}. \tag{2.13}$$

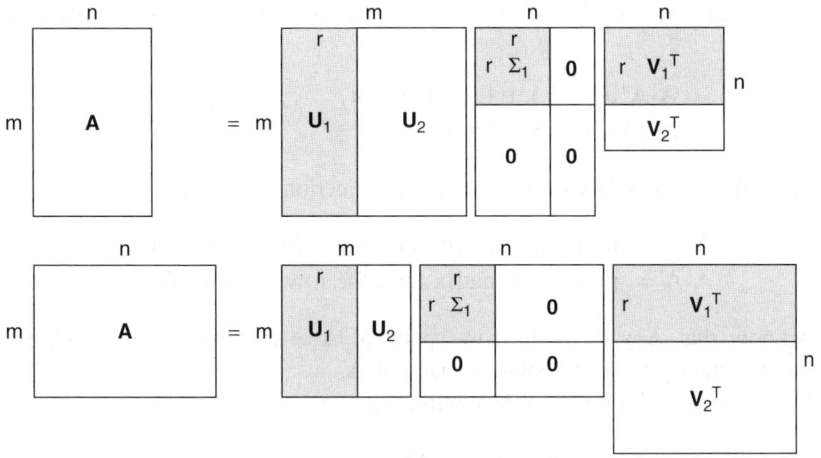

Figure 2.8 Graphical representations of the SVD, $\mathbf{A} = \mathbf{U}\Sigma\mathbf{V}^{\mathrm{T}}$

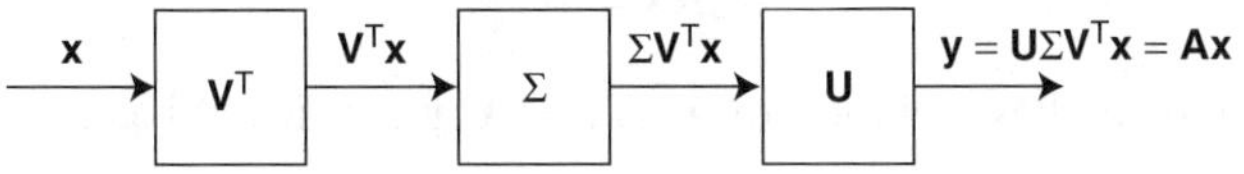

Figure 2.9 Interpretation of $\mathbf{y} = \mathbf{A}\mathbf{x} = \mathbf{U}\Sigma\mathbf{V}^{\mathrm{T}}\mathbf{x}$

This has the very useful property of splitting any matrix **A** into rank-one pieces ordered by their sizes:

$$\mathbf{A} = \sigma_1 \mathbf{u}_1 \mathbf{v}_1^{\mathrm{T}} + \cdots + \sigma_r \mathbf{u}_r \mathbf{v}_r^{\mathrm{T}}. \tag{2.14}$$

Figure 2.8 shows two graphical representations of the SVD. If $\sigma_{t+1}, \ldots, \sigma_r$ are negligibly small, we may approximate **A** by the truncated SVD as

$$\mathbf{A} \approx \sigma_1 \mathbf{u}_1 \mathbf{v}_1^{\mathrm{T}} + \cdots + \sigma_t \mathbf{u}_t \mathbf{v}_t^{\mathrm{T}} \tag{2.15}$$

with $t < r$.

We may interpret the linear transform $\mathbf{y} = \mathbf{A}\mathbf{x} = \mathbf{U}\Sigma\mathbf{V}^{\mathrm{T}}\mathbf{x}$ as shown in Figure 2.9. First, $\mathbf{V}^{\mathrm{T}}\mathbf{x}$ provides coefficients of **x** along the input directions of $\{\mathbf{v}_j\}_{j=1}^r$. Second, $\Sigma\mathbf{V}^{\mathrm{T}}\mathbf{x}$ scales $\mathbf{V}^{\mathrm{T}}\mathbf{x}$ by $\{\sigma_j\}_{j=1}^r$. Third, $\mathbf{U}\Sigma\mathbf{V}^{\mathrm{T}}\mathbf{x}$ reconstructs the output **y** in the directions of $\{\mathbf{u}_j\}_{j=1}^r$. From the relation $\mathbf{A}\mathbf{v}_i = \sigma_i \mathbf{u}_i$, we can interpret $\mathbf{v}_1$ as the most sensitive input direction and $\mathbf{u}_1$ as the most sensitive output direction with the largest gain of σ_1.

2.4.5 Pseudo-inverse

The pseudo-inverse of **A** is

$$\mathbf{A}^{\dagger} = \mathbf{V}\Sigma^{\dagger}\mathbf{U}^{\mathrm{T}} = \frac{1}{\sigma_1}\mathbf{v}_1 \mathbf{u}_1^{\mathrm{T}} + \cdots + \frac{1}{\sigma_r}\mathbf{v}_r \mathbf{u}_r^{\mathrm{T}}. \tag{2.16}$$

Here, in the pseudo-inverse $\Sigma^\dagger$ of the diagonal matrix Σ, each $\sigma \neq 0$ is replaced by $1/\sigma$.

Since

$$\begin{matrix} \mathbf{A}(\mathbf{A}^\dagger\mathbf{u}_j) = \mathbf{A}^\dagger((1/\sigma_j)\mathbf{v}_j) = \mathbf{u}_j \\ \mathbf{A}^\dagger(\mathbf{A}\mathbf{v}_j) = \mathbf{A}^\dagger(\sigma_j\mathbf{u}_j) = \mathbf{v}_j \end{matrix} \quad (j = 1, \ldots, r), \tag{2.17}$$

the products $\mathbf{A}\mathbf{A}^\dagger$ and $\mathbf{A}^\dagger\mathbf{A}$ can be viewed as projection matrices:

$$\begin{aligned} \mathbf{A}\mathbf{A}^\dagger &= \text{ projection matrix onto the column space of } \mathbf{A}, \\ \mathbf{A}^\dagger\mathbf{A} &= \text{ projection matrix onto the row space of } \mathbf{A}. \end{aligned} \tag{2.18}$$

We also know that $\mathbf{A}^\dagger\mathbf{y}$ is in the row space of $\mathbf{A} = \text{span}\{\mathbf{u}_1, \ldots, \mathbf{u}_r\}$ and $\mathbf{A}\mathbf{x} = \mathbf{y}$ is solvable only when $\mathbf{y}$ is in the column space of $\mathbf{A}$.

The least-squares solution $\mathbf{x}^\dagger = \mathbf{A}^\dagger\mathbf{y}$ minimizes

$$\|\mathbf{A}\mathbf{x} - \mathbf{y}\|^2 = (\mathbf{A}\mathbf{x} - \mathbf{y})^{\mathrm{T}}(\mathbf{A}\mathbf{x} - \mathbf{y}). \tag{2.19}$$

Hence, $\mathbf{x}^\dagger$ satisfies the normal equation

$$\mathbf{A}^{\mathrm{T}}\mathbf{A}\mathbf{x} = \mathbf{A}^{\mathrm{T}}\mathbf{y}. \tag{2.20}$$

Moreover, $\mathbf{x}^\dagger$ is the smallest solution of $\mathbf{A}^{\mathrm{T}}\mathbf{A}\mathbf{x} = \mathbf{A}^{\mathrm{T}}\mathbf{y}$ because it has no null components.

2.5 Fourier Transform

Joseph Fourier (1768–1830) introduced the Fourier series to solve the heat equation. Since then, Fourier analysis has been widely used in many branches of science and engineering. It decomposes a general function $f(x)$ into a linear combination of basic harmonics, sines or cosines, that is easier to analyze. For details of the theory, refer to Bracewell (1999) and Gasquet and Witomski (1998). We introduce the Fourier transform as a linear transform since it is also widely used in the inverse problem area.

2.5.1 Series Expansion

Assume that a set $\{\phi_0, \phi_1, \ldots\}$ is an orthonormal basis of a Hilbert space H, that is:

1. $\{\phi_0, \phi_1, \ldots\}$ is orthonormal;
2. every $f \in H$ can be expressed as $f = \sum c_j\phi_j$.

If we denote by P_n the projection map from H to V_n, then

$$P_n f(x) = \sum_{j=0}^{n} \langle f, \phi_j \rangle \phi_j(x).$$

If we denote $P_n^\perp f := f - P_n f$, then $\langle P_n^\perp f, P_n f \rangle = 0$. By the Pythagorean theorem, we have

$$\|f\|^2 = \|P_n f + P_n^\perp f\|^2 = \|P_n f\|^2 + \|P_n^\perp f\|^2$$

and we obtain the Bessel inequality

$$\|f\| \geq \|P_n f\| = \sqrt{\sum_{j=0}^{n} \langle f, \phi_j \rangle^2}. \tag{2.21}$$

By the Bessel inequality, the series

$$P_\infty f := \sum_{j=0}^{\infty} \langle f, \phi_j \rangle \phi_j \text{ converges with respect to the norm } \|\cdot\|.$$

Moreover,

$$\langle f - P_\infty f, \phi_l \rangle = \langle f, \phi_l \rangle - \langle f, \phi_l \rangle = 0 \quad \text{for all } l = 0, 1, 2, \ldots$$

or

$$\langle f - P_\infty f, g \rangle = 0 \quad \text{for all } g \in H.$$

Hence,

$$\|f - P_\infty f\| = 0.$$

This gives us the following series expansion:

$$f(x) = \sum_{j=0}^{\infty} \langle f, \phi_j \rangle \phi_j = \lim_{n\to\infty} P_n f(x).$$

Exercise 2.5.1 *Let* $V_N = \text{span}\{1, \cos 2\pi x, \sin 2\pi x, \ldots, \cos 2\pi Nx, \sin 2\pi Nx\}$. *Prove that* $f \in V_N$ *can be expanded as the following trigonometric series:*

$$f(x) = a_0 + \sum_{j=1}^{N} a_j \cos 2\pi jx + \sum_{j=1}^{N} b_j \sin 2\pi jx,$$

where

$$a_0 = \int_0^1 f(x)\,\mathrm{d}x,$$

$$a_j = 2\int_0^1 f(x) \cos 2\pi jx\,\mathrm{d}x,$$

$$b_j = 2\int_0^1 f(x) \sin 2\pi jx\,\mathrm{d}x.$$

Example 2.5.2 *Let* $V_N = \text{span}\{\mathrm{e}^{2\pi ikx} \mid k = 0, \pm 1, \ldots, \pm N\}$ *be the subspace of* $L^2([0, 1])$ *and let* H *be the closure of* $\cup_{N=1}^{\infty} V_N$. *Then, any complex-valued function* $f \in H$ *can be uniquely represented by*

$$f(x) = \sum_{n=-\infty}^{n=\infty} \alpha_n \mathrm{e}^{2\pi inx}, \quad \alpha_n = \int_0^1 f(y)\,\overline{\mathrm{e}^{2\pi iny}}\,\mathrm{d}y.$$

To see this, note that

$$\int_0^1 e^{2\pi i n x}\,\overline{e^{2\pi i m x}}\,dx = \begin{cases} 1 & \text{if } n = m, \\ 0 & \text{if } n \neq m. \end{cases}$$

The projection map P_N from H to V_N can be expressed as

$$P_N f(x) = \sum_{n=-N}^{N} \alpha_n e^{2\pi i n x}.$$

From the Bessel inequality, we have

$$\sum_{n=-N}^{N} |\alpha_n|^2 = \|P_N f\|^2_{L^2(0,1)} \leq \|f\|^2_{L^2(0.1)}.$$

Hence, we have the Riemann–Lebesgue lemma:

$$\lim_{n\to\pm\infty} \alpha_n = \lim_{n\to\pm\infty} \int_0^1 f(y)\,\overline{e^{2\pi i n y}}\,dy = 0$$

and

$$\lim_{N\to\infty} \|P_N f - f\|^2 = \lim_{N\to\infty} \|P_N^{\perp} f\|^2 = \sum_{n=N+1}^{\infty} |\alpha_n|^2 + |\alpha_{-n}|^2 = 0.$$

2.5.2 Fourier Transform

For each p with $1 \leq p < \infty$, we denote the class of measurable functions f on $\mathbb{R}$ such that $\int_{-\infty}^{\infty} |f(x)|^p\,dx < \infty$ by $L^p(\mathbb{R})$:

$$L^p(\mathbb{R}) = \{f \mid \|f\|_p < \infty\}$$

where the L^p-norm of f is

$$\|f\|_p = \left(\int_{-\infty}^{\infty} |f(x)|^p\,dx\right)^{1/p}.$$

Definition 2.5.3 *We have the following:*

1. *If $f \in L^1(\mathbb{R})$, the* **Fourier transform** *of f is $\widehat{f}$ defined by*

$$\widehat{f}(\xi) = (\mathcal{F}f)(\xi) = \int_{-\infty}^{\infty} f(x)e^{-2\pi i \xi x}\,dx.$$

2. *The* **convolution** *of two functions f and g is given by*

$$f * g(x) = \int_{-\infty}^{\infty} f(x-y)g(y)\,dy.$$

3. *The Dirac* **delta function** *δ is a distribution satisfying*

$$\delta * f(x) = \int_{-\infty}^{\infty} \delta(y-x)f(y)\,dy = f(x) \quad \text{for all } f \in C(\mathbb{R}) \cap L^1(\mathbb{R}).$$

4. *The* **support** *of* f *is the set* $\operatorname{supp}(f) := \overline{\{x \in \mathbb{R} : f(x) \neq 0\}}$.
5. *The function* f *is* **band-limited** *if* $\operatorname{supp}(\widehat{f})$ *is bounded. The function* f *is said to be band-limited with width* L *if* $\operatorname{supp}(\widehat{f}) \subset [-L, L]$.
6. *The* **Gaussian function** *with a mean value b and a standard deviation* σ *is*

$$G_{\sigma,b}(x) := \frac{1}{\sigma\sqrt{2\pi}}\,\mathrm{e}^{-(x-b)^2/(2\sigma^2)}.$$

The Gaussian function has the property that $\int_{\mathbb{R}} G_{\sigma,b}(x)\,\mathrm{d}x = 1$ *and* $\lim_{\sigma\to 0} G_{\sigma,b}(x) = \delta(x-b)$.

Let us summarize the Fourier transform pairs of some important functions.

- Scaling property:

$$\widehat{f(ax)}(\xi) = \int_{\mathbb{R}} f(ax)\,\underbrace{\mathrm{e}^{-2\pi i a x\xi/a}}_{=\mathrm{e}^{-2\pi i x\xi}}\,\mathrm{d}x = \frac{1}{a}\widehat{f}\left(\frac{\xi}{a}\right).$$

- Shifting property:

$$\widehat{f(x-a)}(\xi) = \int_{\mathbb{R}} f(x-a)\,\underbrace{\mathrm{e}^{-2\pi i(x-a)\xi}\,\mathrm{e}^{-2\pi i a\xi}}_{=\mathrm{e}^{-2\pi i x\xi}}\,\mathrm{d}x = \widehat{f}(\xi)\,\mathrm{e}^{-2\pi i a\xi}.$$

- Dirac delta function:

$$\widehat{\delta(x)}(\xi) = 1 \quad \text{and} \quad \widehat{\delta(x-T)}(\xi) = \mathrm{e}^{-2\pi i\xi T}.$$

- Dirac comb:

$$\operatorname{comb}(x) := \sum_{n=-\infty}^{\infty} \delta(x-nT) \quad \Longrightarrow \quad \widehat{\operatorname{comb}}(\xi) = \sum_{n=-\infty}^{\infty} \mathrm{e}^{-2\pi i n T\xi}.$$

- Indicator function of the interval $[a, b]$, denoted by $\chi_{[a,b]}(x)$, is

$$\chi_{[a,b]}(x) = \begin{cases} 1 & \text{if } x \in [a,b], \\ 0 & \text{otherwise.} \end{cases}$$

  We have

$$f(x) = \chi_{[-T/2,T/2]}(x) \quad \Longrightarrow \quad \widehat{f}(\xi) = \int_{-T/2}^{T/2} \mathrm{e}^{-2\pi i x\xi}\,\mathrm{d}x = \frac{\sin(T\pi\xi)}{\pi\xi}.$$

- The Fourier transform of Gaussian function e^{-ax^2} is given by

$$\begin{aligned} \widehat{\mathrm{e}^{-ax^2}}(\xi) &= \int_{\mathbb{R}} \mathrm{e}^{-ax^2}\mathrm{e}^{-2\pi i x\xi}\,\mathrm{d}x = \mathrm{e}^{-\pi^2\xi^2/a}\int_{\mathbb{R}} \mathrm{e}^{-a(x+(\pi/a)i\xi)^2}\,\mathrm{d}x \\ &= \mathrm{e}^{-\pi^2\xi^2/a}\underbrace{\int_{\mathbb{R}} \mathrm{e}^{-ax^2}\,\mathrm{d}x}_{=\sqrt{\pi}/\sqrt{a}} = \frac{\sqrt{\pi}}{\sqrt{a}}\,\mathrm{e}^{-\pi^2\xi^2/a}. \end{aligned}$$

For the above identity $\int_{\mathbb{R}} e^{-a(x+(\pi/a)i\xi)^2}\,dx = \int_{\mathbb{R}} e^{-ax^2}\,dx$, we use the Cauchy integral formula that $\oint_C e^{-az^2}\,dz = 0$ for any closed curve C.

- The Fourier transform of $f * g(x) = \int f(x-y)g(y)\,dy$ is

$$\widehat{f * g}(\xi) = \int_{\mathbb{R}} f * g(x)\, \underbrace{e^{-2\pi i(x-y)\xi}\, e^{-2\pi i y\xi}}_{=e^{-2\pi i x\xi}}\,dx = \widehat{f}(\xi)\,\widehat{g}(\xi).$$

- Modulation: if $g(x) = f(x)\cos(\xi_0 x)$, then

$$\widehat{g}(\xi) = \int_{\mathbb{R}} f(x)\frac{e^{2\pi i\xi_0 x} + e^{-2\pi i\xi_0 x}}{2}\, e^{-2\pi i x\xi}\,dx = \frac{1}{2}[\widehat{f}(\xi+\xi_0) + \widehat{f}(\xi-\xi_0)].$$

- Derivative: if $f \in C_0^1(\mathbb{R}) := \{\phi \in C^1(\mathbb{R}) : \text{supp}(\phi) \text{ is bounded}\}$, then integrating by parts leads to

$$\widehat{\frac{d}{dx}f}(\xi) = \int_{\mathbb{R}} \frac{d}{dx}f(x)\, e^{-2\pi i x\xi}\,dx = -\int_{\mathbb{R}} f(x)\frac{d}{dx}(e^{-2\pi i x\xi})\,dx = 2\pi i\xi\,\widehat{f}(\xi).$$

The following Fourier inversion provides that f can be recovered from its Fourier transform $\widehat{f}$.

Theorem 2.5.4 (Fourier inversion formula) *If $f, \widehat{f} \in L^1(\mathbb{R}) \cap C(\mathbb{R})$, then*

$$f(x) = \int_R \hat{f}(\xi)\, e^{2\pi i\xi x}\,d\xi. \tag{2.22}$$

Proof. The inverse Fourier transform of $\widehat{f}$ can be expressed as

$$\begin{aligned}\int_{\mathbb{R}} \widehat{f}(\xi)\, e^{2\pi i\xi x}\,d\xi &= \int_{\mathbb{R}} \widehat{f}(\xi)\, e^{2\pi i\xi x} \underbrace{\lim_{\epsilon\to 0}[\exp(-\pi\epsilon^2\xi^2)]}_{=1}\,d\xi \\ &= \lim_{\epsilon\to 0}\int_{\mathbb{R}}\left[\int_{\mathbb{R}} e^{2\pi i(x-y)\xi}\exp(-\pi\epsilon^2\xi^2)\,d\xi\right] f(y)\,dy.\end{aligned} \tag{2.23}$$

The last identity can be obtained by interchanging the order of the integration. From the identity $\widehat{e^{-\pi x^2}}(\xi) = e^{-\pi\xi^2}$ and the scaling property $\widehat{f(ax)}(\xi) = (1/a)\widehat{f}(\xi/a)$, the last quantity in (2.23) can be expressed as

$$\left[\int_{\mathbb{R}} e^{2\pi i(x-y)\xi}\exp(-\pi\epsilon^2\xi^2)\,d\xi\right] = \underbrace{\frac{1}{\epsilon}\exp\left(\frac{-\pi(x-y)^2}{\epsilon^2}\right)}_{\phi_\epsilon(x-y)}.$$

Then, the identity (2.23) can be simplified into

$$\int_{\mathbb{R}} \widehat{f}(\xi)\, e^{2\pi i\xi x}\,d\xi = \lim_{\epsilon\to 0}\phi_\epsilon * f(x).$$

Since $\lim_{\epsilon\to 0}\phi_\epsilon(x) = 0$ for all $x \neq 0$ and $\int_{\mathbb{R}} \phi_\epsilon(x)\,\mathrm{d}x = 1$, we have

$$\lim_{\epsilon\to 0} \phi_\epsilon * f(x) = \int_{\mathbb{R}} \delta(x-y) f(y)\,\mathrm{d}y = f(x). \qquad \square$$

Indeed, the Fourier inversion formula holds for general $f \in L^2(\mathbb{R})$ because $C_0(\mathbb{R})$ is dense in $L^2(\mathbb{R})$ and $\int_{\mathbb{R}} |\widehat{f}|^2\,\mathrm{d}\xi = \int_{\mathbb{R}} |f|^2\,\mathrm{d}x$. We omit the proof because it requires some time-consuming arguments of Lebesgue measure theory and some limiting process in L^2. For a rigorous proof, please refer to Gasquet and Witomski (1998).

The following Poisson summation formula indicates that the T-periodic summation of f is expressed as discrete samples of its Fourier transform $\widehat{f}$ with the sampling distance $1/T$.

Theorem 2.5.5 (Poisson summation formula) *For $f \in C(\mathbb{R}) \cap L^1(\mathbb{R})$, we have*

$$\sum_{n=-\infty}^{\infty} f(x-nT) = \frac{1}{T} \sum_{n=-\infty}^{\infty} \widehat{f}\left(\frac{n}{T}\right) \mathrm{e}^{2\pi\mathrm{i}(n/T)x}. \tag{2.24}$$

In particular,

$$\mathrm{supp}(f) \subset [0,T] \implies f(x) = \frac{1}{T} \sum_{n=-\infty}^{\infty} \widehat{f}\left(\frac{n}{T}\right) \mathrm{e}^{2\pi\mathrm{i}(n/T)x} \quad \textit{for } x \in [0,T],$$

$$\mathrm{supp}(f) \subset [0,2T] \implies f(x) + f(x+T) = \frac{1}{T} \sum_{n=-\infty}^{\infty} \widehat{f}\left(\frac{n}{T}\right) \mathrm{e}^{2\pi\mathrm{i}(n/T)x} \quad \textit{for } x \in [0,2T].$$

Proof. Denoting $\mathrm{comb}_T(x) = \sum_{n=-\infty}^{\infty} \delta(x-nT)$, we have

$$\sum_{n=-\infty}^{\infty} f(x-nT) = f * \mathrm{comb}_T(x).$$

The T-periodic function $f * \mathrm{comb}_T(x)$ can be expressed as

$$f * \mathrm{comb}_T(x) = \sum_{n=-\infty}^{\infty} a_n\, \mathrm{e}^{2\pi\mathrm{i}(n/T)x},$$

where

$$a_n = \frac{1}{T} \int_0^T f * \mathrm{comb}_T(x)\, \mathrm{e}^{-2\pi\mathrm{i}(n/T)x}\,\mathrm{d}x.$$

This a_n is given by

$$\begin{aligned} a_n &= \frac{1}{T} \int_0^T \sum_{n=-\infty}^{\infty} f(x-nT)\, \mathrm{e}^{-2\pi\mathrm{i}(n/T)x}\,\mathrm{d}x \\ &= \frac{1}{T} \int_{-\infty}^{\infty} f(x)\, \mathrm{e}^{-2\pi\mathrm{i}(n/T)x}\,\mathrm{d}x \\ &= \frac{1}{T} \widehat{f}\left(\frac{n}{T}\right). \end{aligned} \qquad \square$$

Theorem 2.5.6 *If* $\mathrm{comb}_T(x) = \sum_{n=-\infty}^{\infty} \delta(x - nT)$, *then*

$$\widehat{\mathrm{comb}_T}(\xi) = \sum_{n=-\infty}^{\infty} \mathrm{e}^{-2\pi i n T\xi} = \frac{1}{T}\sum_{n=-\infty}^{\infty} \delta\left(\xi - \frac{n}{T}\right).$$

From this, we again derive the Poisson summation formula:

$$f * \mathrm{comb}_T(x) = \mathcal{F}^{-1}\left(\hat{f}(\xi)\,\underbrace{\left[(1/T)\,\mathrm{comb}_{1/T}(\xi)\right]}_{\widehat{\mathrm{comb}_T}(\xi)}\right)(x) = \frac{1}{T}\sum_{n=-\infty}^{\infty} \hat{f}\left(\frac{n}{T}\right)\mathrm{e}^{2\pi i (n/T)x}. \tag{2.25}$$

Proof. Since $\widehat{\mathrm{comb}_T}(\xi)$ is $1/T$-periodic, it suffices to prove

$$\widehat{\mathrm{comb}_T}(\xi)\chi_{[-1/(2T),1/(2T)]} = \frac{1}{T}\,\delta(\xi).$$

The above identity can be derived from the following facts.

1. $\displaystyle\int_{\mathbb{R}} \widehat{\mathrm{comb}_T}(\xi)\chi_{[-1/(2T),1/(2T)]}\,\mathrm{d}\xi = \frac{1}{T}.$
2. $\displaystyle\widehat{\mathrm{comb}_T}(\xi) = \lim_{N\to\infty}\sum_{n=-N}^{N} \mathrm{e}^{-2\pi i n T\xi} = \lim_{N\to\infty}\frac{\sin[2\pi(N+\frac{1}{2})T\xi]}{\sin(\pi T\xi)}.$
3. For any $\phi \in C(\mathbb{R}) \cap L^1(\mathbb{R})$ with $\phi(0) = 0$,

$$\lim_{N\to\infty}\int_{\mathbb{R}} \frac{\sin[2\pi(N+\frac{1}{2})T\xi]}{\sin(\pi T\xi)}\phi(\xi)\,\mathrm{d}\xi = 0.$$

□

The band-limited function f with bandwidth B, that is, $\mathrm{supp}(\hat{f}) \subset [-B, B]$, does not contain sinusoidal waves at frequencies higher than B. This means that f cannot oscillate rapidly within a distance less than $1/(2B)$. Hence, the band-limited f can be represented by means of its uniformly spaced discrete sample $\{f(n\Delta x) : n = 0, \pm 1, \pm 2, \dots\}$ provided that the sampling interval Δx is sufficiently small. Indeed, the following sampling theorem states that, if $\Delta x \le 1/(2B)$, then the discrete sample $\{f(n\Delta x) : n = 0, \pm 1, \pm 2, \dots\}$ contains the complete information about f. This $2B$ is called the *Nyquist rate*.

Theorem 2.5.7 (Whittaker–Shannon sampling theorem) *Suppose* $f \in C(\mathbb{R}) \cap L^1(\mathbb{R})$ *and* $\mathrm{supp}(\hat{f}) \subset [-B, B]$. *The original data* f *can be reconstructed by the interpolation formula*

$$f(x) = \sum_{n=-\infty}^{\infty} f\left(\frac{1}{2B}n\right)\,\mathrm{sinc}\left(2B\left(x - \frac{1}{2B}n\right)\right),$$

where $\mathrm{sinc}(x) = \sin(\pi x)/(\pi x)$.

Proof. It is enough to prove the above sampling theorem for $B = \frac{1}{2}$. Denoting

$$\mathrm{comb}(x) = \sum_{n=-\infty}^{\infty} \delta(x-n),$$

the sampled version $f(x)\,\mathrm{comb}(x)$ is

$$f(x)\,\mathrm{comb}(x) = f(x) \sum_{n=-\infty}^{\infty} \delta(x-n) = \sum_{n=-\infty}^{\infty} f(n)\delta(x-n).$$

Taking the Fourier transforms of the above identity, we obtain

$$\widehat{f\,\mathrm{comb}}(\omega) = \sum_{n=-\infty}^{\infty} f(n)\,\mathrm{e}^{-2\pi i n\omega} = \sum_{n=-\infty}^{\infty} \widehat{f}(\omega - n).$$

The last equality comes from the Poisson summation formula (2.24). From the assumption $\mathrm{supp}(\widehat{f}\,) \subset [-\frac{1}{2}, \frac{1}{2}]$,

$$\widehat{f\,\mathrm{comb}}(\omega) = \sum_{n=-\infty}^{\infty} \widehat{f}(\omega - n) = \widehat{f}(\omega) \quad \text{for all } -\frac{1}{2} < \xi < \frac{1}{2}.$$

This means that

$$\widehat{f}(\omega) = \widehat{f\,\mathrm{comb}}(\omega)\chi_{[-1/2,1/2]}(\omega).$$

Since $\widehat{\mathrm{sinc}}(\xi) = \chi_{[-1/2,1/2]}(\xi)$, we get

$$\widehat{f}(\omega) = \widehat{(f\,\mathrm{comb}) * \mathrm{sinc}}(\omega).$$

This gives

$$f(x) = (f\,\mathrm{comb}) * \mathrm{sinc}(x) = \sum_{n=-\infty}^{\infty} f(n)\,\mathrm{sinc}(x-n).$$

□

Remark 2.5.8 *When we need to analyze an analog signal, it must be converted into digital form by an analog-to-digital converter (ADC). Assume that $\widehat{f}$ is the analog signal to be analyzed and $1/T$ is the sampling interval. According to the Poisson summation formula, the T-periodic function $f * \mathrm{comb}_T$ is reconstructed from the samples $\widehat{f}(n/T), n = 0, \pm 1, \ldots$. This result is based on the fact that the Fourier transform of $\mathrm{comb}_T(x) = \sum_{n=-\infty}^{\infty} \delta(x - nT)$ is $(1/T)\,\mathrm{comb}_{1/T}(\xi)$. Hence, if $f(x) \neq 0$ for $\frac{1}{2}T < |x| < T$, the formula (2.25) produces aliasing or wrap-around. To avoid aliasing, we require that $\mathrm{supp}(f) \subset [-T/2, T/2]$, that is, the spatial frequency of f should be less than $T/2$.*

2.5.3 Discrete Fourier Transform (DFT)

Assume that f is a continuous signal such that

$$f(x) \approx 0 \text{ for } |x| > \frac{T}{2} \quad \text{and} \quad \widehat{f}(\xi) \approx 0 \text{ for } |\xi| > \frac{\Xi}{2}.$$

Choose a number N so that

$$\frac{T}{N} \le \frac{1}{\Xi} \quad \text{and} \quad \frac{\Xi}{N} \le \frac{1}{T}.$$

With the condition $N \ge T\Xi$, the original signals f and $\hat{f}$ are approximately recovered from the samples $\{f(kT/N - T/2) : k = 0, 1, \ldots, N-1\}$ and $\{\hat{f}(k\Xi/N - \Xi/2) : k = 0, 1, \ldots, N-1\}$ based on either the Poisson summation formula or the Whittaker–Shannon sampling theorem.

Let us convert the continuous signal f into the digital signal $\{f(k\Delta x - T/2) : k = 0, 1, \ldots, N-1\}$ with the sampling spacing $\Delta x = T/N$. The points $x_k = k\Delta x - T/2$, with $k = 0, 1, \ldots, N-1$, are called the *sampling points*.

Writing $f_k = f(k\Delta x - T/2)$, the digital signal corresponding to the continuous signal f can be expressed in the following vector form:

$$\mathbf{f} := \begin{pmatrix} f_0 \\ \vdots \\ f_{N-1} \end{pmatrix} = \begin{pmatrix} f(-T/2) \\ \vdots \\ f((N-1)\Delta x - T/2) \end{pmatrix}.$$

Its Fourier transform $\hat{f}(\xi)$ for $\xi \in [-\Xi/2, \Xi/2]$ is expressed approximately by

$$\hat{f}(\xi) = \int_{-\infty}^{\infty} f(x)\,\mathrm{e}^{-2\pi\mathrm{i}x\xi}\,\mathrm{d}x \approx \int_{-T/2}^{T/2} f(x)\,\mathrm{e}^{-2\pi\mathrm{i}x\xi}\,\mathrm{d}x \approx \sum_{k=0}^{N-1} f_k\,\mathrm{e}^{-2\pi\mathrm{i}x_k\xi}\,\Delta x.$$

Similarly, we denote $\hat{f}_j = \hat{f}(\xi_j)$ where $\xi_j = j\Delta\xi - \Xi/2$ and $\Delta\xi = \Xi/N$. Then, $f(x)$ for $x \in [-T/2, T/2]$ is expressed approximately by

$$f(x) = \int_{-\infty}^{\infty} \hat{f}(\xi)\,\mathrm{e}^{2\pi\mathrm{i}x\xi}\,\mathrm{d}\xi \approx \int_{-\Xi/2}^{\Xi/2} \hat{f}(\xi)\,\mathrm{e}^{2\pi\mathrm{i}x\xi}\,\mathrm{d}\xi \approx \sum_{j=0}^{N-1} \hat{f}_j\,\mathrm{e}^{2\pi\mathrm{i}x\xi_j}\,\Delta\xi.$$

In particular, we have

$$\hat{f}_j \approx \sum_{k=0}^{N-1} f_k\,\mathrm{e}^{-2\pi\mathrm{i}x_k\xi_j}\,\Delta x, \qquad \xi_j = j\Delta\xi - \frac{\Xi}{2},$$

$$f_k \approx \sum_{j=0}^{N-1} \hat{f}_j\,\mathrm{e}^{2\pi\mathrm{i}x_k\xi_j}\,\Delta\xi, \qquad x_k = k\Delta x - \frac{T}{2}.$$

Since

$$\Delta x\,\Delta\xi = \frac{1}{N} \quad \text{and} \quad T\Xi = N \quad \Longrightarrow \quad \mathrm{e}^{-2\pi\mathrm{i}x_k\xi_j} = \mathrm{e}^{-2\pi\mathrm{i}jk/N},$$

we have the following approximations:

$$\hat{f}_j \approx \sum_{k=0}^{N-1} f_k\,\mathrm{e}^{-(2\pi\mathrm{i}/N)jk}\,\Delta x \quad \text{and} \quad f_k \approx \sum_{j=0}^{N-1} \hat{f}_j\,\mathrm{e}^{(2\pi\mathrm{i}/N)jk}\,\Delta\xi.$$

From the above approximations, we can define the discrete Fourier transform.

Definition 2.5.9 *The discrete Fourier transform (DFT) on $\mathbb{C}^N$ is a linear transform $\mathcal{F}_N : \mathbb{C}^N \to \mathbb{C}^N$ given by*

$$\widehat{\mathbf{f}} = \mathcal{F}_N \mathbf{f} = \underbrace{\begin{pmatrix} 1 & 1 & 1 & \cdots & 1 \\ 1 & \gamma_N^1 & \gamma_N^2 & \cdots & \gamma_N^{N-1} \\ 1 & \gamma_N^2 & \gamma_N^4 & \cdots & \gamma_N^{2(N-1)} \\ 1 & \gamma_N^3 & \gamma_N^6 & \cdots & \gamma_N^{3(N-1)} \\ \vdots & \vdots & \vdots & \ddots & \vdots \\ 1 & \gamma_N^{(N-1)} & \gamma_N^{2(N-1)} & \cdots & \gamma_N^{(N-1)^2} \end{pmatrix}}_{=\mathcal{F}_N} \mathbf{f}, \quad \mathbf{f} \in \mathbb{C}^N,$$

where $\gamma_N := \exp(-2\pi \mathrm{i}/N)$ is the Nth principal root of 1. Here, the DFT linear transform $\mathcal{F}_N$ can be viewed as an $N \times N$ matrix.

The DFT matrix $\mathcal{F}_N$ has the following interesting properties.

- Let $\mathbf{a}_{j-1}$ be the jth row (or column) of the Fourier transform matrix $\mathcal{F}_N$. Then,

$$\widehat{f}_j = \langle \mathbf{f}, \mathbf{a}_j \rangle \quad \text{for } j = 0, 1, \ldots, N-1.$$

- The column vectors $\mathbf{a}_0, \ldots, \mathbf{a}_{N-1}$ are eigenvectors of the following matrix corresponding to the Laplace operator:

$$\underbrace{\begin{pmatrix} 2 & -1 & 0 & 0 & \cdots & -1 \\ -1 & 2 & -1 & 0 & \cdots & 0 \\ 0 & -1 & 2 & -1 & \cdots & 0 \\ \vdots & \vdots & \vdots & \vdots & \ddots & \vdots \\ -1 & 0 & 0 & 0 & \cdots & 2 \end{pmatrix}}_{N \times N \text{ matrix}} \mathbf{a}_j = (2 - \gamma_N^j - \gamma_N^{-j}) \mathbf{a}_j.$$

- Hence, $\{\mathbf{a}_0, \ldots, \mathbf{a}_{N-1}\}$ forms an orthogonal basis over the N-dimensional vector space $\mathbb{C}^N$ and

$$\mathcal{F}_N^* \mathcal{F}_N = N \begin{pmatrix} 1 & 0 & \cdots & 0 \\ 0 & 1 & \cdots & 0 \\ \vdots & \vdots & \ddots & \vdots \\ 0 & 0 & \cdots & 1 \end{pmatrix},$$

  where A^* is the complex conjugate of the transpose of the matrix A. The inverse of the DFT linear transform $\mathcal{F}_N$ is simply its transpose:

$$(\mathcal{F}_N)^{-1} = \frac{1}{N} \mathcal{F}_N^*.$$

- If $\mathcal{F}_N(\mathbf{f})$ and $\mathcal{F}_N(\mathbf{g})$ are the DFTs of $\mathbf{f}$ and $\mathbf{g}$, then we have Parseval's identity:

$$\langle \mathbf{f}, \mathbf{g} \rangle = \frac{1}{N} \langle (\mathcal{F}_N^* \mathcal{F}_N) \mathbf{f}, \mathbf{g} \rangle = \frac{1}{N} \langle \mathcal{F}_N(\mathbf{f}), \mathcal{F}_N(\mathbf{g}) \rangle,$$

  where the inner product is defined by $\langle \mathbf{f}, \mathbf{g} \rangle = \sum_{k=0}^{N-1} f_k \, \overline{g_k}$.

2.5.4 Fast Fourier Transform (FFT)

The fast Fourier transform (FFT) is a DFT algorithm that reduces the number of computations from something on the order of N^2 to $N \log N$. Let $N = 2M$ and

$$\mathbf{f} := \begin{pmatrix} f_0 \\ f_1 \\ f_2 \\ \vdots \\ f_{2M-1} \end{pmatrix}, \quad \mathbf{f}_{\text{even}} := \begin{pmatrix} f_0 \\ f_2 \\ \vdots \\ f_{2M-2} \end{pmatrix}, \quad \mathbf{f}_{\text{odd}} := \begin{pmatrix} f_1 \\ f_3 \\ \vdots \\ f_{2M-1} \end{pmatrix}.$$

From the definition of the DFT, we have

$$\mathcal{F}_{2M}\mathbf{f} = \underbrace{\begin{pmatrix} 1 & 1 & 1 & \cdots & 1 \\ 1 & \gamma_{2M}^1 & \gamma_{2M}^2 & \cdots & \gamma_{2M}^{2M-1} \\ 1 & \gamma_{2M}^2 & \gamma_{2M}^4 & \cdots & \gamma_{2M}^{2(2M-1)} \\ 1 & \gamma_{2M}^3 & \gamma_{2M}^6 & \cdots & \gamma_{2M}^{3(2M-1)} \\ \vdots & \vdots & \vdots & \ddots & \vdots \\ 1 & \gamma_{2M}^{(2M-1)} & \gamma_{2M}^{2(2M-1)} & \cdots & \gamma_{2M}^{(2M-1)^2} \end{pmatrix}}_{=\mathcal{F}_{2M}} \left[\begin{pmatrix} f_0 \\ 0 \\ f_2 \\ \vdots \\ f_{2M-2} \\ 0 \end{pmatrix} + \begin{pmatrix} 0 \\ f_1 \\ 0 \\ \vdots \\ 0 \\ f_{2M-1} \end{pmatrix} \right].$$

Since $\gamma_{2M}^M = -1$ and $\gamma_{2M}^2 = \gamma_M$, the $(M+j, 2k)$-component of the matrix $\mathcal{F}_{2M}$ is

$$\gamma_{2M}^{j(2k)} = \gamma_M^{jk} \quad \text{and} \quad \gamma_{2M}^{(M+j)(2k)} = \gamma_M^{jk} \quad \text{for } k = 0, 1, \ldots, M-1,$$

and the $(M+j, 2k-1)$-component of the matrix $\mathcal{F}_{2M}$ is

$$\gamma_{2M}^{(M+j)(2k-1)} = -\gamma_{2M}^j \gamma_{2M}^{j(2k)} = -\gamma_{2M}^j \gamma_M^{jk} \quad \text{for } k = 0, 1, \ldots, M-1.$$

The FFT is based on the following key identity:

$$\mathcal{F}_{2M}\mathbf{f} = \begin{pmatrix} \mathcal{F}_M \mathbf{f}_{\text{even}} \\ \mathcal{F}_M \mathbf{f}_{\text{even}} \end{pmatrix} + \begin{pmatrix} \Theta_m & \begin{matrix} 0 & 0 & \cdots \\ 0 & 0 & \\ \vdots & & \ddots \end{matrix} \\ \begin{matrix} 0 & 0 & \cdots \\ 0 & 0 & \\ \vdots & & \ddots \end{matrix} & -\Theta_m \end{pmatrix} \begin{pmatrix} \mathcal{F}_M \mathbf{f}_{\text{odd}} \\ \mathcal{F}_M \mathbf{f}_{\text{odd}} \end{pmatrix}, \tag{2.26}$$

where $\mathcal{F}_N$ is the $N \times N$ DFT matrix defined in the previous section and

$$\Theta_m = \begin{pmatrix} 1 & 0 & 0 & \cdots & 0 \\ 0 & \gamma_{2M} & 0 & \cdots & 0 \\ 0 & 0 & \gamma_{2M}^2 & \cdots & 0 \\ \vdots & \vdots & \vdots & \ddots & \vdots \\ 0 & 0 & 0 & \cdots & \gamma_{2M}^{M-1} \end{pmatrix} \in \mathbb{C}^M \times \mathbb{C}^M.$$

Remark 2.5.10 *Assume that f is supported in $[-\frac{1}{2}N\Delta x, \frac{1}{2}N\Delta x]$ and that $\mathbf{f} = (f_{-N/2}, \ldots, f_{(N/2)-1})$ is a digital image of $f(x)$, where $f_n = f(n\Delta x)$ and Δx is the sampling interval (or pixel size). The interval $[-\frac{1}{2}N\Delta x, \frac{1}{2}N\Delta x]$ is referred to as the field of view (FOV). For simplicity, assume $N\Delta x = 1$, that is, $FOV = 1$. The number N may be regarded as the number of pixels. Assume that f is band-limited with $[-\frac{1}{2}N\Delta\xi, \frac{1}{2}N\Delta\xi]$, that is, $\max\{|\xi| : \widehat{f}(\xi) \neq 0\} \leq \frac{1}{2}N\Delta\xi$ and denote $\widehat{f}_k = \widehat{f}(k\Delta\xi)$. We can recover f from $\mathbf{f}$ without any loss of information provided that it meets the Nyquist criterion:*

$$\Delta\xi \leq \frac{1}{N\Delta x} = \frac{1}{FOV}.$$

Hence, if $\Delta\xi = 1/(N\Delta x)$, the DFT gives

$$f_n = \sum_{k=-N/2}^{N/2-1} \widehat{f}_k \, e^{2\pi i(kn)/N}, \quad n = -\frac{N}{2}, \ldots, \frac{N}{2} - 1.$$

Extending the sequence $\mathbf{f} = (f_{-N/2}, \ldots, f_{(N/2)-1})$ to the N-periodic sequence in such a way that $f_{n+mN} = f_n$, the discrete version of the Poisson summation formula is

$$f_n + f_{n+N/2} = \sum_{k=-N/4}^{N/4-1} \widehat{f}_{2k} \, e^{2\pi i(2kn)/N}, \quad n = -\frac{N}{4}, \ldots, \frac{N}{4} - 1.$$

2.5.5 Two-Dimensional Fourier Transform

The one-dimensional definition of the Fourier transform can be extended to higher dimensions. The Fourier transform of a two-dimensional function $\rho(x, y)$, denoted by $S(k_x, k_y)$, is defined by

$$S(k_x, k_y) = \int_{-\infty}^{\infty}\int_{-\infty}^{\infty} \rho(x, y) \, e^{-2\pi i(k_x x + k_y y)} \, dx \, dy.$$

We can generalize all the results in the one-dimensional Fourier transform to the two-dimensional case because the two-dimensional Fourier transform can be expressed as two one-dimensional Fourier transforms along x and y variables:

$$S(k_x, k_y) = \int_{-\infty}^{\infty} \underbrace{\left(\int_{-\infty}^{\infty} \rho(x, y) \, e^{-2\pi i k_x x} \, dx \right)}_{=(\mathcal{F}\rho(\cdot, y))(k_x)} e^{-2\pi i k_y y} \, dy.$$

If $\rho(x, y)$ and $S(k_x, k_y)$ are a Fourier transform pair, we have the following properties.

- Scaling property:

$$\rho(ax, by) \quad \longleftrightarrow \quad \frac{1}{|ab|} S\left(\frac{k_x}{a}, \frac{k_y}{b}\right).$$

- Shifting property:

$$\rho(x - a, y - b) \quad \longleftrightarrow \quad e^{-2\pi i(k_x x + k_y y)} S\left(k_x, k_y\right).$$

- Modulation:

$$\rho(x, y)\cos\omega x \quad \longleftrightarrow \quad \frac{1}{2}S\left(k_x + \frac{\omega}{2\pi}, k_y\right) + \frac{1}{2}S\left(k_x - \frac{\omega}{2\pi}, k_y\right).$$

- Fourier inversion formula:

$$\rho(x, y) = \int_{-\infty}^{\infty}\int_{-\infty}^{\infty} S(k_x, k_y)\,\mathrm{e}^{2\pi\mathrm{i}(k_x x + k_y y)}\,\mathrm{d}k_x\,\mathrm{d}k_y.$$

- Sampling theorem: if $\mathrm{supp}(S) \subset [-B, B] \times [-B, B]$, then the original data $\rho(x, y)$ can be reconstructed by the interpolation formula

$$\rho(x, y) = \sum_{n=-\infty}^{\infty}\sum_{m=-\infty}^{\infty} \rho\left(\frac{n}{2B}, \frac{m}{2B}\right)\mathrm{sinc}\left(2B\left(x - \frac{n}{2B}\right)\right)\mathrm{sinc}\left(2B\left(y - \frac{m}{2B}\right)\right).$$

References

Bracewell R 1999 *The Fourier Transform and Its Applications*, 3rd edn. McGraw-Hill, New York.
Gasquet C and Witomski P 1998 *Fourier Analysis and Applications: Filtering, Numerical Computation, Wavelets*. Springer, New York.
Rudin W 1970 *Real and Complex Analysis*. McGraw-Hill, New York.
Strang G 2005 *Linear Algebra and Its Applications*, 4th edn. Thomson Learning, London.
Strang G 2007 *Computational Science and Engineering*. Wellesley-Cambridge, Wellesley, MA.

3

Basics of Forward Problem

To solve an inverse problem, we need to establish a mathematical model of the underlying physical phenomena as a forward problem. Correct formulation of the forward problem is essential to obtain a meaningful solution of the associated inverse problem. Since the partial differential equation (PDE) is a suitable mathematical tool to describe most physical phenomena, we will study the different kinds of PDEs commonly used in physical science.

When we set up a forward problem associated with an inverse problem, we should take account of the well-posedness (Hadamard 1902) as described in Chapter 1. In constructing a mathematical model that transforms physical phenomena into a collection of mathematical expressions and data, we should consider the following three properties.

- Existence: at least one solution exists. For example, the problem $u''(x) = 1$ in $[0, 1]$ has at least one possible solution, $u = x^2/2$, whereas the problem $|u''(x)|^2 = -1$ has no solution.
- Uniqueness: only one solution exists. For example, the boundary value problem

$$u''(x) = u(x) \quad (0 < \forall x < 1), \quad u(0) = 1, \quad u(1) = 1$$

  has the unique solution $u(x) = e^x$, whereas the boundary value problem

$$\left(\frac{u'(x)}{|u'(x)|}\right)' = 0 \quad (0 < \forall x < 1), \quad u(0) = 0, \quad u(1) = 1$$

  has infinitely many solutions, $u(x) = x, x^2, x^3, \ldots$.
- Continuity or stability: a solution depends continuously on the data.

A problem without the above three properties is ill-posed. To formulate a manageable problem dealing with key information, we often go through a simplification process and sacrifice some physical details. There also exist uncertainties in physical conditions and material characterizations. We should, therefore, take account of the well-posedness of the forward problem when we formulate it to seek a solution of a related inverse problem.

Nonlinear Inverse Problems in Imaging, First Edition. Jin Keun Seo and Eung Je Woo.

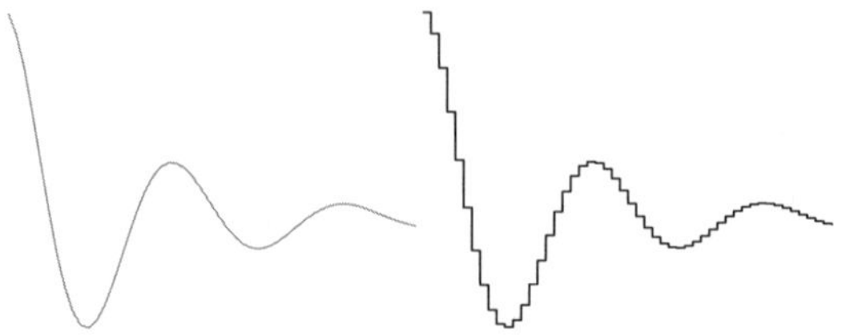

Figure 3.1 Continuous signal $u(x)$ (left) and the corresponding sampled signal $\{u_n\}$ (right)

3.1 Understanding a PDE using Images as Examples

Understanding a PDE and its solution is often hindered by its mathematical expression. Before we deal with it mathematically, we provide an intuitive understanding of basic PDEs through examples of digital images with concrete pictures. A continuous signal will be represented as a function of one variable $u(x)$ or $u(t)$, and the corresponding sampled signal will be written as a sequence $\{u_n\}$ as shown in Figure 3.1. A continuous image will be represented as a function $u(\mathbf{r})$ of two variables $\mathbf{r} = (x, y)$, and the corresponding digital image will be written as a sequence $\{u_{m,n}\}$ as shown in Figure 3.2. The partial derivatives $\partial u/\partial x$, $\partial u/\partial y$ and the Laplacian $\nabla^2 u = \partial^2 u/\partial x^2 + \partial^2 u/\partial y^2$ will be understood in the following discrete sense:

$$\frac{\partial u}{\partial x} \approx u_{m+1,n} - u_{m,n},$$

$$\frac{\partial u}{\partial y} \approx u_{m,n+1} - u_{m,n},$$

$$\nabla^2 u = \frac{\partial^2 u}{\partial x_1^2} + \frac{\partial^2 u}{\partial x_2^2} \approx u_{m+1,n} + u_{m-1,n} + u_{m,n+1} + u_{m,n-1} - 4u_{m,n}.$$

Example 3.1.1 (Poisson's equation) *In Figure 3.3, the image at the top left is represented as u. We can view the image u as a solution of Poisson's equation*

$$\nabla^2 u(\mathbf{r}) = \rho(\mathbf{r}) \ in\ \Omega, \quad u|_{\partial\Omega} = f,$$

where $\Omega = \{\mathbf{r} = (x, y) : 0 < x, y < L\}$ *is the image domain,* ρ *is its Laplacian at the bottom right in Figure 3.3 and* f *is the boundary intensity of* u.

Example 3.1.2 (Wave equation) *In Figure 3.4, we represent the image u as a solution of the wave equation*

$$2\frac{\partial u}{\partial x} + 3\frac{\partial u}{\partial y} = \phi(x, y) \quad in\ \Omega,$$

with the initial condition $u(x, 0) = f_1(x)$, $u(0, y) = f_2(y)$ *for* $0 < x, y < L$. *Here,* ϕ *is plotted on the right-hand side of Figure 3.4 and the initial data are the boundary intensity of the image u on the left.*

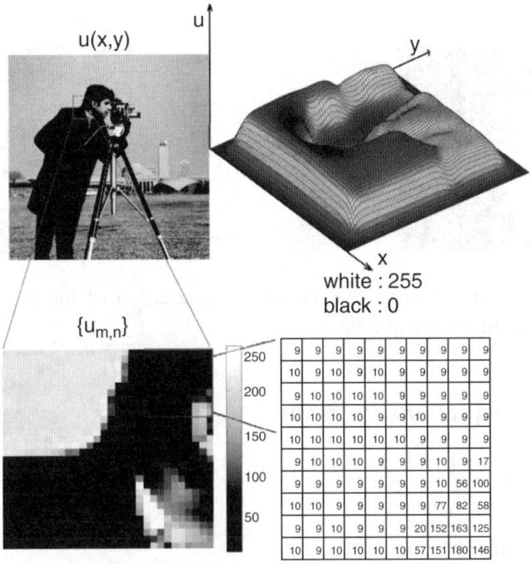

Figure 3.2 Continuous signal $u(\mathbf{r})$ (top left), surface map $z = u(\mathbf{r})$ (top right) and corresponding sampled signals $\{u_{m,n}\}$ (bottom)

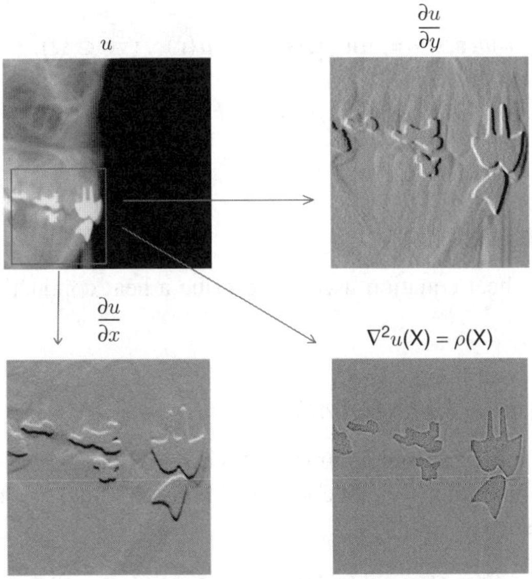

Figure 3.3 The image $u(\mathbf{r})$ (top left) can be viewed as a solution of Poisson's equation $\nabla^2 u = \rho$ (bottom right)

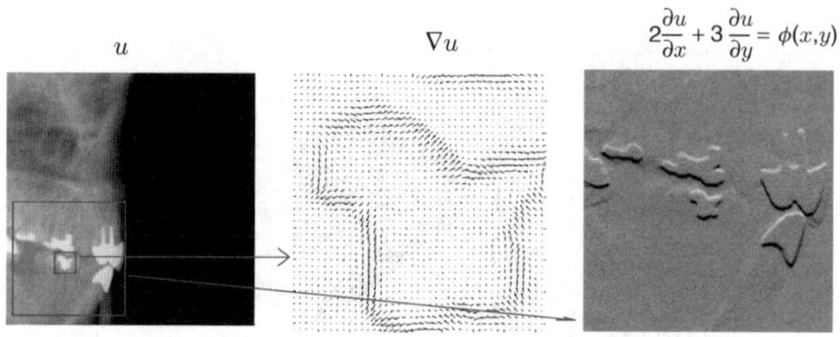

Figure 3.4 The image $u(\mathbf{r})$ (left) can be viewed as a solution of the wave equation $2\,\partial u/\partial x + 3\,\partial u/\partial y = \phi(x, y)$ with the image ϕ (right)

Example 3.1.3 (Heat equation) *We define*

$$w(\mathbf{r}, t) = \int_\Omega \frac{1}{4\pi t} \mathrm{e}^{-|\mathbf{r}-\mathbf{r}'|^2/(4t)} u(\mathbf{r}')\,\mathrm{d}\mathbf{r}' \quad (\forall\, \mathbf{r} \in \Omega,\ t > 0). \tag{3.1}$$

Then, $w(\mathbf{r}, t)$ satisfies the heat equation

$$\left[\frac{\partial}{\partial t} - \nabla^2\right] w(\mathbf{r}, t) = 0 \quad (\forall\, \mathbf{r} \in \Omega,\ t > 0),$$

with the initial condition

$$w(\mathbf{r}, 0) = \lim_{t\to 0+} w(\mathbf{r}, t) = u(\mathbf{r}) \quad (\forall \mathbf{r} \in \Omega).$$

Figure 3.5 shows the images of u, the Gaussian kernel and w.

3.2 Heat Equation

We first introduce the heat equation used to describe a heat conduction process within an object.

3.2.1 Formulation of Heat Equation

Suppose $u(\mathbf{r}, t)$ measures the temperature at time t and position $\mathbf{r} = (x, y, z) \in \Omega$ in a three-dimensional domain $\Omega \subset \mathbb{R}^3$. According to Fourier's law, the heat flow, denoted by $\mathbf{J} = (J_x, J_y, J_z)$, satisfies

$$\mathbf{J}(\mathbf{r}, t) = -k(\mathbf{r})\nabla u(\mathbf{r}, t) \quad (\mathbf{r} \in \Omega,\ t > 0),$$

where k is the thermal conductivity. The temperature u is dictated by the law of conservation of energy. For an arbitrarily small volume $D \subset \Omega$ with its surface area ∂D, heat energy can escape from D only through its surface.

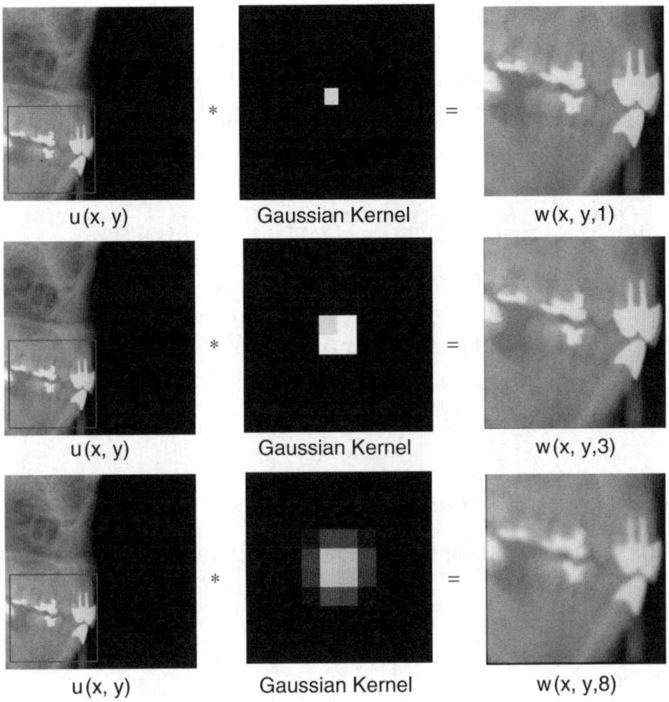

Figure 3.5 The three images in the right-hand column are the images of $w(\mathbf{r}, t) = \int_\Omega [1/(4\pi t)]\,\mathrm{e}^{-|\mathbf{r}-\mathbf{r}'|^2/(4t)} u(\mathbf{r}')\,\mathrm{d}\mathbf{r}'$ at $t = 1$ (top), $t = 3$ (middle) and $t = 8$ (bottom). Here, $w(\mathbf{r}, t)$ represents a solution of the heat equation at time t with the initial data $u(\mathbf{r})$ (left column). The images in the middle column represent the Gaussian kernel $[1/(4\pi t)]\,\mathrm{e}^{-|\mathbf{r}|^2/(4t)}$ with $t = 1$ (top), $t = 3$ (middle) and $t = 8$ (bottom)

The change in the total amount of heat per unit time in D can be expressed as

$$\int_D \rho c \frac{\partial u}{\partial t}\,\mathrm{d}\mathbf{r},$$

where ρ is the density and c is the specific heat of the object, representing the amount of heat required to raise the temperature of one unit mass by one degree. The amount of heat per unit time flowing into D can be expressed by using the divergence theorem as

$$\int_{\partial D} \mathbf{J} \cdot (-\mathbf{n})\,\mathrm{d}S = \int_{\partial D} k\nabla u \cdot \mathbf{n}\,\mathrm{d}S = \int_D \nabla \cdot (k\nabla u)\,\mathrm{d}\mathbf{r},$$

where $\mathbf{n}$ is the unit outward normal vector to ∂D. If $f(\mathbf{r}, t)$ represents the instantaneous volumetric rate of heat generation at $\mathbf{r} \in D$, the total amount of heat produced in D per unit time is

$$\int_D f(\mathbf{r}, t)\,\mathrm{d}\mathbf{r}.$$

From the law of conservation of energy, we set

$$\int_D \rho c \frac{\partial u}{\partial t}\,\mathrm{d}\mathbf{r} = \int_D \nabla\cdot(k\nabla u)\,\mathrm{d}\mathbf{r} + \int_D f(\mathbf{r})\,\mathrm{d}\mathbf{r}.$$

Since D is arbitrary, u satisfies

$$\frac{\partial}{\partial t}u(\mathbf{r},t) - \frac{1}{\rho(\mathbf{r})c(\mathbf{r})}\nabla\cdot(k(\mathbf{r})\nabla u(\mathbf{r},t)) = \frac{1}{\rho(\mathbf{r})c(\mathbf{r})} f(\mathbf{r},t). \tag{3.2}$$

If ρ and c are constants, (3.2) becomes

$$\frac{\partial}{\partial t}u(\mathbf{r},t) - \nabla\cdot(\kappa(\mathbf{r})\nabla u(\mathbf{r},t)) = \tilde{f}(\mathbf{r},t), \tag{3.3}$$

where $\kappa = k/\rho c$ and $\tilde{f} = (1/\rho c)f$.

3.2.2 One-Dimensional Heat Equation

To understand the behavior of the temperature u, we consider the one-dimensional heat equation

$$\left[\frac{\partial}{\partial t} - k\frac{\partial^2}{\partial x^2}\right]u(x,t) = 0 \quad (\forall\, x\in\mathbb{R}, t>0), \tag{3.4}$$

where k is a positive constant.

Theorem 3.2.1 *For $\phi \in C_0(\mathbb{R})$, we define*

$$u(x,t) = \int_{-\infty}^{\infty} K(x-x',t)\phi(x')\,\mathrm{d}x' \quad (\forall\, x\in\mathbb{R}, t>0), \tag{3.5}$$

where K is the one-dimensional heat kernel set by

$$K(x,t) = \frac{1}{\sqrt{4\pi k t}}\,\mathrm{e}^{-|x|^2/(4kt)}. \tag{3.6}$$

Then, u satisfies the heat equation with the initial condition ϕ:

$$\begin{cases} \left[\dfrac{\partial}{\partial t} - k\dfrac{\partial^2}{\partial x^2}\right]u(x,t) = 0 & (\forall\, x\in\mathbb{R}, t>0),\\ u(x,0) = \phi(x) & (\forall\, x\in\mathbb{R}). \end{cases} \tag{3.7}$$

Proof. A direct computation gives

$$\left[\frac{\partial}{\partial t} - k\frac{\partial^2}{\partial x^2}\right]K(x,t) = 0 \quad (\forall\, x\in\mathbb{R}, t>0).$$

Hence, for $t>0$, we have

$$\left[\frac{\partial}{\partial t} - k\frac{\partial^2}{\partial x^2}\right]u(x,t) = \int_{-\infty}^{\infty}\left(\left[\frac{\partial}{\partial t} - k\frac{\partial^2}{\partial x^2}\right]K(x-x',t)\right)\phi(x')\,\mathrm{d}x' = 0.$$

It remains to prove the initial condition

$$\lim_{t\to 0^+} u(x,t) = \phi(x) \quad (\forall\, x \in \mathbb{R}).$$

This comes from the fact that

$$\lim_{t\to 0^+} K(x,t) = \delta(x) \quad (\forall x \in \mathbb{R}).$$

The above approximation holds since

$$\int_{-\infty}^{\infty} K(x,t)\,\mathrm{d}x = \frac{1}{\sqrt{\pi}}\int_{-\infty}^{\infty} \mathrm{e}^{-|x|^2}\,\mathrm{d}x = 1$$

and

$$\lim_{t\to 0^+} K(x,t) = 0 \quad (\forall\, x \neq 0).$$

This completes the proof. □

Theorem 3.2.2 Maximum principle *If a non-constant function $u(x,t)$ satisfies*

$$\left[\frac{\partial}{\partial t} - k\frac{\partial^2}{\partial x^2}\right] u(x,t) = 0 \quad (\forall\, 0 < x < L, 0 < t < T), \tag{3.8}$$

then $u(x,t)$ cannot attain its maximum anywhere in the rectangle $(0,L)\times(0,T]$. In other words, u attains its maximum on the bottom $[0,L]\times\{t=0\}$ or the lateral side $\{0,L\}\times[0,T]$.

Proof. For $\epsilon > 0$, $v_\epsilon(x,t) := u(x,t) + \epsilon x^2$ satisfies

$$\left[\frac{\partial}{\partial t} - k\frac{\partial^2}{\partial x^2}\right] v_\epsilon = -2k\epsilon > 0 \quad (\forall\, (x,t) \in [0,L]\times[0,T]).$$

If v_ϵ attains its maximum inside the rectangle $(0,L)\times(0,T]$, there exists $(x_0,t_0) \in (0,L)\times(0,T]$ at which u attains its maximum, but

$$k\frac{\partial^2}{\partial x^2} v_\epsilon(x_0,t_0) = \frac{\partial}{\partial t} v_\epsilon(x_0,t_0) + 2k\epsilon \geq 0 + 2k\epsilon > 0,$$

which is not possible. Since $\lim_{\epsilon\to 0} v_\epsilon = u$, u also cannot attain its maximum anywhere inside the rectangle $(0,L)\times(0,T]$. □

Theorem 3.2.3 Uniqueness *There is at most one solution of*

$$\begin{cases} \left[\dfrac{\partial}{\partial t} - k\dfrac{\partial^2}{\partial x^2}\right] u(x,t) = f(x,t) & ((\forall\, (x,t) \in (0,L)\times(0,T]), \\ u(x,0) = \phi(x) & (\forall\, x \in [0,L]), \\ u(0,t) = g_0(t), \quad u(L,t) = g_1(t) & (0 < t < T), \end{cases} \tag{3.9}$$

where f, ϕ, g and h are smooth functions. If u_1 and u_2 are two solutions of the above problem, then $u_1 = u_2$.

Proof. The difference $w = u_1 - u_2$ satisfies

$$\begin{cases} \left[\dfrac{\partial}{\partial t} - k\dfrac{\partial^2}{\partial x^2}\right] w(x,t) = 0 & (\forall\ (x,t) \in [0,L] \times [0,T]), \\ w(x,0) = 0 & (\forall\ lx \in [0,L]), \\ w(0,t) = 0, \quad w(L,t) = 0 & (0 < t < T). \end{cases} \tag{3.10}$$

By the maximum principle, $w = 0$ □

3.2.3 Two-Dimensional Heat Equation and Isotropic Diffusion

We consider the heat equation in two dimensions with $\mathbf{r} = (x, y)$.

Theorem 3.2.4 *Let $\phi \in C_0(\mathbb{R}^2)$. If $u(\mathbf{r}, t)$ is a solution of the two-dimensional heat equation*

$$\begin{cases} \left[\dfrac{\partial}{\partial t} - \nabla^2\right] u(\mathbf{r}, t) = 0 & (\mathbf{r} \in \mathbb{R}^2, t > 0), \\ u(\mathbf{r}, 0) = \phi(\mathbf{r}), \end{cases} \tag{3.11}$$

then u can be expressed as

$$u(\mathbf{r}, t) = \int_{\mathbb{R}^2} K(\mathbf{r} - \mathbf{r}', t)\phi(\mathbf{r}')\,\mathrm{d}\mathbf{r}', \tag{3.12}$$

where $K(\mathbf{r}, t) = [1/(4\pi t)]\,\mathrm{e}^{-|\mathbf{r}|^2/(4t)}$ is the heat kernel in two dimensions.

Proof. The proof is exactly the same as that for the one-dimensional case. Indeed, the proof comes from the following facts:

$$\left[\frac{\partial}{\partial t} - \nabla^2\right] K(\mathbf{r}, t) = 0 \quad (\forall\ \mathbf{r} \in \mathbb{R}^2, t > 0)$$

and

$$\lim_{t \to 0^+} K(\mathbf{r}, t) = \delta(\mathbf{r}) \quad (\forall\ \mathbf{r} \in \mathbb{R}^2).$$

□

Exercise 3.2.5 *Let Ω be a bounded smooth domain in $\mathbb{R}^2$. Assume that $u(\mathbf{r}, t)$ is a solution of the two-dimensional heat equation*

$$\left[\frac{\partial}{\partial t} - \nabla^2\right] u(\mathbf{r}, t) = 0 \quad (\mathbf{r} \in \Omega, t > 0). \tag{3.13}$$

1. *State and prove the maximum principle for $u(\mathbf{r}, t)$ as in Theorem 3.2.2.*
2. *State and prove the uniqueness theorem for $u(\mathbf{r}, t)$ as in Theorem 3.2.3.*

Remark 3.2.6 *Uses of PDEs are not limited to describing physical phenomena. In addition to describing a physical heat conduction process, the two-dimensional heat equation has an interesting application of denoising image data. With* $t = \sigma^2/2$*, we can express (3.12) as a convolution form:*

$$u\left(\mathbf{r}, \frac{\sigma^2}{2}\right) = G_\sigma * \phi(\mathbf{r}) = \int_{\mathbb{R}^2} G_\sigma(\mathbf{r} - \mathbf{r}')\phi(\mathbf{r}')\,\mathrm{d}\mathbf{r}', \tag{3.14}$$

where G_σ *is the Gaussian filter defined by*

$$G_\sigma(\mathbf{r}) = \frac{1}{2\pi\sigma^2}\,\mathrm{e}^{-|\mathbf{r}|^2/(2\sigma^2)}. \tag{3.15}$$

When ϕ *in (3.14) is an observed image containing noise, we can view* $G_\sigma * \phi$ *as a denoised image. For a small* σ, $G_\sigma * \phi \approx \phi$ *and, therefore, details in the image are kept. A larger* σ *results in a blurred image* $G_\sigma * \phi$ *with reduced noise. Hence,* σ *determines the local scale of the Gaussian filter that reduces noise while eliminating details of the image* ϕ.

3.2.4 Boundary Conditions

We assume a three-dimensional domain Ω occupying a solid object where heat conduction occurs. If there is no heat source inside Ω, then the temperature $u(\mathbf{r}, t)$ at time t and position $\mathbf{r} = (x, y, z)$ satisfies the heat equation:

$$\underbrace{c(\mathbf{r})\rho(\mathbf{r})\partial_t u(\mathbf{r}, t)}_{\text{rate of change of heat energy}} \quad = \quad \underbrace{\operatorname{div}(k\nabla u)}_{\text{heat flux into voxel region through boundary}}.$$

This PDE tells us the physical law of the rate of change of the temperature inside Ω. To predict future temperature, we need some additional information. Indeed, we need to know the initial temperature distribution

$$u(\mathbf{r}, 0) = u_0(\mathbf{r})$$

and some boundary conditions, which usually will be one of the following:

case 1 $\quad u(\cdot, t)|_{\partial\Omega} = f$ (prescribed temperature);

case 2 $\quad k\mathbf{n} \cdot \nabla u(\cdot, t)|_{\partial\Omega} = g$ (prescribed flux);

case 3 $\quad k\mathbf{n} \cdot \nabla u(\cdot, t)|_{\partial\Omega} = -H(u(\cdot, t) - f(\cdot, t))|_{\partial\Omega}$ (Newton's law of cooling).

Here, H is called the heat transfer coefficient. Case 1 is called the Dirichlet boundary condition, case 2 is the Neumann boundary condition and case 3 is the Robin boundary condition.

Now suppose that the boundary condition is independent of time (steady boundary condition in cases 1 and 2). Then, as $t \to \infty$, the temperature $u(\mathbf{r}, t)$ converges to a function $v(\mathbf{r})$, independent of time, which is called the equilibrium temperature. Then, v satisfies the equation

$$\nabla \cdot (k(\mathbf{r})\nabla v(\mathbf{r})) = 0.$$

For the case of constant k, the above equation becomes the Laplace equation,

$$\nabla^2 v = 0.$$

3.3 Wave Equation

A wave is a spatial disturbance of a medium that propagates at a wave speed depending on the substance of the medium. The wave equation describes wave propagation in the medium. We begin with the following linear wave equation:

$$a(x, y)\frac{\partial u}{\partial x} + b(x, y)\frac{\partial u}{\partial y} = 0. \tag{3.16}$$

We can first understand the structure of its general solution by simple examples.

Example 3.3.1 *When $a = 1, b = 0 = c$, (3.16 becomes $\partial u/\partial x = 0$, which means that u does not depend on x. Hence, the general solution is*

$$u = f(y) \quad \textit{for an arbitrary function } f.$$

For example, $u = y^2 - 5y$ and $u = \mathrm{e}^y$ could be solutions of $\partial u/\partial x = 0$.

Example 3.3.2 *When a and b are constant and $c = 0$, (3.16) becomes $au_x + bu_y = 0$ or $(a, b) \cdot \nabla u = 0$, which means that u does **not** change in the direction (a, b). Hence, u is constant in the direction (a, b), that is, u is constant on any line $bx - ay = constant$, called the characteristic line. We should note that the characteristic line $bx - ay = constant$ satisfies $\mathrm{d}y/\mathrm{d}x = b/a$. Hence, the general solution is*

$$u = f(bx - ay) \quad \textit{for an arbitrary function } f.$$

Example 3.3.3 *Consider $2u_x + yu_y = 0$ or $(2, y) \cdot \nabla u = 0$. Then, u is constant on **any** characteristic curve satisfying $\mathrm{d}y/\mathrm{d}x = y/2$. This means that $u = constant$ for **any** characteristic curve $y\,\mathrm{e}^{-x/2} = C$ that is a solution of $\mathrm{d}y/\mathrm{d}x = y/2$. Hence, the general solution is*

$$u = f(\mathrm{e}^{-x/2}y) \quad \textit{for an arbitrary function } f.$$

Example 3.3.4 *Consider $2u_x + 4xy^2u_y = 0$. Then u is constant along the characteristic curves satisfying*

$$\frac{\mathrm{d}y}{\mathrm{d}x} = \frac{4xy^2}{2} = 2xy^2.$$

Since the characteristic curves are $y = (C - x^2)^{-1}$ or $x^2 + 1/y = C$, the general solution is

$$u = f\left(x^2 + \frac{1}{y}\right) \quad \textit{for an arbitrary function } f.$$

Now, we will derive a wave equation from Hooke's law and Newton's second law of motion by considering a flexible, elastic homogeneous string that is stretched between two points $x = 0$ and $x = L$. Assume that the string undergoes relatively small transverse vibrations and its displacement $u(x, t)$ at time t and position x is perpendicular to the direction of wave propagation. Assume that the string is released from the initial configuration described by the curve $y = f(x), 0 \leq x \leq L$, and that it is at rest when released from this configuration. We also assume that the tension T and the density ρ are constants over the length of the string.

We apply Newton's law to the part of the string over the interval $[x, x + \Delta x]$. The vertical component of the force acting on the string at $[x, x + \Delta x]$ is

$$T\frac{u_x(x+\Delta x, t)}{\sqrt{1+u_x^2}} - T\frac{u_x(x,t)}{\sqrt{1+u_x^2}} \approx T\left(u_x(x+\Delta x, t) - u_x(x,t)\right)$$

$$= T\int_x^{x+\Delta x} u_{xx}(s,t)\,\mathrm{d}s.$$

Since force = mass × acceleration = $\int_x^{x+\Delta x} \rho u_{tt}(s,t)\,\mathrm{d}s$ over the interval $[x, x + \Delta x]$, we obtain

$$\int_x^{x+\Delta x} \rho u_{tt}(s,t)\,\mathrm{d}s = T\int_x^{x+\Delta x} u_{xx}(s,t)\,\mathrm{d}s.$$

Letting $\Delta x \to 0$, we have $\rho u_{tt} - Tu_{xx} = 0$. The displacement function $u(x, t)$ must also satisfy the initial conditions $u(x, 0) = f(x), u_t(x, 0) = 0$ and the boundary conditions $u(0, t) = u(L, t) = 0$. Hence, the displacement $u(x, t)$ satisfies the following PDE approximately:

$$\begin{cases} \rho u_{tt} - Tu_{xx} = 0 & (x,t) \in (0,L) \times (0,\infty), \\ u(x,0) = f(x), \quad u_t(x,0) = 0 & x \in (0,L), \\ u(0,t) = 0 = u(L,t) & t \in (0,\infty). \end{cases}$$

Theorem 3.3.5 *The general solution of the wave equation*

$$\left[\frac{\partial^2}{\partial t^2} - c^2\frac{\partial^2}{\partial x^2}\right]u(x,t) = 0 \quad \text{on } -\infty < x < \infty \tag{3.17}$$

is

$$u(x,t) = f(x+ct) + g(x-ct), \tag{3.18}$$

where f and g are two arbitrary functions of a single variable.

Proof. We can decompose the wave operator into

$$\left[\frac{\partial^2}{\partial t^2} - c^2\frac{\partial^2}{\partial x^2}\right] = \left(\frac{\partial}{\partial t} - c\frac{\partial}{\partial x}\right)\left(\frac{\partial}{\partial t} + c\frac{\partial}{\partial x}\right).$$

Writing

$$v(x,t) = \left(\frac{\partial}{\partial t} + c\frac{\partial}{\partial x}\right)u(x,t),$$

v satisfies

$$\left(\frac{\partial}{\partial t} - c\frac{\partial}{\partial x}\right)v = 0 \quad \text{in } -\infty < x < \infty.$$

The general solution of v is

$$v(x,t) = h(x+ct)$$

for an arbitrary differentiable function h. Then, u satisfies

$$\left(\frac{\partial}{\partial t}+c\frac{\partial}{\partial x}\right)u(x,t)=v(x,t)=h(x+ct), \tag{3.19}$$

where h is an arbitrary function of the single variable. We can express the general solution of (3.19) as

$$u(x,t)=g(x-ct)+\text{particular solution}, \tag{3.20}$$

where $g(x-ct)$ is the general solution of $u_t+cu_x=0$. Now, it remains to determine the particular solution of (3.19). Substituting $f(x+ct)$ into (3.19) yields

$$\left(\frac{\partial}{\partial t}+c\frac{\partial}{\partial x}\right)f(x+ct)=2cf'(x+ct)=h(x+ct). \tag{3.21}$$

Hence, f satisfying $2cf'=h$ is a particular solution. This completes the proof. □

We may adopt a different method of using the characteristic coordinates to prove (3.18). Taking account of

$$\left(\frac{\partial}{\partial t}-c\frac{\partial}{\partial x}\right)\phi(x+ct)=0 \quad\text{and}\quad \left(\frac{\partial}{\partial t}+c\frac{\partial}{\partial x}\right)\phi(x-ct)=0 \quad \forall\,\phi\in C^1(\mathbb{R}),$$

we introduce the characteristic coordinates

$$\xi=x+ct \quad\text{and}\quad \eta=x-ct.$$

With these new coordinates, (3.17) leads to

$$\left[\frac{\partial^2}{\partial t^2}-c^2\frac{\partial^2}{\partial x^2}\right]u=\frac{\partial}{\partial\xi}\frac{\partial}{\partial\eta}u.$$

The above identity comes from the change of variables:

$$\begin{pmatrix}\partial/\partial x\\ \partial/\partial t\end{pmatrix}=\begin{pmatrix}1 & 1\\ c & -c\end{pmatrix}\begin{pmatrix}\partial/\partial\xi\\ \partial/\partial\eta\end{pmatrix} \iff \begin{pmatrix}\partial/\partial\xi\\ \partial/\partial\eta\end{pmatrix}=\frac{1}{2c}\begin{pmatrix}c & 1\\ c & -1\end{pmatrix}\begin{pmatrix}\partial/\partial x\\ \partial/\partial t\end{pmatrix}.$$

Therefore, the solution of this transformed equation is

$$u=f(\xi)+g(\eta)=f(x+ct)+g(x-ct).$$

Theorem 3.3.6 *For a given initial displacement $\phi\in C^1(\mathbb{R})$ and initial velocity $\psi\in C^1(\mathbb{R})$, the solution of the initial value problem*

$$\begin{cases} u_{tt}=c^2u_{xx} & (-\infty<x<\infty, t>0),\\ u(x,0)=\phi(x) & (-\infty<x<\infty),\\ u_t(x,0)=\psi(x) & (-\infty<x<\infty), \end{cases} \tag{3.22}$$

can be expressed as

$$u(x,t)=\frac{1}{2}[\phi(x+ct)+\phi(x-ct)]+\frac{1}{2c}\int_{x-ct}^{x+ct}\psi(s)\,\mathrm{d}s. \tag{3.23}$$

Proof. Since the general solution is $u(x,t) = f(x+ct) + g(x-ct)$, we need to determine f and g using the two initial conditions:

$$f(x) + g(x) = u(x,0) = \phi(x),$$
$$cf'(x) - cg'(x) = u_t(x,0) = \psi(x).$$

Hence, f and g must satisfy

$$f' = \tfrac{1}{2}(\phi' + \psi/c) \quad \text{and} \quad g' = \tfrac{1}{2}(\phi' - \psi/c),$$

which lead to

$$f(x) = \frac{1}{2}\left[\phi(x) + \frac{1}{c}\int_0^x \psi(s)\,\mathrm{d}s\right] + C_1,$$
$$g(x) = \frac{1}{2}\left[\phi(x) - \frac{1}{c}\int_0^x \psi(s)\,\mathrm{d}s\right] + C_2,$$

where C_1 and C_2 are constants. Since $u(x,0) = f(x) + g(x) = \phi(x) + C_1 + C_2$, we have $C_1 + C_2 = 0$. This completes the proof. □

The formula (3.23) clearly explains the effects of an initial position ϕ and an initial velocity ψ that determine how a wave spreads out with speed c in both directions.

Example 3.3.7 (Euler's equation) *Consider the flow of fluid with pressure $p(\mathbf{x},t)$, density $\rho(\mathbf{x},t)$ and the velocity of a particle of fluid $\mathbf{v}(\mathbf{x},t)$. Let Ω be a volume element with boundary $\partial\Omega$ and let $\mathbf{n}(\mathbf{x})$ with $\mathbf{x} \in \partial\Omega$ be the the unit outward normal vector. The conservation of mass in a unit time interval is expressed by the relation*

$$\underbrace{\frac{\mathrm{d}}{\mathrm{d}t}\int_\Omega \rho(\mathbf{x},t)\,\mathrm{d}\mathbf{x}}_{\text{rate of increase of mass in }\Omega} = \underbrace{-\int_{\partial\Omega} \rho\,\mathbf{v}\cdot\mathbf{n}\,\mathrm{d}S.}_{\text{rate at which mass is flowing into }\partial\Omega}$$

From the divergence theorem

$$\int_{\partial\Omega} \rho\mathbf{v}\cdot\mathbf{n}\,\mathrm{d}S = \int_\Omega \nabla\cdot(\rho\mathbf{v})\,\mathrm{d}\mathbf{x},$$

we get

$$\int_\Omega \left(\frac{\partial}{\partial t}\rho(\mathbf{x},t) + \nabla\cdot(\rho\mathbf{v})\right)\mathrm{d}\mathbf{x} = 0,$$

which leads to the continuity equation

$$\frac{\partial}{\partial t}\rho(\mathbf{x},t) + \nabla\cdot(\rho\mathbf{v}) = 0.$$

The total force on Ω is the sum of the forces exerted across the surface $\partial\Omega$, which is given by $\mathbf{F} = -\int_{\partial\Omega} p(\mathbf{x},t)\mathbf{n}(\mathbf{x})\,\mathrm{d}S_{\mathbf{x}}$. The gravitational force on Ω is $\int_\Omega \rho\mathbf{g}\,\mathrm{d}\mathbf{x}$. Newton's law gives

$$\underbrace{-\int_{\partial\Omega} p(\mathbf{x},t)\mathbf{n}(\mathbf{x})\,\mathrm{d}S_{\mathbf{x}} + \int_\Omega \rho\mathbf{g}\,\mathrm{d}\mathbf{x}}_{\text{force of stress + gravitational force}} = \underbrace{\int_\Omega \rho\frac{\mathrm{d}}{\mathrm{d}t}\mathbf{v}(\mathbf{x},t)\,\mathrm{d}\mathbf{x}.}_{\text{mass}\times\text{acceleration}}$$

It follows from the divergence theorem that $-\int_{\partial\Omega} p(\mathbf{x},t)\mathbf{n}(\mathbf{x})\,\mathrm{d}S_{\mathbf{x}} = -\int_{\Omega} \nabla p(\mathbf{r},t)\,\mathrm{d}V$. *Hence,*

$$\int_{\Omega} \left(\rho \frac{\mathrm{d}}{\mathrm{d}t}\mathbf{v}(\mathbf{x},t) + \nabla p(\mathbf{r},t) - \rho \mathbf{g} \right) \mathrm{d}V = 0,$$

which leads to

$$\frac{\mathrm{d}}{\mathrm{d}t}\mathbf{v}(\mathbf{x},t) + \frac{1}{\rho}\nabla p(\mathbf{x},t) = \mathbf{g}(\mathbf{x},t).$$

The time derivative of the fluid velocity $\mathbf{v}(\mathbf{x},t)$ *is the material derivative defined as*

$$\frac{\mathrm{d}}{\mathrm{d}t}\mathbf{v}(\mathbf{x},t) = \frac{\partial}{\partial t}\mathbf{v}(\mathbf{x},t) + (\mathbf{v}\cdot\nabla)\mathbf{v}(\mathbf{x},t).$$

If we denote by $\mathbf{x}(t)$ *the path followed by a fluid particle, then the velocity is* $\mathbf{v}(\mathbf{x}(t),t) = \mathbf{x}'(t)$ *and acceleration* $\mathbf{a}$ *satisfies*

$$\boldsymbol{a}(t) = \frac{\mathrm{d}}{\mathrm{d}t}\mathbf{v}(\mathbf{x}(t),t) = (\mathbf{v}\cdot\nabla)\mathbf{v}(\mathbf{x},t) + \partial_t\mathbf{v}(\mathbf{x},t).$$

Hence, Euler's equation of motion can be expressed as

$$\partial_t\mathbf{v} + (\mathbf{v}\cdot\nabla)\mathbf{v} = -\frac{1}{\rho}\nabla p + \mathbf{g}.$$

3.4 Laplace and Poisson Equations

There are many physical boundary value problems expressed as Laplace or Poisson equations. After introducing the equations, we provide several examples to obtain some insights into their solutions. Then, we will describe the potential technique to find the solutions.

3.4.1 Boundary Value Problem

Let Ω with a smooth connected boundary $\partial\Omega$ be a domain in three-dimensional space $\mathbb{R}^3$. The Laplace equation of a scalar variable u is

$$\nabla^2 u = u_{xx} + u_{yy} + u_{zz} = 0 \quad \text{in } \Omega. \tag{3.24}$$

To determine its solution uniquely, we need to impose a boundary condition. There are three types of boundary conditions, and we need to specify one boundary condition for every point on the boundary $\partial\Omega$.

1. Dirichlet boundary condition: $u|_{\partial\Omega} = f$ on the boundary $\partial\Omega$.
2. Neumann boundary condition: $\nabla u \cdot \mathbf{n}|_{\partial\Omega} = g$ on the boundary $\partial\Omega$.
3. Mixed boundary condition: for two non-overlapping parts Γ_D and Γ_N with $\partial\Omega = \overline{\Gamma_D \cup \Gamma_N}$,

$$u|_{\Gamma_D} = f \quad \text{and} \quad \left.\frac{\partial u}{\partial \mathbf{n}}\right|_{\Gamma_N} = \nabla u \cdot n|_{\Gamma_N} = g.$$

When an internal source exists in Ω, we use the Poisson equation, which is the inhomogeneous form of the Laplace equation:

$$-\nabla^2 u = u_{xx} + u_{yy} + u_{zz} = \rho \quad \text{in}\Omega, \tag{3.25}$$

where ρ represents the source term in Ω. The Poisson equation also requires proper boundary conditions.

For a given $\rho \in C(\bar{\Omega})$ and $h \in C(\partial\Omega)$, consider the following Poisson equation with a Dirichlet boundary condition:

$$\begin{cases} -\nabla^2 u = -(u_{xx} + u_{yy} + u_{zz}) = \rho & \text{in } \Omega, \\ u|_{\partial\Omega} = f. \end{cases} \tag{3.26}$$

By the principle of superposition, we can decompose its solution u as

$$u(\mathbf{r}) = v(\mathbf{r}) + h(\mathbf{r}), \quad \mathbf{r} \in \Omega,$$

where v and h satisfy

$$\begin{cases} -\nabla^2 v = \rho & \text{in } \Omega, \\ v|_{\partial\Omega} = 0, \end{cases} \quad \text{and} \quad \begin{cases} -\nabla^2 h = 0 & \text{in } \Omega, \\ h|_{\partial\Omega} = f. \end{cases}$$

Example 3.4.1 *We consider a one-dimensional special case where* $\rho = 1$, *that is,* $u''(x) = 1$ *in the interval* $(0, 1)$ *and* $u(0) = u(1) = 0$. *The solution is* $u(x) = \frac{1}{2}x(x-1)$.

Example 3.4.2 *We consider a two-dimensional special case of* $\nabla^2 u = 1$ *with* $u|_{\partial\Omega} = 0$, *where the domain* Ω *is the unit disk centered at the origin. The solution is* $u(x, y) = \frac{1}{4}(x^2 + y^2 - 1)$.

Example 3.4.3 *We consider a two-dimensional Laplace equation in the rectangle* $\Omega = (0, \pi) \times (0, \pi)$. *The solution of*

$$\begin{cases} \nabla^2 u = 0 \quad \text{in } \Omega = (0, \pi) \times (0, \pi), \\ u(x, 0) = g(x), \quad 0 = u(x, \pi) = u(0, y) = u(\pi, y), \quad 0 \le x, y \le \pi, \end{cases} \tag{3.27}$$

can be expressed as

$$u(x, y) = \sum_{n=1}^{\infty} c_n \sinh n(\pi - y) \sin nx, \quad \sum_{n=1}^{\infty} c_n \sinh n\pi \sin nx = g(x). \tag{3.28}$$

Here, we use separable variables and Fourier analysis. In short, if

$$V := \{u \in C^2(\Omega) : u \text{ satisfies (3.27) for some } g \in C_0([0, \pi])\},$$

then V *must be contained in the vector space*

$$\text{span}\{\sinh n(\pi - y) \sin nx : n = 1, 2, \dots\}.$$

Hence, $\sinh n(\pi - y) \sin nx (n = 1, 2, \dots)$ *form a basis of the solution space.*

3.4.2 Laplace Equation in a Circle

We derive an explicit representation formula for solutions of the Laplace equation in a circle. We will investigate the mean value property, which is the most important property of solutions of the Laplace equation.

Theorem 3.4.4 *Let $\Omega = B_1(0)$, the unit disk centered at the origin. Suppose u is a solution of the Laplace equation with a Dirichlet boundary condition:*

$$\nabla^2 u = \frac{\partial^2 u}{\partial r^2} + \frac{1}{r}\frac{\partial u}{\partial r} + \frac{1}{r^2}\frac{\partial^2 u}{\partial \theta^2} = 0 \quad in\ \Omega, \qquad u|_{\partial\Omega} = f,$$

where $r = \sqrt{x^2+y^2}$ and $\theta = \tan^{-1}(y/x)$. We can express u as

$$u(r\cos\theta,\ r\sin\theta) = \frac{1}{2\pi}\int_0^{2\pi} f(\cos\tilde{\theta},\ \sin\tilde{\theta})\frac{1-r^2}{1+r^2-2r\cos(\theta-\tilde{\theta})}\,\mathrm{d}\tilde{\theta}.$$

Proof. For convenience, we denote $\hat{u}(r,\theta) = u(r\cos\theta,\ r\sin\theta)$. Then, $\hat{u}(r,\theta)$ satisfies

$$\frac{\partial^2 \hat{u}}{\partial r^2} + \frac{1}{r}\frac{\partial \hat{u}}{\partial r} + \frac{1}{r^2}\frac{\partial^2 \hat{u}}{\partial \theta^2} = 0 \quad \text{in } [0,1)\times[0,2\pi]$$

with

$$\underbrace{u(1,\theta) = f(\theta)}_{\text{boundary condition}} \quad \text{and} \quad \underbrace{u(r,2\pi) = u(r,0), \quad u(0,\theta) = u(0,0)}_{\text{continuity conditions}}.$$

Polar separation of variables is based on the assumption that the solution takes the form $\hat{u}(r,\theta) = v(r)w(\theta)$, which must satisfy

$$\frac{r^2 v''(r) + rv'(r)}{v(r)} = -\frac{w''(\theta)}{w(\theta)} = \lambda = \text{constant}.$$

This comes from the use of $\nabla^2(vw) = 0$, which is equivalent to

$$\frac{\partial^2(vw)}{\partial r^2} + \frac{1}{r}\frac{\partial(vw)}{\partial r} + \frac{1}{r^2}\frac{\partial^2(vw)}{\partial \theta^2} = 0.$$

Hence, the PDE splits into the following pair of ordinary differential equations (ODEs):

$$r^2 v''(r) + rv'(r) - \lambda v(r) = 0 \quad \text{and} \quad w''(\theta) + \lambda w(\theta) = 0.$$

The eigenvalue problem

$$w''(\theta) + \lambda w(\theta) = 0 \quad \text{with periodic boundary condition} \quad w(0) = w(2\pi)$$

has the following non-zero eigenfunctions:

eigenvalue		eigenfunction
$\lambda = 0$	$\longleftrightarrow$	$w = 1$
$\lambda = n^2 (n = 1, 2, \ldots)$	$\longleftrightarrow$	$\sin n\theta, \cos n\theta.$

For each eigenvalue $\lambda = n^2$, the corresponding eigenfunction v must lie in the set

$$V_{n^2} := \{v \in C^2([0,1]) : r^2 v'' + r v' - n^2 v = 0 \text{ and } v(0) < \infty\},$$

which leads to

$$\text{span}\{v(r) = r^n\} = V_{n^2}.$$

Hence, all possible polar separable solutions are

$$v(r)w(\theta) = r^n \cos n\theta \quad \text{or} \quad r^n \sin n\theta \quad (n = 0, 1, 2, 3, \ldots).$$

Now, we are ready to show that $\hat{u}$ can be expressed as a linear combination of polar separable solutions:

$$\hat{u}(r,\theta) = a_0 + \sum_{n=1}^{\infty} (a_n r^n \cos n\theta + b_n r^n \sin n\theta).$$

Since this expression for $\hat{u}$ satisfies all conditions except $\hat{u}(1,\theta) = f(\theta)$, it suffices to show that f can be expressed as a Fourier series. According to Fourier analysis, f can be expressed as

$$f(\theta) = \hat{u}(1,\theta) = \frac{a_0}{2} + \sum_{n=1}^{\infty} (a_n \cos n\theta + b_n \sin n\theta),$$

where

$$a_n = \frac{1}{\pi}\int_0^{2\pi} f(\theta)\cos n\theta \, d\theta, \quad b_n = \frac{1}{\pi}\int_0^{2\pi} f(\theta)\sin n\theta \, d\theta.$$

Hence, the solution $u(r\cos\theta, r\sin\theta) = \hat{u}(r,\theta)$ can be expressed as

$$\hat{u}(r,\theta) = \frac{1}{\pi}\int_0^{2\pi} f(\tilde{\theta}) \underbrace{\left[\frac{1}{2} + \sum_{n=1}^{\infty} r^n (\cos n\theta \cos n\tilde{\theta} + \sin n\theta \sin n\tilde{\theta})\right]}_{=\frac{1}{2}+\sum_{n=1}^{\infty} r^n \cos n(\theta-\tilde{\theta})} d\tilde{\theta}.$$

Denoting $z = r\,\mathrm{e}^{\mathrm{i}(\theta-\tilde{\theta})}$, we have

$$\frac{1}{2} + \sum_{n=1}^{\infty} r^n \cos n(\theta - \tilde{\theta}) = \Re\left(\frac{1}{2} + \sum_{n=1}^{\infty} z^n\right) = \frac{1 - r^2}{2(1 + r^2 - 2r\cos(\theta - \tilde{\theta}))}$$

and

$$\hat{u}(r,\theta) = \frac{1}{\pi}\int_0^{2\pi} f(\tilde{\theta}) \frac{(1 - r^2)}{2(1 + r^2 - 2r\cos(\theta - \tilde{\theta}))} d\tilde{\theta}.$$

□

Theorem 3.4.5 (Mean value property and maximum principle) *Assume that* $\nabla^2 u(\mathbf{r}) = 0$ *in* Ω. *Here,* $\mathbf{r} = (x, y)$. *If* $\overline{B_{r_0}(\mathbf{r}_0)} \subset \Omega$, *then* $u(\mathbf{r}_0)$ *is the average of* u *over the circle* $\partial B_{r_0}(\mathbf{r}_0)$:

$$u(\mathbf{r}_0) = \frac{1}{2\pi r_0}\oint_{\partial B_{r_0}(\mathbf{r}^0)} u(\mathbf{r}) \, d\ell_{\mathbf{r}}.$$

Moreover, u cannot have a strict local maximum or minimum at any interior point of Ω.

Proof. Writing

$$\hat{u}(r,\theta) = u\left(\mathbf{r}_0 + \frac{1}{r_0}(r\cos\theta,\ r\sin\theta)\right)$$

and using Theorem 3.4.4, we have

$$\hat{u}(r,\theta) = \frac{1}{2\pi}\int_0^{2\pi} \hat{u}(1,\tilde{\theta})\frac{1-r^2}{1+r^2-2r\cos(\theta-\tilde{\theta})}\,\mathrm{d}\tilde{\theta}.$$

The above identity directly yields

$$u(\mathbf{r}_0) = \hat{u}(0,\theta) = \frac{1}{2\pi}\int_0^{2\pi}\hat{u}(1,\tilde{\theta})\,\mathrm{d}\tilde{\theta} = \frac{1}{2\pi r_0}\oint_{\partial B_{r_0}(\mathbf{r}_0)} u(\mathbf{r})\,\mathrm{d}\ell_{\mathbf{r}}, \tag{3.29}$$

which means that $u(\mathbf{r}_0)$ is the average of u over the circle $\partial B_{r_0}(\mathbf{r}_0)$.

Now, we prove the maximum principle using the mean value property. To derive a contradiction, suppose u has a strict local maximum at $\mathbf{r}_0 \in \Omega$. Then there exists $s > 0$ such that $u(\mathbf{r}_0) > u(\mathbf{r})$ for $\mathbf{r} \in \partial B_s(\mathbf{r}_0)$, where s is sufficiently small to be $B_\epsilon(\mathbf{r}_0) \subset \Omega$. Then, we have

$$u(\mathbf{r}_0) > \frac{1}{2\pi s}\oint_{\partial B_s(\mathbf{r}_0)} u(\mathbf{r})\,\mathrm{d}\ell_{\mathbf{r}},$$

which contradicts (3.29). □

Theorem 3.4.6 (Uniqueness of Dirichlet problem) *Suppose that $u, v \in C^2(\Omega)\cap C^1(\bar{\Omega})$ satisfy*

$$\nabla^2 u = \nabla^2 v \quad \text{in } \Omega, \qquad u|_{\partial\Omega} = v|_{\partial\Omega}.$$

Then $u = v$ in Ω.

Proof. Denoting $w = u - v$, w satisfies

$$\nabla^2 w = 0 \quad \text{in } \Omega, \qquad w|_{\partial\Omega} = 0.$$

From the maximum principle, w cannot have a strict maximum or minimum inside Ω. Since $w|_{\partial\Omega} = 0$, $w = 0$ in Ω. □

3.4.3 Laplace Equation in Three-Dimensional Domain

Let Ω be a smooth domain in $\mathbb{R}^3$ with a connected boundary $\partial\Omega$. The Green's function $G(\mathbf{r},\mathbf{r}')$ is a unique solution to the following Poisson equation: for each $\mathbf{r}' = (x', y', z') \in \Omega$,

$$\begin{cases} -\nabla^2 G(\mathbf{r},\mathbf{r}') = \delta(\mathbf{r}-\mathbf{r}') & (\forall\ \mathbf{r}\in\Omega), \\ G(\mathbf{r},\mathbf{r}') = 0 & (\forall\ \mathbf{r}\in\partial\Omega). \end{cases} \tag{3.30}$$

Theorem 3.4.7 *The function* $1/(4\pi|\mathbf{r}|)$ *satisfies the following properties:*

$$-\nabla^2\left(\frac{1}{4\pi|\mathbf{r}-\mathbf{r}'|}\right) = \delta(\mathbf{r}-\mathbf{r}'), \tag{3.31}$$

$$-\nabla^2\int_{\mathbb{R}^3}\frac{1}{4\pi|\mathbf{r}-\mathbf{r}'|}\rho(\mathbf{r}')\,d\mathbf{r}' = \rho(\mathbf{r}) \quad (\forall\,\rho\in C_0^2(\mathbb{R}^3)). \tag{3.32}$$

Proof. Let $\rho\in C_0^2(\mathbb{R}^3)$. Straightforward computation gives

$$-\nabla^2\left(\frac{1}{4\pi|\mathbf{r}-\mathbf{r}'|}\right) = 0 \quad \text{if}\,\mathbf{r}\neq\mathbf{r}'. \tag{3.33}$$

Hence, for an arbitrary ball B containing $\mathbf{r}$, we have

$$\nabla^2\int_{\mathbb{R}^3\setminus B}\frac{-1}{4\pi|\mathbf{r}-\mathbf{r}'|}\rho(\mathbf{r}')\,d\mathbf{r}' = 0 \quad (\forall\ B\ni\mathbf{r}). \tag{3.34}$$

This leads to

$$-\nabla^2\int_{\mathbb{R}^3}\frac{1}{4\pi|\mathbf{r}-\mathbf{r}'|}\rho(\mathbf{r}')\,d\mathbf{r}' = -\nabla^2\int_B\frac{1}{4\pi|\mathbf{r}-\mathbf{r}'|}\rho(\mathbf{r}')\,d\mathbf{r}' \quad (\forall B\ni\mathbf{r}).$$

The last term can be decomposed into

$$\begin{aligned}-\nabla^2\int_B\frac{1}{4\pi|\mathbf{r}-\mathbf{r}'|}\rho(\mathbf{r}')\,d\mathbf{r}' &= \lim_{\epsilon\to0+}\int_B\nabla_{\mathbf{r}}^2\left(\frac{-1}{4\pi(|\mathbf{r}-\mathbf{r}'|+\epsilon)}\right)\rho(\mathbf{r}')\,d\mathbf{r}'\\ &= \lim_{\epsilon\to0+}\int_B\nabla_{\mathbf{r}}^2\left(\frac{-1}{4\pi(|\mathbf{r}-\mathbf{r}'|+\epsilon)}\right)(\rho(\mathbf{r}')-\rho(\mathbf{r}))\,d\mathbf{r}'\\ &\quad+\rho(\mathbf{r})\lim_{\epsilon\to0+}\int_B\nabla_{\mathbf{r}}^2\left(\frac{-1}{4\pi(|\mathbf{r}-\mathbf{r}'|+\epsilon)}\right)d\mathbf{r}'.\end{aligned}$$

Since $|\rho(\mathbf{r}')-\rho(\mathbf{r})| = O(|\mathbf{r}'-\mathbf{r}|)$, it is easy to show that

$$\lim_{\epsilon\to0+}\int_B\nabla_{\mathbf{r}}^2\left(\frac{-1}{4\pi(|\mathbf{r}-\mathbf{r}'|+\epsilon)}\right)(\rho(\mathbf{r}')-\rho(\mathbf{r}))\,d\mathbf{r}' = 0$$

and

$$-\nabla^2\int_{\mathbb{R}^3}\frac{1}{4\pi|\mathbf{r}-\mathbf{r}'|}\rho(\mathbf{r}')\,d\mathbf{r}' = \rho(\mathbf{r})\nabla^2\int_B\frac{-1}{4\pi|\mathbf{r}-\mathbf{r}'|}\,d\mathbf{r}' \quad (\forall\ B\ni\mathbf{r}).$$

Hence, it suffices to prove

$$-\nabla^2\int_B\frac{1}{4\pi|\mathbf{r}-\mathbf{r}'|}\,d\mathbf{r}' = 1 \quad (\forall\ B\ni\mathbf{r}). \tag{3.35}$$

The above identity can be shown using a careful use of the divergence theorem:

$$\nabla^2\int_B\frac{-1}{4\pi|\mathbf{r}-\mathbf{r}'|}\,d\mathbf{r}' = \nabla\cdot\int_B\nabla_{\mathbf{r}}\left(\frac{-1}{4\pi|\mathbf{r}-\mathbf{r}'|}\right)d\mathbf{r}'$$

$$
\begin{aligned}
&= -\nabla \cdot \int_B \nabla_{\mathbf{r}'} \left(\frac{-1}{4\pi |\mathbf{r} - \mathbf{r}'|} \right) \mathrm{d}\mathbf{r}' \\
&= -\nabla \cdot \int_{\partial B} \mathbf{n}(\mathbf{r}') \left(\frac{-1}{4\pi |\mathbf{r} - \mathbf{r}'|} \right) \mathrm{d}S_{\mathbf{r}'} \\
&= \int_{\partial B} \mathbf{n}(\mathbf{r}') \cdot \nabla_{\mathbf{r}'} \left(\frac{-1}{4\pi |\mathbf{r} - \mathbf{r}'|} \right) \mathrm{d}S_{\mathbf{r}'} \\
&= \int_{\partial B_r(\mathbf{r})} \mathbf{n}(\mathbf{r}') \cdot \nabla_{\mathbf{r}'} \left(\frac{-1}{4\pi |\mathbf{r} - \mathbf{r}'|} \right) \mathrm{d}S_{\mathbf{r}'} \\
&= \int_{\partial B_r(\mathbf{r})} \frac{\mathbf{r}' - \mathbf{r}}{|\mathbf{r} - \mathbf{r}'|} \cdot \left(\frac{\mathbf{r}' - \mathbf{r}}{4\pi |\mathbf{r} - \mathbf{r}'|^3} \right) \mathrm{d}S_{\mathbf{r}'} = 1,
\end{aligned}
$$

where $B_r(\mathbf{r})$ is the ball with center $\mathbf{r}$ and radius r. Here, we use the fact that, for $\mathbf{r} \in B$,

$$\int_{\partial B} \mathbf{n}(\mathbf{r}') \cdot \nabla_{\mathbf{r}'} \left(\frac{-1}{4\pi |\mathbf{r} - \mathbf{r}'|} \right) \mathrm{d}S_{\mathbf{r}'} = \int_{\partial B_r(\mathbf{r})} \mathbf{n}(\mathbf{r}') \cdot \nabla_{\mathbf{r}'} \left(\frac{-1}{4\pi |\mathbf{r} - \mathbf{r}'|} \right) \mathrm{d}S_{\mathbf{r}'}.$$

This completes the proof. □

Example 3.4.8 *Coulomb's law states that there exists a force between two charged particles, which is proportional to the product of the charges and inversely proportional to the square of the distance between them. Considering a space with charged particles, there exists an electric field with its intensity denoted by* $\mathbf{E}$. *We define* $\mathbf{E}(\mathbf{r})$ *as the force that a stationary unit test charge at* $\mathbf{r}$ *will experience. The electric field intensity from a point charge* q, *located at* $\mathbf{r}_0$ *in an unbounded free space, can be expressed as*

$$\mathbf{E}(\mathbf{r}) = \frac{q}{\epsilon_0} \frac{\mathbf{r} - \mathbf{r}_0}{4\pi |\mathbf{r} - \mathbf{r}_0|^3}, \tag{3.36}$$

where $\epsilon_0 = 8.85 \times 10^{-12}$ *is the permittivity of free space. In general, we can express* $\mathbf{E}$ *subject to a certain charge distribution* $\rho(x)$ *in* Ω *as*

$$\mathbf{E}(\mathbf{r}) = \frac{1}{\epsilon_0} \int_\Omega \frac{\mathbf{r} - \mathbf{r}'}{4\pi |\mathbf{r} - \mathbf{r}'|^3} \rho(\mathbf{r}') \, \mathrm{d}\mathbf{r}'.$$

According to Theorem 3.4.7, $\mathbf{E}$ *in (3.36) can be expressed as*

$$\mathbf{E}(\mathbf{r}) = -\nabla u(\mathbf{r}) \quad \textit{where} \quad u(\mathbf{r}) = \int_{\mathbb{R}^3} \frac{1}{4\pi |\mathbf{r} - \mathbf{r}'|} \rho(\mathbf{r}') \, \mathrm{d}\mathbf{r}'. \tag{3.37}$$

Here, u *is the electric potential and it satisfies the Poisson equation:*

$$-\nabla^2 u(\mathbf{r}) = \rho(\mathbf{r}). \tag{3.38}$$

Theorem 3.4.9 (Two-dimensional version of Theorem 3.4.7) *In two dimensions, the Green's function for the entire space* $\mathbb{R}^2$ *is* $[-1/(2\pi)] \ln |\mathbf{x} - \mathbf{y}|$, *where* $\mathbf{x} = (x_1, x_2)$ *and* $\mathbf{y} = (y_1, y_2)$. *Let* Ω *be a two-dimensional domain in* $\mathbb{R}^2$. *For* $f \in C^1(\bar{\Omega})$, *we have*

$$-\nabla^2 \int_\Omega \frac{-1}{2\pi} \ln |\mathbf{x} - \mathbf{y}| \rho(\mathbf{y}) \, \mathrm{d}\mathbf{y} = \rho(\mathbf{x}) \quad (\forall\, \mathbf{x} \in \Omega, \forall \rho \in C^1(\bar{\Omega})). \tag{3.39}$$

Proof. The proof is exactly the same as that of Theorem 3.4.7 with minor modifications. Let $\mathbf{x} \in \Omega$ be fixed. Straightforward computations give

$$\nabla_{\mathbf{y}}^2 \left(\frac{1}{2\pi} \ln |\mathbf{x} - \mathbf{y}| \right) = 0 \qquad (\forall\ \mathbf{y} \neq \mathbf{x}), \tag{3.40}$$

$$\int_{\partial B} \mathbf{n}(\mathbf{y}) \cdot \nabla_{\mathbf{y}} \left(\frac{1}{2\pi} \ln |\mathbf{x} - \mathbf{y}| \right) d\ell_{\mathbf{y}} = 1 \qquad (\forall\ B \ni \mathbf{x}). \tag{3.41}$$

Here, B is a disk. The identity in (3.40) gives

$$\nabla^2 \int_{\Omega \setminus B} \ln(|\mathbf{x} - \mathbf{y}|)\rho(\mathbf{y})\, d\mathbf{y} = 0 \quad (\forall\ B \ni \mathbf{x}).$$

Since the above identity is true for arbitrarily small $r > 0$, we may understand the left-hand side in (3.39) as

$$\nabla^2 \int_{\Omega} \ln |\mathbf{x} - \mathbf{y}| \rho(\mathbf{y})\, d\mathbf{y} = \rho(\mathbf{x})\, \nabla^2 \int_B \ln(|\mathbf{x} - \mathbf{y}|)\, d\mathbf{y} \quad (\forall\ B \ni \mathbf{x}). \tag{3.42}$$

Hence, (3.39) can be proven by showing that

$$\nabla^2 \int_B \ln(|\mathbf{x} - \mathbf{y}|)\, d\mathbf{y} = 2\pi, \tag{3.43}$$

which is equivalent to

$$\nabla^2 \int_{\Omega} \ln |\mathbf{x} - \mathbf{y}|\, d\mathbf{y} = 2\pi. \tag{3.44}$$

With $\epsilon > 0$, we can move ∇^2 inside the integral (via the Lebesgue dominated convergence theorem):

$$\nabla^2 \int_{\Omega} \ln |\mathbf{x} - \mathbf{y}|\, d\mathbf{y} = \lim_{\epsilon \to 0^+} \int_{\Omega} \nabla_{\mathbf{y}}^2 \ln(|\mathbf{x} - \mathbf{y}| + \epsilon)\, d\mathbf{y}.$$

Using Green's theorem and (3.40) and (3.41),

$$\begin{aligned}
\lim_{\epsilon \to 0^+} \int_{\Omega} \nabla_{\mathbf{y}}^2 \ln(|\mathbf{x} - \mathbf{y}| + \epsilon)\, d\mathbf{y} &= \lim_{\epsilon \to 0^+} \int_{\partial\Omega} \mathbf{n}(\mathbf{y}) \cdot \nabla \ln(|\mathbf{x} - \mathbf{y}| + \epsilon)\, d\ell_{\mathbf{y}} \\
&= \int_{\partial\Omega} \mathbf{n}(\mathbf{y}) \cdot \nabla \ln(|\mathbf{x} - \mathbf{y}|)\, d\ell_{\mathbf{y}} \\
&= \int_{\partial B_r(\mathbf{x})} \mathbf{n}(\mathbf{y}) \cdot \nabla_{\mathbf{y}} (\ln |\mathbf{x} - \mathbf{y}|)\, d\ell_{\mathbf{y}} \quad (\forall\ r > 0) \\
&= 2\pi.
\end{aligned}$$

This completes the proof of (3.44). □

Using Theorem 3.4.7, we now express the Green's function of the Poisson equation with zero Dirichlet boundary condition in a domain Ω.

Definition 3.4.10 *Let Ω be a smooth domain in $\mathbb{R}^3$. The Green's function of the Laplace equation for the domain Ω is*

$$G(\mathbf{r},\mathbf{r}') = \frac{1}{4\pi|\mathbf{r}-\mathbf{r}'|} - H(\mathbf{r},\mathbf{r}'), \quad \mathbf{r},\mathbf{r}' \in \Omega, \tag{3.45}$$

where $H(\mathbf{r},\mathbf{r}')$ satisfies the following: for each $\mathbf{r}' \in \Omega$,

$$\begin{cases} -\nabla^2 H(\mathbf{r},\mathbf{r}') = 0 & \textit{for } \mathbf{r} \in \Omega, \\ H(\mathbf{r},\mathbf{r}') = \dfrac{1}{4\pi|\mathbf{r}-\mathbf{r}'|} & \textit{for } \mathbf{r} \in \partial\Omega. \end{cases} \tag{3.46}$$

According to Theorem 3.4.7 and (3.46), the Green's function satisfies the following in the sense of distribution:

$$\text{for each } \mathbf{r}' \in \Omega, \quad \begin{cases} -\nabla^2 G(\mathbf{r},\mathbf{r}') = \delta(\mathbf{r}-\mathbf{r}') & \text{for } \mathbf{r} \in \Omega, \\ G(\mathbf{r},\mathbf{r}') = 0 & \text{for } \mathbf{r} \in \partial\Omega. \end{cases} \tag{3.47}$$

Exercise 3.4.11 *When $\Omega = \mathbb{R}^3$, $H = 0$ and the Green's function*

$$G(\mathbf{r},\mathbf{r}') = \frac{1}{4\pi|\mathbf{r}-\mathbf{r}'|}$$

can be viewed as a potential subject to a unit positive source located at $\mathbf{r}'$. We can obtain the Green's function for the three-dimensional space $\mathbb{R}^3$ using the symmetry of $v(\mathbf{r}) := G(\mathbf{r}, 0)$, which depends only on the distance $r = |\mathbf{r}| = \sqrt{x^2+y^2+z^2}$ from the origin. Prove the following:

1. *v satisfies*

$$\nabla^2 v(\mathbf{r}) = \frac{d^2 v}{dr^2} + \frac{2}{r}\frac{dv}{dr} = -\delta(r);$$

2. *setting $v(r) = v(x, y)$ with $r = \sqrt{x^2+y^2}$, solve*

$$\nabla^2 v(\mathbf{r}) = \frac{d^2 v}{dr^2} + \frac{1}{r}\frac{dv}{dr} = -\delta(r);$$

3. *the general solution is*

$$v(r) = C_1\frac{1}{r} + C_2;$$

4. *$C_1 = 1/(4\pi)$ and $C_2 = 0$.*

Theorem 3.4.12 *We can express the solution of the Poisson equation*

$$\begin{cases} -\nabla^2 u = \rho & \textit{in } \Omega, \\ u|_{\partial\Omega} = 0, \end{cases} \tag{3.48}$$

as

$$u(\mathbf{r}) = \int_G (\mathbf{r}, \mathbf{r}')\rho(\mathbf{r}')\,\mathrm{d}\mathbf{r}' \quad (\forall\, \mathbf{x} \in \Omega). \tag{3.49}$$

Proof. The property of the Green's function in (3.47) directly gives the proof. □

Example 3.4.13 (Green's function in half-space) *Let $\Omega := \{\mathbf{r} = (x, y, z) : z > 0\}$ be the upper half-space. To find a Green's function, we need to find the function $H(\mathbf{r}, \mathbf{r}')$ such that, for each $\mathbf{r}' \in \Omega$,*

$$\begin{cases} -\nabla^2 H(\mathbf{r}, \mathbf{r}') = 0 & \textit{for } \mathbf{r} \in \Omega, \\ H(\mathbf{r}, \mathbf{r}') = \dfrac{1}{4\pi|\mathbf{r} - \mathbf{r}'|} & \textit{for } \mathbf{r} \in \partial\Omega. \end{cases}$$

The idea is to match the boundary value of the Green's function using the mirror image of a source in Ω. When the mirror is located at $\partial\Omega = \{\mathbf{r} = (x, y, 0) : (\mathbf{x}, y) \in \mathbb{R}^2\}$, the point $\mathbf{r}'_* = (x', y, -z)$ is the mirror image point of $\mathbf{r}$ with respect to the surface $\partial\Omega$. With the aid of the mirror image point $\mathbf{r}'_* = (x', y', -z')$ of $\mathbf{r}'$, the function

$$H(\mathbf{r}, \mathbf{r}') := \frac{1}{4\pi|\mathbf{r} - \mathbf{r}'_*|} \quad (\forall \mathbf{r}, \mathbf{r}' \in \Omega)$$

exactly matches the boundary value

$$\frac{1}{4\pi|\mathbf{r} - \mathbf{r}'|} = H(\mathbf{r}, \mathbf{r}') \quad \forall \mathbf{r} \in \partial\Omega,$$

and for each $\mathbf{r}' \in \Omega$,

$$\nabla^2 H(\mathbf{r}, \mathbf{r}') = 0 \quad \forall \mathbf{r} \in \Omega.$$

Hence, the Green's function of the half-space is given by

$$G(\mathbf{r}; \mathbf{r}') = \frac{1}{4\pi|\mathbf{r} - \mathbf{r}'|} - \frac{1}{4\pi|\mathbf{r} - \mathbf{r}'_*|}, \quad \bar{\mathbf{r}}'_* = (x', y', -z').$$

Example 3.4.14 (Green's function in a ball) *Let $\Omega = \{\mathbf{r} = (x, y, z) : |\mathbf{r}| < 1\}$ be the unit ball in $\mathbb{R}^3$. To find a Green's function in the ball, we need to find the harmonic function $H(\mathbf{r}, \mathbf{r}')$ of $\mathbf{r}$ such that, for each $\mathbf{r}' \in \Omega$,*

$$\begin{cases} -\nabla^2 H(\mathbf{r}, \mathbf{r}') = 0 & \textit{for } |\mathbf{r}| < 1, \\ H(\mathbf{r}, \mathbf{r}') = \dfrac{1}{4\pi|\mathbf{r} - \mathbf{r}'|} & \textit{for } |\mathbf{r}| = 1. \end{cases}$$

In this case, the image point $\mathbf{r}'_*$ with respect to the unit sphere $\partial\Omega$ is given by

$$\mathbf{r}'_* = \frac{\mathbf{r}'}{|\mathbf{r}'|^2}.$$

As a result,

$$H(\mathbf{r}, \mathbf{r}') = \frac{1}{4\pi|\mathbf{r}'|\,|\mathbf{r} - \mathbf{r}'/|\mathbf{r}'|^2|}$$

has the same boundary condition on the unit sphere as $1/(4\pi|\mathbf{r}-\mathbf{r}'|)$*, that is,*

$$\frac{1}{4\pi|\mathbf{r}'|\,|\mathbf{r}-\mathbf{r}'/|\mathbf{r}'|^2|}=\frac{1}{4\pi|\mathbf{r}-\mathbf{r}'|}\quad \textit{for all } |\mathbf{r}|=1.$$

Hence, the Green's function for the unit ball is

$$G(\mathbf{r};\mathbf{r}')=\frac{1}{4\pi|\mathbf{r}-\mathbf{r}'|}-\frac{1}{4\pi|\mathbf{r}'|\,|\mathbf{r}-\mathbf{r}'/|\mathbf{r}'|^2|}.$$

In the case of a solid ball with radius s, we choose the image point given by

$$\mathbf{r}'_*=\frac{s^2\mathbf{r}'}{|\mathbf{r}'|^2}.$$

Then, the Green's function for the solid ball with the radius s is

$$G(\mathbf{r};\mathbf{r}')=\frac{1}{4\pi|\mathbf{r}-\mathbf{r}'|}-\frac{s}{4\pi|\mathbf{r}'|\,|\mathbf{r}-s^2\mathbf{r}'/|\mathbf{r}'|^2|}. \tag{3.50}$$

Lemma 3.4.15 *If G is a Green's function for the domain* Ω*, then it is symmetric:*

$$G(\mathbf{x},\mathbf{y})=G(\mathbf{y},\mathbf{x})\quad(\forall\,\mathbf{x},\mathbf{y}\in\Omega), \tag{3.51}$$

where $\mathbf{x}=(x_1,x_2,x_3)$ *and* $\mathbf{y}=(y_1,y_2,y_3)$.

Proof. Integrating by parts gives

$$\begin{aligned}G(\mathbf{x},\mathbf{y})&=\int_\Omega \delta(\mathbf{z}-\mathbf{x})G(\mathbf{z},\mathbf{y})\,d\mathbf{z}=-\int_\Omega \nabla^2_{\mathbf{z}}G(\mathbf{z},\mathbf{x})G(\mathbf{z},\mathbf{y})\,d\mathbf{z}\\ &=\int_\Omega \nabla_{\mathbf{z}}G(\mathbf{z},\mathbf{x})\cdot\nabla_{\mathbf{z}}G(\mathbf{z},\mathbf{y})\,d\mathbf{z}\\ &=-\int_\Omega G(\mathbf{z},\mathbf{x})\cdot\nabla^2_{\mathbf{z}}G(\mathbf{z},\mathbf{y})\,d\mathbf{z}=\int_\Omega G(\mathbf{z},\mathbf{x})\delta(\mathbf{z}-\mathbf{y})\,d\mathbf{z}\\ &=G(\mathbf{y},\mathbf{x}).\end{aligned}$$

□

Exercise 3.4.16 *Find Green's functions of the following domains* Ω *in* $\mathbb{R}^2$*:*

1. $\Omega=\mathbb{R}^2$;
2. $\Omega=$ *upper half-plane;*
3. $\Omega=$ *unit disk;*
4. $\Omega=\{(x,y)\mid\sqrt{(x-x_0)^2+(y-y_0)^2}<3\}$*, disk with radius* 3.

3.4.4 Representation Formula for Poisson Equation

Using the Green's function, we can find a simple representation formula for the solution of the following Poisson equation:

$$\begin{cases} -\nabla^2 u = \rho & \text{in } \Omega, \\ u|_{\partial\Omega} = f. \end{cases} \tag{3.52}$$

Theorem 3.4.17 *For $f \in C(\partial\Omega)$, the solution of the Poisson equation (3.52) is expressed as*

$$u(\mathbf{r}) = \int_\Omega G(\mathbf{r}; \mathbf{r}')\rho(\mathbf{r}')\,\mathrm{d}\mathbf{r}' + \int_{\partial\Omega} K(\mathbf{r}; \mathbf{r}')f(\mathbf{r}')\,\mathrm{d}S_{\mathbf{r}'}, \quad \mathbf{r} \in \Omega, \tag{3.53}$$

where the Poisson kernel $K(\mathbf{r}, \mathbf{r},)$ is

$$K(\mathbf{r}, \mathbf{r}') = -\mathbf{n}(\mathbf{r}') \cdot \nabla_{\mathbf{r}'} G(\mathbf{r}, \mathbf{r}'), \quad (\forall\, \mathbf{r}' \in \partial\Omega \text{ and } \forall\, \mathbf{r} \in \Omega).$$

Proof. For a fixed $\mathbf{r} \in \Omega$, integration by parts gives

$$\begin{aligned} u(\mathbf{r}) &= \int_\Omega \delta(\mathbf{r} - \mathbf{r}')u(\mathbf{r}')\,\mathrm{d}\mathbf{r}' \\ &= -\int_\Omega \nabla^2_{\mathbf{r}'} G(\mathbf{r}, \mathbf{r}')u(\mathbf{r}')\,\mathrm{d}\mathbf{r}' \quad \text{(distributional sense)} \\ &= \int_{\partial\Omega} K(\mathbf{r}, \mathbf{r}')f(\mathbf{r}')\,\mathrm{d}S_{\mathbf{r}'} + \int_\Omega \nabla_{\mathbf{r}'} G(\mathbf{r}, \mathbf{r}') \cdot \nabla u(\mathbf{r}')\,\mathrm{d}\mathbf{r}' \\ &= \int_{\partial\Omega} K(\mathbf{r}, \mathbf{r}')f(\mathbf{r}')\,\mathrm{d}S_{\mathbf{r}'} + \int_\Omega G(\mathbf{r}, \mathbf{r}')\rho(\mathbf{r}')\,\mathrm{d}\mathbf{r}'. \end{aligned}$$

□

Exercise 3.4.18 *Assume $u \in C^2(\bar{\Omega})$ is a solution of (3.52). As in Theorem 3.4.24 for the two-dimensional case, the potential u minimizes the energy functional $\Phi(u)$ within the class $\mathcal{A} := \{w \in C^2(\bar{\Omega}) : u|_{\partial\Omega} = f\}$, where the energy functional is defined by*

$$\Phi(u) := \int_\Omega \left(\tfrac{1}{2}|\nabla u|^2 - \rho u\right)\mathbf{dr}.$$

The procedure of the proof is as follows.

1. *For all $\phi \in \mathcal{A}$,*

$$0 = \int_\Omega (-\nabla^2 u - \rho)(u - \phi)\,\mathrm{d}x = \int_\Omega \nabla u \cdot \nabla(u - \phi) - \rho(u - \phi).$$

2. *For all $\phi \in \mathcal{A}$,*

$$\int_\Omega |\nabla u|^2 - \rho u\,\mathbf{dr} = \int_\Omega \nabla u \cdot \nabla\phi - \rho\phi\,\mathbf{dr}.$$

3. *Since $|\nabla u \cdot \nabla\phi| \le \frac{1}{2}|\nabla u|^2 + \frac{1}{2}|\nabla u|^2$, then*

$$\Phi(u) \le \Phi(\phi) \quad \textit{for all } \phi \in \mathcal{A}.$$

Theorem 3.4.19 (Mean value property and maximum principle) *Assume that $u \in C^2(\bar{\Omega})$ satisfies $\nabla^2 u(\mathbf{r}) = 0$ in Ω. For each $\mathbf{r}^* \in \Omega$, $u(\mathbf{r}^*)$ is the average of u over its neighboring sphere centered at $\mathbf{r}^*$:*

$$u(\mathbf{r}^*) = \frac{1}{4\pi s^2} \int_{\partial B_s(\mathbf{r}^*)} u(\mathbf{r})\, \mathrm{d}S_{\mathbf{r}} \quad (\forall\ B_s(\mathbf{r}^*) \subset \Omega). \tag{3.54}$$

Moreover, u cannot have a strict local maximum or minimum at any interior point of Ω. In other words,

$$\sup_{\partial\Omega} u = \sup_{\Omega} u \quad \textit{and} \quad \min_{\partial\Omega} u = \min_{\Omega} u. \tag{3.55}$$

Proof. Assume $B_s(\mathbf{r}^*) \subset \Omega$. If $K(\mathbf{r}, \mathbf{r}')$ is the Poisson kernel for the ball $B_s(\mathbf{r}^*)$, a straightforward computation gives

$$K(\mathbf{r}, \mathbf{r}^*)|_{\mathbf{r} \in B_s(\mathbf{r}^*)} = \text{constant} = \frac{1}{4\pi s^2}.$$

Then, the mean value property in (3.54) follows from the representation formula (3.53) with $\rho = 0$ and the above identity. Moreover, the mean value property (3.54) gives the maximum principle (3.55). □

Remark 3.4.20 *We can prove the weak maximum principle (3.55) in a different approach. Set*

$$v_\epsilon^\pm(\mathbf{r}) = u(\mathbf{r}) \pm \epsilon |\mathbf{r}|^2 \quad (\epsilon \approx 0 \textit{ is a small positive number}).$$

Then, $\nabla^2 v_\epsilon^+ > 0$ and $\nabla^2 v_\epsilon^- < 0$ in Ω. Hence, v_ϵ^+ and v_ϵ^- cannot have a local maximum and a local minimum, respectively, at any interior point of Ω. Hence,

$$\sup_{\partial\Omega} v_\epsilon^+ = \sup_{\Omega} v_\epsilon^+ \quad \textit{and} \quad \inf_{\partial\Omega} v_\epsilon^+ = \inf_{\Omega} v_\epsilon^- \quad (\forall\ \epsilon > 0). \tag{3.56}$$

Letting $\epsilon \to 0^+$, we get

$$\sup_{\partial\Omega} u = \sup_{\Omega} u \quad \textit{and} \quad \inf_{\partial\Omega} u = \inf_{\Omega} u.$$

Theorem 3.4.21 *Denote $B_s = \{|\mathbf{r}| < s\}$. If $u \in C^2(\bar{B}_s)$ satisfies $-\nabla^2 u = \rho$ in B_s, then*

$$u(0) = \mathrm{ave}_{\partial B_s} u + \int_{B_s} \frac{1}{4\pi} \left(\frac{1}{|\mathbf{r}|} - \frac{1}{s} \right) \rho(\mathbf{r})\, \mathbf{dr}, \tag{3.57}$$

where

$$\mathrm{ave}_{\partial B_s} u = \frac{1}{4\pi s^2} \int_{\partial B_s} u(\mathbf{r})\, \mathrm{d}S_{\mathbf{r}}.$$

In particular,

$$\begin{aligned} \nabla^2 u = \rho \geq 0 \ \textit{in}\ B_s \quad &\Longrightarrow \quad u(0) \leq \mathrm{ave}_{\partial B_s} u, \\ \nabla^2 u = \rho \leq 0 \ \textit{in}\ B_s \quad &\Longrightarrow \quad u(0) \geq \mathrm{ave}_{\partial B_s} u. \end{aligned} \tag{3.58}$$

Proof. The identity follows from the representation formula (3.53), the mean value theorem (3.54) and the Green's function on the ball (3.50). □

Theorem 3.4.22 (Hopf's lemma) *Let* $\mathbf{r}^* \in \partial B_s$. *If* $u \in C^2(\bar{B}_s(0))$ *satisfies*

$$u(\mathbf{r}^*) > u(\mathbf{r}) \quad (\forall \mathbf{r} \in B_s) \quad \textit{and} \quad \nabla^2 u = 0 \ \textit{in} \ B_s, \tag{3.59}$$

then

$$\frac{\partial u}{\partial \mathbf{n}}(\mathbf{r}^*) = \frac{\mathbf{r}^*}{|\mathbf{r}^*|} \cdot \nabla u(\mathbf{r}^*) > 0. \tag{3.60}$$

Proof. Define

$$v_\alpha(\mathbf{r}) := \mathrm{e}^{-\alpha|\mathbf{r}|^2} - \mathrm{e}^{-\alpha s^2}.$$

Note that $v|_{\partial B_s(0)} = 0$ and

$$\nabla^2 v_\alpha(\mathbf{r}) = \mathrm{e}^{-\alpha|\mathbf{r}|^2}\{4\alpha^2|\mathbf{r}|^2 - 6\alpha\} \quad (\mathbf{r} \in B_s).$$

Using the above identity and $u(\mathbf{r}^*) > \sup_{|\mathbf{r}|=s/2} u$, we can choose a sufficiently small $\epsilon > 0$ and large α such that

$$\nabla^2 v_\alpha(\mathbf{r}) > 0 \quad (\forall \tfrac{1}{2}s < |\mathbf{r}| < s) \quad \text{and} \quad u(\mathbf{r}) + \epsilon v_\alpha(\mathbf{r}) \le u(\mathbf{r}^*) \quad (\forall |\mathbf{r}| = \tfrac{1}{2}s).$$

Since

$$\nabla^2(u(\mathbf{r}) + \epsilon v_\alpha(\mathbf{r})) > 0 \quad (\tfrac{1}{2}s < \forall |\mathbf{r}| < s),$$

$u + \epsilon v_\alpha$ cannot have a local maximum in $\{\frac{1}{2}s < |\mathbf{r}| < s\}$ and therefore $u + \epsilon v_\alpha$ has a maximum at $\mathbf{r}^*$. Hence,

$$\frac{\mathbf{r}^*}{|\mathbf{r}^*|} \cdot \nabla(u + \epsilon v_\alpha)(\mathbf{r}^*) \ge 0$$

or

$$\frac{\mathbf{r}^*}{|\mathbf{r}^*|} \cdot \nabla u(\mathbf{r}^*) \ge -\epsilon \frac{\mathbf{r}^*}{|\mathbf{r}^*|} \cdot \nabla v_\alpha(\mathbf{r}^*) > 0.$$

□

Theorem 3.4.23 (Strong maximum principle) *Assume that* $u \in C^2(\bar{\Omega})$ *satisfies* $\nabla^2 u(\mathbf{r}) = 0$ *in* Ω. *If* u *attains its maximum at* $\mathbf{r}^* \in \partial\Omega$, *then*

$$\textit{either} \quad \frac{\partial}{\partial \mathbf{n}} u(\mathbf{r}^*) > 0 \quad \textit{or} \quad u = \textit{constant}. \tag{3.61}$$

If u *is not constant, then*

$$u(\mathbf{r}^*) > \sup_{\tilde{\Omega}} u \quad (\forall \tilde{\Omega} \subset\subset \Omega). \tag{3.62}$$

Proof. A careful use of Hopf's lemma gives the proof. □

We can view a solution of the Poisson equation as an energy minimizer subject to some constraints.

Theorem 3.4.24 *Let $\rho \in C(\bar{\Omega})$ and $h \in C(\partial\Omega)$. Suppose that $u \in C^2(\Omega) \cap C(\bar{\Omega})$ is the solution of*

$$\begin{cases} -\nabla^2 u = \rho & \text{in } \Omega, \\ u|_{\partial\Omega} = f. \end{cases} \tag{3.63}$$

Then, u is characterized as the unique solution that minimizes the Dirichlet functional

$$\Phi(v) := \int_\Omega \tfrac{1}{2}|\nabla v|^2 - \rho v \, \mathrm{d}\mathbf{x} \tag{3.64}$$

within the set $\mathcal{A} := \{v \in C^2(\Omega) \cap C(\bar{\Omega}) : v|_{\partial\Omega} = f\}$.

Proof. Suppose $w = \arg\min_{v\in\mathcal{A}} \Phi(v)$. Since $\Phi(w) \le \Phi(w + t\phi)$ for any $\phi \in C_0^1(\Omega)$ and any $t \in \mathbb{R}$,

$$0 = \frac{\mathrm{d}}{\mathrm{d}t}\Phi(w + t\phi)\bigg|_{t=0} = \int_\Omega \nabla w \cdot \nabla\phi - \rho\phi \, \mathrm{d}\mathbf{x} \quad \forall\, \phi \in C_0^1(\Omega).$$

Integration by parts yields

$$\int_\Omega (\nabla^2 w + \rho)\phi \, \mathrm{d}\mathbf{x} = 0 \quad \forall\, \phi \in C_0^1(\Omega),$$

which leads to $\nabla^2 w + \rho = 0$ in Ω. Then, it follows from Theorem 3.4.21 that $w = u$ in Ω. □

References

Hadamard J 1902 Sur les problèmes aux dérivées partielles et leur signification physique. *Bull. Univ. Princeton*, **13**, 49–52.

Further Reading

Evans LC 2010 *Partial Differential Equations*. Graduate Studies in Mathematics, no. 19. American Mathematical Society, Providence, RI.

Gilbarg D and Trudinger N 2001 *Elliptic Partial Differential Equations of Second Order*. Springer, Berlin.

John F 1982 *Partial Differential Equations*. Applied Mathematical Sciences, vol. 1. Springer, New York.

Lieb EH and Loss M 2001 *Analysis*, 2nd edn. Graduate Studies in Mathematics, no. 14. American Mathematical Society, Providence, RI.

Marsden JE 1974 *Elementary Classical Analysis*. W. H. Freeman, San Francisco.

Strauss W A 1992 *Partial Differential Equations, An Introduction*. John Wiley & Sons, Inc., New York.

4

Analysis for Inverse Problem

Inverse problems are ill-posed when measurable data are either insufficient for uniqueness or insensitive to perturbations of parameters to be imaged. To solve an inverse problem in a robust way, we should adopt a reasonably well-posed modified model at the expense of a reduced spatial resolution and/or add additional *a priori* information. Finding a well-posed model subject to practical constraints of measurable quantities requires deep knowledge about various mathematical theories in partial differential equations (PDEs) and functional analysis, including uniqueness, regularity, stability, layer potential techniques, micro-local analysis, regularization, spectral theory and others. In this chapter, we present various mathematical techniques that are frequently used for rigorous analysis and investigation of quantitative properties in forward and inverse problems.

4.1 Examples of Inverse Problems in Medical Imaging

Most inverse problems in imaging are to reconstruct cross-sectional images of a material property P from knowledge of input data X and output data Y. We express its forward problem in an abstract form as

$$Y = F(P, X), \tag{4.1}$$

where F is a nonlinear or linear function of P and X. To treat the problem in a computationally manageable way, we need to figure out its sensitivity, explaining how a perturbation $P + \Delta P$ influences the output data $Y + \Delta Y$. In this section, we briefly introduce some examples of forward problem formulations in imaging electrical and mechanical material properties of an imaging object such as the human body.

4.1.1 Electrical Property Imaging

We assume that an imaging object occupies a domain Ω in the three-dimensional space $\mathbb{R}^3$. When the object is a human body, we are interested in visualizing electrical properties of biological tissues inside the body. We denote the admittivity of a biological tissue at a

Nonlinear Inverse Problems in Imaging, First Edition. Jin Keun Seo and Eung Je Woo.

Table 4.1 Time-varying and time-harmonic electromagnetic fields in Maxwell's equations

Name	Time-varying field	Time-harmonic field
Gauss's law	$\nabla \cdot \widetilde{\mathbf{E}} = \rho/\epsilon$	$\nabla \cdot \mathbf{E} = \rho/\epsilon$
Gauss's law for magnetism	$\nabla \cdot \widetilde{\mathbf{B}} = 0$	$\nabla \cdot \mathbf{B} = 0$
Faraday's law of induction	$\nabla \times \widetilde{\mathbf{E}} = -\partial \widetilde{\mathbf{B}}/\partial t$	$\nabla \times \mathbf{E} = -\mathrm{i}\omega \mathbf{B}$
Ampère's circuit law	$\nabla \times \widetilde{\mathbf{H}} = \sigma \widetilde{\mathbf{E}} + \partial \widetilde{\mathbf{D}}/\partial t$	$\nabla \times \mathbf{H} = \sigma \mathbf{E} + \mathrm{i}\omega \mathbf{D}$

position $\mathbf{r}$ and angular frequency ω as $\gamma_\omega(\mathbf{r}) = \sigma_\omega(\mathbf{r}) + \mathrm{i}\omega\epsilon_\omega(\mathbf{r})$, where $\sigma_\omega(\mathbf{r})$ and $\epsilon_\omega(\mathbf{r})$ are the conductivity and permittivity, respectively. Most biological tissues are resistive at low frequencies but capacitive terms may not be negligible at 10 kHz or above. Electrical property imaging aims to image the internal admittivity distribution $P = \gamma_\omega(\mathbf{r})$, which depends on $\mathbf{r} = (x, y, z)$ and ω. To measure the passive material property, we must employ a probing method, which excites the object by externally applying a form of energy and measures its response affected by the admittivity.

As a probing method, we inject a sinusoidal current of $I \sin(\omega t)$ mA through a pair of electrodes that are attached on the surface Ω. This sinusoidally time-varying current produces sinusoidal variations of electric and magnetic fields at every point $\mathbf{r}$ with the same angular frequency ω. For these sinusoidal fields, it is convenient to use the phasor notation. We denote the sinusoidally time-varying electric and magnetic fields by $\widetilde{\mathbf{E}}(\mathbf{r}, t)$ and $\widetilde{\mathbf{H}}(\mathbf{r}, t)$, respectively. We express them using vector field phasors of $\mathbf{E}(\mathbf{r})$ and $\mathbf{H}(\mathbf{r})$ as

$$\widetilde{\mathbf{E}}(\mathbf{r}, t) = \Re\{\mathbf{E}(\mathbf{r})\mathrm{e}^{\mathrm{i}\omega t}\} \quad \text{and} \quad \widetilde{\mathbf{H}}(\mathbf{r}, t) = \Re\{\mathbf{H}(\mathbf{r})\mathrm{e}^{\mathrm{i}\omega t}\}.$$

Each component of $\mathbf{E}(\mathbf{r})$ or $\mathbf{H}(\mathbf{r})$ is a complex-valued function (independent of time) that contains amplitude and phase information. Table 4.1 summarizes important variables and parameters in the *steady-state* or *time-harmonic* electromagnetic field analysis.

For the admittivity imaging, it is imperative to produce a current density $\mathbf{J}$ inside Ω to sense the admittivity by Ohm's law $\mathbf{J} = (\sigma + \mathrm{i}\omega\epsilon)\mathbf{E}$, where $\mathbf{E}$ is the electric field intensity. To create $\mathbf{J}(\mathbf{r})$ and $\mathbf{E}(\mathbf{r})$ inside the body Ω, we can use a pair of electrodes (attached on the surface of the body) to inject sinusoidal current. This produces $\mathbf{J}, \mathbf{E}, \mathbf{B}$ inside Ω, where $\mathbf{B}$ is the magnetic flux density. Alternatively, we may use an external coil to produce eddy currents inside the body Ω. In both cases, we should measure some quantities that enable us to estimate $\mathbf{J}$ and $\mathbf{E}$ (directly or iteratively). Here, we use Maxwell's equations to establish relations among $\mathbf{J}$, $\mathbf{E}$, input data and measured data.

We inject current of $I \sin(\omega t)$ mA with $0 \le \omega/2\pi \le 100$ kHz through a pair of surface electrodes $\mathcal{E}^+$ and $\mathcal{E}^-$ on $\partial\Omega$. The diameter of the imaging object is less than 1 m. Since the Faraday induction is negligibly small in this case, $\mathbf{E}$, $\mathbf{J}$ and $\mathbf{H}$ approximately satisfy

$$\nabla \times \mathbf{E} \approx 0 \qquad \text{in } \Omega, \tag{4.2}$$

$$\nabla \times \mathbf{H} \approx \mathbf{J} = \gamma_\omega \mathbf{E} \qquad \text{in } \Omega, \tag{4.3}$$

with the boundary conditions

$$\mathbf{J} \cdot \mathbf{n} = 0 \qquad \text{on } \partial\Omega \setminus \overline{\mathcal{E}^+ \cup \mathcal{E}^-}, \tag{4.4}$$

$$\mathbf{J} \times \mathbf{n} \approx 0 \qquad \text{on } \mathcal{E}^+ \cup \mathcal{E}^-, \tag{4.5}$$

$$I = -\int_{\mathcal{E}^+} \mathbf{J} \cdot \mathbf{n}\, \mathrm{d}S = \int_{\mathcal{E}^-} \mathbf{J} \cdot \mathbf{n}\, \mathrm{d}S, \tag{4.6}$$

where $\mathbf{n}$ is the outward unit normal vector on $\partial\Omega$ and $\mathrm{d}S$ is the area element on $\partial\Omega$. The first boundary condition (4.4) comes from $\mathbf{J} = \gamma_\omega \mathbf{E} = 0$ in the air, where $\gamma_\omega \approx 0$. The boundary condition (4.5) means that the vector $\mathbf{J}$ on each electrode is parallel or antiparallel to $\mathbf{n}$ since the electrodes are highly conductive. One may adopt the Robin boundary condition to include effects of the electrode–skin contact impedance.

Since $\nabla \times \mathbf{E} \approx 0$, there exists a scalar potential u such that

$$-\nabla u(\mathbf{r}) \approx \mathbf{E}(\mathbf{r}) \quad \text{in } \Omega.$$

Since $0 = \nabla \cdot (\nabla \times \mathbf{H}) = \nabla \cdot \mathbf{J} = -\nabla \cdot (\gamma_\omega \nabla u)$, the potential u is a solution of the following boundary value problem:

$$\begin{cases} \nabla \cdot \left(\gamma_\omega \nabla u\right) = 0 & \text{in } \Omega, \\ \gamma_\omega \dfrac{\partial u}{\partial \mathbf{n}} = 0 & \text{on } \partial\Omega \setminus (\mathcal{E}^+ \cup \mathcal{E}^-), \\ \nabla u \times \mathbf{n} = 0 & \text{on } \mathcal{E}^+ \cup \mathcal{E}^-, \\ I = \displaystyle\int_{\mathcal{E}^+} \gamma_\omega \frac{\partial u}{\partial \mathbf{n}}\, \mathrm{d}s = -\int_{\mathcal{E}^-} \gamma_\omega \frac{\partial u}{\partial \mathbf{n}}\, \mathrm{d}s & \text{on } \mathcal{E}^+ \cup \mathcal{E}^-. \end{cases} \tag{4.7}$$

Example 4.1.1 *Electrical impedance tomography (EIT) aims to visualize (the time or frequency change of) the admittivity distribution $P = \gamma$. One may use multiple electrodes $\mathcal{E}_1, \mathcal{E}_2, \ldots, \mathcal{E}_N$ attached on the boundary $\partial\Omega$ to inject currents and to measure boundary voltages. The input X could be a sequence of injection currents through chosen pairs of electrodes, and the output Y is the resulting boundary voltage data set $u_j[\sigma]|_{\partial\Omega}$, $j = 1, \ldots, N-1$, where the potential $u_j[\sigma]$ is a solution of the Neumann boundary value problem (4.7) with $\mathcal{E}^+ = \mathcal{E}_j$ and $\mathcal{E}^- = \mathcal{E}_{j+1}$. If $\gamma = \gamma_0 + \Delta\gamma$ (or $P = P_0 + \Delta P$) is a small perturbation of a known quantity γ_0, we may deal with a linearized problem:*

$$\Delta Y = F(P_0 + \Delta P, X) - F(P, X) \approx \partial_P F(P_0, X) \Delta P.$$

Example 4.1.2 *Magnetic resonance current density imaging (MRCDI) is an imaging technique that visualizes a current density vector distribution by using a magnetic resonance imaging (MRI) scanner. Measurements are performed by applying an external current to the imaging object during an MRI acquisition. In MRCDI, the quantity to be imaged is the internal current density $P = \mathbf{J}$; the input X is an injection current through a pair of electrodes $\mathcal{E}^+$ and $\mathcal{E}^-$; the output Y is the internal data of $\mathbf{B}$ that is measured by the MRI scanner. Note that measuring all three components of $\mathbf{B} = (B_x, B_y, B_z)$ requires rotating the imaging object inside the MRI scanner, since one can measure only one component of $\mathbf{B}$, which is in the direction of the main magnetic field of the MR scanner. MRCDI uses the relation*

$$\nabla \times \mathbf{B} = \mathbf{J}.$$

Example 4.1.3 *In magnetic resonance electrical impedance tomography (MREIT), the property to be imaged is the conductivity distribution $P = \sigma$ at a low frequency; the input X is an injection current through a pair of electrodes $\mathcal{E}^+$ and $\mathcal{E}^-$; the output Y is the internal data $H_z(\mathbf{r})$ that is measured by an MRI scanner with its main field in the z direction. The relation between H_z and σ is determined by the z component of the Biot–Savart law:*

$$H_z(\mathbf{r}) = \frac{1}{4\pi}\int_\Omega \frac{\langle \mathbf{r}-\mathbf{r}', -\sigma(\mathbf{r}')\nabla u(\mathbf{r}') \times \hat{\mathbf{z}}\rangle}{|\mathbf{r}-\mathbf{r}'|^3}\,d\mathbf{r}' + \mathcal{H}(\mathbf{r}) \quad \textit{for } \mathbf{r} \in \Omega. \tag{4.8}$$

Here, $\mathcal{H}$ is a harmonic function determined by the geometry of lead wire and electrodes and u is the voltage satisfying the Neumann boundary value problem (4.7) with γ replaced by σ.

Example 4.1.4 *Magnetic resonance electrical property tomography (MREPT) aims to visualize $P = \gamma_\omega$ at the Larmor frequency. In MREPT, the input X is a radio-frequency (RF) excitation by an external coil; the output Y is the positive rotating magnetic field $H^+ = \frac{1}{2}(H_x - iH_y)$ that is measured by a B1-mapping technique (Stollberger and Wach 1996; Akoka et al. 2009). The relation between γ_ω and H^+ is determined by*

$$-\nabla^2\mathbf{H}(\mathbf{r}) = \frac{\nabla\gamma_\omega(\mathbf{r})}{\gamma_\omega(\mathbf{r})} \times [\nabla \times \mathbf{H}(\mathbf{r})] - i\omega\gamma_\omega\mathbf{H}(\mathbf{r}). \tag{4.9}$$

Example 4.1.5 *Magnetic induction tomography (MIT) aims to provide the admittivity distribution $P = \gamma_\omega$ in the frequency range of 1–10 MHz. In MIT, multiple transmit coils are used to excite the imaging object and the same or different receive coils are used to measure induced voltages by time-varying magnetic fields produced by eddy currents. The sinusoidal frequency of the external excitation must be high enough to produce a measurable induced voltage:*

$$\oint_C \mathbf{E}(\mathbf{r})\cdot d\mathbf{r} = -i\omega_0\int_S \mu_0\mathbf{H}(\mathbf{r})\cdot d\mathbf{r}, \tag{4.10}$$

where C is the closed path of the receive coil and S is the coil surface. We note that the magnetic field $\mathbf{H}$ outside Ω conveys information on the admittivity γ_ω inside the imaging object Ω. In MIT, the input X is the external excitation; the output Y is the measured induced voltage.

4.1.2 Mechanical Property Imaging

Elasticity is a mechanical property of an elastic object describing how the deformed object, subject to an external force, returns to its original state after the force is removed. Elastography measures the propagation of transverse strain waves in an object. Tissue stiffness is closely related to the velocity of the wave, and the shear modulus (or modulus of rigidity) varies over a wide range, differentiating various pathological states of tissues. Hence, the speed of the harmonic elastic wave provides quantitative information for describing malignant tissues, which typically are known to be much stiffer than normal tissues.

In imaging a mechanical property, we should use Hooke's law, which links the stress and strain tensors:

$$\sigma_{ij} = \sum_{l=1}^{3} \sum_{k=1}^{3} C_{ijkl} \epsilon_{kl},$$

where (C_{ijkl}) is the stiffness tensor, (σ_{ij}) is the strain tensor and (ϵ_{ij}) is the stress tensor. We need to apply a mechanical vibration or stress to create a displacement in Ω and measure some quantity that enables us to estimate the displacement inside Ω. Then, we use an elasticity equation to connect the stress tensor, strain tensor, displacement, input data and measured data.

Assuming that we are trying to image a shear modulus distribution in a linearly elastic and isotropic material, the time-harmonic elastic displacement field denoted by $\mathbf{u} = (u_1, u_2, u_3)$ is dictated by the following PDE:

$$\nabla \cdot (\mu \nabla \mathbf{u}) + \nabla((\lambda + \mu)\nabla \cdot \mathbf{u}) = -\rho\omega^2 \mathbf{u}, \qquad (4.11)$$

where we use the following notation: ρ is the density of the material,

$$\mu = \frac{E}{2(1+\sigma)}$$

is the shear modulus,

$$\lambda = \frac{\sigma E}{(1-2\sigma)(1+\sigma)}$$

is the Lamé coefficient, E is Young's modulus and σ is Poisson's ratio.

Example 4.1.6 *Elastography has been used as a non-invasive technique for the evaluation of the stiffness of the liver, for example. The mechanical property to be imaged is the shear modulus $P = \mu$; the input X is an external low-frequency mechanical vibration applied to the boundary of the imaging object; the output Y is an internal measurement of the displacement vector* $\mathbf{u}$. *The relation between $P = \mu$ and $Y = \mathbf{u}$ is determined by the PDE (4.11). Magnetic resonance elastography (MRE) uses an MR scanner to measure the interior displacement, whereas transient elastography (TE) uses an ultrasound system.*

4.1.3 Image Restoration

The image restoration problem is to recover an original image $P = u$ from a degraded measured (or observed) image $Y = f$ that are related by

$$f = Hu + \eta, \qquad (4.12)$$

where η represents noise and H is a linear operator including blurring and shifting. One may often use *a priori* knowledge about the structure of the true image that can be investigated by looking at geometric structure by level curves.

A typical way of denoising (or image restoration) is to find the best function by minimizing the functional

$$\Phi(u) = \frac{1}{2}\int_\Omega |Hu - f|^2 \,\mathrm{d}\mathbf{r} + \frac{\lambda}{p}\int |\nabla u|^p \,\mathrm{d}\mathbf{r} \quad (p = 1, 2), \tag{4.13}$$

where the first fidelity term $\|Hu - f\|_{L^2(\Omega)}$ forces the residual $Hu - f$ to be small, the second regularization term $\|\nabla u\|_{L^p(\Omega)}$ enforces the regularity of u and the regularization parameter λ controls the tradeoff between the residual norm and the regularity.

4.2 Basic Analysis

To deal with an imaging object with an inhomogeneous material property, we consider the following PDE:

$$\begin{cases} -\nabla\cdot(\sigma(\mathbf{r})\nabla u(\mathbf{r})) = 0 & \text{in } \Omega, \\ u|_{\partial\Omega} = f, \end{cases} \tag{4.14}$$

where σ is differentiable and u is twice differentiable in Ω for $\nabla\cdot(\sigma(\mathbf{r})\nabla u(\mathbf{r})) = 0$ to make sense with the classical derivatives.

In practice, the material property σ may change abruptly. For example, we may consider a conductivity distribution σ inside the human body Ω. Then, σ may have a jump along the boundary of two different organs. Along such a boundary, the electrical field $\mathbf{E} = -\nabla u$, induced by an injection current through a pair of surface electrodes attached on $\partial\Omega$, may not be continuous due to interface conditions of the electric field (like the refractive condition of Snell's law). In this case, there exists no solution $u \in C^2(\Omega)$ in the classical sense and we should seek a practically meaningful solution $u \notin C^2(\Omega)$. Hence, we need to expand the admissible set of solutions of (4.14), which must include all practically meaningful solutions.

This motivates the concept of the generalized derivative called the *weak derivative*, which is a natural extension of the classical derivative. With the use of the weak derivative (reflecting the refractive condition of Snell's law), we can manage the equation $\nabla\cdot(\sigma\nabla u) = 0$ for a discontinuous σ. For a quick understanding of the weaker derivative, we consider the following one-dimensional Dirichlet problem:

$$\begin{cases} -\dfrac{\mathrm{d}}{\mathrm{d}x}\left(\sigma(x)\dfrac{\mathrm{d}}{\mathrm{d}x}u(x)\right) = 0 & \text{in } (-1, 1), \\ u(-1) = -2, \quad u(1) = 1, \end{cases} \quad \text{where } \sigma(x) = \begin{cases} 2 & \text{if } x \geq 0, \\ 1 & \text{if } x < 0. \end{cases} \tag{4.15}$$

Hence, u satisfies

$$u''(x) = 0 \ \text{ in } (-1, 1)\setminus\{0\}, \quad u(-1) = 2, \quad u(1) = 1.$$

We should note that u is different from the solution v of

$$v''(x) = 0 \ \text{ in } (-1, 1), \quad v(-1) = 2, \quad v(1) = 1.$$

This is because v is linear whereas the potential u is piecewise-linear satisfying the following transmission condition (the refractive condition) at $x = 0$ where σ is discontinuous:

$$u(0^+) = u(0^-) \quad \text{and} \quad 2u'(0^+) = u'(0^-). \tag{4.16}$$

Indeed, the practical solution of (4.15) is

$$u(x) = \begin{cases} x & \text{if } 0 \le x < 1, \\ 2x & \text{if } -1 < x < 0. \end{cases}$$

Note that the classical derivative u' does not exist at $x = 0$:

$$u'(x) = \begin{cases} 1 & \text{if } x > 0, \\ \nexists & \text{if } x = 0, \\ 2 & \text{if } x < 0. \end{cases}$$

The difficulty regarding the refraction contained in the PDE

$$\frac{\mathrm{d}}{\mathrm{d}x}\left(\sigma(x)\frac{\mathrm{d}}{\mathrm{d}x}u(x)\right) = 0$$

can be removed by the use of the variational framework:

$$\int_{-1}^{1} \sigma(x)u'(x)\phi'(x)\,\mathrm{d}x = 0 \quad \forall \phi \in C_0^1(-1,1). \tag{4.17}$$

Exercise 4.2.1 *Show that a solution u of (4.17) satisfies the transmission condition (4.16).*

We need to take account of the set of physically meaningful solutions of the variational problem (4.17). A practically meaningful solution u, which is a voltage, should have a finite energy, that is,

$$\Phi(u) := \frac{1}{2}\int_{-1}^{1} \sigma|u'|^2\,\mathrm{d}x < \infty.$$

We solve (4.15) by finding u in the admissible set $\mathcal{A} := \{v : \Phi(v) < \infty, u(-1) = -1, u(1) = 1\}$. Indeed, the solution u of (4.15) is a minimizer of $\Phi(v)$ within the set $\mathcal{A}$:

$$u = \arg\min_{v\in\mathcal{A}} \Phi(v). \tag{4.18}$$

This will be explained in section 4.3.

We return to the three-dimensional problem of (4.14), where $u(\mathbf{r})$ represents an electrical potential at $\mathbf{r}$ in an electrically conducting domain Ω. The physically meaningful solution u must have a finite energy:

$$\Phi(v) = \int_\Omega \sigma(\mathbf{r})|\nabla v(\mathbf{r})|^2\,\mathrm{d}x < \infty.$$

Hence, the solution of (4.14) should be contained in the set $\{v \in L^2(\Omega) : \Phi(v) < \infty\}$. Assuming $0 < \inf_\Omega \sigma < \sup_\Omega \sigma < \infty$, the set $\{v \in L^2(\Omega) : \Phi(v) < \infty\}$ is the same as the following Hilbert space:

$$H^1(\Omega) := \{v : \|v\|_{H^1(\Omega)} < \infty\},$$

where

$$\|u\|_{H^1(\Omega)} = \sqrt{\int_\Omega |\nabla u|^2 + |u|^2 \, d\mathbf{x}}.$$

Indeed, the solution is the minimizer of $\Phi(u)$ within the set $\{v \in H^1(\Omega) : v|_{\partial\Omega} = f\}$. As we mentioned before, the PDE $-\nabla \cdot (\sigma(\mathbf{r})\nabla u(\mathbf{r})) = 0$ should be understood in the variational framework as

$$\int_\Omega \sigma(\mathbf{r})\nabla u(\mathbf{r}) \cdot \nabla \phi(\mathbf{r}) \, dx = 0 \quad (\forall \phi \in C_0^1(\Omega)). \tag{4.19}$$

Remark 4.2.2 *When $u \in C^2(\bar{\Omega})$, there is no difference between the classical and variational problems. However, there exist reasonable situations where the variational problem lacks a smooth solution; there are many practical σ for which the minimization problem has no solution in the class $C^2(\bar{\Omega})$. Indeed, we can consider a problem with $\sigma \notin C(\bar{\Omega})$. Obviously, the classical problem does not have a solution. We can construct a minimizing sequence $\{u_n\}$ in $C^2(\bar{\Omega})$ that is a Cauchy sequence with respect to the norm $\|u\|_{H^1(\Omega)}$. Although the Cauchy sequence $\{u_n\}$ does not converge within $C^2(\bar{\Omega})$, it converges in the Sobolev space $H^1(\Omega)$, the completion of $C^2(\bar{\Omega})$ with respect to the norm $\|u\|_{H^1(\Omega)}$. This means that we can solve the minimization and variational problem within the Sobolev space $H^1(\Omega)$.*

4.2.1 Sobolev Space

To explain the solution of a PDE in the variational framework, it is convenient to introduce the Sobolev space. Let Ω be a bounded smooth domain in $\mathbb{R}^3$ with its smooth boundary $\partial\Omega$ and let $\mathbf{x} = (x_1, \ldots, x_n)$ represent a position. We introduce a Sobolev space $H^1(\Omega)$, which is the closure of the set $C^\infty(\Omega)$ equipped with the norm

$$\|u\| = \sqrt{\int_\Omega |u|^2 + |\nabla u|^2 \, dx}.$$

The finite element method (FEM) for computing an approximate numerical solution and its error analysis can be easily accomplished within the variational framework with the Hilbert space. This Hilbert space $H^1(\Omega)$ is the most widely used Sobolev space in PDE theory.

The generalized derivative can be explained by means of the integration-by-parts formula:

$$\int_\Omega u \partial_{x_i} \phi \, d\mathbf{x} = -\int_\Omega \partial_{x_i} u \phi \, d\mathbf{x} \quad (\forall \phi \in C_0^\infty(\Omega)).$$

In general,

$$\int_\Omega u\partial^\alpha\phi\,\mathrm{d}\mathbf{x} = (-1)^{|\alpha|}\int_\Omega \partial^\alpha u\phi\,\mathrm{d}\mathbf{x} \quad (\forall\phi\in C_0^\infty(\Omega)),$$

where the notions ∂^α and $|\alpha|$ are understood in the following way:

- $\alpha = (\alpha_1,\ldots,\alpha_n)\in\mathbb{N}_0^n$, $\mathbb{N}=\{1,2,\ldots\}$ and $\mathbb{N}_0=\mathbb{N}\cup\{0\}$;
- $\partial^\alpha u = \partial_{x_1}^{\alpha_1}\cdots\partial_{x_n}^{\alpha_n}u$, for example, $\partial^{(2,0,3)}u = \partial_{x_1}^2\partial_{x_3}^3 u$;
- $|\alpha| = \sum_{k=1} n\alpha_k$.

A function v_i satisfying the following equality behaves like the classical derivative $\partial_{x_i}u$:

$$\int_\Omega u\partial_{x_i}\phi\,\mathrm{d}\mathbf{x} = -\int_\Omega v_i\phi\,\mathrm{d}\mathbf{x}, \quad \forall\phi\in C_0^1(\Omega).$$

Definition 4.2.3 *Let $k\in\mathbb{N}_0$ and $0\le\alpha\le 1$. An open set $\Omega\subset\mathbb{R}^n$ is said to be a $C^{k,\alpha}$ domain if, for each $p\in\partial\Omega$, there exists an open neighborhood U_p of p and a $C^{k,\alpha}$ diffeomorphism $\Phi_p : U_p\to B_1(0)$ such that*

$$\Phi_p(U_p\cap\Omega) = \{x\in B_1(0) : x_n>0\} \quad \textit{and} \quad \Phi_p(U_p\cap\partial\Omega) = \{x\in B_1(0) : x_n=0\}.$$

We will assume that Ω is an open subdomain of $\mathbb{R}^n$ with its $C^{k,\alpha}$ boundary $\partial\Omega$, where $k+\alpha\ge 1$. We are now ready to introduce the Sobolev spaces $W^{k,p}(\Omega)$ and $W_0^{k,p}(\Omega)$, where Ω is a domain in $\mathbb{R}^n$ with its boundary $\partial\Omega$ and $1\le p<\infty$:

- $W^{1,p}(\Omega)$ and $W_0^{1,p}(\Omega)$ are the completion (or closure) of $C^\infty(\bar\Omega)$ and $C_0^\infty(\Omega)$, respectively, with respect to the norm

$$\|u\|_{W^{1,p}} := \left(\int_\Omega |u|^p + |\nabla u|^p\,\mathrm{d}\mathbf{x}\right)^{1/p}.$$

  In other words,

$$W_0^{1,p}(\Omega) = \left\{u : \exists\, u_k\in C_0^\infty(\Omega) \text{ s.t. } \lim_{k\to\infty}\|u_k-u\|_{W^{1,p}(\Omega)} = 0\right\}.$$

- $W^{2,p}(\Omega)$ and $W_0^{2,p}(\Omega)$ are the completion (or closure) of $C^\infty(\bar\Omega)$ and $C_0^\infty(\Omega)$, respectively, with respect to the norm

$$\|u\|_{W^{2,p}} := \left(\int_\Omega |u|^p + |\nabla u|^p + |\nabla\nabla u|^p\,\mathrm{d}\mathbf{x}\right)^{1/p}.$$

- We denote $H^1(\Omega)=W^{1,2}(\Omega)$, $H_0^1(\Omega)=W_0^{1,2}(\Omega)$, $H^2(\Omega)=W^{2,2}(\Omega)$ and $H_0^1(\Omega) = W_0^{2,2}(\Omega)$.

Example 4.2.4 *Let $\Omega=(0,1)$ and $u(x)=x(1-x)$ in Ω. We can show that $u\in H_0^1(\Omega)\cap H^2(\Omega)$ but $u\notin H_0^2(\Omega)$. To see this, assume that there exists $u_m\in C_0^\infty(\Omega)$ such that*

$u_m \to u$ in $H^2(\Omega)$ and $\sup_m \|u_m\|_{H^2} \le \|u\|_{H^2} + 1 := M$ without loss of generality. For $0 < s < 1$, it follows from the Schwarz inequality that

$$|u'_m(s)| = \left| \int_0^s u''_n(t)\,\mathrm{d}t \right| \le \sqrt{s}\, \|u_m\|_{H^2}.$$

Hence,

$$\int_0^x |u'_m(s)|^2\,\mathrm{d}x \le \frac{x^2}{2} \|u_m\|_{H^2}^2 \quad (0 < x < 1).$$

An elementary computation shows that

$$\int_0^x |u'(s)|^2\,\mathrm{d}s = \int_0^x |1 - 2x|^2\,\mathrm{d}s \ge x - 2x^2 > \frac{x}{2} \quad (0 < x < \tfrac{1}{4}).$$

Combining the above two inequalities leads to

$$\inf_m \int_0^x |u'(s) - u'_m|^2\,\mathrm{d}s \ge \frac{x}{2} - \frac{x^2}{2} M \quad (0 < x < \tfrac{1}{4}). \tag{4.20}$$

In particular, substituting $x = 1/(4M) < \frac{1}{4}$ into (4.20) yields

$$\inf_m \int_0^x |u'(s) - u'_m|^2\,\mathrm{d}s \ge \frac{1}{8M} - \frac{1}{32M^2} M = \frac{3}{32M} > 0 \quad (0 < x < \tfrac{1}{4})$$

and therefore

$$\inf_m \|u - u_m\|_{H^2} \ge \frac{3}{32M}.$$

This means that $\lim_{m\to\infty} \|u_m - u\|_{H^2} \ne 0$, which contradicts the assumption. For details on this example, see the lecture note by Feldmann and Uhlmann (2003).

Example 4.2.5 *There exists a function $u \in H^1(\mathbb{R}^3)$ that is not bounded on every non-empty open set in $\mathbb{R}^3$. Denoting the set of rational numbers by $\mathbb{Q}$, there exists a sequence $\{\mathbf{q}_m\}_{m=1}^\infty = \mathbb{Q}^3$. For*

$$u(\mathbf{r}) := \frac{1}{1 + |\mathbf{r}|^2} \sum_{m=1}^\infty 2^{-m} \log |\mathbf{r} - \mathbf{q}_m| \quad (\mathbf{r} \in \mathbb{R}^3),$$

we can show by a direct computation that $u \in H^1(\mathbb{R}^3)$ but it is unbounded on every non-empty open set.

Example 4.2.6 *Let $\Omega := \{(r\cos\theta,\ r\sin\theta) : 0 < r < 1, 0 < \theta < 2\pi/3\}$. Let $u_1(x, y) = r^{-3/2}\sin(\frac{3}{2}\theta)$ and $u_2(x, y) = r^{3/2}\sin(\frac{3}{2}\theta)$ with $r = \sqrt{x^2 + y^2}$ and $\tan\theta = y/x$. Show that u_1 and u_2 satisfy*

$$\nabla^2 u_j = 0 \text{ in } \Omega\ (j = 1, 2), \quad u_1|_{\partial\Omega} = u_2|_{\partial\Omega}.$$

Hence, $w = u_1 - u_2$ satisfies

$$\nabla^2 w = 0 \;\; in\ \Omega, \quad w|_{\partial\Omega} = 0.$$

According to the maximum principle, we obtain

$$w = 0 \quad in\ \Omega.$$

But, we know that $w = (r^{-3/2} - r^{3/2})\sin(\frac{3}{2}\theta) \neq 0$. What is wrong with this conclusion? We should note that $u_1 \notin H^1(\Omega)$ and hence u_1 is not a practically meaningful solution. Because of $w \notin H^1(\Omega)$, we cannot apply the maximum principle.

4.2.2 Some Important Estimates

We begin by explaining simplified versions of two important inequalities, the Poincaré and trace inequalities.

Example 4.2.7 *Let $\Omega = \{(x, y) : 0 < x, y < a\}$ be a square with side length $a > 0$.*

- *A simplified version of the Poincaré inequality is*

$$\sup_{u \in C_0^1(\Omega)} \frac{\|u\|_{L^2(\Omega)}}{\|\nabla u\|_{L^2(\Omega)}} \leq C \quad \text{where } C \text{ is a positive constant depending only on } \Omega.$$

- *A simplified version of the trace inequality is*

$$\sup_{u \in C^1(\bar{\Omega})} \frac{\|u\|_{L^2(\partial\Omega)}}{\|u\|_{H^1(\Omega)}} \leq C \quad \text{where } C \text{ is a positive constant depending only on } \Omega.$$

These inequalities are based on the fundamental theorem of calculus:

$$u(x, y) = \begin{cases} u(0, y) + \displaystyle\int_0^x \frac{\partial}{\partial x} u(x', y)\, dx' \\ u(a, y) - \displaystyle\int_x^a \frac{\partial}{\partial x} u(x', y)\, dx' \\ u(x, 0) + \displaystyle\int_0^y \frac{\partial}{\partial y} u(x, y')\, dy' \\ u(x, a) - \displaystyle\int_y^a \frac{\partial}{\partial y} u(x, y')\, dy' \end{cases} \qquad (0 < x, y < a).$$

From the first identity in the above expressions, we have

$$\int_0^a |u(0, y)|^2\, dy = \int_0^a \left| u(x, y) - \int_0^x \frac{\partial}{\partial x} u(x', y)\, dx' \right|^2 dy$$

and therefore

$$\begin{aligned}\int_0^a |u(0,y)|^2\,\mathrm{d}y &\le \int_0^a \left[|u(x,y)| + \int_0^a |\nabla u(x',y)|\,\mathrm{d}x'\right]^2 \mathrm{d}y \\ &\le \int_0^a \left[2|u(x,y)|^2 + \left|\int_0^a |\nabla u(x',y)|\,\mathrm{d}x'\right|^2\right] \mathrm{d}y \\ &\le \int_0^a \left[2|u(x,y)|^2 + \underbrace{\int_0^a 1^2\,\mathrm{d}x'}_{=a} \int_0^a |\nabla u(x',y)|^2\,\mathrm{d}x'\right] \mathrm{d}y \\ &\le 2\int_0^a |u(x,y)|^2\,\mathrm{d}y + 2a\int_\Omega |\nabla u|^2\,\mathrm{d}\mathbf{r} \quad (\mathrm{d}\mathbf{r} = \mathrm{d}x\,\mathrm{d}y).\end{aligned}$$

Integrating over the y variable gives

$$\underbrace{\int_0^a \left(\int_0^a |u(0,y)|^2\,\mathrm{d}y\right)\mathrm{d}x}_{a\int_0^a |u(0,y)|^2\,\mathrm{d}y} \le 2\int_\Omega |u(x,y)|^2\,\mathrm{d}\mathbf{r} + 2a\underbrace{\int_0^a \left(\int_\Omega |\nabla u|^2\,\mathrm{d}\mathbf{r}\right)\mathrm{d}x}_{a\int_\Omega |\nabla u|^2\,\mathrm{d}\mathbf{r}}$$

and therefore

$$\int_0^a |u(0,y)|^2\,\mathrm{d}y \le 2\left(\frac{1}{a} + a\right)\|u\|^2_{H^1}.$$

Application of the above estimate to the three other sides of $\partial\Omega$ leads to the trace inequality.

The Poincaré inequality uses the special property that $u|_{\partial\Omega} = 0$ to get

$$|u(x,y)|^2 \le \left|\int_0^a \left|\frac{\partial}{\partial x}u(x',y)\right|\mathrm{d}x'\right|^2 \le a\int_0^a |\nabla u(x',y)|^2\,\mathrm{d}x'$$

and hence

$$\int_\Omega |u|^2 \le a^2 \int_\Omega |\nabla u|^2,$$

which gives the Poincaré inequality.

Theorem 4.2.8 (Hardy–Littlewood–Sobolev inequality) *Let $p, r > 1$ and $0 < \lambda < n$, with*

$$\frac{1}{p} + \frac{\lambda}{n} + \frac{1}{r} = 2.$$

For $f \in L^p(\mathbb{R}^n)$ and $h \in L^r(\mathbb{R}^n)$,

$$\left|\int_{\mathbb{R}^n}\int_{\mathbb{R}^n} f(\mathbf{x})|\mathbf{x}-\mathbf{y}|^{-\lambda} h(\mathbf{y})\,\mathrm{d}\mathbf{x}\,\mathrm{d}\mathbf{y}\right| \le C(n,\lambda,p)\|f\|_{L^p}\|h\|_{L^r},$$

where

$$C(n, \lambda, p) \le \frac{n}{(n-\lambda)pr}\left(\frac{|S^{n-1}|}{n}\right)^{\lambda^*}\left(\left(\frac{\lambda^*}{1-1/p}\right)^{\lambda^*} + \left(\frac{\lambda^*}{1-1/r}\right)^{\lambda^*}\right)$$

$\lambda^* = \lambda/n$ and S^{n-1} denotes the unit sphere.

For the proof, see Theorem 4.3 in the book by Lieb and Loss (2001).

Theorem 4.2.9 (Sobolev's inequality) *Let $u \in H^1(\mathbb{R}^n)$. Then the following inequalities hold.*

- *For $n \ge 3$,*

$$\|u\|^2_{L^{2n/(n-2)}} \le C_n \|\nabla u\|^2_{L^2(\mathbb{R}^n)}, \tag{4.21}$$

where

$$C_n = \frac{4}{n(n-2)} 2^{-2/n} \pi^{-1-1/n} [\Gamma(\tfrac{1}{2}(n+1))]^{2/n}.$$

- *If $u \in H^1(\mathbb{R})$ (one dimension), then*

$$\|u\|_{L^\infty(\mathbb{R})} \le \tfrac{1}{2}\|\nabla u\|^2_{H^1(\mathbb{R})}, \quad \sup_{x,y} \tfrac{|u(x)-u(y)|}{|x-y|} \le \|u'\|_{L^2}. \tag{4.22}$$

- *If $u \in H^1(\mathbb{R}^2)$ (two dimensions), then, for each $2 \le q < \infty$,*

$$\|u\|_{L^q(\mathbb{R}^2)} \le C_q \|\nabla u\|^2_{H^1(\mathbb{R})}, \quad (\forall 2 \le q < \infty), \tag{4.23}$$

where

$$C_q \le \left[q^{1-2/q} (q-1)^{-1+1/q} \left(\frac{(q-2)}{8\pi}\right)^{1/2-1/q} \right]^2.$$

- *Let Ω be a $C^{0,1}$ domain and $1 \le p \le q$, $m \ge 1$ and $k \le m$. Then,*

$$\|u\|_{L^{np/(n-kp)}(\Omega)} \le C\|u\|_{W^{k,p}(\Omega)} \qquad \text{if } kp < n, \tag{4.24}$$

$$\|u\|_{C^m(\Omega)} \le C\|u\|_{W^{k+m,p}(\Omega)} \qquad \text{if } kp > n, \tag{4.25}$$

where C is independent of u.

For the proof, see Theorem 8.8 in the book by Lieb and Loss (2001).

Lemma 4.2.10 *Let Ω be a convex open set in $\mathbb{R}^n$ with diameter d and let E be an open subset of Ω. For $u \in C^1(\Omega) \cap H^1(\Omega)$, we have*

$$|u(\mathbf{x}) - u_E| \le \frac{d^n}{n|E|} \int_\Omega \frac{1}{|\mathbf{x}-\mathbf{y}|^{n-1}} |\nabla u| \, \mathrm{d}y \quad (\mathbf{x} \in \Omega), \tag{4.26}$$

$$\|u - u_E\|_{L^2(\Omega)} \le C_n \frac{d^n}{|E|} |\Omega|^{1/n} \|\nabla u\|_{L^2(\Omega)}, \tag{4.27}$$

where $u_E = (1/|E|)\int_E u$ *is the average of* u *over* E *and*

$$C_n = \left(\frac{2\pi^{n/2}}{n\Gamma(n/2)}\right)^{1-1/n}.$$

Proof. Let $\mathbf{x} \in \Omega$ be fixed. We have

$$u(\mathbf{x}) - u_E = \frac{1}{|E|}\int_E \left[u(\mathbf{x}) - u(\mathbf{y})\right] \mathrm{d}\mathbf{y} = \frac{1}{|E|}\int_E \left[\int_0^{|\mathbf{x}-\mathbf{y}|} \underbrace{\frac{\mathrm{d}}{\mathrm{d}t}u(\mathbf{x}+t\theta_{\mathbf{y}})}_{=\theta_{\mathbf{y}}\cdot\nabla u}\right] \mathrm{d}y,$$

where $\theta_{\mathbf{y}} = (\mathbf{y}-\mathbf{x})/|\mathbf{y}-\mathbf{x}|$. Hence,

$$|u(\mathbf{x}) - u_E| \le \frac{1}{|E|}\int_\Omega \left[\int_0^{|\mathbf{x}-\mathbf{y}|} |\nabla u(\mathbf{x}+t\theta_y)|\,\mathrm{d}t\right] \mathrm{d}\mathbf{y}.$$

Using the change of the order of integration and $y = \mathbf{x} + \rho\theta$, we obtain

$$\begin{aligned}
|u(\mathbf{x}) - u_E| &\le \frac{1}{|E|}\int_{S^{n-1}}\int_0^d \left[\int_0^\rho \square\,\mathrm{d}t\right]\rho^{n-1}\,\mathrm{d}\rho\,\mathrm{d}\theta \\
&\le \frac{1}{|E|}\int_{S^{n-1}}\int_0^d \left[\int_t^d \rho^{n-1}\,\mathrm{d}\rho\right]\square\,\mathrm{d}t\,\mathrm{d}\theta \\
&\le \frac{1}{|E|}\int_{S^{n-1}}\int_0^d \frac{d^n}{n}\square\,\mathrm{d}t\,\mathrm{d}\theta \\
&= \frac{d^n}{n|E|}\int_{S^{n-1}}\int_0^d t^{1-n}\square\, t^{n-1}\,\mathrm{d}t\,\mathrm{d}\theta \\
&= \frac{d^n}{n|E|}\int_\Omega |\mathbf{x}-\mathbf{y}|^{1-n}|\nabla u(\mathbf{y})|\,\mathrm{d}y,
\end{aligned}$$

where $\square = |\nabla u(\mathbf{x}+t\theta)|\chi_\Omega(x+t\theta)$. The second inequality (4.27) follows from the Hardy–Littlewood–Sobolev inequality. □

Definition 4.2.11 *For* $s \in \mathbb{R}$, *define*

$$H^s(\mathbb{R}^n) := \left\{u \in L^2(\mathbb{R}^n) : \underbrace{\left[\frac{1}{(2\pi)^n}\int (1+|\xi|^2)^s|\hat{u}(\xi)|^2\,\mathrm{d}\xi\right]^{1/2}}_{:=\|u\|_s} < \infty\right\}.$$

This definition is consistent with the previous definition of $H^k(\Omega)$ for $k \in \mathbb{N}$. For the example of $H^1(\mathbb{R}^n)$, it follows from the Plancherel theorem that

$$\int (1+|\xi|^2)|\hat{u}(\xi)|^2\,\mathrm{d}\xi = (2\pi)^n \int |u(\mathbf{x})|^2 + |\nabla u(\mathbf{x})|^2\,\mathrm{d}x.$$

Theorem 4.2.12 *For given* $\ell \in (H^s(\mathbb{R}^n))^*$ *with* $s \ge 0$, *there exists* $v \in H^{-s}(\mathbb{R}^n)$ *such that*

$$\ell(u) = \frac{1}{(2\pi)^n}\int \hat{u}(\xi)\overline{\hat{v}(\xi)}\,\mathrm{d}\xi.$$

Moreover, $\|\ell\| = \|v\|_{H^{-s}}$ *and*

$$\|v\|_{H^{-s}} = \sup_{\|u\|_{H^s}\le 1} \int u\bar{v}\,\mathrm{d}x = \sup_{\|u\|_{H^s}\le 1} \frac{1}{(2\pi)^n}\int \hat{u}\bar{\hat{v}}\,\mathrm{d}\xi.$$

Proof. By the Riesz representation theorem, there exists $w \in H^s(\mathbb{R}^n)$ such that

$$\ell(u) = \langle u, w\rangle_s := \int (1+|\xi|^2)^s \hat{u}(\xi)\overline{\hat{w}(\xi)}\,\mathrm{d}\xi,$$

where $\langle\cdot,\cdot\rangle_s$ is the inner product of H^s. Defining v by $\hat{v}(\xi) := (1+|\xi|^2)^s\hat{w}(\xi) \in H^{-s}$, we have $\|\ell\| = \|w\|_{H^s} = \|v\|_{H^{-s}}$. □

Exercise 4.2.13 *Prove that, for each* $u \in H^{-1}(\mathbb{R}^2)$, *there exists* $\{f_0, f_1, f_2\} \in [L^2(\mathbb{R}^2)]^3$ *such that*

$$u = f_0 + \partial_{x_1} f_1 + \partial_{x_2} f_2 \quad \text{and} \quad C_1\|u\|^2_{H^{-1}} \le \sum_{k=0}^{2}\|f_k\|^2_{L^2} \le C_2\|u\|^2_{H^{-1}},$$

where C_1 *and* C_2 *are positive constants and independent of* u. *Note that we need to choose* $\{f_0, f_1, f_2\} \subset L^2(\mathbb{R}^2)$ *such that*

$$\hat{f}_0(\xi) + (\mathrm{i}\xi_1)\hat{f}_1(\xi) + (\mathrm{i}\xi_2)\hat{f}_2(\xi) = \hat{u}(\xi).$$

There are many choice of $\{f_0, f_1, f_2\}$. *For example, we could choose*

$$\hat{f}_0 = \frac{1}{1+|\xi|^2}\hat{u}, \quad \hat{f}_1 = \frac{-\mathrm{i}\xi_1}{1+|\xi|^2}\hat{u}, \quad \hat{f}_2 = \frac{-\mathrm{i}\xi_2}{1+|\xi|^2}\hat{u}.$$

Exercise 4.2.14 *Let* Ω *be a domain in* $\mathbb{R}^n$. *Prove that* $H^1(\Omega) = H^1_0(\Omega)$ *if and only if* $\Omega = \mathbb{R}^n$. *This means that if* $\Omega \ne \mathbb{R}^n$, *then* $H^1(\Omega) \ne H^1_0(\Omega)$.

Definition 4.2.15 (Negative Sobolev space) *The negative Sobolev space is* $H^{-s}(\Omega) = [H^s_0(\Omega)]^*$ *for* $s > 0$ *where* $\Omega \ne \mathbb{R}^n$ *so that* $H^s_0(\Omega) \subsetneq H^s(\Omega)$.

Theorem 4.2.16 *For each* $\ell \in [H^1_0(\Omega)]^*$, *there exists* $\{f_0, f_1, \ldots, f_n\} \subset L^2(\Omega)$ *such that*

$$\ell(u) = \int_\Omega \left[u\bar{f}_0 + \sum_{p=1}^{n} \partial_{x_k} u\bar{f}_k\right]\mathrm{d}x \quad (\forall u \in H^1_0(\Omega)). \tag{4.28}$$

Moreover, $\|\ell\| = \min\{\sum_{k=0}^{n}\|f_k\|_{L^2} : \{f_0, f_1, \ldots, f_n\}$ *obeys (4.28)*$\}$.

Proof. Let $\Pi : H^1_0(\Omega) \to [L^2(\Omega)]^{n+1}$ be a bounded linear map defined by $\Pi(u) = (u, \partial_{x_1}u, \ldots, \partial_{x_1}u)$. Since $\|u\|_{H^1} = \|\Pi(u)\|_{L^2}$, the map $\Pi : H^1_0(\Omega) \to W := \Pi(H^1_0(\Omega))$ is unitary. Hence, we can define a bounded linear map $\tilde{\ell} : W \to \mathbb{R}$ by

$$\tilde{\ell}(\Pi(u)) = \ell(u).$$

From the Riesz representation theorem, there exists $(f_0, f_1, \ldots, f_n) \in W$ such that

$$\ell(u) = \tilde{\ell}(u, \partial_{x_1} u, \ldots, \partial_{x_1} u) = \int_\Omega \left[u\bar{f}_0 + \sum_{k=1}^{n} \partial_{x_k} u \bar{f}_k \right] \mathrm{d}x \quad (\forall u \in H_0^1(\Omega))$$

and $\|\ell\| = \sum_{k=0}^{n} \|f_k\|_{L^2}$. For the complete proof, see the lecture note by Feldmann and Uhlmann (2003). □

Remark 4.2.17 *For $v \in L^2(\Omega)$, the map $\ell_v : H_0^1(\Omega) \to \mathbb{C}$ is defined by $\ell_v(u) = \int_\Omega vu \,\mathrm{d}x$. Since $|\ell_v(u)| \le \|v\|_{L^2} \|u\|_{H^1}$, we have $\ell_v \in [H_0^1(\Omega)]^*$ and $\|\ell_v\|_{H^{-1}} \le \|v\|_{L^2}$. Moreover, we can show that $L^2(\Omega)$ is dense in $H^{-1}(\Omega)$. By the Riesz representation theorem, for all $\ell \in H^{-1}(\Omega)$, there exists $w \in H_0^1(\Omega)$ such that $\ell(u) = \langle u, w_\ell \rangle_{H^1}$ for all $u \in H_0^1(\Omega)$ and $\|\ell\| = \|w_\ell\|_{H^1}$. Consider the map*

$$\begin{array}{ccccc} v & \to & \ell_v & \to & w_{\ell_v} \\ L^2(\Omega) & \to & H^{-1}(\Omega) & \to & H_0^1(\Omega). \end{array}$$

If $L^2(\Omega)$ is not dense in $H^{-1}(\Omega)$, there exists non-zero $w_0 \in H_0^1(\Omega)$ such that $\langle w_0, w_{\ell_v} \rangle_{H^1} = 0$ for all $v \in L^2(\Omega)$. Since $\langle w_0, w_{\ell_v} \rangle_{H^1} = \ell_v(w_0) = \langle w_0, w_{\ell_v} \rangle_{L^2} = 0$ for all $v \in L^2(\Omega)$, then $w_0 = 0$. Hence, $H^{-1}(\Omega)$ can be viewed as a completion of $L^2(\Omega)$ with the norm

$$\|v\|_{H^{-1}} = \sup_{0 \neq u \in H_0^1(\Omega)} \frac{\langle u, \bar{v} \rangle}{\|u\|_{H^1}}.$$

Remark 4.2.18 *According to the Hahn–Banach theorem in section 4.4.2, we can extend ℓ to $\bar{\ell} \in \mathcal{L}(H^s(\Omega), \mathbb{R})$. But $[H_0^1(\Omega)]^*$ is more useful than $[H^1(\Omega)]^*$ since (4.28) can be written as*

$$\ell(u) = \int_\Omega \left[u\bar{f}_0 - \sum_{p=1}^{n} u \partial_{x_k} \bar{f}_k \right] \mathrm{d}x \quad (\forall u \in H_0^1(\Omega))$$

provided that $f_0, \ldots, f_n$ are differentiable.

Definition 4.2.19 *Let $\Omega \subset \mathbb{R}^n$ be a C^∞ bounded domain. Then, there exist $p_1, \ldots, p_N \in \partial\Omega$, an open neighborhood U_p of p_j and a C^∞ diffeomorphism $\Phi_j : U_{p_j} \to \mathbb{R}^n$ such that $\partial\Omega \cup \bigcup_{j=1}^{N} U_{p_j}$ and*

$$\Phi_j(U_{p_j} \cap \Omega) = \{x \in \mathbb{R}^n : x_n > 0\} \quad \text{and} \quad \Phi_j(U_{p_j} \cap \partial\Omega) = \{x \in \mathbb{R}^n : x_n = 0\}.$$

Choosing $\eta_j \in C_0^\infty(U_{p_j})$ such that $\sum_{j=1}^{N} \eta_j = 1$ in a neighborhood of $\partial\Omega$, we define $H^s(\partial\Omega)$ by the closure of $C^\infty(\partial\Omega)$ with respect to the norm

$$\|u\|_{H^s(\partial\Omega)}^2 = \sum_{j=1}^{n} \|(\eta_j u) \circ \Phi_j^{-1}\|_{H^s(\mathbb{R}^{n-1})}^2.$$

Lemma 4.2.20 (Trace: restriction and extension) *Let $s > \frac{1}{2}$. There is a positive constant C depending only on s such that*

$$\|u\|_{H^{s-1/2}(\partial\mathbb{R}^n_+)} \le C\|u\|_{H^s(\mathbb{R}^n)} \quad (\forall u \in H^s(\mathbb{R}^n)).$$

On the other hand, for each $f \in H^{s-1/2}(\partial\mathbb{R}^n_+)$, there exists $u \in H^s(\mathbb{R}^n)$ such that

$$u|_{\partial\mathbb{R}^n_+} = f \quad \text{and} \quad \|u\|_{H^s(\mathbb{R}^n)} \lesssim \|f\|_{H^{s-1/2}(\partial\mathbb{R}^n_+)}.$$

Proof. From the uniform bounded principle theorem, it suffices to prove that

$$\|u\|_{H^{s-1/2}(\partial\mathbb{R}^n_+)} \le C\|u\|_{H^s(\mathbb{R}^n)} \quad (\forall u \in C_0^\infty(\mathbb{R}^n)).$$

For simplicity of notation, we shall prove only the two-dimensional case $n = 2$. Indeed, we can apply the same idea to prove it for the general case. Let $\Pi : \mathbb{R}^2 \to \mathbb{R}$ be the projection map to the first component. From the Fourier inversion formula, we have

$$\Pi u(x_1) = u(x_1, 0) = \frac{1}{(2\pi)^2} \int e^{ik_1x_1} \hat{u}(k_1, k_2)\, dk_1\, dk_2$$

and

$$\Pi u(x_1) = \frac{1}{2\pi} \int e^{ik_1x_1} \widehat{\Pi u}(k_1)\, dk_1.$$

Comparing the above two identities, we have $\widehat{\Pi u}(k_1) = [1/(2\pi)] \int \hat{u}(k_1, k_2)\, dk_2$ and it follows from the Schwarz inequality that

$$|\widehat{\Pi u}(k_1)| \lesssim \underbrace{\left(\int [1 + |k|^2]^{-s}\, dk \right)^{1/2}}_{\cong (1+k_1^2)^{-s+1/2}} \left(\int |\hat{u}(k)|^2 [1 + |k|^2]^s\, dk_2 \right)^{1/2}.$$

Hence, we have

$$\begin{aligned} \|\Pi u\|^2_{H^{s-1/2}} &= \int (1 + |k_1|^2)^{s-1/2} |\widehat{\Pi u}(k_1)|^2\, dk_1 \\ &\lesssim \int |\hat{u}(k)|^2 [1 + |k|^2]^s\, dk = \|u\|_{H^s(\mathbb{R}^2)}. \end{aligned}$$

For the remaining proof, see the lecture note by Feldmann and Uhlmann (2003). □

4.2.3 Helmholtz Decomposition

The Helmholtz decomposition states that any smooth vector field $\mathbf{F}$ in a smooth bounded domain Ω can be resolved into the sum of a divergence-free (solenoidal) vector field and a curl-free (irrotational) vector field.

Theorem 4.2.21 *Every vector field $\mathbf{F}(\mathbf{r}) = (F_1(\mathbf{r}), F_2(\mathbf{r}), F_3(\mathbf{r})) \in [L^2(\Omega)]^3$ can be decomposed into*

$$\mathbf{F}(\mathbf{r}) = -\nabla u(\mathbf{r}) + \nabla \times \mathbf{A}(\mathbf{r}) + \textit{harmonic} \quad \text{in } \Omega, \tag{4.29}$$

where u is a scalar function, $\nabla \cdot \mathbf{A} = 0$ and harmonic is a vector field whose Laplacian is zero in Ω. Moreover, u and $\mathbf{A}$ are solutions of $\nabla^2 u = \nabla \cdot \mathbf{F}$ and $\nabla^2 \mathbf{A} = \nabla \times \mathbf{F}$, with *appropriate boundary conditions. Hence, these can be uniquely determined up to harmonic functions:*

$$u(\mathbf{r}) = -\int_\Omega \frac{\nabla \cdot \mathbf{F}(\mathbf{r}')}{4\pi|\mathbf{r}-\mathbf{r}'|}\,d\mathbf{r}' + \mathit{harmonic} \tag{4.30}$$

and

$$\mathbf{A}(\mathbf{r}) = \int_\Omega \frac{\nabla \times \mathbf{F}(\mathbf{r}')}{4\pi|\mathbf{r}-\mathbf{r}'|}\,d\mathbf{r}' + \mathit{harmonic}. \tag{4.31}$$

Proof. We write the vector field $\mathbf{F}$ as

$$\mathbf{F}(\mathbf{r}) = \int_\Omega \delta(\mathbf{r}-\mathbf{r}')\mathbf{F}(\mathbf{r}')\,d\mathbf{r}' = -\int_\Omega \nabla^2\left(\frac{1}{4\pi|\mathbf{r}-\mathbf{r}'|}\right)\mathbf{F}(\mathbf{r}')\,d\mathbf{r}'.$$

Integration by parts yields

$$\mathbf{F}(\mathbf{r}) = -\int_\Omega \frac{1}{4\pi|\mathbf{r}-\mathbf{r}'|}\nabla^2\mathbf{F}(\mathbf{r}')\,d\mathbf{r}' + \int_{\partial\Omega} \Psi_1(\mathbf{r},\mathbf{r}')\,dS_{\mathbf{r}'}, \quad \mathbf{r}\in\Omega, \tag{4.32}$$

where

$$\Psi_1(\mathbf{r},\mathbf{r}') = -\frac{\partial}{\partial\mathbf{n}}\left(\frac{1}{4\pi|\mathbf{r}-\mathbf{r}'|}\right)\mathbf{F}(\mathbf{r}') + \frac{1}{4\pi|\mathbf{r}-\mathbf{r}'|}\frac{\partial}{\partial\mathbf{n}}\mathbf{F}(\mathbf{r}').$$

Owing to the property

$$\nabla_{\mathbf{r}}^2\left(\frac{-1}{4\pi|\mathbf{r}-\mathbf{r}'|}\right) = \delta(\mathbf{r}-\mathbf{r}'),$$

we have

$$\nabla_{\mathbf{r}}^2\Psi_1(\mathbf{r},\mathbf{r}') = 0 \quad \text{for } \mathbf{r}\in\Omega,\ \mathbf{r}'\in\partial\Omega.$$

Using the vector identity $-\nabla^2\mathbf{F} = \nabla\times(\nabla\times\mathbf{F}) - \nabla(\nabla\cdot\mathbf{F})$, we can express (4.32) as

$$\mathbf{F}(\mathbf{r}) = \int_\Omega \frac{1}{4\pi|\mathbf{r}-\mathbf{r}'|}[\nabla\times(\nabla\times\mathbf{F}) - \nabla(\nabla\cdot\mathbf{F})]\,dV + \int_{\partial\Omega}\Psi_1(\mathbf{r},\mathbf{r}')\,dS, \quad \mathbf{r}\in\Omega. \tag{4.33}$$

Integrating by parts again, we have

$$\mathbf{F}(\mathbf{r}) = \nabla\times\int \frac{\nabla\times\mathbf{F}(\mathbf{r}')}{4\pi|\mathbf{r}-\mathbf{r}'|}\,dV - \nabla\int\frac{\nabla\cdot\mathbf{F}(\mathbf{r}')}{4\pi|\mathbf{r}-\mathbf{r}'|}\,d\mathbf{r}' + \int_{\partial\Omega}\Psi_1(\mathbf{r},\mathbf{r}') + \Psi_2(\mathbf{r},\mathbf{r}')\,dS,$$

where Ψ_2 is a function satisfying $\nabla_{\mathbf{r}}^2\Psi_2(\mathbf{r},\mathbf{r}') = 0$ for $\mathbf{r}\in\Omega$ and $\mathbf{r}'\in\partial\Omega$. This completes the proof of the Helmholtz decomposition. □

4.3 Variational Problems

4.3.1 Lax–Milgram Theorem

Let Ω be a bounded domain in $\mathbb{R}^n$ with smooth boundary $\partial\Omega$. In this section, we discuss the existence, uniqueness and stability of the boundary value problem (BVP):

$$\begin{cases} -\nabla\cdot(\sigma\nabla u) = f & \text{in } \Omega \\ u|_{\partial\Omega} = 0 \end{cases} \qquad (0 < c_0 < \sigma(\mathbf{r}) < c_1 < \infty), \tag{4.34}$$

where c_0 and c_1 are positive constants. In the case when σ is discontinuous, the fundamental mathematical theories of solutions such as existence, uniqueness and stability can be established within the framework of the Hilbert space $H^1(\Omega)$, which is a norm closure of $C^1(\Omega) \cap C(\bar{\Omega})$ with the norm $\|u\|_{H^1(\Omega)}$.

This problem (4.34) is equivalent to the following variational problem:

$$\text{find } u \in H_0^1(\Omega) = \{u \in H^1(\Omega) : u|_{\partial\Omega} = 0\} \text{ such that}$$

$$\int_\Omega \nabla u \nabla \phi = \int_\Omega f\phi \,\mathrm{d}x, \quad \forall \phi \in H_0^1(\Omega).$$

Define the map $a : H_0^1(\Omega) \times H_0^1(\Omega) \to \mathbb{R}$ by

$$a(u, \phi) = \int_\Omega \sigma \nabla u \nabla \phi \tag{4.35}$$

and define $b : H_0^1(\Omega) \to \mathbb{R}$ by

$$b(\phi) = \int_\Omega f\phi \,\mathrm{d}x. \tag{4.36}$$

Hence, the solvability problem of (4.34) is equivalent to the uniqueness and existence question of finding $u \in X = H_0^1(\Omega)$ satisfying

$$a(u, \phi) = b(\phi), \quad \forall \phi \in X. \tag{4.37}$$

Note that the map $a : H_0^1(\Omega) \times H_0^1(\Omega) \to \mathbb{R}$ satisfies the following:

- both $a(u, \cdot) : X \to \mathbb{R}$ and $a(\cdot, w) : X \to \mathbb{R}$ are linear for all $u, w \in X$;
- $|a(u, v)| < c_1\|u\|\,\|v\|$ and $c_0\|u\|^2 \le a(u, u)$.

Here, the norm $\|u\|$ is $\|u\| = \sqrt{\int_\Omega |u|^2 + |\nabla u|^2 \,\mathrm{d}\mathbf{r}}$. The estimate $|a(u, v)| < c_1\|u\|\,\|v\|$ can be obtained using the Schwarz inequality and the Poincaré inequality gives the estimate $c_0\|u\|^2 \le a(u, u)$.

Assuming that the Hilbert space $H_0^1(\Omega)$ has a basis $\{\phi_k : k = 1, 2, \ldots\}$, the variational problem (4.37) is to determine the coefficient $\{u_k : k = 1, 2, \ldots\}$ of $u = \sum u_k\phi_k$ satisfying

$$a\left(\sum_k u_k\phi_k, \phi_j\right) = b(\phi_j), \quad \forall j = 1, 2, \ldots. \tag{4.38}$$

Taking advantage of the linearity of $a(\cdot, \cdot)$ and $b(\cdot)$, the problem (4.38) is equivalent to solving

$$\underbrace{\begin{pmatrix} a(\phi_1, \phi_1) & a(\phi_1, \phi_2) & \cdots \\ a(\phi_2, \phi_1) & a(\phi_2, \phi_2) & \cdots \\ a(\phi_3, \phi_1) & a(\phi_3, \phi_2) & \cdots \\ \vdots & \vdots & \ddots \end{pmatrix}}_{A} \underbrace{\begin{pmatrix} u_1 \\ u_2 \\ u_3 \\ \vdots \end{pmatrix}}_{\mathbf{u}} = \underbrace{\begin{pmatrix} b(\phi_1) \\ b(\phi_2) \\ b(\phi_3) \\ \vdots \end{pmatrix}}_{\mathbf{b}},$$

where A can be viewed as an $\infty \times \infty$ matrix with coefficients $a_{kj} = a(\phi_k, \phi_j)$. From the fact that $c_0\|u\|^2_{H^1(\Omega)} \leq a(u, u)$, A is positive definite and invertible. Therefore, for given data $\mathbf{b}$, there exists a unique $\mathbf{u}$ satisfying $A\mathbf{u} = \mathbf{b}$.

Now, we will prove the uniqueness and existence in $X = H_0^1(\Omega)$ in a general setting. Let X be a Hilbert space over $\mathbb{R}$ with a norm $\|\cdot\|$. The map $a(\cdot,\cdot) : X \times X \to \mathbb{R}$ is called a bounded bilinear map on the Hilbert space X if

- both $a(u,\cdot) : X \to \mathbb{R}$ and $a(\cdot, w) : X \to \mathbb{R}$ are linear for all $u, w \in X$, and
- there exist M so that $|a(u, v)| < M\|u\|\,\|v\|$.

The solvability of the problem (4.34) can be obtained from the following theorem.

Theorem 4.3.1 (Lax–Milgram theorem) *Suppose the bounded bilinear map $a : X \times X \to \mathbb{R}$ is symmetric and*

$$a(u, u) \geq c\|u\|^2, \quad c > 0. \tag{4.39}$$

Suppose $b(\cdot) : X \to \mathbb{R}$ is linear.

1. *There exists a unique solution $u \in X$ of the following minimization problem:*

$$\text{minimize} \quad \Phi(u) := \tfrac{1}{2}a(u, u) - b(u), \quad u \in X. \tag{4.40}$$

2. *The solution u of the minimization problem (4.40) is the solution of the following variational problem:*

$$a(u, \phi) = b(\phi), \quad \forall \phi \in X. \tag{4.41}$$

Before proving Theorem 4.3.1, we gain its key idea by simple examples.

Example 4.3.2 *Let $X = \mathbb{R}^n$, $\mathbf{b} \in X$ and A be an $n \times n$ symmetric matrix having positive eigenvalues. Define*

$$a(\mathbf{x}, \mathbf{y}) = \langle A\mathbf{x}, \mathbf{y}\rangle := \sum_{i=1}^{n}\sum_{i=1}^{n} a_{ij}x_i y_j, \quad b(\mathbf{x}) := \mathbf{b}\cdot\mathbf{x}.$$

Then, $a(\cdot,\cdot)$ is bounded bilinear and $b(\cdot)$ is bounded linear. Since A has positive eigenvalues, $a(\cdot,\cdot)$ satisfies the condition (4.39). Hence, it meets the requirements of Theorem 4.3.1. Noting that

$$\mathbf{y}\cdot\nabla\Phi(\mathbf{x}) = \langle A\mathbf{x} - \mathbf{b}, \mathbf{y}\rangle \quad (\forall\, \mathbf{y} \in X),$$

we have

$$\nabla\Phi(\mathbf{x}) = 0 \iff A\mathbf{x} - \mathbf{b} = 0.$$

Example 4.3.3 (Laplace equation) *Let Ω be a bounded domain in $\mathbb{R}^n$ with a smooth boundary. We consider the following minimization problem:*

$$\text{minimize} \quad \Phi(v) := \frac{1}{2}\int_\Omega |\nabla v(\mathbf{x})|^2\,\mathrm{d}\mathbf{x} - \int_\Omega f(\mathbf{x})v(\mathbf{x})\,\mathrm{d}\mathbf{x}, \quad v \in X = H_0^1(\Omega).$$

This minimization problem has a unique solution in the Sobolev space $H_0^1(\Omega)$.

- *$X = H_0^1(\Omega)$ is a Hilbert space with the norm $\|v\| = \sqrt{\int_\Omega |\nabla u|^2 + |u|^2\,\mathrm{d}\mathbf{x}}$.*
- *$a(v, w) = \int_\Omega \nabla v \cdot \nabla w\,\mathrm{d}x$ is bounded bilinear on X.*
- *$b(v) = \int_\Omega f v\,\mathrm{d}x$ is bounded linear on X when $f \in L^2(\Omega)$.*
- *$a(\cdot,\cdot)$ is coercive (which will be proved in the next section):*

$$\|u\| = \sqrt{\int_\Omega |\nabla u|^2 + |u|^2\,\mathrm{d}\mathbf{x}} \le C\sqrt{a(u,u)} = C\sqrt{\int_\Omega |\nabla u|^2\,\mathrm{d}\mathbf{x}}.$$

- *From Theorem 4.3.1, there exists a unique solution $u \in H_0^1(\Omega)$ of the above minimization problem.*
- *This minimizer u satisfies the variational problem:*

$$\int_\Omega \nabla u(\mathbf{x}) \cdot \nabla\phi(\mathbf{x})\,\mathrm{d}\mathbf{x} = a(u,\phi) = b(\phi) = \int_\Omega f(\mathbf{x})\phi(\mathbf{x})\,\mathrm{d}\mathbf{x}, \quad \phi \in X.$$

- *This solution u is a solution of the corresponding Euler–Lagrange equation:*

$$-\nabla^2 u = f \quad \text{in } \Omega, \tag{4.42}$$

$$u|_{\partial\Omega} = 0. \tag{4.43}$$

Proof. (Lax–Milgram theorem) We begin by proving the existence of the minimization problem. Denoting

$$\alpha := \inf_{u\in X} \Phi(u) > -\infty,$$

there exists a minimizing sequence $\{u_n\}$, that is,

$$\Phi(u_n) \to \alpha.$$

Existence of the minimizer can be proved by showing that

$$\exists\, u \in X \text{ s.t. } \Phi(u) = \alpha. \tag{4.44}$$

The proof of (4.44) is based on the identity

$$a(u_n, u_n) + a(u_m, u_m) = \tfrac{1}{2}a(u_n - u_m, u_n - u_m) + \tfrac{1}{2}a(u_n + u_m, u_n + u_m). \tag{4.45}$$

This leads to

$$2\Phi(u_n) + 2\Phi(u_m) = \tfrac{1}{2}a(u_n - u_m, u_n - u_m) + 4\Phi(\tfrac{1}{2}(u_n + u_m)). \tag{4.46}$$

Setting $n, m \to \infty$, we have

$$\begin{aligned} 4\alpha \longleftarrow 2\Phi(u_n) + 2\Phi(u_m) &= \tfrac{1}{2}a(u_n - u_m, u_n - u_m) + 4\Phi(\tfrac{1}{2}(u_n + u_m)) \\ &\geq \tfrac{1}{2}c\|u_n - u_m\|^2 + 4\alpha. \end{aligned}$$

Hence $\{u_n\}$ is a Cauchy sequence and there exists $u \in X$ so that $u = \lim_{n\to\infty} u_n$. Since $\Phi(\cdot)$ is continuous, owing to the assumption that $a(\cdot,\cdot)$ and $b(\cdot)$ are bounded and linear, we prove the existence of the minimizer u:

$$\Phi(u) = \lim_{n\to\infty} \Phi(u_n) = \alpha.$$

Now, we prove uniqueness. If u is a minimizer,

$$\frac{\mathrm{d}}{\mathrm{d}t}\Phi(u + t\phi)|_{t=0} = 0 \quad (\forall \phi \in X) \quad \Longrightarrow \quad a(u, \phi) = b(\phi) \quad (\forall \phi \in X).$$

Hence, if u and v are minimizers, then

$$a(u, \phi) = b(\phi) = a(v, \phi) \quad (\forall \phi \in X),$$

which is equivalent to

$$a(u - v, \phi) = 0 \quad (\forall \phi \in X).$$

In particular, $a(u - v,\ u - v) = 0$. Then, it follows from (4.39) that

$$c\|u - v\|^2 \leq a(u - v,\ u - v) = 0,$$

which gives $u = v$. □

Exercise 4.3.4 *Prove that* $\inf_{u\in X} \Phi(u) > -\infty$.

4.3.2 Ritz Approach

Taking account of the finite element method, we continue to study Hilbert space techniques. As before, we assume the following:

- X is a real Hilbert space with norm $\|\cdot\|$;
- $\{X_n\}$ is a sequence of a finite-dimensional subspaces of X such that $X_n \subset X_{n+1}$ and

$$\overline{\bigcup_{n=1}^{\infty} X_n} = X;$$

- $\{\phi_j^n : j = 1, \ldots, N_n\}$ is a basis of X_n;
- $a(\cdot,\cdot) : X \times X \to \mathbb{R}$ is a bounded, symmetric, strongly positive, bilinear map and

$$|a(u, v)| \leq M\|u\|\,\|v\|, \quad c\|u\|^2 \leq a(u, u), \quad \forall u, v \in X;$$

- $b \in X^*$ where X^* is the set of linear functionals on X.

As before, we consider the minimization problem:

$$\text{minimize} \quad \Phi(u_n) := \tfrac{1}{2}a(u_n, u_n) - b(u_n), \quad u_n \in X_n. \tag{4.47}$$

The variational problem is as follows:

$$\text{find} \quad u_n \in X_n \text{ s.t. } a(u_n, \phi_n) = b(\phi_n), \quad \forall \phi_n \in X_n. \tag{4.48}$$

According to the Lax–Milgram theorem, we know the solvability of (4.47) and (4.48). We try to give a rather different analysis by using the new inner product $a(u, v)$. Since $\|u\|^2 \approx a(u, u)$, $\|u\|_a := \sqrt{a(u, u)}$ can be viewed as a norm of the Hilbert space X.

Theorem 4.3.5 (Projection theorem) *The space X equipped with the new inner product and norm*

$$\langle u, v\rangle_a := a(u, v), \quad \|u\|_a := \sqrt{a(u, u)} \quad (\forall u, v \in X) \tag{4.49}$$

is also a Hilbert space. For any closed subspace $V = \bar{V} \subset X$ and for each $u \in X$,

$$\exists\, u_V \in V \quad \text{s.t.} \quad \|u - u_V\|_a = \min_{v \in V} \|u - v\|_a. \tag{4.50}$$

Denoting $V^{\perp} := \{w \in X : a(w, v) = 0, \forall v \in V\}$, if u is decomposed as

$$u = u_1 + u_2 \quad (u_1 \in V,\ u_2 \in V^{\perp}),$$

then $u_1 = u_V$ and $u_2 = u - u_V$.

Proof. Let u be fixed. Since

$$\|u - v\|_a^2 = a(v, v) - 2a(u, v) + a(u, u),$$

the problem (4.50) is equivalent to solving the following minimization problem:

$$\text{minimize} \quad \widetilde{\Phi}(v) = \tfrac{1}{2}a(v, v) - b(v) \quad \text{within the set } V, \tag{4.51}$$

where $b(v) = a(u, v)$. According to the Lax–Milgram theorem, there exists a unique solution $u_V \in V$ of the minimization problem (4.51). Next, we will show $u - u_V \in V^{\perp}$. Since

$$\|u - u_V\|_a^2 \leq \|u - (u_V + tv)\|_a^2 \quad (\forall v \in V,\ \forall t \in \mathbb{R}),$$

we have

$$0 \leq -2ta(u - u_V, v) + t^2 a(v, v) \quad (\forall v \in V, \forall t \in \mathbb{R}).$$

Hence,

$$-ta(v, v) \leq 2a(u - u_V, v) \leq ta(v, v), \quad (\forall v \in V,\ \forall t > 0).$$

Letting $t \to 0$, we get

$$a(u - u_V, v) = 0 \quad (\forall v \in V)$$

and $u - u_V \in V^\perp$. Uniqueness of the decomposition follows from the fact that

$$0 = u - u = \underbrace{(u_1 - u_V)}_{\in V} + \underbrace{(u_2 - u + u_V)}_{\in V^\perp} \quad \Longrightarrow \quad u_1 = u_V \text{ and } u_2 = u - u_V.$$

□

Definition 4.3.6 *Let X be a Hilbert space over the scalar field $\mathbb{R}$. The dual of X, denoted by X^*, is the set of linear maps $\ell : X \to \mathbb{R}$.*

Theorem 4.3.7 (Riesz representation theorem) *For each linear functional $b \in X^*$, there is a unique $u_* \in X$ such that*

$$b(v) = a(u_*, v) \quad (\forall\, v \in X). \tag{4.52}$$

Moreover, $\|b\| = \sqrt{a(u_*, u_*)}$.

Proof. Assume $b \neq 0$, that is, $b(v) \neq 0$ for some $v \in X$. We first prove that the null space $V := \{v \in X : b(v) = 0\}$ is a closed subspace of X of codimension 1, that is,

$$\dim V^\perp = 1.$$

From the assumption of $b \neq 0$ and using the linearity of $b(\cdot)$,

$$\exists\, u \in V^\perp \text{ s.t. } b(u) = 1.$$

Owing to the linearity of $b(\cdot)$ and $b(u) = 1$, we have

$$b(v - b(v)u) = b(v) - b(v)b(u) = 0 \quad (\forall\, v \in X).$$

Hence,

$$v - b(v)u \in V \quad (\forall\, v \in X).$$

This eventually gives a decomposition:

$$v = \underbrace{v - b(v)u}_{\in V} + \underbrace{b(u)u}_{\in V^\perp} \quad (\forall\, v \in X).$$

According to the projection theorem, the above decomposition is unique, and this proves that $\dim V^\perp = 1$. Setting $u_* = u/a(u, u)$, we have

$$a(u_*, v) = a(u_*, v - b(v)u + b(v)u) = a(u_*, b(v)u) = a(u_*, u)b(v) = b(v) \quad \forall\, v \in X.$$

From the above identity, it is easy to see that u_* is unique. Since $|b(v)| \leq \|u_*\|_a \|v\|_a$ and $|b(u_*)| = \|u_*\|_a^2$, we have $\|b\| = \|u_*\|_a$. This completes the proof. □

Theorem 4.3.8 *Assume that $u_n \in X_n$ is a unique solution of the minimization problem (4.47) or the variational problem (4.48). Let u be a solution of the variational problem (4.48) with X_n replaced by X. Then,*

(i) $\lim_{n\to\infty} \|u_n - u\| = 0$,
(ii) $\|u - u_n\| \leq (c/M) \min_{v \in X_n} \|u - v\|$,
(iii) $(c/2)\|u - u_n\|^2 \leq \Phi(u_n) - \Phi(u)$.

Proof. Since $a(u, v) = b(v) = a(u_n, v)$ for all $v \in X_n$, we have

$$a(u - u_n, v) = 0 \quad (\forall\, v \in X_n).$$

Since $a(u - u_n, u - u_n) = a(u - u_n, u)$ and $a(u - u_n, u - v) = a(u - u_n, u)$ for all $v \in X_n$, we have

$$\begin{aligned}\|u - u_n\|^2 &\leq \frac{1}{c}a(u - u_n, u - u_n) = \frac{1}{c}a(u - u_n, u - v) \\ &\leq \frac{M}{c}\|u - u_n\|\,\|u - v\|\end{aligned} \qquad (\forall\, v \in X_n),$$

which gives (ii). Part (i) follows from the assumption that $X_n \to X$ as $n \to \infty$. For all $v \in X$,

$$\begin{aligned}\Phi(u + v) &= \tfrac{1}{2}a(u + v, u + v) - b(u + v) \\ &= \tfrac{1}{2}a(v, v) + (a(u, v) - b(v)) + \Phi(u) \\ &= \tfrac{1}{2}a(v, v) + \Phi(u),\end{aligned}$$

and therefore

$$\Phi(u + v) - \Phi(u) = \tfrac{1}{2}a(v, v) \geq \tfrac{1}{2}c\|v\|^2 \quad (\forall\, v \in X),$$

which proves (iii). □

Exercise 4.3.9 (Contraction mapping) *Let X be a Hilbert space over $\mathbb{R}$. Assume an operator $\mathcal{T} : X \to X$ is a contraction mapping, that is, there exists $0 < \theta < 1$ such that $\|\mathcal{T}u - \mathcal{T}v\| \leq \theta\|u - v\|$ for all $u, v \in X$. Then, there exists a unique $u_* \in X$ such that*

$$\mathcal{T}(u_*) = u_*.$$

Note that, for fixed $u_0 \in X$, we can define a sequence $u_{n+1} = \mathcal{T}(u_n)$ and prove that the sequence is a Cauchy sequence.

Exercise 4.3.10 (Lax–Milgram theorem for non-symmetric bilinear operator) *Let X be a Hilbert space over $\mathbb{R}$. Assume that the map $a(\cdot, \cdot) : X \times X \to \mathbb{R}$ is bounded, coercive and bilinear:*

$$|a(u, v)| < M\|u\|\,\|v\| \quad \text{and} \quad c\|u\|^2 \leq a(u, u).$$

Show that, for each $b \in X^*$, there is a unique $u \in X$ such that

$$a(\phi, u) = b(\phi) \quad \forall\, \phi \in X.$$

Note that, for each $v \in X$, we can define a linear functional $\ell_v \in X^*$ by $\ell_v(\phi) = a(\phi, v)$. We need to prove that there exists a unique $u \in X$ such that $\ell_u = b$. By the Riesz representation theorem, there exists a bounded linear operator $\mathcal{L} : X^* \to X$ such that $\ell(\phi) = \langle \phi, \mathcal{L}(\ell)\rangle$ for $\phi \in X$ and $\ell \in X^*$. We need to show that there exists a unique u such that $\mathcal{L}(\ell_u) = \mathcal{L}(b)$. Define a map $\mathcal{T} : X \to X$ by

$$\mathcal{T}v := v - \beta\mathcal{L}(\ell_v - b).$$

For sufficiently small $\beta > 0$, $\mathcal{T}$ is a contraction map because

$$\|\mathcal{T}v - \mathcal{T}w\|^2 = \|v - w\|^2 - 2\beta\langle v - w, \mathcal{L}(\ell_{v-w})\rangle + \beta^2\|\mathcal{L}(\ell_{v-w})\|^2.$$

With $0 < \beta < c/(M^2 + 1)$, there is a unique $u \in X$ such that $\mathcal{T}u = u$, which is equivalent to $\mathcal{L}(\ell_u) = \mathcal{L}(b)$.

Exercise 4.3.11 (Lax–Milgram theorem on Hilbert space over $\mathbb{C}$) *Let X be a Hilbert space over the complex scalar field $\mathbb{C}$. Assume that the map $a(\cdot,\cdot) : X \times X \to \mathbb{C}$ satisfies the following:*

- *$a(\cdot, u) : X \to \mathbb{R}$ is linear for each $u \in X$;*
- *$a(u, v) = \overline{a(v, u)}$ for all $u, w \in X$;*
- *there exists M so that $|a(u, v)| < M\|u\|\,\|v\|$;*
- *there exists c so that $c\|u\|^2 \le \Re\{a(u, u)\}$ for all u.*

Show that, for each $b \in X^*$, there is a unique $u \in X$ such that

$$a(u, \phi) = b(\phi) \quad \forall \phi \in X.$$

As in the previous exercise, we need to prove that there is a unique $u \in X$ such that $\mathcal{L}(\ell_u) = \mathcal{L}(b)$. Let $X_1 = \{\mathcal{L}\ell_v : v \in X\}$. Existence can be proven by showing that $X_1 = X$. Indeed, if $v \in X_1^\perp$, then

$$c\|v\|^2 \le \Re\{a(v, v)\} = \Re\{\langle \mathcal{L}v, v\rangle\} = 0 \quad \Longrightarrow \quad v = 0.$$

Uniqueness comes from a careful use of the coercivity condition.

4.3.3 Euler–Lagrange Equations

We now study how to compute the gradient of $\Phi(u)$ on a subset X of the Hilbert space $H^1(\Omega)$ using several important examples.

Example 4.3.12 (Minimization problem in one dimension) *Let $\sigma(x) \in C_+([a, b]) = \{v \in C([a, b]) : v > 0\}$. For a given $f \in L^2(a, b)$ and $\alpha, \beta \in \mathbb{R}$, suppose that u_* is a minimizer of the following minimization problem:*

$$\begin{aligned}&\textit{minimize} \quad \Phi(u) := \int_a^b |f(x) - u(x)|^2 + \sigma(x)|u'(x)|^2\,\mathrm{d}x\\ &\textit{subject to the constraint}\\ &u \in X := \{u \in H^1(a, b) \mid u(a) = \alpha,\ u(b) = \beta\}.\end{aligned}$$

Then, u_* satisfies

$$-(\sigma(x)u_*'(x))' + (u_*(x) - f(x)) = 0 \quad (a < \forall x < b)$$

with the boundary conditions $u_*(a) = \alpha$, $u_*(b) = \beta$.

Proof.

1. In this Hilbert space H^1, the inner product and the distance (or metric) between two functions u and v are given respectively by

$$\langle u, v\rangle = \int_a^b uv + u'v'\,\mathrm{d}x \quad \text{and} \quad \|u - v\| = \sqrt{\langle u - v, u - v\rangle}.$$

2. The goal is to investigate a minimizer $u_* \in X$ satisfying

$$\Phi(u_*) \le \Phi(u) \qquad \text{for all } u \in X,$$

which is equivalent to

$$\Phi(u_*) \le \Phi(u_* + tv) \quad \forall\, t \in \mathbb{R},\ \forall\, v \in H_0^1(a, b),$$

where $H_0^1(a, b) := \{u \in H^1(a, b) : u(a) = 0 = u(b)\}$.

3. Hence,

$$\frac{\mathrm{d}}{\mathrm{d}t}\Phi(u_* + tv)|_{t=0} = 0 \quad \forall\, t \in \mathbb{R},\ \forall\, v \in H_0^1.$$

4. By the chain rule and integrating by parts,

$$\begin{aligned}
\frac{\mathrm{d}}{\mathrm{d}t}\Phi(u_* + tv)|_{t=0} &= \lim_{t\to 0}\frac{1}{t}(\Phi(u_* + tv) - \Phi(u_*)) \\
&= \lim_{t\to 0}\frac{1}{t}\int_a^b [2tv(u_* - f) + 2t\sigma u_*' v' + t^2(v^2 + \sigma v'^2)]\,\mathrm{d}x \\
&= \int_a^b [2v(u_* - f) + 2\sigma u_*' v']\,\mathrm{d}x \\
&= 2\int_a^b [(u_* - f) - (\sigma u_*')']v\,\mathrm{d}x.
\end{aligned}$$

This can be viewed as a directional derivative of Φ at u_* in the direction v, that is, the direction where the derivative is computed.

5. Since this holds for all functions $v \in H_0^1$, we have

$$\begin{aligned}
&\frac{\mathrm{d}}{\mathrm{d}t}\Phi(u_* + tv)|_{t=0} = 0 \quad (\forall\, v \in H_0^1) \\
\iff\ & \int_a^b \left[(u_* - f) - (\sigma u_*')'\right] v\,\mathrm{d}x \quad (\forall\, v \in H_0^1) \\
\iff\ & -(\sigma u_*')' + (u_* - f) = 0 \quad (a < \forall\, x < b).
\end{aligned}$$

The last one is the Euler–Lagrange equation of the minimization problem. □

Example 4.3.13 *Consider the curve minimization problem joining two points (a, α) and (b, β):*

$$\text{minimize} \quad \Phi(u) = \int_a^b \sqrt{1 + (u')^2}\, \mathrm{d}x$$
$$\text{within the set} \quad X = \{u \in H^1(a, b) \mid u(a) = \alpha,\ u(b) = \beta\}.$$

If u_* is a minimizer, then

$$u_*''(x) = 0 \quad (a < x < b).$$

Proof. For all $v \in H_0^1$, we have

$$\frac{\mathrm{d}}{\mathrm{d}t}\Phi(u + tv)|_{t=0} = \frac{\mathrm{d}}{\mathrm{d}t}\int_a^b \sqrt{1 + (u' + tv')^2}\, \mathrm{d}x\bigg|_{t=0}$$

$$= \int_a^b \frac{2u'v' + 2tv'^2}{\sqrt{1 + (u' + tv')^2}}\, \mathrm{d}x\bigg|_{t=0} = \int_a^b \frac{2u'v'}{\sqrt{1 + (u')^2}}\, \mathrm{d}x$$

$$= 2\int_a^b \frac{d}{\mathrm{d}x}\left[\frac{u'}{\sqrt{1 + (u')^2}}\right] v\, \mathrm{d}x$$

$$= 2\int_a^b \frac{u''(\sqrt{1 + (u')^2} - u')}{1 + (u')^2} v\, \mathrm{d}x.$$

Since the above identity holds for all $v \in H_0^1$, the minimizer u_* satisfies the Euler–Lagrange equation:

$$u''\frac{\sqrt{1 + (u')^2} - u'}{1 + (u')^2} = 0 \quad (a < x < b).$$

Since

$$\frac{\sqrt{1 + (u')^2} - u'}{1 + (u')^2} \neq 0,$$

we have

$$u_*''(x) = 0 \quad (a < x < b).$$

Hence, the straight line joining (a, α) and (b, β) is the minimizer. □

Example 4.3.14 *Let Ω be a domain in $\mathbb{R}^2$. Consider the minimization problem:*

$$\text{minimize} \quad \Phi(u) := \int_\Omega |\nabla u(x, y)|^2\, \mathrm{d}x\, \mathrm{d}y$$
$$\text{within the set} \quad X := \{u \in H^1(\Omega) : u|_{\partial\Omega} = f(x, y)\}.$$

If u is a minimizer, then u satisfies the Dirichlet problem:

$$\begin{cases} \nabla^2 u(x,y) = 0 & (x,y) \in \Omega, \\ u|_{\partial\Omega} = f & \textit{(prescribed boundary potential)}. \end{cases}$$

Proof. For all $v \in H_0^1(\Omega) := \{v \in H^1(\Omega) : u|_{\partial\Omega} = 0\}$,

$$\begin{aligned} 0 &= \frac{\mathrm{d}}{\mathrm{d}t}\Phi(u+tv)|_{t=0} = 2\int_\Omega \nabla u \cdot \nabla v \,\mathrm{d}x\,\mathrm{d}y \\ &= -2\int_\Omega \nabla^2 u v \,\mathrm{d}x\,\mathrm{d}y. \end{aligned}$$

Since this holds for all $v \in H_0^1(\Omega)$, we have

$$0 = \nabla^2 u \quad \text{in } \Omega.$$

□

Example 4.3.15 *Consider the minimal surface problem:*

$$\begin{aligned} &\textit{minimize} \quad \Phi(u) := \int_\Omega \sqrt{1+u_x^2+u_y^2}\,\mathrm{d}x\,\mathrm{d}y \\ &\textit{within the set} \quad X := \{u \in H^1(\Omega) : u|_{\partial\Omega} = f\}. \end{aligned}$$

If a minimizer u exists, it satisfies the following nonlinear problem with the Dirichlet boundary condition:

$$\begin{cases} \nabla \cdot \left(\dfrac{\nabla u}{\sqrt{1+|\nabla u|^2}}\right) = 0 & \textit{in } \Omega, \\ u|_{\partial\Omega} = f & \textit{(Dirichlet boundary)}. \end{cases}$$

Proof. For all $v \in H_0^1(\Omega)$,

$$\begin{aligned} 0 &= \frac{\mathrm{d}}{\mathrm{d}t}\Phi(u+tv)|_{t=0} = \int_\Omega \frac{\nabla u}{\sqrt{1+|\nabla u|^2}} \cdot \nabla v \;\mathrm{d}x\,\mathrm{d}y \\ &= \int_\Omega \left[-\nabla \cdot \left(\frac{\nabla u}{\sqrt{1+|\nabla u|^2}}\right)\right] v \,\mathrm{d}x\,\mathrm{d}y. \end{aligned}$$

Since this holds for all $v \in H_0^1(\Omega)$, we have

$$0 = \nabla \cdot \left(\frac{\nabla u}{\sqrt{1+|\nabla u|^2}}\right) \quad \text{in } \Omega.$$

□

Example 4.3.16 *For a given $f \in L^2(\Omega)$, consider the total variation minimization problem:*

$$\begin{aligned} &\textit{minimize} \quad \Phi(u) := \int_\Omega |\nabla u(x,y)| + |u-f|^2 \,\mathrm{d}x\,\mathrm{d}y \\ &\textit{within the set} \quad X := \{u \in W^{1,1}(\Omega) : \mathbf{n} \cdot \nabla u|_{\partial\Omega} = 0\}. \end{aligned}$$

If a minimizer u exists, it satisfies the following nonlinear problem with zero Neumann boundary condition:

$$\begin{cases} \nabla \cdot \left(\dfrac{\nabla u}{|\nabla u|} \right) = 2(u - f) & \textit{in } \Omega, \\ \mathbf{n} \cdot \nabla u|_{\partial\Omega} = 0 & (\textit{insulating boundary}). \end{cases}$$

Proof. For all $v \in X$,

$$\begin{aligned} 0 = \frac{\mathrm{d}}{\mathrm{d}t} \Phi(u + tv)|_{t=0} &= \int_\Omega \frac{\nabla u}{|\nabla u|} \cdot \nabla v + 2(u - f)v \, \mathrm{d}x \, \mathrm{d}y \\ &= \int_\Omega \left[-\nabla \cdot \left(\frac{\nabla u}{|\nabla u|} \right) + 2(u - f) \right] v \, \mathrm{d}x \, \mathrm{d}y. \end{aligned}$$

Since this holds for all $v \in X$, we have

$$0 = -\nabla \cdot \left(\frac{\nabla u}{|\nabla u|} \right) + 2(u - f) \quad \text{in } \Omega.$$

□

4.3.4 Regularity Theory and Asymptotic Analysis

We briefly discuss the regularity theory for the elliptic PDE. We know that a solution of the Laplace equation has the mean value property and therefore it is analytic ("having the ability to analyze"): if we have knowledge of all derivatives of a harmonic function at one fixed point, we can get full knowledge of the solution in its neighborhood. This mean value type property can be applied in some sense to solutions of elliptic equations. Basically, Harnack's inequality comes from a weighted mean value property. For details, please refer to the book by Gilbarg and Trudinger (2001).

Let Ω be a bounded smooth domain in $\mathbb{R}^3$. Consider the following divergence-form elliptic operator:

$$Lu(\mathbf{r}) := -\nabla \cdot (\sigma(\mathbf{r}) \nabla u(\mathbf{r})) + c(\mathbf{r}) u(\mathbf{r}), \tag{4.53}$$

where $\|c\|_{L^\infty(\Omega)} \leq c_0$ and

$$\sigma = \begin{pmatrix} \sigma_{11} & \sigma_{12} & \sigma_{13} \\ \sigma_{21} & \sigma_{22} & \sigma_{23} \\ \sigma_{31} & \sigma_{32} & \sigma_{33} \end{pmatrix}$$

is symmetric, bounded and positive definite. Assume that there exist positive constants c_1 and c_2 such that

$$c_1 |\mathbf{r}|^2 \leq \langle \sigma \mathbf{r}, \mathbf{r} \rangle \leq c_2 |\mathbf{r}|^2 \quad (\forall \mathbf{r} \in \mathbb{R}^3). \tag{4.54}$$

The bounded linear form $a(\cdot, \cdot) : H^1(\Omega) \times H^1(\Omega) \to \mathbb{R}$ associated with a divergence-form elliptic operator L is

$$a(u, \phi) = \int_\Omega \sigma(\mathbf{r}) \nabla u(\mathbf{r}) \cdot \nabla \phi(\mathbf{r}) + c(\mathbf{r}) u(\mathbf{r}) \phi(\mathbf{r}) \, \mathrm{d}\mathbf{r}. \tag{4.55}$$

Theorem 4.3.17 *Let $f \in L^2(\Omega)$, $\|c\|_{L^\infty(\Omega)} \le c_0$ and $\|\sigma\|_{C^1(\bar\Omega)} \le c_1$. Assume that $u \in H_0^1(\Omega)$ is a weak solution of $Lu = f$, that is,*

$$a(u,\phi) = \int_\Omega f(\mathbf{r})\phi(\mathbf{r})\,d\mathbf{r} \quad (\forall \phi \in H_0^1(\Omega)). \tag{4.56}$$

Then, for each $\Omega_0 \subset\subset \Omega$,

$$\|u\|_{H^2(\Omega_0)} \le C(\|f\|_{L^2(\Omega)} + \|u\|_{L^2(\Omega)}), \tag{4.57}$$

where the constant C depends only on Ω_0, $\widetilde{\Omega}$, Ω, c_0 and c_1. Here by $\Omega_0 \subset\subset \Omega$ we mean $\overline{\Omega_0} \subset \Omega$.

Proof. Set $\tilde f = f - cu$. Choose an open set Ω_1 such that $\Omega_0 \subset\subset \Omega_1 \subset\subset \Omega$. Take a cut-off function $\zeta \in C_0^1(\Omega_1)$ such that $0 \le \zeta \le 1$ and $\zeta|_{\Omega_0} = 1$. Substituting $\phi = u\zeta^2$ into (4.56) gives

$$\int_\Omega \langle \sigma\nabla u, \nabla u\rangle \zeta^2\,d\mathbf{x} + 2\int_\Omega \langle \sigma\nabla u\zeta, \nabla\zeta u\rangle\,d\mathbf{r} = \int_\Omega \tilde f u\zeta^2\,d\mathbf{x}. \tag{4.58}$$

From (4.54), we have

$$\|\nabla u\zeta\|^2_{L^2(\Omega)} \le \frac{1}{c_1}\int_\Omega \langle \sigma\nabla u, \nabla u\rangle\zeta^2\,d\mathbf{x}. \tag{4.59}$$

Using $ab < \epsilon a^2 + (1/\epsilon)b^2$, we have

$$\left|\int_\Omega \langle \sigma\nabla u\zeta, \nabla\zeta u\rangle\,d\mathbf{x}\right| \le \epsilon\|\nabla u\zeta\|^2_{L^2(\Omega)} + \frac{c_2}{\epsilon}\|\nabla\zeta\|_{L^\infty(\Omega)}\|u\|^2_{L^2(\Omega)}. \tag{4.60}$$

Similarly, using the Schwarz inequality, we have

$$\left|\int_\Omega \tilde f u\zeta^2\,d\mathbf{x}\right| \le (c_0+1)\|u\|^2_{L^2(\Omega)} + \|f\|^2_{L^2(\Omega)}. \tag{4.61}$$

By taking sufficiently small $\epsilon > 0$, the estimates (4.59)–(4.61) lead to

$$\|\nabla u\|^2_{L^2(\Omega_0)} \le \|\nabla u\zeta\|^2_{L^2(\Omega)} \le C_1(\|u\|^2_{L^2(\Omega)} + \|f\|^2_{L^2(\Omega)}) \tag{4.62}$$

for some positive constant C_1. □

Now, we will estimate $\|\nabla\nabla u\|^2_{L^2(\Omega_0)}$. From now on, we shall use the notation $\mathbf{x} = (x_1, x_2, x_3) = \mathbf{r} = (x, y, z)$ just for simple expressions. For the estimate of $\|\nabla\nabla u\|^2_{L^2(\Omega_0)}$, substitute $\phi = -D_k^{-h}(D_k^h u\zeta^2)$ into (4.56), where $D_k^h u$ denotes the difference quotient

$$D_k^h u(\mathbf{x}) = \frac{u(\mathbf{x} + h\mathbf{e}_k) - u(\mathbf{x})}{h} \approx \frac{\partial u}{\partial x_k},$$

where h is very small and $0 < h < \mathrm{dist}(\Omega_0, \partial\Omega)$. Then,

$$\underbrace{\int_\Omega \langle \sigma\nabla u, \nabla(-D_k^{-h}(D_k^h u\zeta^2))\rangle\,d\mathbf{x}}_{=\mathrm{I}} = \underbrace{\int_\Omega \tilde f(-D_k^{-h}(D_k^h u\zeta^2))\,d\mathbf{x}}_{=\mathrm{II}}.$$

We write I as

$$\mathrm{I} = \underbrace{\int_\Omega \sum_{j=1}^{3} \sigma_{ij} \left(D_k^h \frac{\partial u}{\partial x_j}\right)\left(D_k^h \frac{\partial u}{\partial x_j}\right) \zeta^2 \, d\mathbf{x}}_{=I_1} + \underbrace{\int_\Omega \sum_{j=1}^{3} D_k^h \sigma_{ij} \left(\frac{\partial u}{\partial x_j}\right)\left(D_k^h \frac{\partial u}{\partial x_j}\right) \zeta^2 \, d\mathbf{x}}_{=I_2}.$$

From the assumption of ellipticity,

$$C_1 \|\zeta D_k^h \nabla u\|^2_{L^2(\Omega)} \le I_1.$$

Using the Hölder inequality and Poincaré inequality,

$$|I_2| + |\mathrm{II}| \le \epsilon \|\zeta D_k^h \nabla u\|^2_{L^2(\Omega)} + C_\epsilon (\|u\|_{H^1(\Omega)} + \|f\|^2_{L^2(\Omega)}),$$

where ϵ can be chosen arbitrarily small, $C_{\epsilon,2}$ depends on ϵ, $\|\sigma\|_{C_1(\Omega)}$ and $\|c\|_{L^\infty(\Omega)}$.

Then, we have the estimate

$$\|\zeta D_k^h \nabla u\|^2_{L^2(\Omega)} \le C_* (\|u\|_{H^1(\Omega)} + \|f\|^2_{L^2(\Omega)}).$$

Since this is true for arbitrarily small h, we have the estimate

$$\|\nabla\nabla u\|^2_{L^2(\Omega_0)} \le C_* (\|u\|_{H^1(\Omega)} + \|f\|^2_{L^2(\Omega)}).$$

Here, we may use a convergence theorem in real analysis. Combining this estimate into (4.59) with an appropriate choice of ζ leads to (4.57).

Remark 4.3.18 *For beginners in this area, we recommend starting with the equation $Lu = -\nabla^2 u = f$ to prove Theorem 4.3.17. With this simpler model, a much simpler computation gives the estimate (4.57). Then, it is easy to see that the use of the condition (4.54) leads to an extension to the general case.*

Next, we study some important techniques for asymptotic analysis. We will use a nice simple model in the book by Chipot (2009) describing asymptotic analysis for problems in a large cylinder.

For a given $f(y) \in C([-a, a])$, let u_L and u_∞, respectively, be solutions of

$$\begin{cases} -\nabla^2 u_L(x, y) = f(y) & (\forall\, (x, y) \in \Omega_L := \{(x, y) : |x| < L,\ |y| < a\}), \\ u_L|_{\partial\Omega_L} = 0 \end{cases} \tag{4.63}$$

and

$$\begin{cases} -\dfrac{\partial^2}{\partial y^2} u_\infty(y) = f(y) & (\forall\, y \in (-a, a)), \\ u_\infty(\pm a) = 0. \end{cases} \tag{4.64}$$

Then, we have the following estimates.

Theorem 4.3.19 *For $0 < \ell < L$, we have*

$$\|\nabla(u_L - u_\infty)\|_{L^2(\Omega_l)} \le 4 \left(\frac{4a}{1 + 4a}\right)^{[L-\ell]} \|u_\infty\|_{H^1(-a,\, a)}, \tag{4.65}$$

where $[L - \ell]$ is the maximal integer less than $L - \ell$.

Proof. The proof of (4.65) is based on the identity

$$\int_{\Omega_L} \nabla(u_L(\mathbf{r}) - u_\infty(y)) \cdot \nabla[(u_L(\mathbf{r}) - u_\infty(y))\phi_\ell^2(x)]\,\mathbf{dr} = 0 \quad (\forall\, 0 < \ell < L), \tag{4.66}$$

where $\mathbf{r} = (x, y)$ and $\phi_\ell \in H_0^1(-L, L)$ is a linear function such that

$$0 \le \phi_\ell \le 1, \quad \phi_\ell|_{(-\ell,\ell)} = 1, \quad \sup_x |\phi_\ell'(x)| \le \frac{1}{L-\ell}.$$

The identity (4.66) can be expressed as

$$\int_{\Omega_L} |\nabla(u_L - u_\infty)|^2 \phi_\ell^2\,\mathbf{dr} = -2\int_{\Omega_L} [\phi_\ell \nabla(u_L - u_\infty)] \cdot [(u_L - u_\infty)\nabla\phi_\ell]\,\mathbf{dr}. \tag{4.67}$$

Since $\nabla\phi_\ell = 0$ in Ω_ℓ,

$$\int_{\Omega_L} |\nabla(u_L - u_\infty)|^2 \phi_\ell^2\,\mathbf{dr} \le 2\int_{\Omega_L\setminus\Omega_\ell} \phi_\ell |\nabla(u_L - u_\infty)|\,|u_L - u_\infty|\,|\nabla\phi_\ell|\,\mathbf{dr}. \tag{4.68}$$

Since $|\nabla\phi_\ell| \le 1/(L-\ell)$,

$$\int_{\Omega_L} |\nabla(u_L - u_\infty)|^2 \phi_\ell^2\,\mathbf{dr} \le \frac{2}{L-\ell}\int_{\Omega_L\setminus\Omega_\ell} \phi_\ell |\nabla(u_L - u_\infty)|\,|u_L - u_\infty|\,\mathbf{dr} \tag{4.69}$$

and the left-hand side of (4.69) can be estimated by

$$\begin{aligned}&\int_{\Omega_L} |\nabla(u_L - u_\infty)|^2 \phi_\ell^2\,\mathbf{dr} \\ &\quad \le \frac{2}{L-\ell}\left[\int_{\Omega_L\setminus\Omega_\ell} |\nabla(u_L - u_\infty)|^2 \phi_\ell^2\,\mathbf{dr}\right]^{1/2}\left[\int_{\Omega_L\setminus\Omega_\ell} |(u_L - u_\infty)|^2\,\mathbf{dr}\right]^{1/2}.\end{aligned} \tag{4.70}$$

Dividing both sides of (4.70) by $\left[\int_{\Omega_L\setminus\Omega_\ell} |\nabla(u_L - u_\infty)|^2\phi_\ell^2\,\mathbf{dr}\right]^{1/2}$, we have

$$\left[\int_{\Omega_L} |\nabla(u_L - u_\infty)|^2 \phi_\ell^2\,\mathbf{dr}\right]^{1/2} \le \frac{2}{L-\ell}\left[\int_{\Omega_L\setminus\Omega_\ell} |(u_L - u_\infty)|^2\,\mathbf{dr}\right]^{1/2}. \tag{4.71}$$

From the Poincaré inequality, the right-hand term in (4.71) can be estimated by

$$\int_{\Omega_L\setminus\Omega_\ell} |(u_L - u_\infty)|^2\,\mathbf{dr} \le a\int_{\Omega_L\setminus\Omega_\ell} \left|\frac{\partial}{\partial y}(u_L - u_\infty)\right|^2 \mathbf{dr} \le a\int_{\Omega_L\setminus\Omega_\ell} |\nabla(u_L - u_\infty)|^2\,\mathbf{dr}. \tag{4.72}$$

From (4.71) and (4.72), we have

$$\Phi(\ell) \le \frac{4a}{(L-\ell)^2}(\Phi(L) - \Phi(\ell)) \quad (\forall\, 0 < \ell < L), \tag{4.73}$$

where

$$\Phi(\ell) := \int_{\Omega_\ell} |\nabla(u_L - u_\infty)|^2\,\mathbf{dr}.$$

The estimate (4.73) can be simplified as

$$\Phi(\ell) \leq \theta(L-\ell)\Phi(L), \quad \theta(L-\ell) = \frac{4a/(L-\ell)^2}{1+4a/(L-\ell)^2} \quad (\forall\, 0 < \ell < L). \tag{4.74}$$

Similar arguments for getting (4.74) with changing ϕ_ℓ give

$$\Phi(\ell_1) \leq \theta(\ell_2 - \ell_1)\Phi(\ell_2) \quad (\forall\, 0 < \ell_1 < \ell_2 \leq L). \tag{4.75}$$

Hence, we have

$$\Phi(\ell) \leq \theta\Phi(\ell+1) \leq \theta^{[L-\ell]}\Phi(L) \quad (\forall\, 0 < \ell < L), \tag{4.76}$$

where $\theta = \theta(1) = 4a/(1+4a) < 1$. Now, we will estimate $\Phi(L)$. Since

$$\int_{\Omega_L} \nabla(u_L - u_\infty)\cdot\nabla(u_L - \phi_{L-1}u_\infty)\,\mathrm{d}\mathbf{r} = 0,$$

we have

$$\int_{\Omega_L} |\nabla(u_L - u_\infty)|^2\,\mathrm{d}\mathbf{r} = -\int_{\Omega_L} \nabla(u_L - u_\infty)\cdot\nabla((1-\phi_{L-1})u_\infty)\,\mathrm{d}\mathbf{r}$$

and, therefore,

$$\begin{aligned}\int_{\Omega_L} |\nabla(u_L - u_\infty)|^2\,\mathrm{d}\mathbf{r} &\leq \int_{\Omega_L\setminus\Omega_{L-1}} |\nabla((1-\phi_{L-1})u_\infty)|^2\,\mathrm{d}\mathbf{r}\\ &= \int_{\Omega_L\setminus\Omega_{L-1}} \left|-\frac{\mathrm{d}}{\mathrm{d}y}\phi_{L-1}u_\infty + (1-\phi_{L-1})\frac{\mathrm{d}}{\mathrm{d}y}u_\infty\right|^2\,\mathrm{d}\mathbf{r}\\ &\leq 2\int_{\Omega_L\setminus\Omega_{L-1}} |u_\infty(y)|^2 + \left|\frac{\mathrm{d}}{\mathrm{d}y}u_\infty(y)\right|^2\,\mathrm{d}\mathbf{r}\\ &\leq 4\|u_\infty\|^2_{H^1(-a,a)}.\end{aligned}$$

This completes the proof. □

We can generalize this simple asymptotic analysis (4.76) to an elliptic PDE:

$$\sum_{i,j=1}^{3} \nabla\cdot(A(\mathbf{r})\nabla u(\mathbf{r})) = f(y,z), \quad A(\mathbf{r}) = \begin{pmatrix} a_{11}(\mathbf{r}) & a_{12}(y,z) & a_{13}(y,z)\\ a_{21}(\mathbf{r}) & a_{22}(y,z) & a_{23}(y,z)\\ a_{31}(\mathbf{r}) & a_{32}(y,z) & a_{33}(y,z)\end{pmatrix}.$$

For details, please see section 6.3 in the book by Chipot (2009).

4.4 Tikhonov Regularization and Spectral Analysis

In this section, we briefly present a functional analytic approach for the Tikhonov regularization method, which is probably the most commonly used method when the problem $T\mathbf{x} = \mathbf{b}$ is ill-posed. Regularization is an important tool to deal with an ill-posed problem by imposing *a priori* information on the solution. We discuss the regularization techniques for solving $T^*T\mathbf{x} = T^*\mathbf{y}$, where T^*T is an $n \times n$ matrix having a large condition number. For more detailed aspects, the reader may refer to the book by Engl *et al.* (1996).

Let X and Y be Hilbert spaces. We denote by $\mathcal{L}(X, Y)$ the set of linear operators $T : X \to Y$. For each $T \in \mathcal{L}(X, Y)$, the Hilbert spaces X and Y, respectively, can be decomposed into

$$X = N(T) \oplus N(T)^{\perp} \quad \text{and} \quad Y = \overline{R(T)} \oplus R(T)^{\perp},$$

where

- $N(T) := \{\mathbf{x} \in X : T\mathbf{x} = 0\}$ is the kernel of T,
- $R(T) := \{T\mathbf{x} : \mathbf{x} \in X\}$ is the range of T,
- $N(T)^{\perp} := \{\mathbf{x} \in X : \langle \mathbf{x}, \mathbf{z}\rangle = 0 \text{ for all } \mathbf{z} \in N(T)\}$ is the orthogonal space of $N(T)$,
- $R(T)^{\perp} := \{\mathbf{y} \in Y : \langle \mathbf{y}, \mathbf{z}'\rangle = 0 \text{ for all } \mathbf{z}' \in R(T)\}$ is the orthogonal space of $R(T)$.

Definition 4.4.1 *For each $T \in \mathcal{L}(X, Y)$, its generalized inverse, denoted by $T^{\dagger}$, is defined as follows:*

- $D(T^{\dagger}) = R(T) \oplus R(T)^{\perp}$,
- $T^{\dagger} : D(T^{\dagger}) \to N(T)^{\perp}$ *is a linear operator such that* $N(T^{\dagger}) = R(T)^{\perp}$ *and*

$$T^{\dagger}T(\mathbf{x}) = \mathbf{x} \quad \textit{for all } \mathbf{x} \in X.$$

Theorem 4.4.2 *Let $T \in \mathcal{L}(X, Y)$ and let $\mathbf{x}^{\dagger} = T^{\dagger}\mathbf{y}$. Then,*

$$\mathbf{x} - \mathbf{x}^{\dagger} \in N(T) \quad \textit{if and only if } T^*T\mathbf{x} = T^*\mathbf{y},$$

where T^* is the dual operator of T, which requires $\langle T\mathbf{x}, \mathbf{y}\rangle = \langle \mathbf{x}, T^*\mathbf{y}\rangle$ for all $x \in X$ and $\mathbf{y} \in R(T)$.

Proof. The proof goes as follows:

$$\begin{aligned} T^*T\mathbf{x} = T^*\mathbf{y} \quad &\Longleftrightarrow \quad T\mathbf{x} - \mathbf{y} \in N(T^*) = R(T)^{\perp} \\ &\Longleftrightarrow \quad T\mathbf{x} \text{ is the closest element in } R(T) \text{ to } \mathbf{y} \\ &\Longleftrightarrow \quad T\mathbf{x} = T\mathbf{x}^{\dagger} \\ &\Longleftrightarrow \quad \mathbf{x} - \mathbf{x}^{\dagger} \in N(T). \end{aligned}$$

□

Theorem 4.4.3 *Let $T \in \mathcal{L}(X, Y)$. The generalized inverse $T^{\dagger} : D(T^{\dagger}) \to N(T)^{\perp}$ is bounded if and only if $R(T)$ is closed, that is, $R(T) = \overline{R(T)}$.*

Proof. The proof follows from the closed graph theorem in Theorem 4.4.12. □

4.4.1 Overview of Tikhonov Regularization

This section is based on a lecture by Bastian von Harrach. To get insight into Tikhonov regularization, we consider the simplest operator $T \in \mathcal{L}(\mathbb{R}^n, \mathbb{R}^m)$, which is an $m \times n$

matrix. Assume that the problem $T\mathbf{x} = \mathbf{b}$ is ill-posed. For ease of explanation, we assume that $T : \mathbb{R}^n \to \mathbb{R}^m$ satisfies the following:

- $T : \mathbb{R}^n \to \mathbb{R}^m$ is injective and bounded,
- $\|(T^*T)^{-1}\| \approx \infty$ (very large),

where T^* is the transpose of T. Consider the ill-posed problem:

$$T\mathbf{x} = \mathbf{b}$$

or the corresponding least-squares problem

$$T^*T\mathbf{x} = T^*\mathbf{b}.$$

In practice, the data $\mathbf{b}$ are always contaminated by noise; hence we may consider the following problem:

$$\min_{\mathbf{x}} \|T\mathbf{x} - \mathbf{b}^\delta\| \quad \text{with some constraints imposed on } \mathbf{x},$$

where $\mathbf{b}^\delta$ are the noisy data. Here, the norm $\|\cdot\|$ is the standard Euclidean distance.

From the assumption that $\|(T^*T)^{-1}\| \approx \infty$, unregularized solutions $(T^*T)^{-1}T^*\mathbf{b}^\delta$ are prone to magnify noise in the data $\mathbf{b}^{\text{true}}$, where $\mathbf{b}^{\text{true}}$ are the true data. Hence, $(T^*T)^{-1}T^*\mathbf{b}^\delta$ may be very different from the true solution $\mathbf{x}^{\text{true}} := (T^*T)^{-1}T^*\mathbf{b}^{\text{true}}$ even if $\delta \approx 0$. Hence, the goal is to find a good operator $G_\delta : X \to X$ so that $G_\delta \approx (T^*T)^{-1}$ and $G_\delta T^*\mathbf{b}^\delta$ is a reasonably good approximation of $\mathbf{x}^{\text{true}}$ when $\mathbf{b}^\delta \approx \mathbf{b}^{\text{true}}$.

The motivation of Tikhonov regularization is to minimize not only $\|T\mathbf{x} - \mathbf{b}^\delta\|$ but also $\|\mathbf{x}\|$. By adding a regularization parameter, α, we consider the following minimization problem:

$$\min_{\mathbf{x}} \left\{ \underbrace{\|T\mathbf{x} - \mathbf{b}^\delta\|^2}_{\text{fitting term}} \quad + \quad \underbrace{\alpha\|\mathbf{x}\|^2}_{\text{regularization term}} \right\}. \tag{4.77}$$

Lemma 4.4.4 *For $\alpha > 0$, $T^*T + \alpha I$ is invertible and its inverse is bounded by*

$$\|(T^*T + \alpha I)^{-1}\| \le \frac{1}{\alpha}. \tag{4.78}$$

More precisely,

$$\|(T^*T + \alpha I)^{-1}T^*\| \le \frac{1}{\sqrt{\alpha}}. \tag{4.79}$$

Proof. Using the property of the adjacent operator T^*, we have

$$\begin{aligned}
\|(T^*T + \alpha I)\mathbf{x}\|^2 &= \langle (T^*T + \alpha I)\mathbf{x},\ (T^*T + \alpha I)\mathbf{x}\rangle \\
&= \|T^*T\mathbf{x}\|^2 + 2\underbrace{\langle T^*T\mathbf{x}, \alpha\mathbf{x}\rangle}_{\alpha\|T\mathbf{x}\|^2} + \alpha^2\|\mathbf{x}\|^2 \\
&\ge \alpha^2\|\mathbf{x}\|^2.
\end{aligned}$$

Hence $T^*T+\alpha I$ is injective. We can prove that $T^*T+\alpha I$ is surjective by the Riesz lemma or Lax–Milgram theorem. Therefore $T^*T+\alpha I$ is invertible and

$$\|(T^*T+\alpha I)^{-1}\| \le \sup_{\|\mathbf{x}\|=1} \frac{\|\mathbf{x}\|}{\|(T^*T+\alpha I)\mathbf{x}\|} \le \frac{1}{\alpha}.$$

Next, we will prove (4.79). Using the fact that

$$\underbrace{(T^*T+\alpha I)^{-1}T^*}_{A} = \underbrace{T^*(TT^*+\alpha I)^{-1}}_{B}$$

(see Exercise 4.4.5), we have

$$\begin{aligned}
\|\underbrace{(T^*T+\alpha I)^{-1}T^*}_{A}\|^2 &= \|\underbrace{T(T^*T+\alpha I)^{-1}}_{A^*}\ \underbrace{(T^*T+\alpha I)^{-1}T^*}_{A}\| \\
&= \|\underbrace{T(T^*T+\alpha I)^{-1}}_{A^*}\ \underbrace{T^*(TT^*+\alpha I)^{-1}}_{B}\| \\
&\le \|T(T^*T+\alpha I)^{-1}T^*\|\ \underbrace{\|(T^*T+\alpha I)^{-1}\|}_{\le 1/\alpha} \\
&\le \frac{1}{\alpha}\|T(T^*T+\alpha I)^{-1}T^*\| \le \frac{1}{\alpha}.
\end{aligned}$$

This proves (4.79). Here, the last inequality in the above estimate comes from

$$\underbrace{T(T^*T+\alpha I)^{-1}}_{C} = \underbrace{(TT^*+\alpha I)^{-1}T}_{D}$$

(see Exercise 4.4.5) and the following estimate:

$$\|\underbrace{T(T^*T+\alpha I)^{-1}}_{C}\ T^*\| = \|\underbrace{(TT^*+\alpha I)^{-1}T}_{D}\ T^*\| \le \|(TT^*)^{-1}TT^*\| = 1.$$

□

Exercise 4.4.5 *Prove the following.*

1. *Prove that* $(T^*T+\alpha I)^{-1}T^* = T^*(TT^*+\alpha I)^{-1}$ *by showing the identity*

$$\begin{aligned}
&\underbrace{(T^*T+\alpha I)T^*(TT^*+\alpha I)^{-1}(TT^*+\alpha I)}_{(T^*T+\alpha I)T^*} \\
&\quad = \underbrace{(T^*T+\alpha I)(T^*T+\alpha I)^{-1}T^*(TT^*+\alpha IT^*)}_{T^*(TT^*+\alpha I)}.
\end{aligned}$$

2. *Prove that* $T(T^*T+\alpha I)^{-1} = (TT^*+\alpha I)^{-1}T$.

Theorem 4.4.6 *The minimization problem (4.77) has a unique solution, and the minimizer of (4.77) is given by*

$$\mathbf{x}^\alpha = (T^*T+\alpha I)^{-1}T^*\mathbf{b}^\delta.$$

Proof. The minimization problem (4.77) can be written as

$$\min \left\| \begin{pmatrix} T\mathbf{x} - \mathbf{b}^\delta \\ \sqrt{\alpha}\, I\mathbf{x} \end{pmatrix} \right\|^2,$$

which is the same as

$$\min \left\| \begin{pmatrix} T \\ \sqrt{\alpha}\, I \end{pmatrix} \mathbf{x} - \begin{pmatrix} \mathbf{b}^\delta \\ \mathbf{0} \end{pmatrix} \right\|^2.$$

The corresponding Euler–Lagrange equation is

$$\underbrace{\begin{pmatrix} T^* & \sqrt{\alpha}\, I \end{pmatrix} \begin{pmatrix} T \\ \sqrt{\alpha}\, I \end{pmatrix}}_{T^*T+\alpha I} \mathbf{x} = \underbrace{\begin{pmatrix} T^* & \sqrt{\alpha}\, I \end{pmatrix} \begin{pmatrix} \mathbf{b}^\delta \\ \mathbf{0} \end{pmatrix}}_{T^*\mathbf{b}^\delta}.$$

Owing to the invertibility of $T^*T + \alpha I$ in Lemma 4.4.4, the minimizer $\mathbf{x}^\alpha$ must satisfy $\mathbf{x}^\alpha = (T^*T + \alpha I)^{-1} T^* \mathbf{b}^\delta$. □

To maximize the effect of the regularization, it is important to choose an appropriate regularization parameter α. The strategy is to choose an appropriate parameter $\alpha = \alpha(\delta)$, which can be viewed as a function of δ, such that

$$\mathbf{x}^\alpha = (T^*T + \alpha I)^{-1} T^* \mathbf{b}^\delta \underset{\text{stably}}{\longrightarrow} \mathbf{x}^{\text{true}} \quad \text{as } \delta \to 0.$$

Hence, we must have $\alpha(\delta) \to 0$ as $\delta \to 0$ because $\mathbf{x}^{\text{true}} = (T^*T)^{-1} T^* \mathbf{b}^{\text{true}}$. We need to make an appropriate choice of $\alpha(\delta)$ so that it provides a "stable" convergence $\mathbf{x}^\alpha = (T^*T + \alpha I)^{-1} T^* \mathbf{b}^\delta \to \mathbf{x}^{\text{true}}$ as $\delta \to 0$.

Theorem 4.4.7 *Let* $T\mathbf{x}^{\text{true}} = \mathbf{b}^{\text{true}}$ *and denote* $G_\alpha := (T^*T + \alpha I)^{-1}$. *Let* $\mathbf{x}^\alpha = G_\alpha T^* \mathbf{b}^\delta$ *with* $\|\mathbf{b}^\delta - \mathbf{b}^{\text{true}}\| \le \delta$. *Then, for* $\alpha > 0$, *we have*

$$\|\mathbf{x}^\alpha - \mathbf{x}^{\text{true}}\| \le \frac{\delta}{\sqrt{\alpha}} + \|G_\alpha T^* \mathbf{b}^{\text{true}} - \mathbf{x}^{\text{true}}\|.$$

Proof. Since $\mathbf{x}^\alpha = G_\alpha T^* \mathbf{b}^\delta$, we have

$$\begin{aligned} \|G_\alpha T^* \mathbf{b}^\delta - \mathbf{x}^{\text{true}}\| &\le \|G_\alpha T^* \mathbf{b}^\delta - G_\alpha T^* \mathbf{b}^{\text{true}}\| + \|G_\alpha T^* \mathbf{b}^{\text{true}} - \mathbf{x}^{\text{true}}\| \\ &= \|G_\alpha T^* (\mathbf{b}^\delta - \mathbf{b}^{\text{true}})\| + \|G_\alpha T^* \mathbf{b}^{\text{true}} - \mathbf{x}^{\text{true}}\| \\ &\le \frac{\delta}{\sqrt{\alpha}} + \|G_\alpha T^* \mathbf{b}^{\text{true}} - \mathbf{x}^{\text{true}}\|, \end{aligned}$$

where the last inequality comes from Lemma 4.4.4. □

In the proof of the previous theorem for a fixed δ, note the estimate

$$\|\mathbf{x}^\alpha - \mathbf{x}^{\text{true}}\| \le \underbrace{\|G_\alpha T^* (\mathbf{b}^\delta - \mathbf{b}^{\text{true}})\|}_{\psi_1(\alpha)} + \underbrace{\|G_\alpha T^* \mathbf{b}^{\text{true}} - \mathbf{x}^{\text{true}}\|}_{\psi_2(\alpha)}.$$

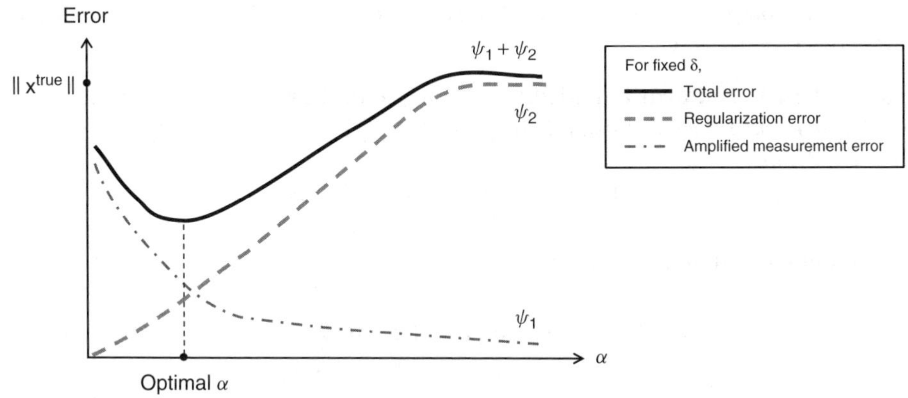

Figure 4.1 Error estimation

The term $\psi_1(\alpha) := \|G_\alpha T^*(\mathbf{b}^\delta - \mathbf{b}^{\text{true}})\|$ indicates the amplified measurement error related to the difference between the exact data and the measurement data, and the term $\psi_2(\alpha) := \|G_\alpha T^*\mathbf{b}^{\text{true}} - \mathbf{x}^{\text{true}}\|$, called the regularization error, illustrates how well $G_\alpha T^*$ approximates T^{-1}. Roughly speaking, the optimal α would be a quantity near the intersecting point of $\psi_1(\alpha) = \psi_2(\alpha)$ as shown in Figure 4.1.

4.4.2 Bounded Linear Operators in Banach Space

Definition 4.4.8 *A normed linear space X equipped with a norm $\|\cdot\|$ is called a Banach space if it is complete.*

The following are important examples of Banach spaces in PDE theory:

- $X = L^p(\mathbb{R}^n) := \{u : \mathbb{R}^n \to \mathbb{R} : \|u\| = \left(\int_\Omega |u|^p\right)^{1/p} < \infty\}$ for $1 \leq p \leq \infty$;
- $X = W^{1,p}(\mathbb{R}^n) :=$ the closure of $C_0^1(\mathbb{R}^n)$ with norm $\|u\| = \left(\int_\Omega |u|^p + |\nabla u|^p\right)^{1/p}$;
- $X = \ell^p := \{\mathbf{a} = (a_1, a_2, \ldots) : \|\mathbf{a}\| = \left(\sum_{j=1}^\infty |a_j|^p\right)^{1/p} < \infty\}$.

Note that the vector space $C_0(\mathbb{R}^n)$ equipped with the L^p-norm is a normed space but not a Banach space.

Theorem 4.4.9 (Baire category) *Let X be a Banach space. If $X = \bigcup_{k=1}^\infty X_k$ and $\overline{X_k} = X_k$, then one of the X_k contains an open ball.*

Proof. If every X_n does not contain an open set, we can choose the following Cauchy sequences $\{x_n\}$ and $\{r_n\}$.

1. Choose $B_{r_1}(x_1) \subset X \setminus X_1$ (because $X \setminus X_1$ is open).
2. For $n = 1, 2, \ldots,$ there exists $B_{r_{n+1}}(x_{n+1}) \subset (X \setminus X_{n+1}) \cap B_{r_n}(x_n)$ with $r_{n+1} < \frac{1}{2}r_n$ (because $(X \setminus X_{n+1}) \cap B_{r_n}(x_n)$ is open).

Since X is complete, $x_n \to x \in X$. However, $x \notin X_n$ for all n, which contradicts the assumption $X = \bigcup_{k=1}^{\infty} X_k$. □

Theorem 4.4.10 (Uniform boundedness principle or Banach–Steinhaus) *Let X be a Banach space and Y be a normed linear space. Let $\{T_\alpha\} \subset \mathcal{L}(X, Y)$. If $\sup_\alpha \|T_\alpha(x)\| < \infty$ for all $x \in X$, then*

$$\sup_\alpha \|T_\alpha\| = \sup_\alpha \sup_{\|x\|\leq 1} \|T_\alpha x\| < \infty.$$

Proof. From the assumption, we have

$$X = \bigcup_{n=1}^{\infty} X_n, \quad X_n := \{x \in X : \sup_\alpha \|T_\alpha(x)\| \leq n\}.$$

From the Baire category theorem, there exists X_n that contains an open ball $B_r(x^*)$:

$$\|x\| < r \quad \Longrightarrow \quad \sup_\alpha \|T_\alpha(x + x^*)\| \leq n.$$

This implies that

$$\sup_\alpha \|T_\alpha\| = \sup_\alpha \sup_{\|x\|\leq 1} \|T_\alpha(x)\| \leq \frac{n}{r} + \sup_\alpha \|T_\alpha(x^*)\| < \infty.$$

□

Theorem 4.4.11 *Let $T : X \to Y$ be linear, where X and Y are Banach spaces. If $T(X) = Y$, then T is open. Moreover, if $T(X) = Y$ and T is one-to-one, then T^{-1} is continuous.*

Proof. Denote $B_r := \{x \in X : \|x\| < r\}$ and $\widetilde{B}_r := \{y \in Y : \|y\| < r\}$. From the Baire category theorem and the assumption $Y = \bigcup_n T(B_n)$, there exists N such that $T(B_N)$ contains an open ball $y^* + \widetilde{B}_r$. This proves that T is open because $T(x + B_a) \subset T(x) + y^* + \widetilde{B}_r$ for all $x \in X$ and $a > 0$. If T is one-to-one, it is obvious that T^{-1} is continuous. □

Theorem 4.4.12 *Let $T : X \to Y$ be linear, where X and Y are Banach spaces. Then, T is bounded if and only if the graph $\{(x, Tx) : x \in X\}$ is closed with respect to the product norm $\|(x, y)\| = \|x\|_X + \|y\|_Y$.*

Proof. If the graph $G := \{(x, Tx) : x \in X\}$ is closed, G can be viewed as a Banach space and hence the projection map $\Pi_1 : G \to X$ defined by $\Pi_1((x, y)) = x$ has a bounded inverse due to the open mapping theorem. Since the projection map $\Pi_2 : G \to Y$ defined by $\Pi_2((x, y)) = y$ is continuous, then $T = \Pi_2 \circ \Pi_1^{-1}$ is bounded. Conversely, if T is bounded, then clearly G is closed because $x_n \to x$ implies $Tx_n \to Tx$. □

Theorem 4.4.13 *Let X be a Hilbert space. If an operator $T : X \to X$ is linear and symmetric, then T is bounded.*

Proof. From the closed graph theorem, it suffices to prove that the graph $G := \{(x, Tx) : x \in X\}$ is closed. We need to show that if $x_n \to x$ and $Tx_n \to y$, then $y = Tx$. From the assumption that $T^* = T$, we have

$$\langle \phi, Tx - y \rangle = \langle T\phi, x \rangle - \langle \phi, y \rangle = \lim_{n\to\infty} \left(\langle T\phi, x_n \rangle - \langle \phi, Tx_n \rangle \right) = 0, \quad \forall \phi \in X.$$

Since $Tx - y$ is orthogonal to all $\phi \in X$, we must have $Tx = y$ and hence the graph G is closed. □

We should note that, in order to apply the closed graph theorem, the domain of T should be a Banach space. To see this, consider the Hilbert space $X = L^2([0, 1])$ equipped with L^2-norm. We know that $C^2([0, 1])$ is a dense subset of the Hilbert space $X = L^2([0, 1])$. It is easy to see that the Laplace operator $T : C^2([0, 1]) \to X$ defined by

$$T(u) = \nabla \cdot \nabla u$$

is unbounded because

$$\int_0^1 |T(\cos 2\pi nx)|^2 \,\mathrm{d}x = (2\pi n)^2 \int_0^1 |\cos 2\pi nx|^2.$$

Defining $V := \{u \in X : \exists\, u_n \in C^2([0, 1]) \text{ s.t. } u_n \to u \;\&\; Tu_n \to w\}$, we can extend the unbounded operator T to V by defining the extension operator $\bar{T} : V \to X$ in such a way that $\lim_{n\to\infty} \bar{T}(u_n) = \bar{T}(\lim_{n\to\infty} u_n)$ whenever it makes sense. Clearly $\bar{T}$ is a closed operator. However, $\bar{T}$ is an unbounded operator, but it does not contract to the closed graph theorem because V is not complete (not a Banach space). At this point, we should note that V is a dense subset in $L^2([0, 1])$ since $C^2([0, 1]) \subsetneq V$.

Theorem 4.4.14 (Hahn–Banach) *Let X be a normed linear space over $\mathbb{R}$ (or $\mathbb{C}$) equipped with a norm $\|\cdot\|$ and let Y be a subspace of X. Let $\ell \in Y^* := \mathcal{L}(Y, \mathbb{R}$ or $\mathbb{C})$ with $\ell(x) \le \|x\|$ for $x \in Y$. Then, there exists an extension $\bar{\ell} \in X^*$ such that $\bar{\ell}|_Y = \ell$ and $\bar{\ell}(x) \le \|x\|$, where $X^* := \mathcal{L}(X, \mathbb{R}$ or $\mathbb{C})$.*

Proof. We will only prove the case where X is a normed linear space over $\mathbb{R}$. Choose $x_1 \in X \setminus Y$ and set $X_1 := \operatorname{span}(Y, x_1)$, the vector space spanned by Y and x_1. From the usual induction process and Zohn's lemma, it suffices to show that there exists an extension $\bar{\ell} \in (X_1)^*$ such that $\bar{\ell}|_Y = \ell$ and $\ell(x) \le \|x\|$. Since for $\bar{\ell}$ we must have $\bar{\ell}(y + \alpha x_1) = \ell(y) + \alpha\bar{\ell}(x_1)$ for all $\alpha \in \mathbb{R}$ and $y \in Y$, we only need to show the existence of $\bar{\ell}(x_1)$ satisfying

$$|\ell(y) + \alpha\bar{\ell}(x_1)| \le \|y + \alpha x_1\| \quad (\forall\, y \in X,\ \forall \alpha \in \mathbb{R})$$

or

$$\frac{-\|y + \alpha x_1\| + \ell(y)}{\alpha} \le \bar{\ell}(x_1) \le \frac{\|y + \alpha x_1\| - \ell(y)}{\alpha} \quad (\forall\, y \in X,\ \forall \alpha > 0).$$

Hence, the existence of $\bar{\ell}(x_1)$ requires

$$\sup_{\alpha > 0, y \in Y} \frac{-\|y - \alpha x_1\| + \ell(y)}{\alpha} \le \inf_{\alpha > 0, y \in Y} \frac{\|y + \alpha x_1\| - \ell(y)}{\alpha} \quad (\forall\, y \in X,\ \forall \alpha > 0),$$

which is equivalent to

$$\frac{-\|y - \alpha x_1\| + \ell(y)}{\alpha} \le \frac{\|\tilde{y} + \tilde{\alpha} x_1\| - \ell(\tilde{y})}{\tilde{\alpha}} \quad (\forall\, y, \tilde{y} \in X,\ \forall \alpha, \tilde{\alpha} > 0).$$

Rearranging the above inequality, we need to show that

$$\alpha\ell(\tilde{y}) + \tilde{\alpha}\ell(y) \le \tilde{\alpha}\|y - \alpha x_1\| + \alpha\|\tilde{y} + \tilde{\alpha}x_1\| \quad (\forall\, y, \tilde{y} \in X,\ \forall \alpha, \tilde{\alpha} > 0).$$

The above inequality holds owing to the convexity of the norm $\|\cdot\|$ and the fact that

$$\alpha\ell(\tilde{y}) + \tilde{\alpha}\ell(y) \le (\alpha + \tilde{\alpha})\left\|\frac{\tilde{\alpha}(y - \alpha x_1) + \alpha(\tilde{y} + \tilde{\alpha}x_1)}{\alpha + \tilde{\alpha}}\right\| \quad (\forall\, y, \tilde{y} \in X, \quad \forall \alpha, \tilde{\alpha} > 0).$$

□

Theorem 4.4.15 (Dual of $L^p(\mathbb{R}^n)$: Riesz representation theorem) *Let $1 \le p < \infty$ and $1/p + 1/q = 1$. For each $g \in L^q(\mathbb{R}^n)$, we define a linear functional $\ell_g : L^p(\mathbb{R}^n) \to \mathbb{R}$ by*

$$\ell_g(f) = \int fg\,\mathrm{d}x \quad (\forall\, f \in L^p(\mathbb{R}^n)).$$

Then, $\|\ell_g\| = \|g\|_{L^q}$. On the other hand, for each $\ell \in (L^p)^* = \mathcal{L}(L^p, \mathbb{R})$, there exists $g \in L^q(\mathbb{R}^n)$ such that $\ell(f) = \int fg\,\mathrm{d}x$ for all $f \in L^p(\mathbb{R}^n)$.

Proof. From the Hölder inequality, $\|\ell_g\| \le \|g\|_{L^q}$. Taking $f = \|g\|_{L^q}^{1-q} g|g|^{q-2}$, we get $\|f\|_{L^p} = 1$ and $\ell_g(f) = \|g\|_{L^q}$, which lead to $\|\ell_g\| \ge \|g\|_{L^q}$. Hence, $\|\ell_g\| = \|g\|_{L^q}$.

For $\ell \in (L^p)^*$, we can define a kind of Radon–Nikodym derivative:

$$g(x) := \lim_{r\to 0} \frac{\ell(B_r(x))}{|B_r(x)|} \quad (\forall x \in \mathbb{R}^n).$$

For the proof of $g \in L^q$, please refer to the book by Lieb and Loss (2001). □

Theorem 4.4.16 (Dual of Hilbert space: Riesz representation theorem) *Let H be a Hilbert space equipped with an inner product $\langle\cdot,\cdot\rangle$. For each $\ell \in H^* = \mathcal{L}(H, \mathbb{R}$ or $\mathbb{C})$, there exists $g \in H$ such that $\ell(f) = \langle f, g\rangle$ for all $f \in H$.*

The proof is left as an exercise for the reader.

4.4.3 Regularization in Hilbert Space or Banach Space

Regularization is an important tool to deal with an ill-posed problem by approximating it as a well-posed problem using *a priori* information on solutions. We discuss regularization techniques for solving $K^*K\mathbf{x} = K^*\mathbf{y}$, where K^*K is an $n \times n$ matrix having a large condition number.

Definition 4.4.17 *Let $K \in \mathcal{L}(X, Y)$.*

- *The map $K : X \to Y$ is said to be a compact operator if $K(\overline{B_1})$ is compact, where B_1 is the unit ball in X. Equivalently, K is compact if, for any bounded sequence $\{\mathbf{x}_n\}$ in X, $\{K\mathbf{x}_n\}$ has a convergent subsequence.*
- *The operator norm of K is $\|K\| = \sup_{\|\mathbf{x}\|\le 1} \|K\mathbf{x}\|$.*
- *For an eigenvalue $\sigma \in \mathbb{R}$, $X_\sigma := \{\mathbf{x} \in X : K^*K\mathbf{x} = \sigma\mathbf{x}\}$ is called a σ-eigenspace of K^*K.*

Throughout this section, we assume that $K \in \mathcal{L}(X, Y)$ is a compact operator. Note that K^*K maps from X to X and KK^* maps from Y to Y. Both K^*K and KK^* are compact operators and self-adjoint, that is, $\langle K^*K\mathbf{x}, \mathbf{x}'\rangle = \langle \mathbf{x}, K^*K\mathbf{x}'\rangle$ for all $\mathbf{x}, \mathbf{x}' \in X$.

Exercise 4.4.18 *Show that*

$$\|K^*K\| = \sup_{\|\mathbf{x}\|\leq 1} \langle K^*K\mathbf{x}, \mathbf{x}\rangle.$$

Theorem 4.4.19 *If $K \in \mathcal{L}(X, Y)$ is a compact operator, the self-adjoint compact operator $K^*K : X \to X$ satisfies the following.*

- *There exist eigenvalues $\sigma_1 \geq \sigma_2 \geq \cdots \geq 0$ and the corresponding orthonormal eigenfunctions $\{v_n : n = 1, 2, \ldots\}$ of K^*K such that*

$$K^*K\mathbf{x} = \sum_n \lambda_n^2 \langle \mathbf{x}, v_n\rangle v_n.$$

- *For each $\sigma_n > 0$, $\dim(X_{\sigma_n^2}) < \infty$ and*

$$\sigma_1^2 = \sup_{\mathbf{x}\in X} \frac{\langle K^*K\mathbf{x}, \mathbf{x}\rangle}{\|\mathbf{x}\|^2} \quad \textit{and} \quad \sigma_2^2 = \sup_{\mathbf{x}\in X^{\perp}_{\sigma_1^2}} \frac{\langle K^*K\mathbf{x}, \mathbf{x}\rangle}{\|\mathbf{x}\|^2}$$

  up to n.
- *If K^*K has infinitely many different eigenvalues, then $\lim_{n\to\infty} \sigma_n = 0$.*
- $\bigoplus_n X_{\sigma_n^2} = X$.
- *Writing $u_n = Kv_n/\|v_n\|$, we have*

$$Kv_n = \sigma_n u_n \quad \textit{and} \quad K^*u_n = \sigma_n v_n.$$

Proof. Since K is compact, it is easy to prove that $K^*K : X \to X$ is compact, that is, for any bounded sequence $\{\mathbf{x}_n\}$ in X, there exists a convergent subsequence $K^*K\mathbf{x}_{n_k}$. Hence, we have a singular system $(\sigma_n; v_n, u_n)$. Moreover, $\dim(X_{\sigma^2}) < \infty$; if not, K^*K is not compact. The remaining proofs are elementary. □

Theorem 4.4.20 *Let $K^\dagger : R(K) \oplus R(K)^\perp \to N(K)^\perp$ be a generalized inverse of a compact operator $K \in \mathcal{L}(X, Y)$. Then,*

$$K^\dagger\mathbf{y} = \sum_n \frac{\langle \mathbf{y}, u_n\rangle}{\sigma_n} v_n \quad \textit{for all } \mathbf{y} \in D(K^\dagger) = R(K) \oplus R(K)^\perp$$

and

$$\mathbf{y} \in R(K) \oplus R(K)^\perp \quad \textit{if and only if} \quad \sum_n \frac{|\langle \mathbf{y}, u_n\rangle|^2}{\sigma_n^2} < \infty.$$

Equivalently,

$$\mathbf{y} \in \overline{R(K)} \setminus R(K) \quad \textit{if and only if} \quad \sum_n \frac{|\langle \mathbf{y}, u_n\rangle|^2}{\sigma_n^2} = \infty.$$

Remark 4.4.21 *Assume that* $\dim R(K) = \infty$. *Then,* $\lim_{n\to\infty} \sigma_n = 0$ *and hence, for any arbitrarily small* $\epsilon > 0$,

$$\lim_{n\to\infty} \|K^{\dagger}(\epsilon u_n)\| = \lim_{n\to\infty} \frac{\langle \epsilon u_n, u_n\rangle}{\sigma_n}\|v_n\| = \lim_{n\to\infty} \frac{\epsilon}{\sigma_n} = \infty.$$

This means that any small error in the data $\mathbf{y}$ *may result in a large error in a reconstructed solution* $\mathbf{x} = K^{\dagger}\mathbf{y}$.

To see this effect clearly, let us consider the backward diffusion equation. For a given time $t > 0$, *let* $K_t : L^2([0,\pi]) \to L^2([0,\pi])$ *be the compact operator defined by*

$$K_t(f) = u(\cdot, t),$$

where $u(x,t)$ *is a solution of the diffusion equation* $u_t - u_{xx} = 0$ *in* $[0,\pi] \times (0,\infty)$ *with the initial condition* $u(x,0) = f(x)$ *and the homogeneous boundary condition* $u(0,t) = u(\pi,t) = 0$. *The standard method of separable variables gives the explicit representation of* $K_t f$:

$$K_t f(x) = \sum_{n=1}^{\infty} \frac{\langle f, v_n\rangle}{e^{tn^2}} v_n(x), \quad x \in [0, 2\pi],$$

where $v_n(x) = u_n(x) = \sqrt{2/\pi}\,\sin nx$ *and*

$$\langle f, v_n\rangle = \sqrt{\frac{2}{\pi}} \int_0^{\pi} \sin(nx) f(x)\,\mathrm{d}x.$$

Hence,

$$\|K_t f\|^2 = \sum_{n=1}^{\infty} \frac{\langle f, v_n\rangle^2}{e^{tn^4}} \le \|f\|^2$$

and its generalized inverse is

$$K_t^{\dagger}[u(\cdot,t)] = \sum_{n=1}^{\infty} \frac{\langle u(\cdot,t), v_n\rangle}{e^{-tn^2}} v_n.$$

Though any small perturbation of f *results in a small perturbation of* $K_t f$, *the generalized inverse is severely ill-posed when* t *is not small. To be precise, suppose that we try to find* $u(x,0)$ *from knowledge of noisy data* $\tilde{u}(x,2) = u(x,2) + e^{-20}v_{10}(x)$ *that is a small perturbation of the true data* $u(x,2) = v_1(x)$. *Then, the computed solution*

$$K_2^{\dagger}[\tilde{u}(\cdot,2)] = \sum_{n=1}^{\infty} \frac{\langle \tilde{u}(\cdot,t), v_n\rangle}{e^{-2n^2}} v_n = e^2 v_1 + 10^{-20} e^{200} v_{10} \approx e^{180} v_{10}$$

is very different from the true solution $u(x,0) = e^2 v_1(x) = K_2^{\dagger}[\tilde{u}(\cdot,2)]$. *This means that the true solution* $u(x,0) = e^2 v_1(x)$ *is negligibly small in the computed solution* $\tilde{u}(x,0) = e^2 v_1(x) + 10^{-20} e^{200} v_{10}(x)$.

For a compact operator K, the following are equivalent:

- $K^{\dagger}$ is unbounded;
- $R(K)$ is not closed, that is, $\overline{R(K)} \neq R(K)$;

- $\dim R(K) = \infty$;
- if $\{(\sigma_n, v_n, u_n) \in \mathbb{R}_+ \times X \times Y : n = 1, 2, \dots\}$ is a singular system for K such that $K^*Kv_n = \sigma_n K^* u_n = \sigma_n^2 v_n$, then there are infinitely many $\sigma_n > 0$ such that $\sigma_n \to 0$ as $n \to \infty$.

Hence, if $K^\dagger$ is unbounded and K is compact, the problem $K\mathbf{x} = \mathbf{y}$ is ill-posed, that is, a small error in the data $\mathbf{y}$ may result in a large error in the solution $\mathbf{x} = K^\dagger\mathbf{y}$. Recall that $K^\dagger = (K^*K)^\dagger K^*$. Numerous regularization techniques have been used to deal with this ill-posedness.

We now introduce the Tikhonov regularization method. Let $K \in \mathcal{L}(X, Y)$ be a compact operator. To explain the regularization method, it will be convenient to express the compact operator K^*K in the form

$$K^*K = \int_0^\infty \lambda \, \mathrm{d}P_\lambda = \sum_n \sigma_n^2 \langle \cdot, v_n \rangle v_n,$$

where P_λ is the projection map from X to the set $\bigoplus\{X_{\sigma_n^2} : \sigma_n^2 < \lambda\}$. Note that $D(K^*K) = \{\mathbf{x} \in X : \|K^*K\mathbf{x}\| < \infty\}$. Hence, if $\mathbf{x} \in D(K^*K)$, then

$$\int_0^\infty \lambda^2 \, \mathrm{d}\|P_\lambda \mathbf{x}\|^2 = \sum_n \sigma_n^4 |\langle \mathbf{x}, v_n \rangle|^2 = \|K^*K\mathbf{x}\|^2 < \infty.$$

Using the above property, we can show that

$$R(K^*) = \{\mathbf{x} \in N(K)^\perp : K\mathbf{x} \in D((KK^*)^\dagger)\} = R(\sqrt{K^*K}),$$

where $\sqrt{K^*K}$ is defined by

$$\sqrt{K^*K} = \int_0^\infty \sqrt{\lambda} \, \mathrm{d}P_\lambda = \sum_n \sigma_n \langle \cdot, v_n \rangle v_n.$$

For $\mathbf{y} \in D(K^\dagger)$, the solution $\mathbf{x}^\dagger = K^\dagger \mathbf{y}$ can be expressed as

$$\mathbf{x}^\dagger = (K^*K)^\dagger K^*\mathbf{y} = \int_0^\infty \frac{1}{\lambda} \, \mathrm{d}P_\lambda \, K^*\mathbf{y} = \sum_n \frac{1}{\sigma_n^2} \langle K^*\mathbf{y}, v_n \rangle v_n. \tag{4.80}$$

If $K \in \mathcal{L}(X, Y)$ is compact and $R(K)$ is not closed, then the operator $K^\dagger$ is unbounded and the problem $K\mathbf{x} = \mathbf{y}$ is ill-posed. If $\mathbf{y} \in \overline{R(K)} \setminus R(K)$, then

$$\mathbf{x} = \int_0^\infty \frac{1}{\lambda} \, \mathrm{d}P_\lambda \, K^*\mathbf{y} \quad \text{does not exist.}$$

On the other hand,

$$\mathbf{x}_\alpha = \int_0^\infty \frac{1}{\lambda + \alpha} \, \mathrm{d}P_\lambda \, K^*\mathbf{y} \quad \text{does exist for any } \alpha > 0.$$

This reconstruction method is called Tikhonov regularization. The solution

$$\mathbf{x}_\alpha = (K^*K + \alpha I)^{-1} K^*\mathbf{y} = \sum_n \frac{\sigma_n}{\sigma_n^2 + \alpha} \langle \mathbf{y}, u_n \rangle v_n$$

can also be interpreted as a minimizer of the energy functional:

$$\mathbf{x}_\alpha = \arg\min_{\mathbf{x}\in X}\{\|K\mathbf{x}-\mathbf{y}\|^2 + \alpha\|\mathbf{x}\|^2\}. \tag{4.81}$$

Denoting $K_\alpha^\dagger = (K^*K + \alpha I)^{-1}K^*$, the regularization $K_\alpha^\dagger$ can be viewed as an approximation of $K^\dagger$ in the sense that

$$\lim_{\alpha\to 0} K_\alpha^\dagger \mathbf{y} = \lim_{\alpha\to 0}(K^*K+\alpha I)^{-1}K^*\mathbf{y} = K^\dagger\mathbf{y}, \quad \forall\, \mathbf{y}\in D(K^\dagger) = R(K)\oplus N(K^*).$$

Here, we should note that $N(K^*) = R(K)^\perp$. The most widely used iteration method for solving $\mathbf{y} = K\mathbf{x}$ is based on the identity $\mathbf{x} = \mathbf{x} + \alpha K^*(\mathbf{y} - K\mathbf{x})$ for some $\alpha > 0$. This method is called the Landweber iteration, which expects $\mathbf{x}_k \to \mathbf{x}$ by setting

$$\mathbf{x}_{k+1} = \mathbf{x}_k + \alpha K^*(\mathbf{y} - K\mathbf{x}_k).$$

With this Landweber iteration, $\mathbf{y}\in D(K^\dagger)$ if and only if $\mathbf{x}_k$ converges. Indeed,

$$\mathbf{x}_{k+1} = (I-\alpha K^*K)\mathbf{x}_k + \alpha K^*\mathbf{y} = (I-\alpha K^*K)^2(\mathbf{x}_{k-1}) + \sum_{j=0}^{1}(I-\alpha K^*K)^j(\alpha K^*\mathbf{y})$$

which leads to

$$\mathbf{x}_{k+1} = (I-\alpha K^*K)^{k+1}(\mathbf{x}_0) + \sum_{j=0}^{k}(I-\alpha K^*K)^j(\alpha K^*\mathbf{y}).$$

Regularization is to replace the ill-conditioned problem $K\mathbf{x} = \mathbf{y}$ by its neighboring well-posed problem in which we impose a constraint on desired solution properties. The problem of choosing the regularization parameter α in (4.81) is critical. In MR-based medical imaging, we often need to deal with the tradeoff between spatial resolution and temporal resolution. Accelerating the MR imaging speed requires the phase encoding lines in k-space to be skipped, and results in insufficient k-space data for MR image reconstruction via inverse Fourier transformation. To deal with the missing data, one may use the minimization problem (4.81), which consists of the fidelity term (ensuring that the k-space data $\mathbf{y}$ are consistent with the image $\mathbf{x}$) and the regularization term (which incorporates the desired solution properties). Nowadays many research areas use regularization techniques.

4.5 Basics of Real Analysis

We now briefly review the real analysis techniques that are used in this chapter. For more details, we refer to the books by Lieb and Loss (2001), Marsden (1974), Rudin (1970) and Wheeden and Zygmund (1977). Throughout this section, we assume that Ω is a bounded domain in $\mathbb{R}^n$ ($n = 1$, 2 or 3) and $f:\Omega\to\mathbb{R}$ is a function.

4.5.1 Riemann Integrability

For a quick overview of the Riemann integral, we will explain its concept only in two-dimensional space. Let $\Omega\subset\mathbb{R}^2$ be a bounded domain and let $f:\Omega\to\mathbb{R}$ be a bounded function. We enclose Ω in a rectangle and extend f to the whole rectangle $B = [a_1, b_1] \times [a_2, b_2]$ by defining it to be zero outside Ω. Let $\mathcal{P}$ be a partition of B obtained by dividing $a_1 = x_0 < x_1 < \cdots < x_n = b_1$ and $a_2 = y_0 < y_1 < \cdots < y_m = b_2$:

$$\mathcal{P} = \{\underbrace{[x_i, x_{i+1}] \times [y_j, y_{j+1}]}_{=\text{subrectangle}\, R} : i = 0, 1, \ldots, n-1;\ j = 0, 1, \ldots, m-1\}.$$

- Define the upper sum of f by

$$U(f, \mathcal{P}) := \sum_{R \in \mathcal{P}} \sup\{f(x, y) \mid (x, y) \in R\} \times (\text{volume of } R)$$

  and the lower sum of f by

$$L(f, \mathcal{P}) := \sum_{R \in \mathcal{P}} \inf\{f(x, y) \mid (x, y) \in R\} \times (\text{volume of } R).$$

- Define the upper integral of f on Ω by

$$\overline{\int_\Omega} f = \inf\{U(f, \mathcal{P}) : \mathcal{P} \text{ is a partition of } B\}$$

  and the lower integral of f on Ω by

$$\underline{\int_\Omega} f = \sup\{L(f, \mathcal{P}) : \mathcal{P} \text{ is a partition of } B\}.$$

We say that f is *Riemann integrable or integrable* if

$$\overline{\int_\Omega} f = \underline{\int_\Omega} f.$$

If f is *integrable* on Ω, we denote

$$\int_\Omega f = \overline{\int_\Omega} f = \underline{\int_\Omega} f.$$

4.5.2 *Measure Space*

For a domain E in $\mathbb{R}^2$, we will denote the area of the domain E by $\mu(E)$. We can measure the area $\mu(E)$ based on the area of rectangles. If $Q = (a, b) \times (c, d)$ and Q_j for $j = 1, 2, \ldots$ is a sequence of rectangles with $Q_j \cap Q_k = \emptyset$ for $j \neq k$, then we measure them in the following ways:

$$\mu(Q) = (b-a)(d-c) \quad \text{and} \quad \mu\left(\bigcup_{j=1}^{\infty} Q_j\right) = \sum_{j=1}^{\infty} \mu(Q_j).$$

It seems that any open domain E is measurable since any open set can be expressed as a countable union of rectangles. Then, can we measure any subset in $\mathbb{R}^2$? For example, consider the set $E = \{\mathbf{x} = (x_1, x_2) \in [0,1] \times [0,1] : x_1 \in \mathbb{Q}, x_2 \in \mathbb{R} \setminus \mathbb{Q}\}$, where $\mathbb{Q}$ is the set of rational numbers. What is its area $\mu(E)$? It is a bit complicated. Measure theory provides a systematic way to assign to each suitable subset E a positive quantity $\mu(E)$ representing its area.

The triple $(X, \mathcal{M}, \mu)$ is said to be a measure space if the following are true:

1. X is a set;
2. $\mathcal{M}$ is a σ-algebra over the set X, that is, closed under a countable union and complementations;
3. μ is a measure on $\mathcal{M}$, that is,
   (a) $\mu(E) \geq 0$ for any $E \in \mathcal{M}$,
   (b) for all countable collections $\{E_j : j = 1, 2, \ldots\}$ of pairwise disjoint sets in $\mathcal{M}$, $\mu(\bigcup_j E_j) = \sum_j \mu(E_j)$, and
   (c) $\mu(\emptyset) = 0$.

If the σ-algebra $\mathcal{M}$ includes all null sets (a null set is a set N such that $\mu(N) = 0$), then μ is said to be *complete*. Let $\mathcal{P}(X)$ denote the collection of all subsets of X.

Definition 4.5.1 (Outer measure) *An outer measure on a non-empty set X is a set function μ^* defined on $\mathcal{P}(X)$ that is non-negative, monotone and countably subadditive.*

With the use of the outer measure, we can describe a general constructive procedure for obtaining the complete measure. Let $\mathcal{M}_{\text{pre}} \subset \mathcal{P}(X)$ be an algebra of sets and $\mu_{\text{pre}} : \mathcal{M}_{\text{pre}} \to \mathbb{R}^+ \cup \{0\}$ a set-valued function such that $\mu_{\text{pre}}(\emptyset) = 0$. For $A \subset X$, we define

$$\mu^*(A) = \inf \left\{ \sum_{j=1}^{\infty} \mu_{\text{pre}}(E_j) : A \subset \bigcup_{j=1}^{\infty} E_j, E_j \in \mathcal{M}_{\text{pre}} \right\}.$$

Then, μ^* is an outer measure.

Example 4.5.2 *Let $X = \mathbb{R}^2$ and let $\mathcal{M}_{\text{pre}}$ be the σ-algebra generated by the set of all open rectangles in $\mathbb{R}^2$. Define*

$$\mu_{\text{pre}}(E) = \textit{area of } E, \quad E \in \mathcal{M}_{\text{pre}}.$$

Then $(X, \mathcal{M}_{\text{pre}}, \rho)$ is a measure space but it may not be complete. This ρ is called the pre-measure. For $A \subset X$, we define

$$\mu^*(A) = \inf\{\mu_{\text{pre}}(E) : A \subset E, \ E \in \mathcal{M}_{\text{pre}}\}.$$

Then, μ^* is an outer measure.

The following Caratheodory definition provides a method of constructing a complete measure space $(X, \mathcal{M}, \mu^*)$.

Definition 4.5.3 *Let μ^* be an outer measure on a set X. A subset $A \subset X$ is said to be μ^*-measurable if*

$$\forall E \subset X, \quad \mu^*(E) = \mu^*(E \cap A) + \mu^*(E \setminus A).$$

The Lebesgue measure on $X = \mathbb{R}$ is an extension of the pre-measure defined by $\mu_{\text{pre}}((a, b]) = b - a$.

4.5.3 Lebesgue-Measurable Function

Throughout this section, we restrict ourselves to the standard Lebesgue measure μ in X ($X = \mathbb{R}^3$ or $\mathbb{R}^N$) that is generated by the outer measure,

$$\mu^*(A) = \inf\{\mu_{\text{pre}}(U) : A \subset U,\ U \text{ open }\},$$

where μ_{pre} is a pre-measure defined on open sets in X. For example, in $X = \mathbb{R}^3$, $\mu_{\text{pre}}(U)$ is the volume of U.

A function $f : X \to \mathbb{R}$ is said to be measurable if $f^{-1}(U)$ is measurable for every open set U.

Exercise 4.5.4 *Given two functions f and g, we define*

$$f \vee g = \max\{f, g\}, \quad f \wedge g = \min\{f, g\},$$
$$f^+ = f \vee 0, \qquad f^- = (-f) \vee 0.$$

1. *Then $f : X \to \mathbb{R}$ is measurable if $\{x \in X : f(x) > a\} \in \mathcal{M}$ for all $a \in \mathbb{R}$.*
2. *Prove that, if f and g are measurable, then so are $f + g$, fg, $f \vee g$, $f \wedge g$, f^+, f^- and $|f|$.*
3. *If $\{f_j\}$ is a sequence of measurable functions, then $\limsup_j f_j$ and $\limsup_j f_j$ are measurable.*

From now on, we assume that E, E_j are measurable sets. The characteristic function of E, denoted by χ_E, is the function defined by

$$\chi_E(x) = \begin{cases} 1 & \text{if } x \in E, \\ 0 & \text{otherwise.} \end{cases}$$

Let $\mathcal{S}_{imple}$ be the set of all simple functions, a finite linear combination of characteristic functions:

$$\mathcal{S}_{imple} = \left\{ \phi = \sum_{j=1}^{n} c_j \chi_{E_j} : E_j \in \mathcal{M},\ c_j \in \mathbb{R} \right\}. \tag{4.82}$$

Let $\mathcal{M}_{able}$ be the set of measurable functions and let

$$\mathcal{M}_{able}^+ = \{f \in \mathcal{M}_{able} : f \geq 0\}.$$

For given $f \in \mathcal{M}_{able}$, we say that $f = 0$ holds *almost everywhere* (abbreviated a.e.) if $\mu(\{x : f(x) \neq 0\}) = 0$.

Theorem 4.5.5 *Any $f \in \mathcal{M}_{able}^+$ can be approximated by a sequence of simple functions*

$$\phi_n = \sum_{k=0}^{2^{2n}-1} k 2^{-n} \chi_{E_{n,k}} + 2^n \chi_{F_n} \nearrow f, \tag{4.83}$$

where $E_{n,k} = f^{-1}((k2^{-n}, (k+1)2^{-n}])$ *and* $F_n = f^{-1}((2^n, \infty])$. *Moreover, each* ϕ_n *satisfies*

$$\phi_n \le \phi_{n+1} \quad and \quad 0 \le f(x) - \phi_n(x) \le 2^{-n} \quad \forall x \in X \setminus F_n.$$

The proof is straightforward by drawing a diagram of ϕ_n. From the above theorem, we can prove that, for any measurable function f, there is a sequence of simple functions ϕ_n such that $\phi_n \to f$ on any set on which f is bounded. The reason is that f can be decomposed into $f = f^+ - f^-$, where $f = \max\{f, 0\}$ and $f^- = \max\{-f, 0\}$.

Now, we are ready to give the definition of the Lebesgue integral using the integral of a measurable simple function $\phi = \sum_{j=1}^{n} c_j \chi_{E_j}$, which is defined as

$$\int \phi \, d\mu = \sum_{j=1}^{n} c_j \mu(E_j). \tag{4.84}$$

We use the convention that $0 \cdot \infty = 0$. Let $\mathcal{S}_{imple}$ be a vector space of measurable simple functions. Then the integral $\int \Box \, d\mu$ (as a function of $\Box$) can be viewed as a linear functional on $\mathcal{S}_{imple}$, that is, the operator $\int \Box \, d\mu : \mathcal{S}_{imple} \to \mathbb{R}$ is linear.

Lemma 4.5.6 *Let* $(X, \mathcal{M}, \mu)$ *be a measure space. For a non-negative* $\phi \in \mathcal{S}_{imple}$ *and* $E \in \mathcal{M}$, *define*

$$\nu(E) = \int_E \phi \, d\mu = \int_X \phi \chi_E \, d\mu. \tag{4.85}$$

Then, $(X, \mathcal{M}, \nu)$ *is also a measure space.*

Proof. It suffices to prove this for $\phi = \chi_A$ where $A \in \mathcal{M}$. The proof follows from the following identity for $E = \bigcup_j E_j$ of mutually disjoint $E_j \in \mathcal{M}$:

$$\nu(E) = \int \chi_A \chi_E \, d\mu = \int \chi_{A \cap E} \, d\mu = \sum_j \int_{E_j} \phi \, d\mu = \sum_j \nu(E_j).$$

□

Using the definition (4.84) of the integral of a measurable simple function, we can define the integral of a measurable function $f \ge 0$.

Definition 4.5.7 *The integral of* $f \in \mathcal{M}^+_{able}$ *is defined by*

$$\int f \, d\mu = \sup \left\{ \int \phi \, d\mu : \phi \le f \quad and \quad \phi \in \mathcal{S}_{imple} \right\}. \tag{4.86}$$

From the definition, we obtain that $f \le g$ implies $\int f \, d\mu \le \int g \, d\mu$.

Theorem 4.5.8 (Monotone convergence theorem) *For a non-decreasing sequence* $\{f_n\} \subset \mathcal{M}^+_{able}$, *we have*

$$\lim_n \int f_n \, d\mu = \int \lim_n f_n \, d\mu.$$

Proof. Since $f_n \nearrow$, there exists $f = \lim_n f_n \in M_{able}$. Since $\int f_n \, \mathrm{d}\mu \nearrow$ and $f_n \leq f$, we have

$$\lim_n \int f_n \, \mathrm{d}\mu \leq \int f \, \mathrm{d}\mu.$$

It remains to prove $\lim_n \int f_n \, \mathrm{d}\mu \geq \int f \, \mathrm{d}\mu$. Since $\int f \, \mathrm{d}\mu = \sup\{\int \phi \, \mathrm{d}\mu : \phi \leq f, \phi \in \mathcal{S}_{imple}\}$, it suffices to prove that, for any α, $0 < \alpha < 1$, and any $\phi \in \mathcal{S}_{imple}$ with $\phi \leq f$,

$$\lim_n \int f_n \, \mathrm{d}\mu \geq \alpha \int \phi \, \mathrm{d}\mu.$$

Let $E_n = \{f_n \geq \alpha\phi\}$. Then,

$$\int f_n \, \mathrm{d}\mu \geq \int_{E_n} f_n \, \mathrm{d}\mu \geq \int_{E_n} \alpha\phi \, \mathrm{d}\mu \overset{\text{define}}{:=} \alpha\nu(E_n).$$

Since ν is a measure and $E_n \nearrow X$, $\lim_n \nu(E_n) = \nu(X) = \int \phi \, \mathrm{d}\mu$. Thus, $\lim_n \int f_n \, \mathrm{d}\mu \geq \alpha \int \phi \, \mathrm{d}\mu$. □

Exercise 4.5.9 *Prove that, if $\phi_n \in \mathcal{S}_{imple}$ and $\phi_n \nearrow f$ for some $f \in \mathcal{M}_{able}$, then*

$$\lim_n \int \phi_n \, \mathrm{d}\mu = \int f \, \mathrm{d}\mu.$$

Proposition 4.5.10 *Let $f \in \mathcal{M}^+_{able}$. Then, $\int f \, \mathrm{d}\mu = 0$ if and only if $f = 0$ a.e.*

Proof. If $f \in \mathcal{S}^+_{imple}$, then the statement is immediate. If $f = 0$ a.e. and $\phi \leq f$, then $\phi = 0$ a.e., and hence $\int f = \sup\{\int \phi : \phi \leq f, \phi \in \mathcal{S}^+_{imple}\} = 0$.

Conversely, if $\int f \, \mathrm{d}\mu = 0$, then

$$0 = \int f \, \mathrm{d}\mu \geq \frac{1}{n}\mu(\{f > 1/n\}), \quad n = 1, 2, \ldots,$$

and therefore $f = 0$ a.e. because

$$\mu(\{f > 0\}) = \mu\left(\bigcup_{n=1}^{\infty}\{f > 1/n\}\right) \leq \bigcup_{n=1}^{\infty}\mu(\{f > 1/n\}) = 0.$$

□

Lemma 4.5.11 (Fatou's lemma) *For any sequence $f_n \in \mathcal{M}^+_{able}$, we have*

$$\int \liminf_n f_n \, \mathrm{d}\mu \leq \liminf_n \int f_n \, \mathrm{d}\mu.$$

Proof. Since $g_k = \inf_{j \geq k} f_j \nearrow$, we have

$$\liminf_n \int f_n \, \mathrm{d}\mu = \sup_{k \geq 1} \inf_{j \geq k} \int f_j \, \mathrm{d}\mu \geq \sup_{k \geq 1} \int \underbrace{\inf_{j \geq k} f_j}_{g_k} \, \mathrm{d}\mu = \lim_{k \to \infty} \int g_k \, \mathrm{d}\mu = \lim_{k \to \infty} \int g_k \, \mathrm{d}\mu.$$

□

Definition 4.5.12 ($L^1(X, \mathrm{d}\mu)$) *A function $f : X \to \mathbb{R}$ is integrable if $f \in \mathcal{M}_{able}$ and $\int |f| \,\mathrm{d}\mu < \infty$. We denote by $L^1(X, \mathrm{d}\mu)$ the class of all integrable functions. For $f \in L^1(X, \mathrm{d}\mu)$, we define*

$$\int f \,\mathrm{d}\mu = \int f^+ \,\mathrm{d}\mu - \int f^- \,\mathrm{d}\mu.$$

It is easy to see that $L^1(X, \mathrm{d}\mu)$ is a vector space equipped with the norm $\|f\| = \int |f| \,\mathrm{d}\mu$ satisfying the following:

1. $\|f\| \geq 0$, for all $f \in L^1$;
2. $\|f\| = 0$ if and only if $f = 0$ a.e.;
3. $\|\lambda f\| = |\lambda| \, \|f\|$, for all $f \in L^1$ and every scalar λ;
4. $\|f + g\| \leq \|f\| + \|g\|$, for all $f, g \in L^1$.

Now, we will prove that $L^1(X, \mathrm{d}\mu)$ is a complete normed space (or Banach space). To prove this, we need to study several convergence theorems.

Theorem 4.5.13 (Lebesgue dominated convergence theorem (LDCT)) *If $\{f_n\} \subset L^1$ with $f_n \to f$ a.e. and there exists $g \in L^1$ so that $|f_n| \leq g$ a.e. for all n, then*

$$f \in L^1 \quad \textit{and} \quad \int f \,\mathrm{d}\mu = \lim_n \int f_n \,\mathrm{d}\mu.$$

Proof. Since $g + f_n \geq 0$, it follows from Fatou's lemma that

$$\int \liminf_n (g + f_n) \,\mathrm{d}\mu \leq \liminf_n \int (g + f_n) \,\mathrm{d}\mu.$$

This leads to $\int f \,\mathrm{d}\mu \leq \lim_n \inf \int f_n \,\mathrm{d}\mu$.

On the other hand, applying the same argument to the sequence $g - f_n \geq 0$ yields

$$-\int f \,\mathrm{d}\mu \leq \liminf_n \int (-f_n) \,\mathrm{d}\mu = -\limsup_n \int f_n \,\mathrm{d}\mu.$$

□

Exercise 4.5.14 *Let*

$$G(x, t) = \frac{1}{\sqrt{4\pi t}} \mathrm{e}^{-x^2/4t}$$

denote the fundamental solution of the heat equation in one dimension. Let $f_n(x) = G(x, 1/n)$. Then, $f_n \to f = 0$ a.e. and

$$\int f \,\mathrm{d}\mu = 0 \quad \textit{and} \quad \lim_n \int f_n \,\mathrm{d}\mu = 1.$$

The above exercise explains the reason why the LDCT requires the assumption that $\{f_n\}$ is dominated by a fixed L^1 function g.

Exercise 4.5.15 *Let $\{f_j\} \subset L^1$ so that $\sum_j \int |f_j| \, d\mu < \infty$. Show that $g_n = \sum_{j=1}^{n} |f_j| \nearrow g = \sum_{j=1}^{\infty} |f_j|$. Using the LDCT, prove that there exists $f \in L^1$ such that*

$$\lim_{n\to\infty} \sum_{j=1}^{n} f_j = f \text{a.e.} \quad \textit{and} \quad \int f \, d\mu = \sum_j \int f_j \, d\mu.$$

From (4.83), for any function $f \in L^1$, there exists a sequence $\phi_n \in \mathcal{S}_{imple}$ such that

$$\phi_n \to f \text{a.e.} \quad \text{and} \quad |\phi_n| < |f| \text{a.e.}$$

From the LDCT, $\|\phi_n - f\| = \int |\phi_n - f| \, d\mu \to 0$. Hence, $\mathcal{S}_{imple}$ is dense in L^1, that is, every element in L^1 is an L^1-limit of a sequence of elements in $\mathcal{S}_{imple}$.

Exercise 4.5.16 *Show that a sequence of continuous functions*

$$f_n(x) := \begin{cases} 1 & \textit{if } 0 \le x \le 1, \\ 0 & \textit{if } 0 < -1/n, \\ 0 & \textit{if } x > 1 + 1/n, \\ \textit{linear} & \textit{otherwise} \end{cases}$$

$$\textit{converges to} f = \chi_{(0,1)} \textit{in } L^1 \textit{ sense.}$$

Using this idea, show that $C_0(\mathbb{R})$ is dense in $L^1(\mathbb{R}, \mu)$.

Exercise 4.5.17 *Let $f(x,t) : X \times [a,b] \to \mathbb{R}$ be a mapping. Suppose that f is differentiable with respect to t and that*

$$g(x) := \sup_{t\in[a,b]} \left| \frac{\partial}{\partial t} f(x,t) \right| \in L^1(X, d\mu).$$

Then, $F(t) = \int f(x,t) \, d\mu$ is differentiable on $a \le t \le b$ and

$$\frac{\partial}{\partial t} \int f(x,t) \, d\mu = \int \frac{\partial}{\partial t} f(x,t) \, d\mu.$$

Note that, for each $t \in (a,b)$, we can apply the LDCT to the sequence

$$h_n(x) = \frac{f(x,t_n) - f(x,t)}{t_n - t}, \quad t_n \to t,$$

because $|h_n| \le g$ from the mean value theorem.

4.5.4 Pointwise, Uniform, Norm Convergence and Convergence in Measure

Let Ω be a subdomain in $\mathbb{R}^N$. The sequence of functions $f_k \in C(\Omega)$ is said to *converge pointwise* to f if, for each $\mathbf{r} \in \mathbb{R}^N$, $f_k(\mathbf{r}) \to f(\mathbf{r})$. We say that the sequence of functions $f_k \in C(\Omega)$ *converges uniformly* to f if $\sup_{\mathbf{r}\in\Omega} |f_k(\mathbf{r}) - f(\mathbf{r})| \to 0$.

Example 4.5.18 *Consider the following.*

1. *The function* $f_k(x) = x^k$ *converges pointwise to*
$$f(x) = \begin{cases} 0, & 0 \le x < 1, \\ 1, & x = 1, \end{cases}$$
on the interval $[0, 1]$, *but* $\sup_{x\in[0,1]} |f_k(x) - f(x)| = 1$ *for all* k. *Hence, the convergence is not uniform.*
2. *The function*
$$g_n(x) = \sum_{k=0}^{n} \frac{(-1)^k x^{2k+1}}{(2k+1)!}$$
converges uniformly to $\sin x$ *in* $[0, 1]$.
3. *The following functions* $\phi_n \in L^1(\mathbb{R}, \mathrm{d}\mu)$ *satisfy* $\|\phi_n\| = \int_{\mathbb{R}} |\phi_n|\,\mathrm{d}\mu = 1 \nrightarrow 0$ *in some sense:*
   (a) $\phi_n = \frac{1}{n}\chi_{(0,n)} \to 0$ *uniformly;*
   (b) $\phi_n = \chi_{(n,n+1)} \to 0$ *pointwise;*
   (c) $\phi_n = n\chi_{(0,1/n)} \to 0$ *a.e.*
   (d) *Define a sequence* $\phi_{2^k+j} = \chi_{(j2^{-k},(j+1)2^{-k})}$ *for* $j = 0, \ldots, 2^k - 1$ *and* $k = 0, 1, 2, \ldots$. *Then,* $\phi_1 = \chi_{(0,1)}, \phi_2 = \chi_{(0,2^{-1})}, \phi_3 = \chi_{(2^{-1},1)}, \cdots$ *and* $\|\psi_{k,j} - 0\| = 2^{-k} \to 0$, *while* $\phi_{2^k+j}(x) \nrightarrow 0$ *for any* x.

Definition 4.5.19 *The sequence* $\{f_n\}$ *is said to converge in measure to* f *if*
$$\forall \epsilon > 0, \quad \lim_{n\to\infty} \mu(\{|f_n - f| \ge \epsilon\}) = 0;$$
and $\{f_n\}$ *is said to be a Cauchy sequence in measure if*
$$\forall \epsilon > 0, \quad \lim_{n,m\to\infty} \mu(\{|f_n - f_m| \ge \epsilon\}) = 0.$$

For example, the sequences $\phi_n = (1/n)\chi_{(0,n)}$, $\chi_{(0,1/n)}$ and $\phi_{2^k+j} = \chi_{(j2^{-k},(j+1)2^{-k})}$ converge to 0 in measure, but the sequence $\phi_n = \chi_{(n,n+1)}$ does not converge to 0 in measure.

Theorem 4.5.20 *Suppose* $\{f_n\}$ *is a Cauchy sequence in measure. Then,*

- *there exists* $f \in \mathcal{M}_{able}$ *such that* $f_n \to f$ *in measure,*
- *there exists* f_{n_k} *such that* $f_{n_k} \to f$ *a.e.,*
- f *is uniquely determined a.e.*

Proof. Choose a subsequence n_k such that $g_k = f_{n_k}$,
$$\mu(E_k) \le 2^{-k}, \quad E_k = \{|g_k - g_{k+1}| \ge 2^{-k}\}.$$
Then, denoting $Z_k := \bigcup_{j=k+1}^{\infty} E_j$ and $Z = \bigcap_{k=1}^{\infty} Z_k$, we have
$$\mu(Z_k) \le 2^{-k} \quad \text{and} \quad \mu(Z) = 0.$$

For $x \in X \setminus Z_M$ and $j > i$, we have

$$|g_{M+i}(x) - g_{M+j}(x)| \le \sum_{k=i}^{j-1} |g_{M+k}(x) - g_{M+k+1}(x)| \le \sum_{k=i}^{j-1} 2^{-M-k} \le 2^{-M-i+1}.$$

Therefore, $\lim_{k\to\infty} g_k(x) = f(x)$ exists for $x \in X \setminus Z$. By letting $f(x) = 0$ for $x \in Z$, we have $g_k \to f$ a.e. It is easy to prove that the sequence g_k converges to f uniformly on $X \setminus Z_M (M = 1, 2, \dots)$ because $\sup_{X\setminus Z_M} |g_{M+k} - f| \le 2^{-M-k+1}$. We can prove $f_n \to f$ in measure from

$$\mu(\{|f - f_n| \ge \epsilon\}) = \mu(\{|f - g_k| \ge \tfrac{1}{2}\epsilon\}) + \mu(\{|g_k - f_n| \ge \tfrac{1}{2}\epsilon\}) \to 0 \quad \text{as } k, n \to \infty.$$

□

The statement [$f_n \to f$ in L^1] implies [$f_n \to f$ in measure] since

$$\forall \epsilon > 0, \quad \mu(\{|f_n - f| > \epsilon\}) \le \frac{1}{\epsilon} \int |f_n - f| \, d\mu \to 0.$$

From the proof of the previous theorem, if [$f_n \to f$ in L^1], then we can choose a subsequence n_k such that $g_k = f_{n_k}$,

$$\mu(E_k) \le 2^{-k}, \quad E_k = \{|g_k - g_{k+1}| \ge 2^{-k}\}.$$

Hence, according to Theorem 4.5.20, [$f_n \to f$ in L^1] implies the statement [$f_{n_k} = g_k \to f$ a.e.].

Theorem 4.5.21 (Egorov's theorem) *Let $\mu(E) < \infty$ and let $f_n \to f$ a.e. Then, for any $\epsilon > 0$, there exists $F \subset E$ so that $\mu(E \setminus F) < \epsilon$ and $f_n \to f$ uniformly on F.*

Proof. Since $f_n \to f$ a.e., there exists $Z \subset E$ so that $\mu(Z) = 0$ and $f_n(x) \to f(x)$ for $x \in E \setminus Z$. Hence, it suffices to prove the theorem for the case when $Z = \emptyset$.

Let $E_{m,n} = \bigcap_{j=m}^{\infty} \{|f_j - f| < 1/n\}$. Then, $\lim_{m\to\infty} \mu(E_{m,n}) = E$ for $n = 1, 2, \dots$. Hence, there exists m_n such that $\mu(E \setminus E_{m_n,n}) \le \epsilon 2^{-n}$. Let $F = \bigcap_{n=1}^{\infty} E_{m_n,n}$. Then, $\mu(E \setminus F) < \epsilon$. Moreover, if $j > m_n$, then $\sup_{x\in F} |f_j(x) - f(x)| < 1/n$. Hence, $f_n \to f$ uniformly on F.

4.5.5 *Differentiation Theory*

Throughout this section, we consider a bounded function $f : [a, b] \to \mathbb{R}$. We will study a necessary and sufficient condition that f' exists almost everywhere and

$$f(y) - f(x) = \int_x^y f' \, d\mu, \quad \mu((x, y)) = |y - x|.$$

Recall that the derivative of f at x exists if all four of the following numbers have the same finite value:

$$\liminf_{h\to 0^\pm} \frac{f(x+h) - f(x)}{h}, \quad \limsup_{h\to 0^\pm} \frac{f(x+h) - f(x)}{h}.$$

According to Lebesgue's theorem, every monotonic function $f : [a, b] \to \mathbb{R}$ is differentiable almost everywhere.

Example 4.5.22 *If f is a Cantor function, then $f' = 0$ almost everywhere but*

$$1 = f(1) - f(0) \neq 0 = \int_0^1 f'(x)\,d\mu.$$

To understand why $f(1) - f(0) \neq \int_0^1 f'(x)\,d\mu$ for a Cantor function f, we need to understand the concept of absolute continuity.

Definition 4.5.23 *A measure ν is absolutely continuous with respect to μ if and only if $\mu(E) = 0 \Rightarrow \nu(E) = 0$. A non-negative $f \in BV[a, b]$ (the space of bounded variation functions in $[a, b]$) is absolutely continuous with respect to μ if and only if the measure $\nu(E) = \int_E f\,d\mu$ is absolutely continuous with respect to μ. The measures μ and ν are mutually singular, written $\mu \perp \nu$, if and only if*

$$\exists\, E, F \in \mathcal{M} \text{ s.t. } X = E \cup F,\ E \cap F = \emptyset,\ \mu(E) = 0 = \nu(F).$$

Theorem 4.5.24 (Absolute continuity) *If f' exists almost everywhere, $f' \in L^1(d\mu)$ and*

$$f(x) = \int_a^x f'(x)\,d\mu, \quad x \in (a, b],$$

then f is absolutely continuous.

Proof. We want to prove that, for a given $\epsilon > 0$, there exists δ so that $\sum_j (y_j - x_j) < \delta \Rightarrow \sum_j |f(y_j) - f(x_j)| < \epsilon$.

If $|f'|$ is bounded, then we choose $\delta = \epsilon/\|f'\|_\infty$ and

$$\sum_j |f(y_j) - f(x_j)| \leq \sum_j \int_{x_j}^{y_j} |f'|\,d\mu \leq C \sum_j (y_j - x_j) < \|f'\|_\infty \delta = \epsilon.$$

If $f' \in L^1(d\mu)$ but not bounded, then we decompose

$$f' = g + h \quad \text{where } g \text{ is bounded and } \int |h|\,d\mu < \tfrac{1}{2}\epsilon.$$

This is possible because the bounded functions are dense in $L^1(d\mu)$. The result follows by choosing $\delta = \epsilon/(2\|g\|_\infty)$. □

From this theorem, the Cantor function is not absolutely continuous.

Every continuous and non-decreasing function f can be decomposed into the sum of an absolutely continuous function and a singular function, both monotone:

$$f(x) = \underbrace{\int_0^x f'(s)\,d\mu_s}_{g} + (f(x) - g(x)),$$

where g is an absolutely continuous function.

Theorem 4.5.25 (Radon–Nikodym, Riesz representation) *Let ν be a finite measure on $[a, b]$, and suppose that ν is absolutely continuous with respect to μ. Then, there exists $f \in L^1(X, \mathrm{d}\mu)$ such that*

$$\nu(E) = \int_E f \,\mathrm{d}\mu \quad \textit{for all } \nu(E) < \infty (E \in \mathcal{M}).$$

References

Akoka S, Franconi F, Seguin F and le Pape A 2009 Radiofrequency map of an NMR coil by imaging. *Magn. Reson. Imag.* **11**, 437–441.

Chipot M 2009 *Elliptic Equations: An Introductory Course*. Birkhäuser Advanced Texts. Birkhäuser, Basel.

Engl HW, Hanke M and Neubauer A 1996 *Regularization of Inverse Problems*. Mathematics and Its Applications. Kluwer Academic, Dordrecht.

Feldman J and Uhlmann G 2003 *Inverse Problems*. Lecture Note. See http://www.math.ubc.ca/~feldman/ibook/.

Gilbarg D and Trudinger N 2001 *Elliptic Partial Differential Equations of Second Order*. Springer, Berlin.

Lieb EH and Loss M 2001 *Analysis*, 2nd edn. Graduate Studies in Mathematics, no. 14. American Mathematical Society, Providence, RI.

Marsden JE 1974 *Elementary Classical Analysis*. W. H. Freeman, San Francisco.

Rudin W 1970 *Real and Complex Analysis*. McGraw-Hill, New York.

Stollberger R and Wach P 1996 Imaging of the active B_1 field in vivo. *Magn. Reson. Med.* **35**, 246–251.

Wheeden RL and Zygmund A 1977 *Measure and Integral: An Introduction to Real Analysis*. Monographs and Textbooks in Pure and Applied Mathematics, vol. 43. Marcel Dekker, New York.

Further Reading

Evans LC 2010 *Partial Differential Equations*. Graduate Studies in Mathematics, no. 19. American Mathematical Society, Providence, RI.

Folland G 1976 *Introduction to Partial Differential Equations*. Princeton University Press, Princeton, NJ.

Giaquinta M 1983 *Multiple Integrals in the Calculus of Variations and Non-linear Elliptic Systems*. Princeton University Press, Princeton, NJ.

Grisvard P 1985 *Elliptic Problems in Nonsmooth Domains*. Monographs and Studies in Mathematics, no. 24. Pitman, Boston, MA.

John F 1982 *Partial Differential Equations*. Applied Mathematical Sciences, vol. 1. Springer, New York.

Kellogg OD 1953 *Foundations of Potential Theory*. Dover, New York.

Reed M and Simon B 1980 *Functional Analysis*. Methods of Modern Mathematical Physics, vol. I. Academic Press, San Diego, CA.

Rudin W 1970 *Functional Analysis*. McGraw-Hill, New York.

Zeidler E 1989 *Nonlinear Functional Analysis and Its Applications*. Springer, New York.

5

Numerical Methods

Quantitative analyses are essential elements in solving forward and inverse problems. To utilize computers, we should devise numerically implementable algorithms to solve given problems, which include step-by-step procedures to obtain final answers. Formulation of a forward as well as an inverse problem should be done bearing in mind that we will adopt a certain numerical algorithm to solve them. After reviewing the basics of numerical computations, we will introduce various methods to solve a linear system of equations, which are most commonly used to obtain numerical solutions of both forward and inverse problems. Considering that most forward problems are formulated by using partial differential equations, we will study numerical techniques such as the finite difference method and finite element method. The accuracy or consistency of a numerical solution as well as the convergence and stability of an algorithm to obtain the solution need to be investigated.

5.1 Iterative Method for Nonlinear Problem

Recall the abstract form of the inverse problem in section 4.1, where we tried to reconstruct a material property P from knowledge of the input data X and output data Y using the forward problem (4.1). It is equivalent to the following root-finding problem:

$$G(P) = Y - F(P, X) = 0,$$

where X and Y are given data.

We assume that both the material property P and $F(P, X)$ are expressed as vectors in N-dimensional space $\mathbb{C}^N$. Imagine that P^* is a root of $G(P) = 0$ and P^0 is a reasonably good guess in the sense that $P^0 + \Delta P = P^*$ for a small perturbation ΔP. From the tangent line approximation, we can approximate

$$0 = G\left(P^0 + \Delta P\right) \approx G\left(P^0\right) + [J\left(P^0\right)]\Delta P,$$

where $J(P^0)$ denotes the Jacobian of $G(P)$ at P^0. Then, the solution $P^* = P^0 + \Delta P$ can be approximated by

$$P^* = P^0 + \Delta P \approx \Phi\left(P^0\right) := P^0 - [J\left(P^0\right)]^{-1} G\left(P^0\right).$$

Nonlinear Inverse Problems in Imaging, First Edition. Jin Keun Seo and Eung Je Woo.

The Newton–Raphson algorithm is based on the above idea, and we can start from a good initial guess P^0 to perform the following recursive process:

$$P^{n+1} = \underbrace{P^n - [J(P^n)]^{-1} G(P^n)}_{\Phi(P^n)}. \tag{5.1}$$

The convergence of this approach is related to the condition number of the Jacobian matrix $J(P^*)$. If $J(P^*)$ is nearly singular, the root-finding problem $G(P) = 0$ is ill-posed. According to the *fixed point theorem*, the sequence P^n converges to a fixed point P^* provided that $\Phi(P) := P - [J(P)]^{-1}G(P)$ is a contraction mapping: for some $0 < \alpha < 1$,

$$\|\Phi(P) - \Phi(Q)\| \le \alpha \|P - Q\| \quad \left(\forall P, Q \in \mathbb{C}^N\right). \tag{5.2}$$

To be precise, if Φ is a contraction mapping satisfying (5.2), then

$$\begin{aligned} \|P^{n+1} - P^n\| &= \|\Phi(P^n) - \Phi(P^{n-1})\| \\ &\le \alpha \|P^n - P^{n-1}\| \le \cdots \le \alpha^n \|P^1 - P^0\| \to 0. \end{aligned}$$

Hence, $\{P^n\}$ is a Cauchy sequence, which guarantees the convergence $P^n \to P^*$. Here, P^* is a fixed point, $\Phi(P^*) = P^*$. Owing to the convergence of the recursive process (5.1), we can compute an approximate solution of the root-finding problem $G(P) = 0$.

Example 5.1.1 *Let $\Phi(x) = \cos x$. With the initial point $x_0 = 0.7$, let $x_{n+1} = \Phi(x_n)$, $n = 0, 1, 2, \ldots$. Prove that the sequence x_n converges to a point x_*, where x_* is a fixed point, $x_* = \phi(x_*)$.*

5.2 Numerical Computation of One-Dimensional Heat Equation

We begin with a simple problem of solving the following one-dimensional heat equation with the thermal diffusivity coefficient $\kappa > 0$:

$$\begin{cases} \partial_t u(x,t) = \kappa \partial_x^2 u(x,t) & (0 < x < 1, t > 0) \quad \text{(diffusion equation)}, \\ u(x,0) = f(x) & (0 < x < 1) \quad \text{(initial condition)}, \\ u(0,t) = 0, u(1,t) = 0 & (t > 0) \quad \text{(boundary condition)}. \end{cases} \tag{5.3}$$

To compute an approximate numerical solution of $u(x,t)$ in (5.3), we discretize xt-space into a collection of grid points and express u in the following discrete form:

$$U_k^n \overset{\text{def}}{=} u(x_k, t_n) \quad \left(x_k = k\Delta x,\ t_n = n\Delta t\right),$$

where $\Delta x = 1/N$ and the superscript n and subscript k, respectively, specify the time step and the space step. The quantity $\partial_t u$ can be approximated by the first-order time derivative using a finite difference scheme of the two-point Euler difference:

$$\partial_t u|_{k,n} \approx \frac{U_k^{n+1} - U_k^n}{\Delta t} \quad \text{(forward difference scheme)},$$

where $\partial_t u|_{k,n} = \partial_t u(k\Delta x, n\Delta t)$.

The expressions for the space derivatives can be obtained by Taylor's expansion.

- Forward difference:

$$U_{k+1}^n = U_k^n + \frac{\partial}{\partial x}u|_{k,n}\frac{\Delta x}{1} + \frac{\partial^2}{\partial x^2}u|_{k,n}\frac{\Delta x^2}{2} + \frac{\partial^3}{\partial x^3}u|_{k,n}\frac{\Delta x^3}{6} + O\left(\Delta x^4\right),$$

$$\frac{\partial}{\partial x}u|_{k,n} = \frac{1}{\Delta x}\left(U_{k+1}^n - U_k^n\right) - \frac{1}{\Delta x}\left(\frac{\partial^2}{\partial x^2}u|_{k,n}\frac{\Delta x^2}{2} + \frac{\partial^3}{\partial x^3}u|_{k,n}\frac{\Delta x^3}{6} + O\left(\Delta x^4\right)\right).$$

- Backward difference:

$$U_{k-1}^n = U_k^n - \frac{\partial}{\partial x}u|_{k,n}\frac{\Delta x}{1} + \frac{\partial^2}{\partial x^2}u|_{k,n}\frac{\Delta x^2}{2} - \frac{\partial^3}{\partial x^3}u|_{k,n}\frac{\Delta x^3}{6} + O\left(\Delta x^4\right),$$

$$\frac{\partial}{\partial x}u|_{k,n} = \frac{1}{\Delta x}\left(U_k^n - U_{k-1}^n\right) - \frac{1}{\Delta x}\left(-\frac{\partial^2}{\partial x^2}u|_{k,n}\frac{\Delta x^2}{2} + \frac{\partial^3}{\partial x^3}u|_{k,n}\frac{\Delta x^3}{6} - O\left(\Delta x^4\right)\right).$$

- Central difference:

$$\frac{\partial^2}{\partial x^2}u|_{k,n} = \frac{U_{k+1}^n - 2U_k^n + U_{k-1}^n}{\Delta x^2} + O\left(\Delta x^2\right).$$

Here, the "big O" notation $\phi(x) = O(\eta(x))$ is used to describe the limiting behavior of the function ϕ such that

$$\limsup_{\eta(x)\to 0}\frac{O(\eta(x))}{\eta(x)} < \infty.$$

Using three-point central difference schemes to approximate the second-order space derivative, we get the following recurrence relations.

- Explicit scheme or forward Euler method: U_k^{n+1} (the temperature at time $n+1$ and position k) depends explicitly on $U_{k-1}^n, U_k^n, U_{k+1}^n$ at the previous time n using the relation

$$\frac{U_k^{n+1} - U_k^n}{\Delta t} \approx \kappa\left[\frac{U_{k+1}^n + U_{k-1}^n - 2U_k^n}{\Delta x^2}\right].$$

- Implicit scheme or backward Euler method: the Laplacian $(\partial^2 u/\partial x^2)|_{k,n+1}$ and U_k^{n+1} at time $n+1$ are evaluated from the implicit relation

$$\frac{U_k^{n+1} - U_k^n}{\Delta t} \approx \kappa\left[\frac{U_{k+1}^{n+1} + U_{k-1}^{n+1} - 2U_k^{n+1}}{\Delta x^2}\right].$$

- Crank–Nicolson scheme: we use the average of the explicit and implicit schemes

$$\frac{U_k^{n+1} - U_k^n}{\Delta t} \approx \frac{\kappa}{2}\left[\frac{U_{k+1}^n + U_{k-1}^n - 2U_k^n}{\Delta x^2}\right] + \frac{\kappa}{2}\left[\frac{U_{k+1}^{n+1} + U_{k-1}^{n+1} - 2U_k^{n+1}}{\Delta x^2}\right].$$

With the explicit scheme, the diffusion equation in (5.3) can be expressed as

$$\frac{U_k^{n+1} - U_k^n}{\Delta t} = \kappa\left(\frac{U_{k+1}^n + U_{k-1}^n - 2U_k^n}{\Delta x^2}\right) + \underbrace{O(\Delta t) + O(\Delta x^2)}_{\text{local truncation error}}. \tag{5.4}$$

Here, the last term can be viewed as the truncation error (or discretization error), which is the difference between the solution of the explicit scheme equation and the exact analytic solution.

Exercise 5.2.1 *Denote the truncation error by*

$$E_k^n := \frac{U_k^{n+1} - U_k^n}{\Delta t} - \kappa\left(\frac{U_{k+1}^n + U_{k-1}^n - 2U_k^n}{\Delta x^2}\right) \quad (0 \le k \le N-1, n > 1). \tag{5.5}$$

Show that the truncation error E_k^n can be expressed as

$$E_k^n = \left[\frac{\partial}{\partial t}u(x_k, \tilde{t}_n) - \kappa\frac{\partial^2}{\partial x^2}u(\tilde{x}_k, t_n)\right], \tag{5.6}$$

where $t_n \le \tilde{t}_n \le t_{n+1}$ and $x_k \le \tilde{x}_k \le x_{k+1}$. Show that the principal part of the local truncation error is

$$E_k^n \approx \frac{\Delta t}{2}\frac{\partial^2}{\partial t^2}u|_{k,n} - \frac{\kappa\Delta x^2}{2}\frac{\partial^4}{\partial x^4}u|_{k,n}.$$

When using a specific computational scheme, we should check the following fundamental conditions.

- Convergence. Does the numerical solution u_k^n converge to the true solution u?
- Consistency. Does the discretization error go to zero as $\Delta t, \Delta x \to 0$?
- Stability. Do small errors in the initial and boundary conditions cause small errors in the numerical solution?

5.2.1 Explicit Scheme

In the explicit scheme, it is easy to compute numerical solutions because it is explicit. However, computational experience shows that, depending on the size of $\kappa(\Delta x)^2/\Delta t$, numerical errors (round-off and truncation errors) can be magnified as the iterative computations progress in time. It is well known that this explicit method requires $\kappa(\Delta t/\Delta x^2) < \frac{1}{2}$ in order to be numerically stable. Owing to this requirement, it is computationally expensive.

If we substitute the approximations into $\partial_t u = \kappa\partial_x^2 u$ and ignore the higher-order terms $O(\Delta t)$ and $O(\Delta x^2)$, we obtain the following system of difference equations for interior grid points:

$$\frac{u_k^{n+1} - u_k^n}{\Delta t} = \kappa\left(\frac{u_{k+1}^n + u_{k-1}^n - 2u_k^n}{\Delta x^2}\right).$$

This formation is called the explicit forward Euler method because the time derivative is represented with a forward Euler approximation. The solution at time $t = n\Delta t$ can be solved explicitly as

$$u_k^{n+1} = (1 - 2r)\,u_k^n + r\left(u_{k+1}^n + u_{k-1}^n\right) \quad \text{where} \quad r = \kappa\frac{\Delta t}{\Delta x^2} \tag{5.7}$$

or

$$\mathbf{u}^{n+1} = \underbrace{\begin{pmatrix} 1-2r & r & 0 & \cdots & 0 & 0 \\ r & 1-2r & r & \cdots & 0 & 0 \\ \vdots & \vdots & \vdots & \ddots & \vdots & \vdots \\ 0 & 0 & 0 & \cdots & r & 1-2r \end{pmatrix}}_{\mathbf{A}} \mathbf{u}^n + r \underbrace{\begin{pmatrix} u_0^n \\ 0 \\ \vdots \\ u_N^n \end{pmatrix}}_{\text{boundary condition}}, \tag{5.8}$$

where $\mathbf{u}^n = [u_1^n, \ldots, u_{N-1}^n]^{\mathrm{T}}$. Since u in (5.3) has homogeneous boundary conditions,

$$\mathbf{u}^n = \mathbf{A}^n \mathbf{u}^0. \tag{5.9}$$

From the above formula, we can directly compute $\mathbf{u}^n$, an approximate solution $u(x, \Delta t)$, from the discrete version of the initial condition $\mathbf{u}^0$ (a discrete form of $u(x, 0) = f(x)$). This step-by-step time advancing process (5.8) is called *time marching* since the unknown values $\mathbf{u}^{n+1}$ at the next time $n + 1$ are directly computed from the computed values $\mathbf{u}^n$ at the previous time n. According to (5.4), the computed solution by the time marching scheme (5.8) can be $O(\Delta t)$ accurate in time and $O(\Delta x^2)$ accurate in space.

We should note that the exact solution u_k^n of the discretized problem (5.7) may not be an accurate approximate solution of the original continuous heat equation even if Δx and Δt are sufficiently small. The stability of a numerical scheme is closely associated with numerical errors. Hence, we need to check the error $e_k^n := U_k^n - u_k^n$ between the numerical solution u_k^n of the finite difference equation (5.7) and the true solution $U_k^n = u(x_k, t_n)$, with u being the solution of the continuous equation (5.3). From (5.7) and (5.5), the error e_k^n can be expressed as

$$e_k^{n+1} = re_{k-1}^n + (1 - 2r)e_k^n + re_{k+1}^n + E_k^n \Delta t.$$

If $r < \frac{1}{2}$, the error e_k^{n+1} is estimated by

$$|e_k^{n+1}| \le \sup_k |e_k^n| + \sup_k |E_k^n| \Delta t.$$

Since the above estimate holds for all k and n, we obtain

$$\sup_k |e_k^{n+1}| \le \sup_k |e_k^0| + \sum_{m=1}^{n} \sup_k |E_k^m| \Delta t. \tag{5.10}$$

We should note that, as $\Delta t \to 0$, $\sum_{m=1}^{n} \sup_k |E_k^m| \Delta t \to 0$. The estimate (5.10) shows that u_k^n converges to the true solution u as $\Delta t \to 0$ provided $r < 0.5$. The numerical errors are proportional to the time step and to the square of the space step. For details, please refer to Haberman (1987).

Theorem 5.2.2 *The explicit scheme (5.7) is stable provided that the value of the gain parameter r defined in (5.7) satisfies*

$$r = \kappa \frac{\Delta t}{\Delta x^2} < \frac{1}{2}.$$

Exercise 5.2.3 *The Von Neumann stability criterion is a method to check the stability condition by inserting the Fourier mode* $u_k^n = \lambda^n \mathrm{e}^{\mathrm{i}mk\pi\Delta x}$ *into the finite difference scheme (5.7):*

$$\underbrace{\lambda^{n+1}\mathrm{e}^{\mathrm{i}mk\pi\Delta x}}_{u_k^{n+1}} = \underbrace{\lambda^n\mathrm{e}^{\mathrm{i}mk\pi\Delta x}[(1-2r)+r(\mathrm{e}^{-\mathrm{i}m\pi\Delta x}+\mathrm{e}^{\mathrm{i}m\pi\Delta x})]}_{(1-2r)u_k^n+r(u_{k+1}^n+u_{k-1}^n)}.$$

A necessary condition for stability can be obtained by restricting Δx *and* Δt *such that* $|\lambda| \le 1$*, that is,* λ^n *will not grow without bound.*

1. *Derive an expression for* λ *for stability:*

$$\lambda = \left[(1-2r)+r\left(\mathrm{e}^{-\mathrm{i}m\pi\Delta x}+\mathrm{e}^{\mathrm{i}m\pi\Delta x}\right)\right] = 1-4r\sin^2\left(\frac{m\pi\Delta x}{2}\right).$$

2. *Show that* $|r| \le \frac{1}{2}$ *implies* $|\lambda| \le 1$.

Exercise 5.2.4 *We can obtain a stability condition using the discrete Fourier transform (DFT). Recall that the DFT of a vector* $\mathbf{u}^n = (u_0^n, u_1^n, \ldots, u_{N-1}^n)^{\mathrm{T}}$ *is a vector* $\hat{\mathbf{u}}^n = (\hat{u}_0^n, \ldots, \hat{u}_{N-1}^n)^{\mathrm{T}}$ *with* $\hat{u}_m^n$ *being*

$$\hat{u}_m^n = \sum_{k=0}^{N-1} u_k^n \mathrm{e}^{(-\mathrm{i}2\pi/N)km}, \quad m = 0, 1, \ldots, N-1.$$

We can apply the same DFT to

$$\partial_x^2 u|_{k,n} \approx \frac{u_{k+1}^n + u_{k-1}^n - 2u_k^n}{\Delta x^2} = N^2\left(u_{k+1}^n + u_{k-1}^n - 2u_k^n\right).$$

1. *Show that*

$$\begin{aligned}
\widehat{\partial_x^2 u}|_{m,n} &\approx N^2 \sum_k \mathrm{e}^{(-\mathrm{i}2\pi/N)km}(u_{k+1}^n - 2u_k^n + u_{k-1}^n) \\
&= N^2 \sum_k \mathrm{e}^{(-\mathrm{i}2\pi/N)km}(\mathrm{e}^{(-\mathrm{i}2\pi/N)m} - 2 + \mathrm{e}^{(\mathrm{i}2\pi/N)m})u_k^n \\
&= N^2 \sum_k \mathrm{e}^{(-\mathrm{i}2\pi/N)km}\left[2\cos(\underbrace{2m\pi/N}_{\xi_m}) - 2\right]u_k^n.
\end{aligned}$$

2. *Denote* $\xi_m = 2m\pi/N$ *and show that*

$$\widehat{\partial_x^2 u}|_{m,n} \approx -N^2 4\sin^2\left(\xi_m/2\right)\widehat{u}_m^n.$$

3. *Show that application of DFT to (5.7) leads to*

$$\hat{u}_m^{n+1} = \left(1 - 4r\sin^2\left(\xi_m/2\right)\right)\hat{u}_m^n, \quad m = 0, 1, \ldots, N,$$

   where $r = \kappa N^2 \Delta t = \kappa \Delta t/\Delta x^2$.
4. *For stability, we need*

$$|1 - 4r\sin^2\left(\xi_m/2\right)| < 1.$$

   Hence, a sufficient condition for stability is $r < \frac{1}{2}$.

Unfortunately, the numerical solution is unstable when $r > \frac{1}{2}$ no matter how small Δx and Δt are. If $r > \frac{1}{2}$, the error e_k^n may grow with the time step n. This is the major shortcoming of the explicit method. Now, we present the Lax equivalence theorem (Lax and Richtmyer 1956), which plays an important role in determining the convergence of a solution of the finite difference method to a solution of the partial differential equation (PDE).

Theorem 5.2.5 Lax equivalence theorem *For a consistent difference scheme of a well-posed linear initial value problem, the method is convergent if and only if it is stable.*

5.2.2 *Implicit Scheme*

The major shortcoming of the explicit method is the time step constraint $\Delta t < [1/(2\kappa)]\Delta x^2$ to guarantee the stability so that the computed solution u_k^n is a good approximation to the true solution $u(x_k, t_n)$. Using the explicit scheme or forward Euler method, the numerical solution u_k^n may blow up if the time step does not satisfy the constraint $\Delta t < [1/(2\kappa)]\Delta x^2$.

We can remove the time step constraint $\Delta t < [1/(2\kappa)]\Delta x^2$ of the forward Euler method by the use of the implicit scheme or backward Euler method:

$$u_k^{n+1} = u_k^n + r \underbrace{\left[u_{k+1}^{n+1} - 2u_k^{n+1} + u_{k-1}^{n+1}\right]}_{\approx \partial_x^2 u|_{k,n+1} \Delta x^2}, \tag{5.11}$$

where we evaluate the approximation for u_{xx} at $t_{n+1} = (n+1)\Delta t$ rather than at t_n. This implicit backward Euler method leads one to solve the following implicit linear system:

$$\mathbf{A}\mathbf{u}^{n+1} = \mathbf{u}^n, \tag{5.12}$$

where

$$\mathbf{A} := \begin{pmatrix} 1+2r & -r & 0 & \cdots & 0 & 0 \\ -r & 1+2r & -r & \cdots & 0 & 0 \\ \vdots & \vdots & \vdots & \ddots & \vdots & \vdots \\ 0 & 0 & 0 & \cdots & -r & 1+2r \end{pmatrix}.$$

Exercise 5.2.6 *Prove that the backward Euler method (5.11) is unconditionally stable. Inserting the Fourier mode $u_k^n = \lambda^n e^{ijk\pi\Delta x}$ into the implicit scheme, find the following expression for λ:*

$$\lambda = \frac{1}{1 + 2r[1 - \cos(j\pi\Delta x)]} < 1.$$

Since the matrix $\mathbf{A}$ in (5.12) is diagonally dominant, the implicit scheme is always numerically stable and convergent regardless of the choice of the time step. This is the major advantage of the implicit backward Euler method over the explicit forward Euler method (5.7). On the other hand, the implicit method requires a computation of the inverse matrix $\mathbf{A}$ at each time step and, therefore, it is computationally more intensive.

5.2.3 Crank–Nicolson Method

The Crank–Nicolson method combines the forward and backward Euler methods as

$$u_k^{n+1} = u_k^n + r\left[\beta\left(u_{k+1}^{n+1} - 2u_k^{n+1} + u_{k-1}^{n+1}\right) + (1-\beta)\left(u_{k+1}^n - 2u_k^n + u_{k-1}^n\right)\right].$$

When $\beta = \frac{1}{2}$, the scheme is called the Crank–Nicolson scheme.

Exercise 5.2.7 *Show that the Crank–Nicolson scheme is unconditionally stable.*

We can summarize the following features.

- For $\beta = 0$, the scheme is explicit. It is stable under the condition $\Delta t \leq [1/(2\kappa)]\Delta x^2$. Owing to this constraint, the time step has to be chosen small enough.
- For $\beta \neq 0$, the scheme is implicit. The scheme for $\beta = 1$ is unconditionally stable.
- When $\beta = \frac{1}{2}$, the scheme is the Crank–Nicolson scheme, which is unconditionally stable.

5.3 Numerical Solution of Linear System of Equations

Finding a numerical solution of a forward problem and also an inverse problem often requires solving a system of linear algebraic equations,

$$\mathbf{Au} = \mathbf{b}$$

for $\mathbf{u}$. We can view a solution $\mathbf{u}$ of the system as a minimizer of the following functional:

$$\Phi(\mathbf{u}) = \tfrac{1}{2}\langle \mathbf{Au}, \mathbf{u}\rangle - \mathbf{b}\cdot\mathbf{u}.$$

Assuming that $\mathbf{A}$ is invertible, we consider a quantitative bound for the error in the computed solution of the linear system. We define the norm of the matrix $\mathbf{A}$ as

$$\|\mathbf{A}\| = \max_{\mathbf{u}\neq 0}\frac{\|\mathbf{Au}\|}{\|\mathbf{u}\|}.$$

With $\mathbf{u}$ being the solution of $\mathbf{Au} = \mathbf{b}$ and $\mathbf{u} + \Delta\mathbf{u}$ being the solution of $\mathbf{A}[\mathbf{u} + \Delta\mathbf{u}] = \mathbf{b} + \Delta\mathbf{b}$, we have

$$\frac{\|\Delta\mathbf{u}\|}{\|\mathbf{u}\|} = \frac{\|\mathbf{A}^{-1}\Delta\mathbf{b}\|}{\|\mathbf{A}^{-1}\mathbf{b}\|} \leq \underbrace{\|\mathbf{A}^{-1}\|\,\|\mathbf{A}\|}_{\mathrm{cond}(\mathbf{A})}\frac{\|\Delta\mathbf{b}\|}{\|\mathbf{b}\|},$$

where cond($\mathbf{A}$) is called the condition number of the matrix $\mathbf{A}$, indicating an amplification factor that bounds the maximum relative change in the solution due to a given relative change in the vector on the right-hand side.

5.3.1 Direct Method using LU Factorization

The most common direct method of solving $\mathbf{Au} = \mathbf{b}$ is LU-factorization with forward and backward substitutions. By applying the Gaussian elimination process, we can express the matrix **A** as

$$\mathbf{A} = \mathbf{LU},$$

where **L** is a unit lower triangular matrix and **U** is an upper triangular matrix. We summarize the process to find **L** and **U** as follows:

(i) $\mathrm{A}^{[1]} = [a_{ij}]$.

(ii)
$$\underbrace{\begin{bmatrix} 1 & & & & \\ -a_{21}/a_{11} & 1 & & & \\ \vdots & 0 & \ddots & & \\ \vdots & \vdots & & \ddots & \\ -a_{N1}/a_{11} & 0 & \dots & 0 & 1 \end{bmatrix}}_{\mathrm{L}_1} \mathrm{A}^{[1]}$$

$$= \underbrace{\begin{bmatrix} a_{11}^{[1]} & a_{12}^{[1]} & \dots & \dots & a_{1N}^{[1]} \\ 0 & a_{22} - (a_{21}/a_{11})a_{12} & & & \\ \vdots & \vdots & \ddots & & \\ \vdots & a_{i2} - (a_{i1}/a_{11})a_{12} & & a_{ij} - (a_{i1}/a_{11})a_{1j} & \\ 0 & \vdots & & & \ddots \end{bmatrix}}_{\mathrm{A}^{[2]}}.$$

(iii) Given $\mathrm{A}^{[k]} = [a_{ij}^{[k]}]$ of the form

$$\mathrm{A}^{[k]} = \begin{bmatrix} a_{11}^{[k]} & & & \\ 0 & \ddots & & \\ \vdots & & a_{kk}^{[k]} & \cdots & a_{kN}^{[k]} \\ \vdots & & \vdots & & \vdots \\ 0 & & a_{Nk}^{[k]} & \dots & a_{NN}^{[k]} \end{bmatrix},$$

$\mathrm{A}^{[k+1]}$ is determined from

$$\mathrm{L}_k \mathrm{A}^{[k]} = \mathrm{A}^{[k+1]},$$

where

$$\mathrm{L}_k = \begin{bmatrix} 1 & & & & & \\ 0 & \ddots & & & & \\ \vdots & & \ddots & & & \\ \vdots & 0 & 1 & \leftarrow k\text{th row} & & \\ \vdots & \vdots & -a_{ik}^{[k]}/a_{kk}^{[k]} & \ddots & & \\ \vdots & \vdots & \uparrow & & \ddots & \\ 0 & 0 & k\text{th column} & & & 1 \end{bmatrix}.$$

Then,

$$
\begin{aligned}
A^{[N]} &= U, \\
L &= L_1^{-1} \cdots L_{N-1}^{-1} \\
&= \begin{bmatrix} 1 & 0 & \dots & 0 \\ \vdots & \ddots & & \\ l_{ik} = -a_{ik}^{[k]}/a_{kk}^{[k]} & & \ddots & 0 \\ \vdots & \dots & \dots & 1 \end{bmatrix}.
\end{aligned}
$$

Given such an LU-factorization, we can solve the system $\mathbf{Au} = \mathbf{b}$ via the forward and backward substitutions. Such direct methods are usually applied to the cases where the size of the matrix $\mathbf{A}$ is relatively small.

5.3.2 *Iterative Method using Matrix Splitting*

When the problem $\mathbf{Au} = \mathbf{b}$ is too large, one may use an iterative method by generating a sequence $\mathbf{u}^n \to \mathbf{u}$ with a low computational cost. The most commonly used iterative methods for solving $\mathbf{Au} = \mathbf{b}$ are based on the decomposition of the matrix $\mathbf{A}$ into two parts $\mathbf{A} = \mathbf{B} + (\mathbf{A} - \mathbf{B})$ so that the linear system $\mathbf{Au} = \mathbf{b}$ can be rewritten as

$$\mathbf{Bu} = \mathbf{b} - (\mathbf{A} - \mathbf{B})\mathbf{u}.$$

Then, we set the iteration scheme as follows:

$$\mathbf{u}^{n+1} = \underbrace{\mathbf{B}^{-1}(\mathbf{B} - \mathbf{A})}_{\mathbf{G}} \mathbf{u}^n + \underbrace{\mathbf{B}^{-1}\mathbf{b}}_{\tilde{\mathbf{b}}}. \tag{5.13}$$

An iterative method expressed in the form of (5.13) is called the stationary iterative method. For the choice of $\mathbf{B}$ in the stationary iterative method, we use the decomposition

$$\mathbf{A} = \mathbf{L} + \mathbf{D} + \mathbf{U},$$

where $\mathbf{D}$ is the diagonal part of $\mathbf{A}$ and $\mathbf{L}$ and $\mathbf{U}$ are the lower and upper triangular matrices with zeros on their diagonals, respectively. The most common stationary iterative schemes are as follows.

- Jacobi method: with $\mathbf{B} = \mathbf{D}$ and $\mathbf{G} = \mathbf{D}^{-1}(\mathbf{L} + \mathbf{U})$, the iterative scheme is

$$\mathbf{u}^{n+1} = \underbrace{\mathbf{D}^{-1}(\mathbf{L} + \mathbf{U})}_{\mathbf{G}} \mathbf{u}^n + \underbrace{\mathbf{D}^{-1}\mathbf{b}}_{\tilde{\mathbf{b}}}.$$

- Damped Jacobi method: use $\mathbf{G} = (\mathbf{I} - \omega\mathbf{D}^{-1}\mathbf{A})$ with the damping factor ω to get the iterative scheme

$$\mathbf{u}^{n+1} = \left(\mathbf{I} - \omega\mathbf{D}^{-1}\mathbf{A}\right)\mathbf{u}^n + \omega\mathbf{D}^{-1}\mathbf{b}.$$

- Gauss–Seidel method: use $\mathbf{G} = -(\mathbf{L}+\mathbf{D})^{-1}\mathbf{U}$ to get

$$\mathbf{u}^{n+1} = -\left(\mathbf{L}+\mathbf{D}\right)^{-1}\mathbf{U}\mathbf{u}^{n} + \left(\mathbf{L}+\mathbf{D}\right)^{-1}\mathbf{b}.$$

- Successive over-relaxation (SOR) method: use $\mathbf{G} = (\omega\mathbf{D}+\mathbf{L})^{-1}\left[(\omega-1)\mathbf{D}-\mathbf{U}\right]$ to get

$$\mathbf{u}^{n+1} = \left(\omega\mathbf{D}+\mathbf{L}\right)^{-1}\left[(\omega-1)\,\mathbf{D}-\mathbf{U}\right]\mathbf{u}^{n} + \left(\omega\mathbf{D}+\mathbf{L}\right)^{-1}\mathbf{b}$$

with the relation factor $\omega > 1$.

Definition 5.3.1 *The set $\sigma(\mathbf{G})$ of all eigenvalues of the matrix $\mathbf{G}$ is said to be the spectrum of $\mathbf{G}$. The value $\rho(\mathbf{G}) = \max_{\lambda\in\sigma(\mathbf{G})}|\lambda|$ is called the spectral radius of $\mathbf{G}$.*

Throughout this section, we assume that the stationary iterative scheme (5.13) is consistent with $\mathbf{Au} = \mathbf{b}$, that is,

$$\mathbf{u} = \mathbf{G}\mathbf{u} + \tilde{\mathbf{b}}. \tag{5.14}$$

From (5.13) and (5.14), we have

$$\mathbf{u} - \mathbf{u}^{n} = \mathbf{G}\left(\mathbf{u}-\mathbf{u}^{n-1}\right) = \cdots = \mathbf{G}^{n}\left(\mathbf{u}-\mathbf{u}^{0}\right). \tag{5.15}$$

The above identity leads directly to the following theorem.

Theorem 5.3.2 *For any initial guess $\mathbf{u}^0$,*

$$\lim_{n\to\infty}\|\mathbf{u}-\mathbf{u}^{n}\| \to 0 \quad \textit{if and only if } \rho\left(\mathbf{G}\right) < 1.$$

To be precise, assume that $\mathbf{G}$ has N linearly independent eigenvectors $\mathbf{v}_1, \ldots, \mathbf{v}_N$ with associated eigenvalues $\lambda_1, \ldots, \lambda_N$ and that $|\lambda_N| = \rho(\mathbf{G})$. Then, from (5.15), we have

$$\mathbf{u} - \mathbf{u}^{n} = \mathbf{G}^{n}(\mathbf{u}-\mathbf{u}^{0}) = \lambda_N^n \sum_{k=1}^{N}\left(\frac{\lambda_k}{\lambda_N}\right)^{n}\left\langle \mathbf{u}-\mathbf{u}^{0}, \frac{\mathbf{v}_k}{\|\mathbf{v}_k\|^2}\right\rangle \mathbf{v}_k$$

and, therefore,

$$\|\mathbf{u}-\mathbf{u}^{n}\| \le [\rho(\mathbf{G})]^{n}\,\|\mathbf{u}-\mathbf{u}^{0}\|.$$

Hence, the spectral radius $\rho(\mathbf{G})$ measures the speed of convergence when $\rho(\mathbf{G}) < 1$.

Exercise 5.3.3 *Show that the Jacobi method $\mathbf{u}^{n+1} = (\mathbf{I}-\mathbf{D}^{-1}\mathbf{A})\mathbf{u}^{n} + \mathbf{D}^{-1}\mathbf{b}$ converges whenever $\mathbf{A}$ is strictly diagonally dominant, that is, $\mathbf{u}^n \to \mathbf{u}$ if*

$$\|\mathbf{D}^{-1}(\mathbf{L}+\mathbf{U})\|_{\infty} := \max_{k}\frac{1}{|a_{kk}|}\sum_{j\neq k}|a_{kj}| < 1.$$

Exercise 5.3.4 *Let $\mathbf{u}^{n+1/2} = (\mathbf{I}-\mathbf{D}^{-1}\mathbf{A})\mathbf{u}^{n} + \mathbf{D}^{-1}\mathbf{b}$ denote the result of the Jacobi method. Show that the damped Jacobi method is*

$$\mathbf{u}^{n+1} = \mathbf{u}^{n} + \omega\left[\mathbf{u}^{n+1/2}-\mathbf{u}^{n}\right],$$

where ω is a damping factor.

Exercise 5.3.5 *Derive the Gauss–Seidel method using*

$$\mathbf{u} - \mathbf{A}\mathbf{u} = \mathbf{u} - \mathbf{b}.$$

The successive relaxation (SR) method is

$$(\mathbf{D} + \omega\mathbf{L})\,\mathbf{u}^{n+1} = \mathbf{b} + [(1-\omega)\,\mathbf{D} - \omega\mathbf{U}]\,\mathbf{u}^{n}.$$

The condition $0 < \omega < 1$ is called *under-relaxation*, whereas the condition $\omega > 1$ is called *over-relaxation*. The choice of ω for faster convergence depends on the structure of $\mathbf{A}$.

5.3.3 Iterative Method using Steepest Descent Minimization

Throughout this section, we assume that $\mathbf{A}$ is a real symmetric $N \times N$ matrix. We can iteratively solve $\mathbf{Au} = \mathbf{b}$ by searching for a solution of the following equivalent minimization problem:

$$\Phi(\mathbf{v}) = \tfrac{1}{2}\langle \mathbf{Av}, \mathbf{v}\rangle - \mathbf{b}\cdot\mathbf{v}.$$

Then, $\mathbf{Au} = \mathbf{b}$ if and only if

$$\Phi(\mathbf{v}) = \Phi(\mathbf{u}) + \tfrac{1}{2}\langle \mathbf{A}(\mathbf{v}-\mathbf{u}), (\mathbf{v}-\mathbf{u})\rangle \quad \left(\forall\ \mathbf{v} \in \mathbb{R}^N\right).$$

From the above equivalence relation, we have the following: if $\mathbf{A}$ is positive definite, then $\mathbf{Au} = \mathbf{b}$ if and only if $\mathbf{u}$ minimizes the energy functional $\Phi(\mathbf{v})$. The minimum value of Φ is $\Phi(\mathbf{A}^{-1}\mathbf{b}) = -\frac{1}{2}\mathbf{A}^{-1}\mathbf{b}\cdot\mathbf{b}$.

Exercise 5.3.6 *Consider*

$$\Phi(x,y) = \frac{1}{2}\left\langle \underbrace{\begin{pmatrix} 3 & 1 \\ 1 & 4 \end{pmatrix}}_{\mathbf{A}} \begin{pmatrix} x \\ y \end{pmatrix}, \begin{pmatrix} x \\ y \end{pmatrix} \right\rangle - \left\langle \begin{pmatrix} 7 \\ 8 \end{pmatrix}, \begin{pmatrix} x \\ y \end{pmatrix} \right\rangle.$$

Show that the minimizer of Φ in $\mathbb{R}^2$ is the solution of the linear system

$$\begin{pmatrix} 3 & 1 \\ 1 & 4 \end{pmatrix}\begin{pmatrix} x \\ y \end{pmatrix} = \begin{pmatrix} 7 \\ 8 \end{pmatrix}.$$

The steepest descent method to find $\mathbf{u} = \mathbf{A}^{-1}\mathbf{b}$ is to search for a minimum point of Φ by traveling in the steepest descent direction of Φ. It is important to note that the residual vector $\mathbf{r} = \mathbf{b} - \mathbf{Av}$ associated with any vector $\mathbf{v}$ points in the steepest descent direction of Φ. Taking an iterative step in the steepest descent direction leads us to approach the minimum point $\mathbf{u} = \mathbf{A}^{-1}\mathbf{b}$. The following updating steps describe the general procedure:

1. Start with an initial guess $\mathbf{u}_0$.
2. Compute the steepest direction $\mathbf{r}_0 = -\nabla\Phi(\mathbf{u}_0) = \mathbf{b} - \mathbf{A}\mathbf{u}_0$.
3. Search for the lowest point $\mathbf{u}_1 = \mathbf{u}_0 + \kappa_0\mathbf{r}_0$ along the line defined by $\mathbf{u}_0 + \kappa\mathbf{r}_0$, with κ variable. To be precise,

$$\kappa_0 := \arg\min_{\kappa} \Phi(\mathbf{u}_0 + \kappa\mathbf{r}_0) = \frac{\mathbf{r}_0\cdot\mathbf{r}_0}{\langle \mathbf{A}\mathbf{r}_0, \mathbf{r}_0\rangle}.$$

4. Compute a new search direction $\mathbf{r}_1 = -\nabla\Phi(\mathbf{u}_1)$.
5. Repeat the previous steps to get a minimizing sequence $\{\mathbf{u}_m\}$ in such a way that

$$\mathbf{u}_{m+1} = \mathbf{u}_m + \underbrace{\frac{\mathbf{r}_m \cdot \mathbf{r}_m}{\langle \mathbf{A}\mathbf{r}_m, \mathbf{r}_m \rangle}}_{\kappa_m} \underbrace{(-\nabla\Phi(\mathbf{u}_m))}_{\mathbf{r}_m := \mathbf{b} - \mathbf{A}\mathbf{u}_m}. \tag{5.16}$$

For the convergence analysis of the steepest descent method, note that

$$\mathbf{r}_m = -\nabla\Phi(\mathbf{u}_m) = -\mathbf{A}\mathbf{u}_m + \mathbf{b} = -\mathbf{A}(\mathbf{u}_m - \mathbf{u}).$$

Hence, (5.16) can be expressed as

$$\mathbf{u}_{m+1} = \mathbf{u}_m - \kappa_m \mathbf{A}\left(\mathbf{u}_m - \mathbf{u}\right). \tag{5.17}$$

Denoting $\mathbf{e}_m := \mathbf{u}_m - \mathbf{u}$, (5.16) can be written as

$$\mathbf{e}_{m+1} = \left(I - \kappa_m \mathbf{A}\right)\mathbf{e}_m = \left[\left(I - \kappa_m \mathbf{A}\right) \cdots \left(I - \kappa_0 \mathbf{A}\right)\right]\mathbf{e}_0. \tag{5.18}$$

For an intuitive understanding of the rate of convergence of the steepest descent method, we should note that the error $\mathbf{e}_m = \mathbf{u}_m - \mathbf{u}$ measures the difference in the domain space, whereas the residual $\mathbf{r}_m = \mathbf{A}(\mathbf{u} - \mathbf{u}_m) = -\nabla\Phi(\mathbf{u}_m)$, indicating the steepest descent direction, measures the difference in the range space.

We can get a convergence analysis using eigenvectors $\mathbf{v}_1, \ldots, \mathbf{v}_N$ of $\mathbf{A}$ (assumed to be an orthonormal basis in $\mathbb{R}^N$) with associated eigenvalues $\lambda_1, \ldots, \lambda_N$, ordered so that $|\lambda_1| \leq \cdots \leq |\lambda_N|$. Assume that

$$\mathbf{e}_m = \sum_j c_j^m \mathbf{v}_j.$$

From (5.18), we have

$$\mathbf{r}_m = -\sum_j \lambda_j c_j^m \mathbf{v}_j \quad \text{and} \quad \mathbf{e}_{m+1} = \sum_j \underbrace{c_j^m \beta_j^m}_{c_j^{m+1}} \mathbf{v}_j,$$

where

$$\beta_j^m = 1 - \lambda_j \kappa_m = 1 - \lambda_j \frac{\sum_k \left(c_k^m \lambda_k\right)^2}{\sum_k \left(c_k^m\right)^2 \left(\lambda_k\right)^3}.$$

Now, consider the following two special cases.

- In the special case of $\mathbf{e}_0 = \mathbf{v}_j$,

$$\mathbf{r}_0 = -\mathbf{A}\mathbf{e}_0 = -\lambda_j \mathbf{v}_j \quad \text{and} \quad \kappa_0 = \frac{1}{\lambda_j}.$$

  From (5.18), we have

$$\mathbf{e}_1 = \left(I - \kappa_0 \mathbf{A}\right)\mathbf{e}_0 = \left(1 - \kappa_0 \lambda_j\right)\mathbf{v}_j = 0.$$

  Therefore, $\mathbf{u}_1$ is the exact solution.
- If $\lambda_1 = \cdots = \lambda_N$, then $\beta_1^0 = 0$ and hence $\mathbf{u}_1$ is the exact solution.

The following exercise explains the rate of convergence of the steepest descent method.

Exercise 5.3.7 *Prove the identity*

$$\langle \mathbf{A}\mathbf{e}_1, \mathbf{e}_1 \rangle = \sum_j \left(c_j^0\right)^2 \lambda_j \left(\beta_j^0\right)^2 = \alpha \langle \mathbf{A}\mathbf{e}_0, \mathbf{e}_0 \rangle,$$

where

$$\alpha = 1 - \frac{\left(\sum_k \left(c_k^0\right)^2 \left(\lambda_k\right)^2\right)^2}{\left(\sum_k \left(c_k^0\right)^2 \left(\lambda_k\right)^3\right)\left(\sum_k \left(c_k^0\right)^2 \lambda_k\right)}.$$

Using the above identity, show that

$$|\langle \mathbf{A}\mathbf{e}_m, \mathbf{e}_m \rangle| \le |\alpha|^m \, |\langle \mathbf{A}\mathbf{e}_0, \mathbf{e}_0 \rangle|.$$

The above exercise explains that the steepest descent method can converge very quickly when $\alpha \approx 0$. Note that $\alpha \approx 0$ if either $\lambda_N/\lambda_1 \approx 1$ or $\mathbf{e}_0$ is close to an eigenvector.

However, the rate of convergence would be very slow when $|\lambda_N|/|\lambda_1|$ is large. In the case where $|\lambda_N|/|\lambda_1| \approx \infty$, it is possible that the steepest descent direction $\mathbf{r}_m$ from a given point $\mathbf{u}_m$ is very different from the direction pointing to the true solution $\mathbf{u}$. The worst case is when $\mathbf{e}_0 \approx \lambda_N \mathbf{v}_1 + \lambda_1 \mathbf{v}_n$, and this causes a sluggish performance of the steepest descent method in which the iterative scheme takes many short and inefficient switchbacks down to the valley floor. It may include steps with the same directions as those in earlier steps.

Remark 5.3.8 *We revisit the steepest descent method by considering an iterative method to solve the minimization problem:*

$$\min_{\mathbf{u}\in\mathbb{R}^N} \Phi(\mathbf{u}),$$

where Φ is a convex function. Given an initial approximation $\mathbf{u}^0 \in \mathbb{R}^N$ of the exact solution $\mathbf{u}$, we find successive approximations $\mathbf{u}^k \in \mathbb{R}^N$, $k = 1, 2, \ldots,$ of the form

$$\mathbf{u}^{k+1} = \mathbf{u}^k + \alpha_k \mathbf{d}^k, \quad k = 0, 1, 2, \ldots, \tag{5.19}$$

where $\mathbf{d} \in \mathbb{R}^N$ is a search direction and $\alpha_k > 0$ is a step length. By Taylor's theorem,

$$\Phi\left(\mathbf{u}^{k+1}\right) = \Phi\left(\mathbf{u}^k\right) + \alpha_k \nabla\Phi\left(\mathbf{u}^k\right)\cdot \mathbf{d}^k + \frac{\alpha_k^2}{2}\mathbf{d}^k \cdot \nabla\nabla\Phi\left(\mathbf{u}\right)\mathbf{d}^k,$$

where $\mathbf{u}$ lies on the line segment between $\mathbf{u}^k$ and $\mathbf{u}^{k+1}$. This implies that

$$\Phi\left(\mathbf{u}^{k+1}\right) = \Phi\left(\mathbf{u}^k\right) + \alpha_k \nabla\Phi\left(\mathbf{u}^k\right)\cdot \mathbf{d}^k + O\left(\alpha_k^2\right), \quad as \alpha_k \to 0.$$

If we choose $\mathbf{d}^k = -\nabla\Phi(\mathbf{u}^k) \neq 0$, then for a sufficiently small α_k we have

$$\nabla\Phi\left(\mathbf{u}^k\right)\cdot \mathbf{d}^k < 0 \quad and \quad \Phi\left(\mathbf{u}^{k+1}\right) < \Phi\left(\mathbf{u}^k\right).$$

To choose the step length α_k, we may determine α_k so that

$$\Phi\left(\mathbf{u}^k + \alpha_k \mathbf{d}^k\right) = \min_{\alpha\ge 0} \Phi\left(\mathbf{u}^k + \alpha \mathbf{d}^k\right),$$

in which case α_k must satisfy

$$\nabla\Phi\left(\mathbf{u}^k + \alpha_k \mathbf{d}^k\right)\cdot \mathbf{d}^k = 0. \tag{5.20}$$

5.3.4 Conjugate Gradient (CG) Method

To deal with inefficient switchbacks in finding $\mathbf{u} = \mathbf{A}^{-1}\mathbf{b}$ using the steepest descent method, the conjugate gradient (CG) method was proposed by choosing new search directions taking account of the progress in previous ones. Throughout this section, we assume that $\mathbf{A}$ is symmetric and positive definite. Two vectors $\mathbf{v}$ and $\mathbf{w}$ are conjugate with respect to $\mathbf{A}$ if they are $\mathbf{A}$-orthogonal:

$$\langle \mathbf{A}\mathbf{v}, \mathbf{w} \rangle = 0.$$

To understand the key idea of the CG method, we pick orthogonal search directions $\mathbf{d}_0, \mathbf{d}_1, \ldots, \mathbf{d}_{N-1}$ and choose

$$\mathbf{u}_{m+1} = \mathbf{u}_m + \underbrace{\frac{\mathbf{e}_m \cdot \mathbf{d}_m}{\|\mathbf{d}_m\|^2}}_{\kappa_m} \mathbf{d}_m.$$

Then, $\mathbf{e}_N = 0$ and $\mathbf{u}_N$ must be the exact solution of $\mathbf{A}\mathbf{u} = \mathbf{b}$. This procedure requires at most N steps to compute $\mathbf{u}$ provided that κ_m is computable. Unfortunately, we cannot compute $\mathbf{u}_N$ since the computation of κ_m requires the unknown quantity $\mathbf{u}$.

To deal with this problem, we use the fact that $\mathbf{A}\mathbf{e}_m$ is computable though $\mathbf{e}_m$ is not, and make the search direction $\mathbf{A}$-orthogonal instead of orthogonal. Suppose that $\mathbf{a}_1, \ldots, \mathbf{a}_{N-1}$ are $\mathbf{A}$-orthogonal:

$$\langle \mathbf{A}\mathbf{a}_i, \mathbf{a}_k \rangle = 0, \quad \forall i \neq k.$$

Our new requirement is that

$$\mathbf{u}_{m+1} = \mathbf{u}_m + \kappa_m \mathbf{a}_m \quad \text{s.t.} \quad \frac{\mathrm{d}}{\mathrm{d}\kappa} \Phi\left(\mathbf{u}_m + \kappa \mathbf{a}_m\right)|_{\kappa=\kappa_m} = 0.$$

It is crucial to observe that κ_m is computable as

$$\kappa_m = -\frac{\langle \mathbf{A}\mathbf{e}_m, \mathbf{a}_m \rangle}{\langle \mathbf{A}\mathbf{a}_m, \mathbf{a}_m \rangle} = \frac{\langle \mathbf{r}_m, \mathbf{a}_m \rangle}{\langle \mathbf{A}\mathbf{a}_m, \mathbf{a}_m \rangle}.$$

Now, we need to determine the $\mathbf{A}$-orthogonal basis set $\{\mathbf{a}_0, \ldots, \mathbf{a}_{N-1}\}$. It is not desirable to use the Gram–Schmidt method since the process is roughly equivalent to performing Gaussian elimination. In the CG method, the search directions are constructed by conjugation of the residuals. We choose the search direction of the vector form

$$\mathbf{p}_m = \mathbf{r}_{m-1} + \beta_{m-1}\mathbf{p}_{m-1} \quad \text{s.t.} \quad \langle \mathbf{A}\mathbf{p}_m, \mathbf{p}_{m-1} \rangle = 0,$$

and hence

$$\mathbf{u}_{m+1} = \mathbf{u}_m + \kappa_m \mathbf{p}_m, \quad \kappa_m = \frac{\mathbf{r}_m \cdot \mathbf{p}_m}{\|\mathbf{p}_m\|_A^2}.$$

This choice makes sense for many reasons.

1. Residual $\mathbf{r}_m = -\nabla\Phi(\mathbf{u}_m)$ and $\mathbf{p}_m = \mathbf{r}_m + \cdots$ contains the steepest descent direction at $\mathbf{u}_m$.
2. We have $\mathbf{r}_m \cdot \mathbf{p}_{m-1} = 0$ because of the choice of $\kappa_{m-1} = 0$, that is,

$$0 = \frac{\mathrm{d}}{\mathrm{d}\kappa} \Phi\left(\mathbf{u}_{m-1} + \kappa \mathbf{p}_{m-1}\right)|_{\kappa_{m-1}} = \mathbf{p}_{m-1} \cdot \nabla\Phi\left(\mathbf{u}_m\right).$$

3. We require that the minimization along the line $\mathbf{u}_m + \kappa \mathbf{p}_m$ does not undo the progress in searching the direction $\mathbf{p}_{m-1}$. Hence, we choose the parameter β_m so that $\mathbf{p}_{m+1}$ is $\mathbf{A}$-conjugate to $\mathbf{p}_m$:

$$\langle \mathbf{A} \underbrace{(\mathbf{r}_m + \beta_m \mathbf{p}^m)}_{\mathbf{p}_{m+1}}, \ \mathbf{p}_m \rangle = 0 \quad \Longrightarrow \quad \beta_m := -\frac{\langle \mathbf{A}\mathbf{r}^{m+1}, \mathbf{p}_m \rangle}{\langle \mathbf{A}\mathbf{p}_m, \mathbf{p}_m \rangle}.$$

4. Assume $\mathbf{u}_0 = 0$. Then,

$$\|\mathbf{e}_m\|_A^2 = \min_{\mathbf{v} \in \mathcal{K}_m} \langle \mathbf{A}(\mathbf{u} - \mathbf{v}), \mathbf{u} - \mathbf{v} \rangle,$$

where $\mathcal{K}_m = \text{span}\{\mathbf{p}_0, \mathbf{p}_1, \ldots, \mathbf{p}_{m-1}\}$ is called a Krylov subspace.

Now, we are ready to explain the basic CG algorithm. Let $\mathbf{A} \in \mathbb{R}^{N \times N}$ be symmetric and positive definite. The following algorithm solves $\mathbf{A}\mathbf{u} = \mathbf{b}$ starting with the initial guess $\mathbf{u}_0$:

1. $\mathbf{r}_0 := \mathbf{b} - \mathbf{A}\mathbf{u}_0$;
2. $\mathbf{p}_0 := \mathbf{r}_0$;
3. for $m = 1, 2, \ldots, n$,
   (a) $\kappa_{m-1} := \dfrac{\mathbf{r}_{m-1} \cdot \mathbf{p}_{m-1}}{\|\mathbf{p}_{m-1}\|_A^2}$,
   (b) $\mathbf{u}_m := \mathbf{u}_{m-1} + \kappa_{m-1}\mathbf{p}_{m-1}$,
   (c) $\mathbf{r}_m := \mathbf{r}_{m-1} + \kappa_{m-1}\mathbf{A}\mathbf{p}_{m-1}$,
   (d) $\mathbf{p}_m := \mathbf{r}_m + \dfrac{\langle \mathbf{A}\mathbf{r}^m, \mathbf{p}_{m-1} \rangle}{\langle \mathbf{A}\mathbf{p}_{m-1}, \mathbf{p}_{m-1} \rangle} \mathbf{p}_{m-1}$.

For detailed explanation for this method, see Allen *et al.* (1988). We may view the CG method as a method between the steepest descent and Newton's methods. Since each search direction $\mathbf{p}^m$ is $\mathbf{A}$-conjugate to all previous search directions, the CG algorithm requires at most N iterations to converge so that $\mathbf{u}_N$ must be the exact solution from the theoretical point of view. However, in practice, when N is large, we need to take account of numerical round-off errors, and it is possible that the exact solution is never obtained.

Since the error $\mathbf{e}_m = \mathbf{u} - \mathbf{u}_m$ satisfies

$$\|\mathbf{e}_m\|_A \le 2\|\mathbf{e}_0\|_A \left[\frac{\sqrt{\text{cond}(\mathbf{A})} - 1}{\sqrt{\text{cond}(\mathbf{A})} + 1} \right]^m,$$

the basic CG method may show poor performance for a large condition number cond(A). These observations motivated the use of a preconditioner. The idea is to replace

$$\mathbf{A}\mathbf{u} = \mathbf{b} \quad \Longrightarrow \quad \underbrace{\mathbf{B}^{-\mathrm{T}}\mathbf{A}\mathbf{B}^{-1}}_{:=\widetilde{\mathbf{A}}} \tilde{\mathbf{u}} = \underbrace{\mathbf{B}^{-\mathrm{T}}\mathbf{b}}_{:=\tilde{\mathbf{b}}} \quad \text{s.t. cond}(\widetilde{\mathbf{A}}) \ll \text{cond}(\mathbf{A}),$$

where $\mathbf{B}^{-\mathrm{T}}$ is the transpose of $\mathbf{B}^{-1}$. To be precise, define $G(\mathbf{x}) := \Phi(\mathbf{B}\mathbf{v})$ with $\mathbf{x} = \mathbf{B}\mathbf{v}$. Then, it is easy to see that

$$G(\mathbf{x}) = \tfrac{1}{2}\langle \mathbf{x}, \widetilde{A}\mathbf{x} \rangle - \tilde{\mathbf{b}} \cdot \mathbf{x}$$

and

$$\tilde{\mathbf{u}} = \arg\min G(\mathbf{x}) \quad \text{and} \quad \tilde{\mathbf{u}} = \mathbf{B}\mathbf{u},$$

which is equivalent to

$$\widetilde{\mathbf{A}}\tilde{\mathbf{u}} = \tilde{\mathbf{b}} \quad \text{and} \quad \tilde{\mathbf{u}} = \mathbf{B}\mathbf{u}.$$

Example 5.3.9 (Preconditioned conjugate gradient algorithm) *Let* $\mathbf{A} \in \mathbb{R}^{N\times N}$ *be a symmetric and positive definite matrix. Let* $\tau > 0$ *be a convergence tolerance on* $\|\mathbf{r}_m\| = \|\mathbf{b} - \mathbf{A}\mathbf{u}_m\|$*. The preconditioned CG algorithm is as follows:*

1. $\mathbf{r}_0 = \mathbf{b} - \mathbf{A}\mathbf{u}_0$;
2. $\mathbf{p}_0 := \mathbf{r}_0$;
3. *for* $m = 1, 2, \ldots,$
   (a) $\kappa_{m-1} := \dfrac{\mathbf{r}_{m-1}\cdot\mathbf{p}_{m-1}}{\|\mathbf{p}_{m-1}\|_A^2}$,
   (b) $\mathbf{u}_m := \mathbf{u}_{m-1} + \kappa_{m-1}\mathbf{p}_{m-1}$,
   (c) $\mathbf{r}_m := \mathbf{r}_{m-1} + \kappa_{m-1}\mathbf{A}\mathbf{p}_{m-1}$,
   (d) *if* $\|\mathbf{r}_m\| > \tau$, *then*
      i. *solve* $\mathbf{B}^{\mathrm{T}}\mathbf{B}\mathbf{y}_m = \mathbf{r}_m$ *for* $\mathbf{y}_m$,
      ii. $\mathbf{p}_m := \mathbf{y}^m + \dfrac{\langle \mathbf{A}\mathbf{y}^m, \mathbf{p}_{m-1}\rangle}{\langle \mathbf{A}\mathbf{p}_{m-1}, \mathbf{p}_{m-1}\rangle}\mathbf{p}_{m-1}$,
      iii. $m := m + 1$,
      iv. *go to 3(a).*

5.4 Finite Difference Method (FDM)

5.4.1 Poisson Equation

We examine numerical techniques to find a solution for the Poisson equation in a two-dimensional rectangular domain $\Omega = \{(x, y) : 0 < x < a, 0 < y < b\}$:

$$\begin{cases} \nabla^2 u = u_{xx} + u_{yy} = f, & \text{for } (x, y) \in \Omega, \\ u(x, y) = g(x, y), & \text{for } (x, y) \in \Omega. \end{cases} \tag{5.21}$$

The simplest way to solve (5.21) is to convert it to an equivalent system of difference equations using a finite difference approximation. Let $\Delta x = a/m$ and $\Delta y = b/n$ denote the step size in the x and y directions, respectively. Let

$$x_k = k\Delta x, k = 1, \ldots, m \quad \text{and} \quad y_j = j\Delta y, j = 1, \ldots, n.$$

Let $u_{k,j}$ denote the computed value of $u(x_k, y_j)$. To convert the Poisson equation $u_{xx} + u_{yy} = f$ into a difference equation, we use finite difference approximations for the second derivatives. The three-point central difference approximations for the second derivatives are

$$u_{xx}(x_k, y_j) = \frac{u_{k+1,j} - 2u_{k,j} + u_{k-1,j}}{\Delta x^2} + O(\Delta x^2),$$

$$u_{yy}(x_k, y_j) = \frac{u_{k,j+1} - 2u_{k,j} + u_{k,j-1}}{\Delta y^2} + O(\Delta y^2).$$

Ignoring the higher-order terms, we get the following system of difference equations for the interior grid points:

$$\begin{cases} \dfrac{u_{k+1,j} - 2u_{k,j} + u_{k-1,j}}{\Delta x^2} + \dfrac{u_{k,j+1} - 2u_{k,j} + u_{k,j-1}}{\Delta y^2} = f_{k,j}, \\ \qquad\qquad 1 \le k \le n-1,\ 1 \le j \le m-1, \\ u_{0,j} = g_{0,j}, \quad u_{m,j} = g_{m,j}, \quad u_{k,0} = g_{k,0}, \quad u_{k,n} = g_{k,n}, \end{cases} \tag{5.22}$$

where $f_{k,j} = f(x_k, y_j)$ and $g_{k,j} = g(x_k, y_j)$.

We often take the square grid as $\Delta x = \Delta y = h$. Then, we may conveniently write the structure of the five-point approximation in terms of the following stencil as

$$\begin{pmatrix} 0 & 1 & 0 \\ 1 & -4 & 1 \\ 0 & 1 & 0 \end{pmatrix} \begin{pmatrix} * & u_{k,j+1} & * \\ u_{k-1,j} & u_{k,j} & u_{k+1,j} \\ * & u_{k,j-1} & * \end{pmatrix} = h^2 f_{k,j}.$$

The problem of the system of difference equations can be written as the following linear system.

- Let $\mathbf{u} = [u_{1,1}, \ldots, u_{1,n-1}, u_{2,1}, \ldots, u_{m-1,n-1}]$.
- Let $\mathbf{A}$ be the $(m-1)(n-1) \times (m-1)(n-1)$ square matrix:

$$\mathbf{A} = \begin{pmatrix} \mathbf{B} & \mathbf{I} & & & \\ \mathbf{I} & \mathbf{B} & \mathbf{I} & & \\ & & \cdots & & \\ & & & \cdots & \\ & & & \mathbf{I} & \mathbf{B} \end{pmatrix},$$

where $\mathbf{I}$ is the $(n-1) \times (n-1)$ identity matrix and $\mathbf{B}$ is the $(n-1) \times (n-1)$ matrix

$$\mathbf{B} = \begin{pmatrix} -4 & 1 & & & \\ 1 & -4 & 1 & & \\ & & \cdots & & \\ & & & \cdots & \\ & & & 1 & -4 \end{pmatrix}.$$

- The finite difference system (5.22) can be written as

$$\mathbf{Au} = \mathbf{f} + \mathbf{BC}, \tag{5.23}$$

where $\mathbf{f} = [f_{1,1}, \ldots, f_{1,n-1}, f_{2,1}, \ldots, f_{m-1,n-1}]$ and $\mathbf{BC}$ is the corresponding boundary condition.

5.4.2 Elliptic Equation

Assume that Ω is a square region in $\mathbb{R}^2$. We will explain a discretized version of the elliptic equation:

$$\nabla \cdot (\sigma \nabla u(\mathbf{r})) = 0 \quad (\forall \mathbf{r} = (x, y) \in \Omega).$$

We divide Ω uniformly into $N \times N$ subsquares $\Omega_{i,j}$ with the center point (x_i, y_j), where $i, j = 0, \ldots, N-1$. We assume that the conductivity σ is constant on each subsquare $\Omega_{i,j}$, say $\sigma_{i,j}$:

$$\sigma_{i,j} = \sigma|_{\Omega_{i,j}} \quad (\forall i, j = 0, 1, \ldots, N-1).$$

The solution u can be approximated by a vector $\mathbf{u} = (u_0, u_1, \ldots, u_{N^2-1})$ such that each interior voltage u_k where $k = i + jN$ is determined by the weighted average (depending on the conductivity σ) of the four neighboring potentials. To be precise, the conductivity equation $\nabla \cdot (\sigma \nabla u(\mathbf{r})) = 0$ can be written in the following discretized form:

$$a_{k,k}u_k + [a_{k,k_N}u_{k_N} + a_{k,k_S}u_{k_S} + a_{k,k_E}u_{k_E} + a_{k,k_W}u_{k_W}] = 0, \tag{5.24}$$

with

$$a_{k,k} = -\sum_d a_{k,k_d} \quad \text{and} \quad a_{k,k_d} = \frac{\sigma_k \sigma_{k_d}}{\sigma_k + \sigma_{k_d}} \quad \text{for } d = \mathrm{N},\ \mathrm{S},\ \mathrm{E},\ \mathrm{W}, \tag{5.25}$$

where $k_\mathrm{N}, k_\mathrm{S}, k_\mathrm{E}, k_\mathrm{W}$ denote north, south, east and west neighboring points of the kth point. The discretized conductivity equation (5.24) with the Neumann boundary condition g can be written as a linear system $\mathbf{Au} = \mathbf{g}$, where $\mathbf{g}$ is the injection current vector associated with g.

5.5 Finite Element Method (FEM)

The finite element method (FEM) is a useful tool to compute an approximate numerical solution of a PDE (Johnson 2009). This section summarizes the FEM to find an approximate solution of a PDE.

5.5.1 *One-Dimensional Model*

We explain Galerkin's method for solving the following elliptic equation as an example:

$$\begin{cases} -\dfrac{\mathrm{d}}{\mathrm{d}x}\left(c(x)\dfrac{\mathrm{d}}{\mathrm{d}x}u(x)\right) = f(x), & 0 < x < 1, \\ u(0) = 0, \quad u'(1) = 0. \end{cases} \tag{5.26}$$

The weak form of (5.26) is

$$\int_0^1 c(x)\frac{\mathrm{d}}{\mathrm{d}x}u(x)\frac{\mathrm{d}}{\mathrm{d}x}\phi(x)\,\mathrm{d}x = \int_0^1 f(x)\phi(x)\,\mathrm{d}x, \tag{5.27}$$

where ϕ satisfies $\phi(0) = 0$. Galerkin's method discretizes the weak form (5.27). Let $h = 1/N$ and define V_h as follows:

$$V_h = \{\phi \in C[0, 1] : \phi(0) = 0, \phi|_{[ih,(i+1)h]} \text{is linear for} i = 0, \ldots, N-1\}.$$

As shown in Figure 5.1, we choose $\phi_1, \ldots, \phi_N \in V_h$ such that

$$\phi_i(jh) = \begin{cases} 1 & \text{if } i = j, \\ 0 & \text{otherwise} . \end{cases}$$

Then, the finite element space is $V_h = \mathrm{span}\{\phi_1, \ldots, \phi_N\}$.

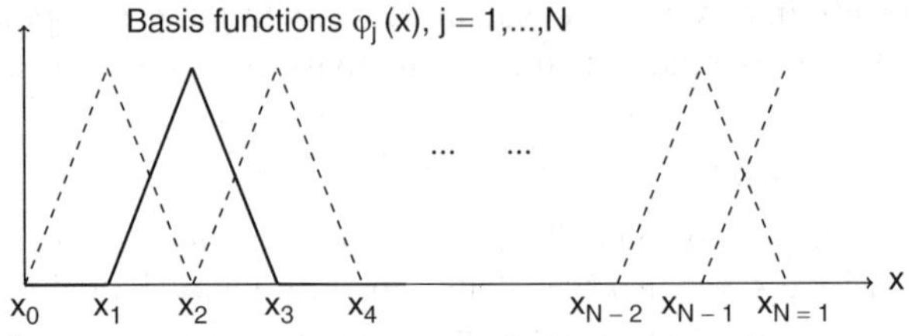

Figure 5.1 Basis functions

We look for an approximate solution

$$u_h = \sum_{i=1}^{N} u_i \phi_i. \tag{5.28}$$

Substituting $u_h = \sum_{i=1}^{N} u_i \phi_i$ into (5.27) yields

$$\int_0^1 c(x) \sum_{i=1}^{N} u_i \phi_i'(x) \phi_j'(x)\, \mathrm{d}x = \int_0^1 f(x)\phi_j(x)\, \mathrm{d}x, \quad j = 1, \ldots, N. \tag{5.29}$$

The identity (5.29) can be written as the following matrix form:

$$\mathbf{Au} = \mathbf{f}, \tag{5.30}$$

where $\mathbf{A}$ is an $N \times N$ matrix given by

$$\mathbf{A} = \begin{bmatrix} a_{11} & \cdots & a_{1N} \\ \vdots & \ddots & \vdots \\ a_{N1} & \cdots & a_{NN} \end{bmatrix} \quad \text{with} \quad a_{ij} = \int_0^1 c(x)\phi_i'(x)\phi_j'(x)\, \mathrm{d}x,$$

$$\mathbf{u} = \begin{bmatrix} u_1 \\ \vdots \\ u_N \end{bmatrix} \quad \text{and} \quad \mathbf{f} = \begin{bmatrix} \vdots \\ \int_0^1 f(x)\phi_j(x)\, \mathrm{d}x \\ \vdots \end{bmatrix}.$$

The matrix $\mathbf{A}$ is sparse:

- If $|i - j| > 1$, then $\int_0^1 c(x)\phi_i'(x)\phi_j'(x)\, \mathrm{d}x = 0$, and hence

$$\mathbf{A} = \begin{bmatrix} * & * & 0 & \cdots & & & 0 \\ * & * & * & & & & \\ 0 & * & * & * & & & \\ \vdots & & \ddots & \ddots & \ddots & & \\ & & & * & * & * & \\ & & & & * & * & * \\ 0 & \cdots & & & 0 & * & * \end{bmatrix},$$

where $*$ denotes a non-zero value.

- If $j - i = 1$, then $\int_0^1 c(x)\phi_i'(x)\phi_j'(x)\,\mathrm{d}x = -\frac{1}{h^2}\int_{ih}^{jh} c(x)\,\mathrm{d}x$.
- If $i = j \neq N$, then $\int_0^1 c(x)\phi_i'(x)\phi_j'(x)\,\mathrm{d}x = \frac{2}{h^2}\int_{(i-1)h}^{(i+1)h} c(x)\,\mathrm{d}x$.
- If $i = j = N$, then $\int_0^1 c(x)\phi_i'(x)\phi_j'(x)\,\mathrm{d}x = \frac{1}{h^2}\int_{(N-1)h}^{1} c(x)\,\mathrm{d}x$.

5.5.2 Two-Dimensional Model

Let Ω be a domain in the two-dimensional space $\mathbb{R}^2$ with smooth boundary $\partial\Omega$ and let σ be a positive bounded function in Ω. For a given $g \in H_\diamond^{-1/2}(\partial\Omega) := \{\phi \in H^{-1/2}(\partial\Omega) : \int_{\partial\Omega}\phi\,\mathrm{d}S = 0\}$, we consider the Neumann boundary value problem:

$$\begin{cases} \nabla\cdot(\sigma(\mathbf{x})\nabla u(\mathbf{x})) = 0 \quad \text{for } \mathbf{x}\in\Omega, \\ \sigma\dfrac{\partial u}{\partial \mathbf{n}}\Big|_{\partial\Omega} = g, \quad \displaystyle\int_{\partial\Omega} u\,\mathrm{d}S = 0. \end{cases} \tag{5.31}$$

We try to find an approximate solution of (5.31) within the set $H_\diamond^1(\Omega) = \{\phi \in H^1(\Omega) : \int_{\partial\Omega}\phi\,\mathrm{d}S = 0\}$. The FEM uses the weak formulation for the solution u:

$$\int_\Omega \sigma\nabla u\cdot\nabla\phi\,\mathrm{d}\mathbf{x} = \int_{\partial\Omega} g\phi\,\mathrm{d}S, \quad \forall\phi\in H_\diamond^1(\Omega). \tag{5.32}$$

We perform triangulation of the domain Ω by subdividing Ω into triangular subdomains $K_1, K_2, \ldots, K_M$ as in Figure 5.2. They are pairwise disjoint and no vertex of one triangle lies on the edge of another triangle within a typical triangular element. The corresponding approximate domain, denoted by Ω_h, is the domain whose closure is the closure of $K_1 \cup \cdots \cup K_M$.

Next, we can construct the finite element space $V \subset H_\diamond^1(\Omega_h)$:

$$V = \left\{v \in C(\Omega_h) : v|_{K_j} \text{ is linear for } j = 1, 2, \ldots, M\right\} \cap H_\diamond^1(\Omega_h).$$

Let $\{\mathbf{z}_j : j = 1, \ldots, N\}$ be the set of nodes of the triangular elements. For each k, we define the function $\varphi_k \in V$ by

$$\varphi_k(\mathbf{z}_j) = \begin{cases} 1 & \text{if } k = j, \\ 0 & \text{otherwise.} \end{cases}$$

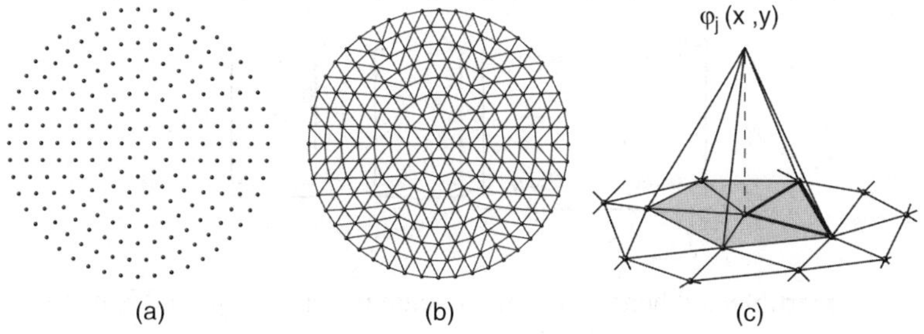

Figure 5.2 (a) Node points, (b) finite element triangulation and (c) basis function ϕ_j

Then, it is easy to see that

$$V = \text{span}\{\varphi_1, \varphi_2, \ldots, \varphi_N\} \cap H^1_\diamond(\Omega_h) \quad \text{and} \quad \dim V = N - 1.$$

According to the Lax–Milgram theorem, there exists a unique $u \in V$ that solves

$$\int_{\Omega_h} \sigma \nabla u \cdot \nabla \varphi_j \, \mathrm{d}\mathbf{x} = \int_{\partial\Omega_h} g \varphi_j \, \mathrm{d}S, \quad \forall\, j = 1, \ldots, N. \tag{5.33}$$

Substituting $u = \sum_{i=1}^N u_i \varphi_i$ (with u_i being unknown constants) into (5.33) leads to the linear system

$$\underbrace{\begin{bmatrix} a_{11} & \ldots & a_{1N} \\ \vdots & \ddots & \vdots \\ a_{N1} & \ldots & a_{NN} \end{bmatrix}}_{\mathbf{A}} \underbrace{\begin{bmatrix} u_1 \\ \vdots \\ u_N \end{bmatrix}}_{\mathbf{u}} = \underbrace{\begin{bmatrix} b_1 \\ \vdots \\ b_N \end{bmatrix}}_{\mathbf{b}}, \tag{5.34}$$

where

$$a_{ij} = \int_{\Omega_h} \sigma \nabla \varphi_i \cdot \nabla \varphi_j \, \mathrm{d}\mathbf{x} \quad \text{and} \quad b_j = \int_{\partial\Omega_h} g \varphi_j \, \mathrm{d}S.$$

The Lax–Milgram theorem provides that the system has a unique solution $\mathbf{u} = (u_1, \ldots, u_N)^{\mathrm{T}}$ satisfying

$$\mathbf{A}\mathbf{u} = \mathbf{b} \quad \text{and} \quad \sum_{k=1}^N u_k \int_{\partial\Omega_h x} \varphi_k \, \mathrm{d}S = 0.$$

Exercise 5.5.1 *Consider the following.*

1. *Show that, if u is the solution of*

$$\begin{cases} \nabla \cdot (\sigma(\mathbf{x}) \nabla u(\mathbf{x})) = 0 \quad \text{for } \mathbf{x} \in \Omega, \\ \sigma \dfrac{\partial u}{\partial \mathbf{n}}\Big|_{\partial\Omega} = g, \end{cases} \tag{5.35}$$

   then $u + c$ is also a solution of (5.35) for any constant c. Hence, $\mathbf{A}(\mathbf{u} + \mathbf{c}) = \mathbf{b}$ for any constant vector $\mathbf{c}$.
2. *Show that $\mathbf{A}$ is not invertible and that the rank of $\mathbf{A}$ is $N - 1$.*

Since the rank of $\mathbf{A}$ is $N - 1$, we drop the first row and column of $\mathbf{A}$ and consider

$$\underbrace{\begin{bmatrix} a_{22} & \ldots & a_{2N} \\ \vdots & \ddots & \vdots \\ a_{N2} & \ldots & a_{NN} \end{bmatrix}}_{\mathbf{A}_{N-1}} \underbrace{\begin{bmatrix} \tilde{u}_2 \\ \vdots \\ \tilde{u}_N \end{bmatrix}}_{\tilde{\mathbf{u}}} = \underbrace{\begin{bmatrix} b_2 \\ \vdots \\ b_N \end{bmatrix}}_{\tilde{\mathbf{b}}}. \tag{5.36}$$

Now, $\mathbf{A}_{N-1}$ is invertible and hence we can compute $\tilde{\mathbf{u}} = [\mathbf{A}_{N-1}]^{-1}\tilde{\mathbf{b}}$. Using the knowledge of $\tilde{\mathbf{u}}$, we can obtain the solution $u = \sum u_j \varphi_j$ of (5.33) by

$$u = \sum_{j=2}^{N} \tilde{u}_j \varphi_j - c = -c\varphi_1 + \sum_{j=2}^{N} (\tilde{u}_j - c)\varphi_j,$$

where c is a constant chosen so that $\int_{\partial\Omega_h} u \, \mathrm{d}S = 0$.

5.5.2.1 Computation of Element Matrix

Let $\mathcal{T} = \{K_1, \ldots, K_M\}$ be the set of the triangular elements. The ijth element of the matrix $\mathbf{A}$ can be decomposed into

$$a_{ij} = \sum_{K \in \mathcal{T}} \underbrace{\int_K \sigma \nabla\varphi_i \cdot \nabla\varphi_j \mathbf{dx}}_{a_{ij}^K}.$$

Fixing the triangle K, we let the nodes of K be $\mathbf{z}^1, \mathbf{z}^2$ and $\mathbf{z}^3$. Let $\varphi_1(\mathbf{x})$ be a linear function on K with the value 1 at the node $\mathbf{z}^1$ and 0 at the other nodes. We will derive formulas for φ_1 and $\nabla\varphi_1$ in terms of $\mathbf{z}^1, \mathbf{z}^2$ and $\mathbf{z}^3$. The function φ_1 in K can be written in the form

$$\varphi_1(\mathbf{x}) = \alpha_1 + \alpha_2 x_1 + \alpha_3 x_2 \quad \text{on } K$$

where $\alpha_1, \alpha_2, \alpha_3$ satisfy

$$\begin{pmatrix} 1 \\ 0 \\ 0 \end{pmatrix} = \begin{pmatrix} 1 & z_1^1 & z_2^1 \\ 1 & z_1^2 & z_2^2 \\ 1 & z_1^3 & z_2^3 \end{pmatrix} \begin{pmatrix} \alpha_1 \\ \alpha_2 \\ \alpha_3 \end{pmatrix}.$$

Therefore, the coefficients $\alpha_1, \alpha_2, \alpha_3$ can be found from

$$\begin{pmatrix} \alpha_1 \\ \alpha_2 \\ \alpha_3 \end{pmatrix} = \begin{pmatrix} 1 & z_1^1 & z_2^1 \\ 1 & z_1^2 & z_2^2 \\ 1 & z_1^3 & z_2^3 \end{pmatrix}^{-1} \begin{pmatrix} 1 \\ 0 \\ 0 \end{pmatrix}.$$

Since we can express φ_1 as

$$\varphi_1(\mathbf{x}) = \left(1, x_1, x_2\right) \begin{pmatrix} 1 & z_1^1 & z_2^1 \\ 1 & z_1^2 & z_2^2 \\ 1 & z_1^3 & z_2^3 \end{pmatrix}^{-1} \begin{pmatrix} 1 \\ 0 \\ 0 \end{pmatrix} \quad \text{for } \mathbf{x} \in K,$$

$$\varphi_1(\mathbf{x}) = \frac{1}{2|K|} \left(\mathbf{z}^2 \times \mathbf{z}^3 - \mathbf{x} \times \mathbf{z}^3 - \mathbf{z}^2 \times \mathbf{x}\right), \quad \forall \mathbf{x} \in K,$$

where

$$\mathbf{z} \times \mathbf{x} = \begin{vmatrix} z_1 & z_2 \\ x_1 & x_2 \end{vmatrix} = z_1 x_2 - z_2 x_1$$

and $|K| = \frac{1}{2}|(\mathbf{z}^2 - \mathbf{z}^1) \times (\mathbf{z}^3 - \mathbf{z}^1)|$ is the area of the triangle. Note that $\varphi_1(\mathbf{x})$ is linear and $\varphi_1(\mathbf{z}^2) = 0 = \varphi_1(\mathbf{z}^3)$.

Similarly, we have

$$\varphi_2(\mathbf{x}) = \frac{1}{2|K|}\left(\mathbf{z}^3 \times \mathbf{z}^1 - \mathbf{x} \times \mathbf{z}^1 - \mathbf{z}^3 \times \mathbf{x}\right),$$

$$\varphi_3(\mathbf{x}) = \frac{1}{2|K|}\left(\mathbf{z}^1 \times \mathbf{z}^2 - \mathbf{x} \times \mathbf{z}^2 - \mathbf{z}^1 \times \mathbf{x}\right).$$

Simple computation yields

$$\nabla\varphi_1(\mathbf{x}) = \frac{1}{2|K|}\left(\mathbf{z}^2 - \mathbf{z}^3\right)^{\perp},$$

$$\nabla\varphi_2(\mathbf{x}) = \frac{1}{2|K|}\left(\mathbf{z}^3 - \mathbf{z}^1\right)^{\perp},$$

$$\nabla\varphi_3(\mathbf{x}) = \frac{1}{2|K|}\left(\mathbf{z}^1 - \mathbf{z}^2\right)^{\perp},$$

where $\mathbf{z}^{\perp} = (z_2, -z_1)$. If $\sigma = \sigma_K$ is a constant in K, then

$$a_{12}^K = \int_K \sigma\nabla\varphi_1 \cdot \nabla\varphi_2 \, \mathrm{d}\mathbf{x} = \frac{1}{4|K|^2}\left(\mathbf{z}^2 - \mathbf{z}^3\right) \cdot \left(\mathbf{z}^3 - \mathbf{z}^1\right)\sigma_K.$$

Hence, the element matrix for the triangular element K is

$$B_K := \begin{pmatrix} a_{11}^K & a_{12}^K & a_{13}^K \\ a_{21}^K & a_{22}^K & a_{23}^K \\ a_{31}^K & a_{32}^K & a_{33}^K \end{pmatrix}$$

$$= \frac{\sigma_K}{4|K|^2}\begin{bmatrix} |\mathbf{z}^2 - \mathbf{z}^3|^2 & (\mathbf{z}^2 - \mathbf{z}^3)\cdot(\mathbf{z}^3 - \mathbf{z}^1) & (\mathbf{z}^2 - \mathbf{z}^3)\cdot(\mathbf{z}^1 - \mathbf{z}^2) \\ (\mathbf{z}^2 - \mathbf{z}^3)\cdot(\mathbf{z}^3 - \mathbf{z}^1) & |\mathbf{z}^3 - \mathbf{z}^1|^2 & (\mathbf{z}^3 - z^1)\cdot(\mathbf{z}^1 - \mathbf{z}^2) \\ (\mathbf{z}^2 - \mathbf{z}^3)\cdot(\mathbf{z}^1 - \mathbf{z}^2) & (\mathbf{z}^3 - \mathbf{z}^1)\cdot(\mathbf{z}^1 - \mathbf{z}^2) & |\mathbf{z}^1 - \mathbf{z}^2|^2 \end{bmatrix}.$$

5.5.2.2 Assembly of Two Element Matrices

We can combine two element matrices from two connected triangles sharing two nodes together. Globally numbering all four nodes, we may express each element matrix as

$$B_{K_1} = \begin{bmatrix} & & & 0 \\ & B_{K_1} & & 0 \\ & & & 0 \\ 0 & 0 & 0 & 0 \end{bmatrix} \quad \text{and} \quad B_{K_2} = \begin{bmatrix} 0 & 0 & 0 & 0 \\ 0 & & & \\ 0 & & B_{K_2} & \\ 0 & & & \end{bmatrix}.$$

Assembly of these two element matrices is straightforward as

$$\widetilde{B}_{K_1 \cup K_2} = B_{K_1} + B_{K_2}.$$

5.5.2.3 Assembly of Master Matrix

We first globally number all nodes. Assembly of all element matrices into a master matrix is more complicated because we must take care of connected or disjoint elements. Since the ijth element of the matrix $\mathbf{A}$ is

$$a_{ij} = \sum_{K \in \mathcal{T}} \int_K \sigma \nabla \varphi_i \cdot \nabla \varphi_j \, \mathrm{d}\mathbf{x},$$

assembly of all element matrices is required to form the global master matrix $\mathbf{A}$. An efficient way of assembling the global master matrix proceeds recursively, building up the finite element representation of one element at a time, as follows.

1. Set the $N \times N$ matrix

$$\mathbf{A}^0 = \begin{bmatrix} 0 & \dots & 0 \\ \vdots & \ddots & \vdots \\ 0 & \dots & 0 \end{bmatrix}.$$

   Here, N is the number of nodes in the finite element model or mesh.

2. For $m = 1, \dots, M$ (M is the number of elements),

$$\begin{aligned} a_{ij}^m &= ij\text{th element of } \mathbf{A}^m \\ &= a_{ij}^{m-1} + b_{rs}^m, \end{aligned}$$

   where $i = \max(g_r, g_s)$ and $j = \min(g_r, g_s)$. Here r and s are local node numbers of the element K^m, and g_r and g_s are global node numbers corresponding to r and s, respectively. We can find values of b_{rs}^m from

$$\mathbf{B}_{K^m} = \begin{bmatrix} b_{11}^m & b_{12}^m & b_{13}^m \\ b_{21}^m & b_{22}^m & b_{23}^m \\ b_{31}^m & b_{32}^m & b_{33}^m \end{bmatrix}.$$

   This procedure generates a lower triangular matrix

$$\mathbf{A}^M = \begin{bmatrix} a_{11}^M & & & 0 \\ a_{21}^M & \ddots & & \\ \vdots & & \ddots & \\ a_{N1}^M & \dots & \dots & a_{NN}^M \end{bmatrix}.$$

Since our master matrix $\mathbf{A}$ is symmetric and $a_{ij}^M = a_{ij}$ when $i \geq j$,

$$\mathbf{A} = \begin{bmatrix} a_{11}^M & a_{21}^M & \dots & a_{N1}^M \\ a_{21}^M & \ddots & & \vdots \\ \vdots & & \ddots & \vdots \\ a_{N1}^M & \dots & \dots & a_{NN}^M \end{bmatrix}.$$

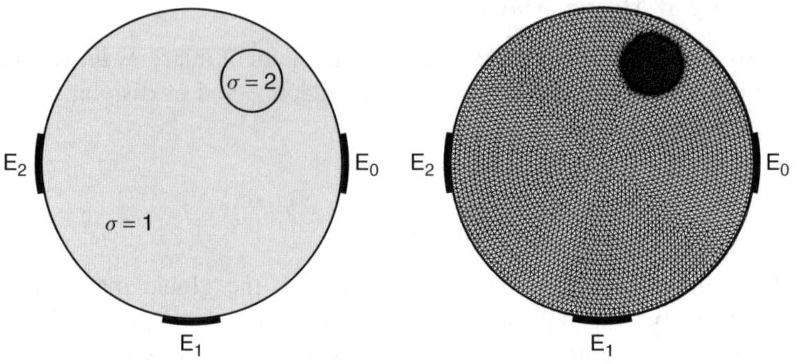

Figure 5.3 Conductivity distribution and mesh

5.5.2.4 Boundary Condition

Since $\mathbf{A}$ is a singular matrix, it is not invertible and we cannot solve the linear system of equations $\mathbf{Au} = \mathbf{b}$. Since the rank of $\mathbf{A}$ is $N - 1$, we set a reference node and modify the master matrix so that the modified matrix is not singular, as follows.

1. For a chosen reference node, set $u_1 = 0$.
2. Modify the matrix and vector as

$$\mathbf{A}_{N-1} = \begin{bmatrix} 1 & 0 & \dots & 0 \\ 0 & a_{22} & \dots & a_{2N} \\ \vdots & \vdots & \ddots & \vdots \\ 0 & a_{N2} & \dots & a_{NN} \end{bmatrix} \quad \text{and} \quad \tilde{\mathbf{b}} = \begin{bmatrix} 0 \\ b_2 \\ \vdots \\ b_N \end{bmatrix}.$$

3. Modify the linear system of equations as

$$\mathbf{A}_{N-1}\tilde{\mathbf{u}} = \tilde{\mathbf{b}},$$

where vectors $\tilde{\mathbf{u}}$ and $\tilde{\mathbf{b}}$ are regarded as the modified potential and current, respectively.

5.5.2.5 Solution of Linear System of Equations

The matrix $\mathbf{A}_{N-1}$ is symmetric, positive definite and sparse. Hence, we can solve $\mathbf{A}_{N-1}\tilde{\mathbf{u}} = \tilde{\mathbf{b}}$ using various techniques introduced in the previous sections.

5.5.3 Numerical Examples

5.5.3.1 Elliptic PDE

Let Ω be a unit disk with the conductivity distribution σ given in Figure 5.3. We attach three electrodes $\mathcal{E}_0, \mathcal{E}_1, \mathcal{E}_2$ as shown in Figure 5.3. If we apply voltage V_0 between a pair

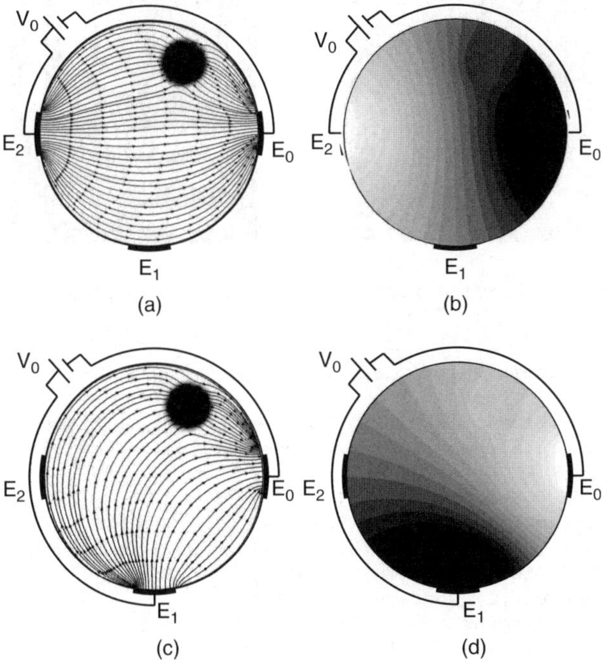

Figure 5.4 (a, c) Current flux and (b, d) equipotential lines

of the electrodes $\mathcal{E}_j$ and $\mathcal{E}_0$, then the induced potential is dictated approximately by the following mixed boundary value problem:

$$\begin{cases} \nabla \cdot (\sigma \nabla u) = 0 & \text{in } \Omega, \\ u|_{\mathcal{E}_0} = V_0, \quad u|_{\mathcal{E}_j} = 0, \\ \sigma \dfrac{\partial u}{\partial \mathbf{n}} = 0, & \partial\Omega \backslash \mathcal{E}_j \cup \mathcal{E}_0, \end{cases} \tag{5.37}$$

where σ is the conductivity distribution. Figure 5.4 shows the images of the current flux and equipotential lines for the solution of (5.37). The following suggests an example code for solving (5.37).

```
clc; clear; close all
h = 1/32; Node = [0 0];
for cir = 1:1/h
Node = [Node;cir*h*cos((0:round(2*pi*cir)-1)*2*pi/round(2*pi*cir))'
...cir*h*sin((0:round(2*pi*cir)-1)*2*pi/round(2*pi*cir))']; end
Element = delaunayn(Node); Sol_u = zeros(length(Node),1);
Sen = zeros(length(Node)); in =length(Node)-ceil(1*round(2*pi*cir/16)):
...length(Node)-ceil(0*round(2*pi*cir/16)); Sol_u(in) = 1;
out = length(Node)-floor(9*round(2*pi*cir/16)):length(Node)-
...floor(8*round(2*pi*cir/16)); Sol_u(out) = -1;
B = [in out]; I = [];
for ele = 1:length(Element)
    Cof=[Node(Element(ele,:),:) ones(3,1)]eye(3);sigma = 1;
```

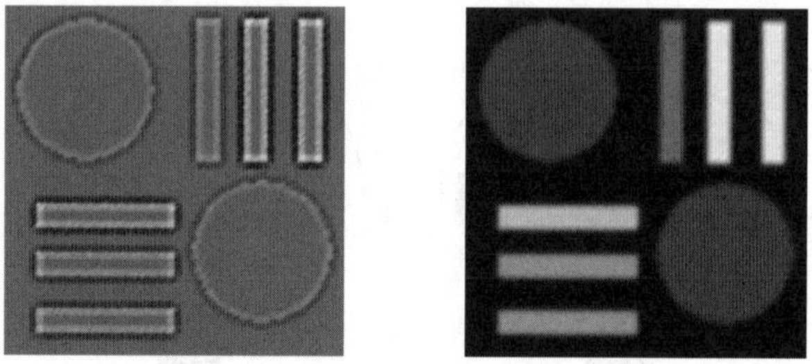

Figure 5.5 Image of ρ and solution of the Poisson equation in (5.38)

```
    if norm(mean(Node(Element(ele,:),:))-[0.5 0.5])<0.2+10^(-2)
        sigma = 2;      end
Sen(Element(ele,:),Element(ele,:)) = Sen(Element(ele,:),Element(ele,:))+sigma*
...abs(1/2*det([Node(Element(ele,:),:) ones(3,1)]))*(Cof(1:2,:)'*Cof(1:2,:));
end
Sen(B,:) = []; RHS =-Sen*Sol_u; Sen(:,B) = [];
Sol_u(setdiff(1:length(Node), find(Sol_u =0))) = Sen\RHS;
figure; plot(Node(:,1),Node(:,2),'k.'); axis equal; axis off;
 hold on; plot(Node(in,1),Node(in,2),'b*');
hold on; plot(Node(out,1),Node(out,2),'r*');
hold on; plot(Node(length(Node)-ceil(9*round(2*pi*cir/16)):length(Node)
...-ceil(8*round(8*pi*cir/16)),1),Node(length(Node)-
ceil(9*round(2*pi*cir/16))
...:length(Node)-ceil(8*round(8*pi*cir/16)),2),'ms');
figure,triplot(Element, Node(:,1), Node(:,2),'black'); axis equal;axis off;
figure, trimesh(Element, Node(:,1), Node(:,2), Sol_u);
```

5.5.3.2 Poisson Equation

We consider the Poisson equation

$$\begin{cases} -\nabla^2 u = \rho & \text{in } \Omega = (0,1) \times (0,1), \\ u|_{\partial\Omega} = 1. \end{cases} \tag{5.38}$$

Figure 5.5 shows the image of ρ in (5.38) and the solution u. The following also suggests an example code to solve (5.38).

```
[X Y] = meshgrid(0:0.005:0.5,0:0.005:0.5); Node = [X(:) Y(:)];
Element = delaunayn(Node); U = ones(length (Node),1); App_U = U;
Sen = sparse(length(Node), length(Node));
U(sum((Node-ones(size(U))*[0.12 0.38])'.^2)<0.1^2+0.1^4
...|sum((Node-ones(size(U))*[0.380.15])'.^2)<0.1^2+0.1^4) = 4;
U((abs(Node(:,1)-0.15)<0.1+0.1^4&abs(Node(:,2)-0.05)<0.015+0.1^4)
...|(abs(Node(:,1)-0.15)<0.1+0.1^4&abs(Node(:,2)-0.13)<0.015+0.1^4)) = 8;
U((abs(Node(:,1)-0.15)<0.1+0.1^4&abs(Node(:,2)-0.2)<0.015+0.1^4)) = 10;
U((abs(Node(:,1)-0.45)<0.015+0.1^4&abs(Node(:,2)-0.38)<0.1+0.1^4)
...|(abs(Node(:,1)-0.37)<0.015+0.1^4&abs(Node(:,2)-0.38)<0.1+0.1^4)) = 12;
U((abs(Node(:,1)-0.3)<0.015+0.1^4&abs(Node(:,2)-0.38)<0.1+0.1^4)) = 6;
On=zeros(length(Node),10);
index=ones(length(Node),1);
for ele=1:length(Element)
```

```
    p= Element(ele,:);
    On(p(1),index(p(1)))=ele;index(p(1))=index(p(1))+1;
    On(p(2),index(p(2)))=ele;index(p(2))=index(p(2))+1;
    On(p(3),index(p(3)))=ele;index(p(3))=index(p(3))+1;
end
grad_u=gradp(Node,Element,U);grad=ele2node(On,grad_u);clear grad_u;
...grad_ux=gradp(Node,Element,grad(:,1));gradx=ele2node(On,grad_ux);
grad_uy=gradp(Node,Element, grad(:,2)); grady=ele2node(On,grad_uy);
f=-(gradx(:,1) + grady(:,2));RHS=zeros(size(U)); figure;
surfc(X,Y,reshape(f,size(X,1),size(X,2)),'FaceColor','interp',
...'EdgeColor','none');view(2);colormap gray; axis equal; axis off;
for ele = 1:length(Element)
Cof = [Node(Element(ele,:),:) ones(3,1)]\eye(3);
Sen(Element(ele,:),Element(ele,:)) = Sen(Element(ele,:),Element(ele,:))+abs
...(1/2*det([Node(Element(ele,:),:) ones(3,1)]))*(Cof(1:2,:)'*Cof(1:2,:));
RHS(Element(ele,:)) = RHS(Element(ele,:))+f(Element(ele,:))
...*abs(1/factorial(2)*det([Node(Element(ele,:),:) ones(3,1)]))/3;
end
B = find((Node(:,1)-0.25)>0.246 & (Node(:,2)-0.25)>0.246); clear Node
Sen(B,:) = [];RHS(B) = [];RHS=RHS-Sen*App_U; Sen(:,B) = [];
App_U(setdiff(1:length(X(:)),B)) = Sen\RHS;
figure;surfc(X,Y,reshape(App_U,size(X,1),size(X,2)),'FaceColor',
...'interp','EdgeColor','none');view(2); colormap gray; axis equal; axis off;
function grad_u=gradp(Node,ele_con,fwd_potential)
%%%how to get gradient potential~~~
num=length(ele_con);
for ele = 1:num
        vertex=ele_con(ele,:);%%%four vertex of tetrahedron
        Jacobian(1,:) = Node(vertex(1),:)-Node(vertex(3),:);
        Jacobian(2,:) = Node(vertex(2),:)-Node(vertex(3),:);
        %%%%Jacobian is the Jacobian Matrix
        %%%%pd=potential difference
        PD(1)=fwd_potential(vertex(1))-fwd_potential(vertex(3));
        PD(2)=fwd_potential(vertex(2))-fwd_potential(vertex(3));
        grad_u(ele,:) = Jacobian\PD';
end
function grad=ele2node(On,gradp)
%%  this function is just for 2D,
% transfer the gradient in each element to gradient in each Node
grad=zeros(size(On,1),2);
for i=1:size(On,1)
    if length(find(On(i,:)>0))==1
       grad(i,:)=gradp(On(i,find(On(i,:)>0)),:);
    else
    grad(i,:)=mean(gradp(On(i,find(On(i,:)>0)),:));
    end
end
```

References

Allen MB, Herrera I and Pinder GF 1988 *Numerical Modeling in Science and Engineering*. John Wiley & Sons, Inc., New York.

Haberman R 1987 *Elementary Applied PDEs*, 2nd edn. Prentice-Hall, Englewood Cliffs, NJ.

Johnson C 2009 *Numerical Solution of Partial Differential Equations by the Finite Element Method*. Reprint of 1987 edition. Dover, Mineola, NY.

Lax PD and Richtmyer RD 1956 Survey of the stability of linear finite difference equations. *Commun. Pure Appl. Math*. **9**, 267–293.

Further Reading

Axelsson O and Barker VA 1984 *Finite Element Solution of Boundary Value Problems: Theory and Computation*. Academic Press, Orlando, FL.

Bank RE 1998 *PLTMG: A Software Package for Solving Elliptic Partial Differential Equations. Users' Guide 8.0*. Software, Environments, Tools, vol. 5. SIAM, Philadelphia, PA.

Braess D 2007 *Finite Elements: Theory, Fast Solvers, and Applications in Elasticity Theory*, 3rd edn. Translated from German by LL Schumaker. Cambridge University Press, Cambridge.

Brenner SC and Scott LR 2008 *The Mathematical Theory of Finite Element Methods*, 3rd edn. Texts in Applied Mathematics, vol. 15. Springer, New York.

Brezzi F and Fortin M 1991 *Mixed and Hybrid Finite Element Methods*. Springer Series in Computational Mathematics, vol. 15. Springer, New York.

Brezzi F, Marini LD, Markowich P and Pietra P 1991 On some numerical problems in semiconductor device simulation. In *Mathematical Aspects of Fluid and Plasma Dynamics* (eds G Toscani, V Boffi and S Rionero). Lecture Notes in Mathematics, vol. 1460, pp. 31–42. Springer, Berlin.

Calhoun D and LeVeque RJ 2005 An accuracy study of mesh refinement on mapped grids. In *Adaptive Mesh Refinement – Theory and Applications*. Lecture Notes in Computer Science and Engineering, no. 41, pp. 91–101. Springer, Berlin.

Carstensen C and Klose R 2003 A posteriori finite element error control for the p-Laplace problem. *SIAM J. Sci. Comput*. **25**(3), 792–814.

Carstensen C, Liu W and Yan N 2006 A posteriori error estimates for finite element approximation of parabolic p-Laplacian. *SIAM J. Numer. Anal*. **43**(6), 2294–2319.

Ciarlet PG 1980 *Metod Konechnykh Elementov dlya Ellipticheskikh Zadach*. Translated from English by BI Kvasov. Mir, Moscow.

Fehrenbach J, Gournay F, Pierre C and Plouraboue F 2012 The generalized Graetz problem in finite domains. *SIAM J. Appl. Math*. **72**(1), 99–123.

Gockenbach MS 2006 *Understanding and Implementing the Finite Element Method*. SIAM, Philadelphia, PA.

Higham DJ and Higham NJ 2005 *MATLAB Guide*, 2nd edn. SIAM, Philadelphia, PA.

Iserles A 1996 *A First Course in the Numerical Analysis of Differential Equations*. Cambridge Texts in Applied Mathematics. Cambridge University Press, Cambridge.

LeVeque RJ 2002 *Finite Volume Methods for Hyperbolic Problems*. Cambridge Texts in Applied Mathematics. Cambridge University Press, Cambridge.

LeVeque RJ 2007 *Finite Difference Methods for Ordinary and Partial Differential Equations: Steady-State and Time-Dependent Problems*. SIAM, Philadelphia, PA.

Moler CB 2004 *Numerical Computing with MATLAB*. SIAM, Philadelphia, PA.

Quarteroni A and Valli A 1994 *Numerical Approximation of Partial Differential Equations*. Springer Series in Computational Mathematics, vol. 23. Springer, Berlin.

Quarteroni A and Saleri F 2003 *Scientific Computing with MATLAB*. Texts in Computational Science and Engineering, vol. 2. Springer, Berlin.

Richtmyer RD and Morton KW 1967 *Difference Methods for Initial-Value Problems*, 2nd edn. Interscience Tracts in Pure and Applied Mathematics, no. 4. Interscience, New York.

Strang G and Fix G 2008 *An Analysis of the Finite Element Method*, 2nd edn. Wellesley-Cambridge, Wellesley, MA.

Strikwerda JC 1989 *Finite Difference Schemes and Partial Differential Equations*. Wadsworth & Brooks/Cole Mathematics Series. Wadsworth & Brooks/Cole, Pacific Grove, CA.

Thomas JW 1995 *Numerical Partial Differential Equations: Finite Difference Methods*. Texts in Applied Mathematics, vol. 22. Springer, New York.

Trefethen LN *Spectral Methods in MATLAB* Software, Environments, Tools, vol. 10. SIAM, Philadelphia, PA.

6

CT, MRI and Image Processing Problems

Image processing has become one of the most important components in medical imaging modalities such as magnetic resonance imaging, computed tomography, ultrasound and other functional imaging modalities. Image processing techniques such as image restoration and sparse sensing are being used to deal with various imperfections in the data acquisition processes of the imaging modalities. Image segmentation, referring to the process of partitioning an image into multiple segments, has numerous applications, including tumor detection, quantification of tissue volume, computer-guided surgery, study of anatomical structure and so on. In this chapter, we review the basic mathematics behind X-ray computed tomography (CT) and magnetic resonance imaging (MRI), and then discuss some image processing techniques.

6.1 X-ray Computed Tomography

X-ray computed tomography (CT) is the most widely used tomographic imaging technique, which uses X-rays passing through the body at different angles. It visualizes the internal structures of the human body by assigning an X-ray attenuation coefficient to each pixel, which characterizes how easily a medium can be penetrated by an X-ray beam Hounsfield (1973). The idea is to visualize the imaging object in a slice by taking X-ray data at all angles around the object based on mathematical methods suggested by Cormack (1963). They shared the 1979 Nobel Prize. Indeed, some of the ideas of CT (reconstructing cross-sectional images of an object from its integral values along lines in all directions) were previously developed by Radon (1917). Bracewell (1956) had applied his theory to radioastronomy, but unfortunately little attention was paid to it at that time. Cormack was unaware of Radon's earlier work, and Radon himself did not know the even earlier work by the Dutch physicist Lorentz, who had already proposed a solution of the mathematical problem for the three-dimensional case (Cormack 1992). We refer to Kalender (2006) for an excellent review of CT.

Nonlinear Inverse Problems in Imaging, First Edition. Jin Keun Seo and Eung Je Woo.

6.1.1 Inverse Problem

The corresponding inverse problem to X-ray CT can be described roughly as follows.

- **Quantity to be imaged.** The distribution of linear attenuation coefficients, denoted by a function $f(\mathbf{x})$ at point $\mathbf{x} = (x_1, x_2)$ or $\mathbf{x} = (x_1, x_2, x_3)$. See Figure 6.1.
- **Input data.** An incident X-ray beam is passed through a patient placed between an X-ray source and a detector. These beams are transmitted in all directions $\Theta := (\cos\theta, \sin\theta), 0 \le \theta \le 2\pi$. Assuming a fixed angle θ, a number of X-ray photons are transmitted through the body along projection lines $L_{\theta,s} := \{\mathbf{x} \in \mathbb{R}^2 : \Theta \cdot \mathbf{x} = s\}$, $s \in \mathbb{R}$. See Figure 6.1.
- **Output data.** Detectors measure X-ray intensity attenuation $I_\theta(s)$ along the individual projection lines $L_{\theta,s}$ at all angles. These measured data $I_\theta(s)$ provide an X-ray image $P_\theta(s)$ that can be expressed roughly by

$$P_\theta(s) = \mathcal{R}_\theta f(s) + \text{nonlinear effects}, \tag{6.1}$$

where $\mathcal{R}_\theta f(s)$, called the Radon transform of f, is the integral along the line $L_{\theta,s} := \{\mathbf{x} \in \mathbb{R}^2 : \Theta \cdot \mathbf{x} = s\}$ in the direction $\Theta = (\cos\theta, \sin\theta)$:

$$\mathcal{R}_\theta f(s) := \int_{\mathbb{R}^2} f(\mathbf{x})\delta(\Theta \cdot \mathbf{x} - s)\,\mathrm{d}\mathbf{x} = \int_{L_{\theta,s}} f(\mathbf{x})\,\mathrm{d}\ell_{\mathbf{x}}, \tag{6.2}$$

where $\mathrm{d}\ell$ is the length element. See Figure 6.1.
- **Inverse problem.** Recover f from the series of X-ray data P_{θ_n}, $n = 1, 2, \ldots, N$, where $\theta_n = 2n\pi/N$.

6.1.2 Basic Principle and Nonlinear Effects

We begin with understanding the relationship between X-ray intensity attenuation $I_\theta(s)$ and the X-ray data $P_\theta(s)$ in (6.1). For simplicity, we ignore scattering and metal shadowing effects for the moment.

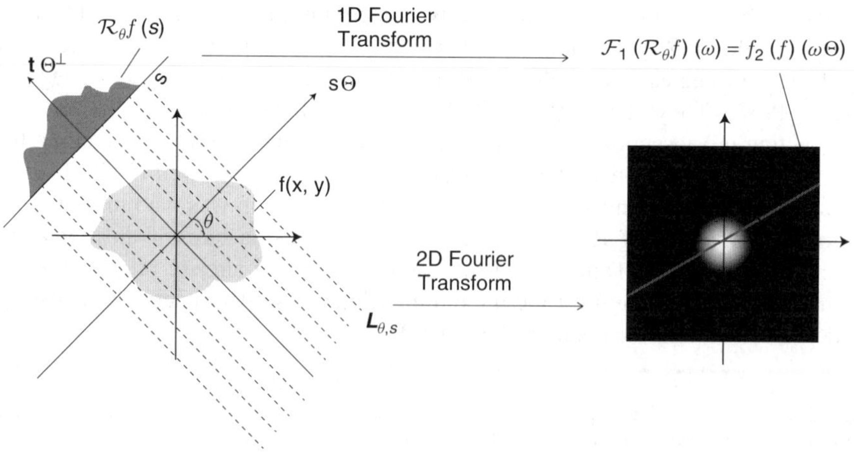

Figure 6.1 Illustration of the Fourier slice theorem

Fix θ and s. The incident X-ray beam is composed of a number of photons of different energies. Hence, the incident intensity of the X-ray beam, denoted by I^0, can be viewed as a function of the photon energy level E (Brooks and Chiro 1976). Imagine that an X-ray passes through a patient along $L_{\theta,s}$. Let $I_\theta(E, s)$ be the measured attenuated intensity along the line $L_{\theta,s}$. Then $I^0(E)$ indicates the source X-ray at energy level E, and $I_\theta(E, s)$ indicates the detected X-ray after passing through the body along the line $L_{\theta,s}$. Denoting by $f_E(\mathbf{x})$ the attenuation coefficient at point $\mathbf{x}$ and at energy level E, the relation between $I_\theta(E, s)$ and f_E is dictated by the Beer–Lambert law (Beer 1852; Lambert 1760):

$$I_\theta(E, s) = I^0(E) \exp\{-\mathcal{R}_\theta f_E(s)\} \quad (E_{\min} \leq E \leq E_{\max}), \tag{6.3}$$

where $E_{\max}$ and $E_{\min}$, respectively, are the maximum and minimum energy levels of the X-ray beam. Soft tissues have roughly $f_E = 0.38$ at $E = 30\,\mathrm{keV\,cm^{-1}}$ and $f_E = 0.21$ at $E = 60\,\mathrm{keV\,cm^{-1}}$. If the beam does not interact with any medium, then the unattenuated beam is $I_\theta(E, s) = I^0(E)$.

Remark 6.1.1 (Beam hardening) *The lower-energy photons tend to be absorbed more rapidly than higher-energy photons. As a result, the mean energy of the incident X-ray beam is lower than the mean energy of the X-rays reaching the detectors after passing through an object, that is,*

$$\frac{\int E I_\theta(E, s)\, \mathrm{d}E}{\int I_\theta(E, s)\, \mathrm{d}E} \geq \frac{\int E I^0(E)\, \mathrm{d}E}{\int I^0(E)\, \mathrm{d}E}.$$

This effect is called beam hardening, since the mean energy of I_θ (after passing through the object) becomes greater than that of I^0 (before passing through the object). In a disk-like object, as shown in Figure 6.2, the reconstructed CT image of this homogeneous

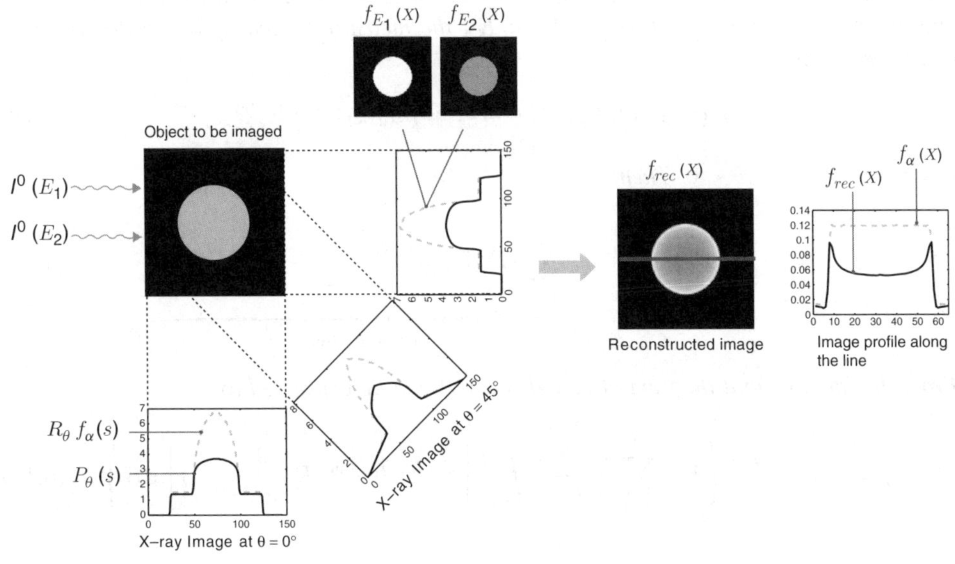

Figure 6.2 Illustration of metal artifacts

object appears brighter near the boundary than in its interior as a result of the beam hardening process.

The measured attenuation intensity $I_\theta(s)$ along the projection line $L_{\theta,s}$ is given by

$$I_\theta(s) = \int_{E_{\min}}^{E_{\max}} I_\theta(E, s)\,\mathrm{d}E = \int_{E_{\min}}^{E_{\max}} I^0(E)\exp\{-\mathcal{R}_\theta f_E(s)\}\,\mathrm{d}E. \tag{6.4}$$

This $I_\theta(s)$ provides the X-ray data $P_\theta(s)$ given by

$$P_\theta(s) = \ln\left(\frac{I^0}{I_\theta(s)}\right), \quad I^0 := \int_{E_{\min}}^{E_{\max}} I^0(E)\,\mathrm{d}E. \tag{6.5}$$

We try to reconstruct an image f such that

$$f = \arg\min_f \int_0^{2\pi}\int_{\mathbb{R}} \left|I_\theta(s) - I^0\exp\{-\mathcal{R}_\theta f(s)\}\right|^2 \mathrm{d}s\,\mathrm{d}\theta. \tag{6.6}$$

We hope that f is related to f_{E_0} for some energy level E_0.

Example 6.1.2 *In the special case when the X-ray beam is monochromatic, the relation between $f = f_{E_0}$ and the measured intensity along the line $L_{\theta,s}$ can be expressed by Beer's law,*

$$I_\theta(s) = I^0(E_0)\exp\{-\mathcal{R}_\theta f(s)\} \quad \textit{and} \quad P_\theta(s) = \mathcal{R}_\theta f(s). \tag{6.7}$$

In this case, there is no nonlinear effect in (6.1), and f can be reconstructed directly by the inverse Radon transform algorithm described in the next section.

Example 6.1.3 *In the case when the X-ray beam is bichromatic, having two dominant energy levels E_1 and E_2 with $E_1 < E_0 < E_2$, the measured intensity along the line $L_{\theta,s}$ can be expressed by*

$$I_\theta(s) = I^0(E_1)\,\mathrm{e}^{-\mathcal{R}_\theta f_{E_1}(s)} + I^0(E_2)\,\mathrm{e}^{-\mathcal{R}_\theta f_{E_2}(s)}. \tag{6.8}$$

Assume that $\partial f_E/\partial E < 0$. Then,

$$I_\theta(s) = I^0\,\mathrm{e}^{-\mathcal{R}_\theta f(s)} + \underbrace{\sum_{j=1}^{2}\left[I^0(E_j)\left(\mathrm{e}^{-\mathcal{R}_\theta f_{E_j}(s)} - \mathrm{e}^{-\mathcal{R}_\theta f(s)}\right)\right]}_{\text{nonlinear effect}}. \tag{6.9}$$

From the fundamental theorem of calculus, (6.9) can be expressed as

$$I_\theta(s) = I^0\,\mathrm{e}^{-\mathcal{R}_\theta f(s)}\left(1 - \sum_{j=1}^{2}\frac{I^0(E_j)}{I^0}\int_{E_0}^{E_j}\left[\mathrm{e}^{-\mathcal{R}_\theta[f_E - f](s)}\mathcal{R}_\theta\frac{\partial}{\partial E}f_E(s)\right]\mathrm{d}E\right). \tag{6.10}$$

Then, the difference between the Radon transform of the tomographic image f and the X-ray image $P_\theta(s) = \ln[I^0/I_\theta(s)]$ is given by

$$P_\theta(s) - \mathcal{R}_\theta f(s) = \ln\left(1 - \sum_{j=1}^{2} \frac{I^0(E_j)}{I^0} \int_{E_0}^{E_j} \left[\mathrm{e}^{-\mathcal{R}_\theta[f_E - f](s)} \mathcal{R}_\theta \frac{\partial}{\partial E} f_E(s) \right] \mathrm{d}E \right). \tag{6.11}$$

The right-hand side of the above identity is the nonlinear effect described in (6.1).

6.1.3 Inverse Radon Transform

Recall that the Radon transform $\mathcal{R}_\theta f$ of an attenuation coefficient $f(\mathbf{x})$ is the one-dimensional projection of $f(\mathbf{x})$ taken at an angle θ. Let $\mathcal{R}f$ denote the projection map in (6.2) as a function of θ and s, which can be expressed as

$$\mathcal{R}_\theta f(s) = \int_{-\infty}^{\infty} f(s\Theta + t\Theta^\perp)\,\mathrm{d}t, \quad \big(\Theta = (\cos\theta, \sin\theta),\ \Theta^\perp = (-\sin\theta, \cos\theta)\big). \tag{6.12}$$

The back-projection operator associated with the projection map $\mathcal{R}f$ is defined by

$$\mathcal{B}(\mathcal{R}f)(\mathbf{x}) = \int_0^\pi \mathcal{R}_\theta f(\mathbf{x}\cdot\Theta)\,\mathrm{d}\theta. \tag{6.13}$$

The Hilbert transform of a function $\phi(t)$ is defined as

$$\mathcal{H}\phi(t) = \frac{1}{\pi}\int_{-\infty}^{\infty} \frac{\phi(s)}{t-s}\,\mathrm{d}s. \tag{6.14}$$

Theorem 6.1.4 (Inverse Radon transform and back-projection) *Given $\mathcal{R}f$, its inverse Radon transform is*

$$f(\mathbf{x}) = \frac{1}{2\pi^2}\int_0^\pi \int_{\mathbb{R}} \frac{\partial \mathcal{R}_\theta f(s)/\partial s}{\mathbf{x}\cdot\Theta - s}\,\mathrm{d}s\,\mathrm{d}\theta \quad (\Theta = (\cos\theta, \sin\theta)), \tag{6.15}$$

which can be expressed as

$$f(\mathbf{x}) = \frac{1}{2\pi}\mathcal{B}\left[\mathcal{H}\left[\frac{\partial}{\partial s}\mathcal{R}_\theta f\right]\right](\mathbf{x}).$$

On the other hand, the back-projection $\mathcal{B}(\mathcal{R}f)$ provides a blurred image of $f(x, y)$ given by

$$\mathcal{B}(\mathcal{R}f)(\mathbf{x}) = f * \eta(\mathbf{x}), \quad \eta(\mathbf{x}) = \frac{1}{|\mathbf{x}|}, \tag{6.16}$$

where $*$ denotes the two-dimensional convolution in Cartesian coordinates.

In order to prove this theorem, we need to know the following Fourier slice theorem.

Theorem 6.1.5 (Fourier slice theorem) *For a fixed θ, the one-dimensional Fourier transform with respect to s of the projection $\mathcal{R}_\theta f(s)$ is equal to the line in the two-dimensional Fourier transform of f taken at θ:*

$$\mathcal{F}_1(\mathcal{R}_\theta f)(\omega) = \mathcal{F}_2(f)(\omega\Theta) \quad (\Theta = (\cos\theta, \sin\theta)), \tag{6.17}$$

where $\mathcal{F}_1$ and $\mathcal{F}_2$ are the one- and two-dimensional Fourier transforms, respectively.

Proof. The proof goes as follows:

$$\begin{aligned}
\mathcal{F}_1(\mathcal{R}_\theta f)(\omega) &= \int_{\mathbb{R}} \mathcal{R}_\theta f(s)\, \mathrm{e}^{-\mathrm{i}2\pi\omega s}\, \mathrm{d}s \\
&= \iint f(s\Theta + t\Theta^\perp)\, \mathrm{e}^{-\mathrm{i}2\pi\omega s}\, \mathrm{d}s\, \mathrm{d}t \\
&= \int_{\mathbb{R}^2} f(\mathbf{x}) \exp[-\mathrm{i}2\pi\omega\mathbf{x}\cdot\Theta]\, \mathrm{d}\mathbf{x} \\
&= \mathcal{F}_2(f)(\omega\Theta).
\end{aligned} \tag{6.18}$$

□

Now, we are ready to prove the inverse Radon transform.

Proof. (Inverse Radon transform) The inverse Fourier transform

$$f(\mathbf{x}) = \frac{1}{(2\pi)^2} \int_{-\infty}^{\infty}\int_{-\infty}^{\infty} \mathcal{F}_2 f(\omega_1, \omega_2)\, \mathrm{e}^{\mathrm{i}2\pi\mathbf{x}\cdot(\omega_1,\omega_2)}\, \mathrm{d}\omega_1\, \mathrm{d}\omega_2$$

□

can be changed into the polar coordinate form

$$f(\mathbf{x}) = \frac{1}{(2\pi)^2} \int_0^{2\pi}\int_0^{\infty} \mathcal{F}_2 f(\omega\Theta)\, \mathrm{e}^{\mathrm{i}2\pi\omega\mathbf{x}\cdot\Theta}\, \omega\, \mathrm{d}\omega\, \mathrm{d}\theta.$$

Using (6.17), we can change the form again into the following:

$$\begin{aligned}
f(\mathbf{x}) &= \frac{1}{(2\pi)^2} \int_0^{\pi}\int_{-\infty}^{\infty} \mathcal{F}_2 f(\omega\Theta)\, \mathrm{e}^{\mathrm{i}2\pi\omega\mathbf{x}\cdot\Theta} |\omega|\, \mathrm{d}\omega\, \mathrm{d}\theta \\
&= \frac{1}{(2\pi)^2} \int_0^{\pi} \left[\int_{-\infty}^{\infty} \omega\mathcal{F}_1\left(\mathcal{R}_\theta f\right)(\omega) \frac{\omega}{|\omega|}\, \mathrm{e}^{\mathrm{i}2\pi\omega\mathbf{x}\cdot\Theta}\, \mathrm{d}\omega\right] \mathrm{d}\theta.
\end{aligned} \tag{6.19}$$

Denoting $g_\theta(s) = \partial\mathcal{R}_\theta f(s)/\partial s$, we obtain

$$\frac{1}{2\pi}\mathcal{F}_1(\mathcal{H}g_\theta)(\omega) = \omega\mathcal{F}_1(\mathcal{R}_\theta f)(\omega)\frac{\omega}{|\omega|} \tag{6.20}$$

because

$$\mathcal{F}_1(\mathcal{H}g_\theta)(\omega) = -\mathrm{i}\frac{\omega}{|\omega|}\mathcal{F}_1 g_\theta(\omega) \quad \text{and} \quad \mathcal{F}_1(g_\theta)(\omega) = 2\pi\mathrm{i}\omega\mathcal{F}_1(\mathcal{R}_\theta f)(\omega).$$

Hence, it follows from (6.19) and (6.20)) that

$$f(\mathbf{x}) = \frac{1}{(2\pi)^2}\int_0^{\pi}\left[\frac{1}{2\pi}\int_{-\infty}^{\infty}\mathcal{F}_1(\mathcal{H}g_\theta)(\omega)\,\mathrm{e}^{\mathrm{i}2\pi\omega\mathbf{x}\cdot\Theta}\,\mathrm{d}\omega\right]\mathrm{d}\theta$$
$$= \frac{1}{2\pi}\int_0^{\pi}\underbrace{\left[\frac{1}{\pi}\int_{\mathbb{R}}\frac{\partial\mathcal{R}_\theta f(s)/\partial s}{\mathbf{x}\cdot\Theta - s}\,\mathrm{d}s\right]}_{\mathcal{H}g_\theta(\Theta\cdot\mathbf{x})}\mathrm{d}\theta. \tag{6.21}$$

This completes the proof of the inverse Radon transform.

Remark 6.1.6 *The back-projection operator $\mathcal{B}$ is not the inverse of the Radon transform $\mathcal{R}$, but the adjoint of $\mathcal{R}$ in the following sense:*

$$\begin{aligned}\langle\mathcal{R}f, g\rangle_{s,\theta} &:= \int_0^{2\pi}\left[\int_{\mathbb{R}}\mathcal{R}_\theta f(s)g(s\Theta)\,\mathrm{d}s\right]\mathrm{d}\theta\\ &= \int_0^{2\pi}\left[\int_{\mathbb{R}}\left\{\int_{\mathbb{R}}f(s\Theta + t\Theta^{\perp})\,\mathrm{d}t\right\}g(s\Theta)\,\mathrm{d}s\right]\mathrm{d}\theta\\ &= \int_{\mathbb{R}^2}f(\mathbf{x})\underbrace{\left[\int_0^{2\pi}g((\mathbf{x}\cdot\Theta)\Theta)\,\mathrm{d}\theta\right]}_{\mathcal{B}g(\mathbf{x})}\mathrm{d}\mathbf{x}\\ &:= \langle f, \mathcal{B}g\rangle_{\mathbb{R}^2}.\end{aligned} \tag{6.22}$$

Since $\mathcal{B}(\mathcal{R}f)(\mathbf{x}) = f * \eta(\mathbf{x})$, the image of $\mathcal{B}(\mathcal{R}f)$ can be viewed as a blurred version of f.

Remark 6.1.7 *There is also the inverse Radon transform without derivative, which is called the filtered back-projection. To be precise, let $\mathcal{F}_1\left(\mathcal{R}_\theta f\right)(\omega)$ be supported in the interval $[-\Gamma, \Gamma]$ (called band-limited). Then, according to (6.19), we have*

$$f(\mathbf{x}) = \int_0^{\pi}\mathrm{d}\theta\int_{-\Gamma}^{\Gamma}\mathcal{F}_1(\mathcal{R}_\theta f)(\omega)|\omega|\,\mathrm{e}^{\mathrm{i}2\pi\omega\mathbf{x}\cdot\Theta}\,\mathrm{d}\omega.$$

Using the fact that $\mathcal{F}(\phi * \psi) = \mathcal{F}(\phi)\mathcal{F}(\psi)$, we have

$$\int\mathcal{F}_1(\mathcal{R}_\theta f)(\omega)|\omega|\,\mathrm{e}^{\mathrm{i}2\pi\omega\mathbf{x}\cdot\Theta}\,\mathrm{d}\omega = (\mathcal{R}_\theta f) * h(\mathbf{x}\cdot\Theta),$$

where h is the inverse Fourier transform of $|\omega|$:

$$h(t) = \int_{-\Gamma}^{\Gamma}|\omega|\,\mathrm{e}^{\mathrm{i}2\pi\omega t}\,\mathrm{d}\omega = \frac{\Gamma^2}{2}\left(\frac{\sin 2\pi\Gamma t}{2\pi\Gamma t}\right) - \Gamma^2\left(\frac{\sin\pi\Gamma t}{\pi\Gamma t}\right)^2.$$

Hence, f can be computed from

$$f(\mathbf{x}) = \int_0^{\pi}(\mathcal{R}_\theta f) * h(\mathbf{x}\cdot\Theta)\,\mathrm{d}\theta.$$

This is called the filtered back-projection method, *which was first described by Bracewell and Riddle (1967). From the Nyquist sampling criterion, the bandwidth Γ and projection sampling interval $\Delta\omega$ should satisfy the relation, $\Gamma = 1/(2\Delta\omega)$.*

Exercise 6.1.8 *Consider the inverse Radon transform of $P_\theta(s)$ in (6.11) described in Example 6.1.3:*

$$f(\mathbf{x}) = \frac{1}{2\pi^2} \int_0^\pi \int_{\mathbb{R}} \frac{\partial P_\theta(s)/\partial s}{\mathbf{x} \cdot \Theta - s} \, \mathrm{d}s \, \mathrm{d}\theta \quad (\Theta = (\cos\theta, \sin\theta)). \tag{6.23}$$

Show that the relation between f and P_θ is $\mathcal{R}_\theta f(s) = P_\theta(s)$ for all θ and s.

We refer to Faridani (2003) and Natterer (2008) for detailed explanations on the Radon inversion formula, and refer to Tuy (1983), Feldkamp *et al.* (1984), Kak and Slaney (1988) and Grangeat (1991) for the inversion algorithm in cone-beam CT, which uses a cone-shaped X-ray beam rather than a conventional linear fan beam.

6.1.4 Artifacts in CT

In X-ray CT, the incident intensity $I^0(E)$ and the attenuated intensity $I_\theta(E, s)$ are functions of the energy level E, which ranges over 10–150 keV, and the attenuation coefficient f_E differs with the energy level E. Mostly, $\partial f_E/\partial E < 0$, which means that the lower-energy photons are absorbed more rapidly than higher-energy photons. Hence, there exist fundamental artifacts in the reconstructed image $f(\mathbf{x})$ as in Figure 6.2 by the inverse Radon transform since f can be regarded as a distribution of the linear attenuation coefficient $f(\mathbf{x}) \approx f_(\mathbf{x})$ at a mean energy level E_0.

The presence of metal objects in the scan field can lead to severe streaking artifacts owing to the above-mentioned inconsistencies by beam hardening and severely high contrast in the attenuation coefficient distribution between the metal and the surrounding subjects. Figure 6.3 shows such metal artifacts. Additional artifacts due to beam hardening, partial volume and aliasing are likely to compound the problem when scanning very dense objects. We refer to Kalender *et al.* (1987), Meyer *et al.* (2010) and Wang *et al.* (1996) for metal artifacts reduction.

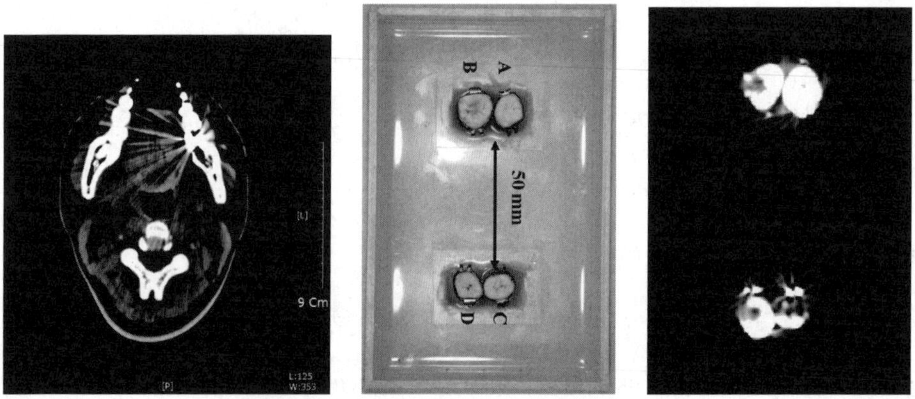

Figure 6.3 Examples of metal artifacts

Example 6.1.9 *Figure 6.2 illustrates how beam hardening produces artifacts in CT images in the simplest case of the Example 6.1.3 with* $I^0 = I^0(E_1) + I^0(E_2)$, $I^0(E_1) = I^0(E_2) = 1$ *and*

$$f_{E_1}(\mathbf{x}) = \begin{cases} 0.2 & \mathbf{x} \in D, \\ 0.01 & \textit{otherwise}, \end{cases} \qquad f_{E_2}(\mathbf{x}) = \begin{cases} 0.04 & \mathbf{x} \in D, \\ 0.02 & \textit{otherwise}. \end{cases}$$

In Figure 6.2,

- $I_\theta(s) = \mathrm{e}^{-R_\theta f_{E_1}(s)} + \mathrm{e}^{-R_\theta f_{E_2}(s)}$,
- $P_\theta(s) = -\ln[I_\theta(s)/I^0] = -\ln\left(\frac{1}{2}\mathrm{e}^{-R_\theta f_{E_1}(s)} + \frac{1}{2}\mathrm{e}^{-R_\theta f_{E_2}(s)}\right)$,
- $f_{\mathrm{rec}}(\mathbf{x}) = \displaystyle\int_0^\pi \left[-\ln\left(\tfrac{1}{2}\mathrm{e}^{-R_\theta f_{E_1}} + \tfrac{1}{2}\mathrm{e}^{-R_\theta f_{E_2}}\right) * h\right](\mathbf{x}\cdot\Theta)\,\mathrm{d}\theta.$

We try to reconstruct $f_\alpha = \alpha f_{E_1} + (1-\alpha) f_{E_2}$, *but the reconstructed image* f *using the standard inversion formula can be very different from* $f_\alpha = \alpha f_{E_1} + (1-\alpha) f_{E_2}$ *for any* α.

6.2 Magnetic Resonance Imaging

MRI uses magnetic fields to visualize the quantity of hydrogen atoms inside biological tissues by creating magnetization; hydrogen atoms can interact with the external magnetic field **B** because a nucleus with spin has a local magnetic field around it. In the human body, the amount of hydrogen would be the major factor of the net magnetization vector **M**. In MRI, we use various techniques to localize **M** in such a way that we provide a cross-sectional image of the density of **M** inside the human body. These techniques are based on the nuclear magnetic resonance (NMR) phenomenon, which is determined by the interaction of a nuclear spin **M** with the external magnetic field **B** and its local environments, including relaxation effects.

The idea behind this imaging modality was published by Lauterbur (1973), and the first cross-sectional image of a living mouse was published by Lauterbur (1974). Damadian (1971) discovered that NMR can distinguish tumors from normal tissues due to their relaxation time. Lauterbur and Mansfield were awarded the Nobel Prize for developing the mathematical framework and some MRI techniques, but the award was denied by Damadian, who claimed that his work was earlier than those of Lauterbur and Mansfield. We refer to Filler (2010) for the history of MRI. We will describe the NMR phenomenon in the next section, taking account of measurable signals in an MRI system.

6.2.1 Basic Principle

We consider a human body occupying a domain Ω inside an MRI scanner with its main magnetic field $\mathbf{B}_0 + \delta\mathbf{B}$. Throughout this section, we shall assume that the field inhomogeneity $\delta\mathbf{B}$ is negligible, that is, $\delta\mathbf{B} = 0$ and $\mathbf{B}_0 = (0, 0, B_0) = B_0\hat{\mathbf{z}}$ is constant. The strong uniform main field produces a distribution of net magnetization $\mathbf{M}(\mathbf{r}, t)$ in Ω by aligning protons inside the human body, where $\mathbf{M}(\mathbf{r}, t) = (M_x(\mathbf{r}, t), M_y(\mathbf{r}, t), M_z(\mathbf{r}, t))$ depends on time t and position $\mathbf{r}$. A stronger $\mathbf{B}_0$ produces a larger $\mathbf{M}$. The interaction of $\mathbf{M}$ with the external magnetic field $\mathbf{B}_0$ is dictated by the Bloch equation:

$$\frac{\partial}{\partial t}\mathbf{M} = -\gamma \mathbf{B}_0 \times \mathbf{M} = -(\gamma B_0 \hat{\mathbf{z}}) \times \mathbf{M}, \tag{6.24}$$

where γ is the gyromagnetic ratio[1]. From (6.24), we have

$$\frac{\partial}{\partial t}\mathbf{M} \perp \mathbf{M} \quad \text{and} \quad \frac{\partial}{\partial t}\mathbf{M} \perp \mathbf{B}_0, \tag{6.25}$$

which means that the vector $\partial\mathbf{M}/\partial t$ is perpendicular to both $\mathbf{M}$ and $\mathbf{B}_0$. To be precise, (6.24) can be written as

$$\frac{\partial}{\partial t}\begin{pmatrix} M_x \\ M_y \\ M_z \end{pmatrix} = \gamma \begin{vmatrix} \hat{\mathbf{x}} & \hat{\mathbf{y}} & \hat{\mathbf{z}} \\ M_x & M_y & M_z \\ 0 & 0 & B_0 \end{vmatrix} = \begin{pmatrix} \omega_0 M_y \\ -\omega_0 M_x \\ 0 \end{pmatrix}. \tag{6.26}$$

Writing $M_\perp = M_x + \mathrm{i}M_y$, we express the above identity as

$$\frac{\partial}{\partial t}M_\perp = -\mathrm{i}\omega_0 M_\perp. \tag{6.27}$$

The solution of the above ODE is

$$M_\perp(\mathbf{r}, t) = M_\perp(\mathbf{r}, 0)\,\mathrm{e}^{-\mathrm{i}\omega_0 t}, \tag{6.28}$$

which means that the transverse component $\mathbf{M}_\perp$ rotates clockwise at the angular frequency $\omega_0 = \gamma B_0$. The above expression explains how $\mathbf{B}_0$ causes $\mathbf{M}$ to precess around the z axis at the angular frequency of $\omega_0 = \gamma B_0$. For a 1.5 T MRI system, the frequency of $\omega_0/2\pi$ is approximately 64 MHz. According to (6.28), $\mathbf{B}_0 = B_0\hat{\mathbf{z}}$ causes $\mathbf{M}$ to precess clockwise about the $\mathbf{B}_0 = B_0\hat{\mathbf{z}}$ direction at the angular frequency $\omega_0 = \gamma|\mathbf{B}_0|$. If we apply a magnetic field $\mathbf{B}(\mathbf{r})$, then $\mathbf{M}(\mathbf{r}, t)$ satisfies the Bloch equation

$$\frac{\partial}{\partial t}\mathbf{M}(\mathbf{r}, t) = -\gamma\mathbf{B}(\mathbf{r}) \times \mathbf{M}(\mathbf{r}, t)$$

and $\mathbf{M}(\mathbf{r}, t)$ precesses at the angular frequency $\omega_0 = \gamma|\mathbf{B}_0(\mathbf{r})|$.

6.2.2 k-Space Data

If we could produce a magnetic flux density $\mathbf{B}$ that is localized at each position $\mathbf{r}$ (in such a way that $\mathbf{r} \neq \mathbf{r}' \Rightarrow B_z(\mathbf{r}) \neq B_z(\mathbf{r}')$), we could encode the positions independently. However, it is impossible to localize the magnetic field $\mathbf{B}$ because $\nabla \cdot \mathbf{B} = 0$. Instead of a point localization, we could make a slice localization by applying a magnetic field gradient that varies with respect to one direction. The following is the one-dimensional magnetic field gradient along the z axis with the magnetic field increasing in the z direction:

$$\mathbf{B}(\mathbf{r}) = \mathbf{B}_0 + G_z z\hat{\mathbf{z}} = B_0\hat{\mathbf{z}} + G_z z\hat{\mathbf{z}}, \tag{6.29}$$

where B_0 and G_z are constants. The first component $\mathbf{B}_0$ is the main magnetic field, which causes individual $\mathbf{M}(\mathbf{r})$ to precess around the z axis, and the second term $G_z z\hat{\mathbf{z}}$ is the gradient field. This G_z is said to be the magnetic field gradient in the z direction. With this $\mathbf{B}$, the vector $\mathbf{M}$ in the body is essentially vertically aligned and rotates at the Larmor frequency $\gamma(B_0 + G_z z)$ around the z axis. Note that the frequency changes with z.

[1] For hydrogen ^{1}H, $\gamma = 2\pi \times 42.576 \times 10^6\ \mathrm{rad\,s^{-1}\,T^{-1}}$.

To extract a signal of $\mathbf{M}$, we flip $\mathbf{M}$ toward the transverse direction to produce its xy component. Flipping $\mathbf{M}$ toward the xy plane requires a second magnetic field $\mathbf{B}_1$ perpendicular to $\mathbf{B}_0 = (0, 0, B_0)$. We can flip $\mathbf{M}$ over the xy plane by using a radio-frequency (RF) magnetic field $\mathbf{B}_1$ that is generated by RF coils through which we inject sinusoidal current at the Larmor frequency $\omega_0 = \gamma B_0$.

After terminating the RF pulse, we apply a phase encoding gradient in the y direction, which makes the spin phase change linearly in the phase encoding direction. Then we apply the frequency encoding gradient field as

$$\mathbf{B} = \mathbf{B}_0 + G_x x\hat{\mathbf{x}},$$

and the signal during frequency encoding in the x direction becomes

$$S(t, k_y) = \int_{\{z=z_*\}} m(x, y)\, \mathrm{e}^{\mathrm{i}k_y y}\, \mathrm{e}^{\mathrm{i}\gamma G_x x t}\, \mathrm{d}S_{xy}. \tag{6.30}$$

Through multiple phase encodings to get multiple signals of (6.30) for different values of k_y, we may collect a set of k-space data

$$S(k_x, k_y) = \int_{\{z=z_*\}} m(x, y)\, \mathrm{e}^{\mathrm{i}(k_x x + k_y y)}\, \mathrm{d}S_{xy} \tag{6.31}$$

for various k_x and k_y. See Figure 6.4 (top left) for an image of $S(k_x, k_y)$. We refer the reader to the book by Haacke *et al.* (1999) for detailed explanations on MRI.

6.2.3 Image Reconstruction

In the Cartesian sampling pattern, we select a cross-sectional slice in the z direction using the gradient coils and get spatial information in the xy plane using the frequency encoding (x direction) and phase encoding (y direction). Let m be a spin density supported in the region $\{(x, y) : -FOV/2 < x, y < FOV/2\}$, where FOV is the field of view, and let $\mathbf{m}$ be the N^2 column vector representing the corresponding $N \times N$ image matrix with pixel size $\Delta x \times \Delta y = FOV/N \times FOV/N$. The inverse Fourier transform provides the image of $m(x, y)$ (see Figure 6.4 (top right)). Let us quickly review the inverse discrete Fourier transform (DFT). Assume that the image of $m(x, y)$ is approximated by an $N \times N$ matrix $\mathbf{m}$:

$$\mathbf{m} = \begin{pmatrix} m(x_0, y_0) & \cdots & m(x_{N-1}, 0) \\ \vdots & \ddots & \vdots \\ m(0, y_{N-1}) & \cdots & m(x_{N-1}, y_{N-1}) \end{pmatrix}, \quad x_j = j\Delta x,\, y_j = j\Delta y.$$

Then, the relation (6.31) between the MR data and the image of $m(x, y)$ can be expressed as

$$S(j\Delta k_x, \ell\Delta k_x) = c \sum_{\tilde{\ell}=0}^{N-1} \sum_{\tilde{j}=0}^{N-1} m(x_{\tilde{j}}, y_{\tilde{\ell}}) \exp\left\{\mathrm{i}\left(j\tilde{j}\Delta x \Delta k_x + \ell\tilde{\ell}\Delta y \Delta k_y\right)\right\} \tag{6.32}$$

for some scaling constant c and $j, \ell = 0, 1, \ldots, N-1$. Here, the inverse DFT with a k-space sampling that is designed to meet the Nyquist criterion provides a discrete

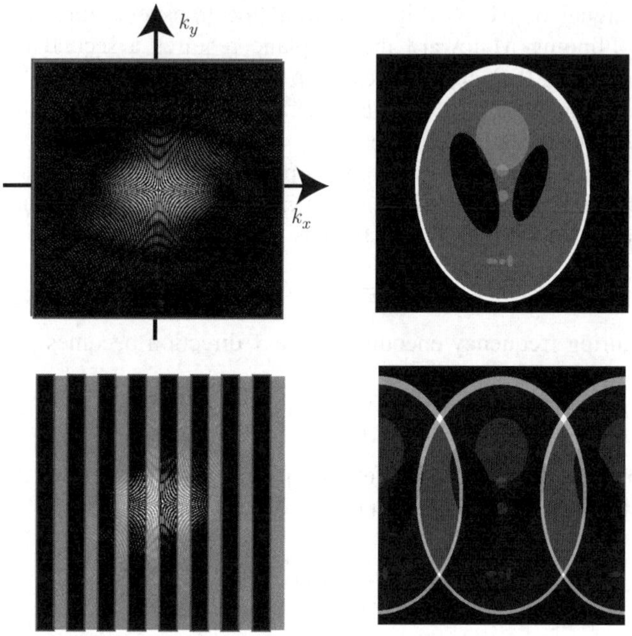

Figure 6.4 The two images at the top are fully sampled k-space data and the inverse Fourier transform. The two images at the bottom are subsampled k-space data by the factor of 2 and the inverse Fourier transform

version of the image $m(x, y)$. Using the two-dimensional inverse DFT, we have

$$m(x_j, y_\ell) = \tilde{c} \sum_{\tilde{\ell}=0}^{N-1} \sum_{\tilde{j}=0}^{N-1} S(\tilde{j}\Delta k_x, \tilde{\ell}\Delta k_x) \exp\{-\mathrm{i}(j\tilde{j}\Delta x \Delta k_x + \ell\tilde{\ell}\Delta y \Delta k_y)\} \tag{6.33}$$

for some scaling constant $\tilde{c}$.

In MRI, the data acquisition speed is roughly proportional to the number of phase encoding lines due to the time-consuming phase encoding, which separates signals from different y positions within the image $m(x, y)$. Hence, to accelerate the acquisition process, we need to skip phase encoding lines in k-space. Definitely, reduction in the k-space data violating the Nyquist criterion is associated with aliasing in the image space. If we use subsampled k-space data by a factor of R in the phase encoding direction (y direction), according to the Poisson summation formula, the corresponding inverse Fourier transform produces the following fold-over artifacts (as shown in Figure 6.4 (bottom right) and Figure 6.5):

$$\sum_{j=0}^{(N-1)/R} S(x_a, k_j)\,\mathrm{e}^{-\mathrm{i}k_j y_b} = \sum_{k=1}^{R} m\left(x, y + k\frac{FOV}{R}\right).$$

Parallel MRI (pMRI) is a way to deal with this fold-over artifact by using multiple receiver coils and supplementary spatial information in the image space. Parallel imaging

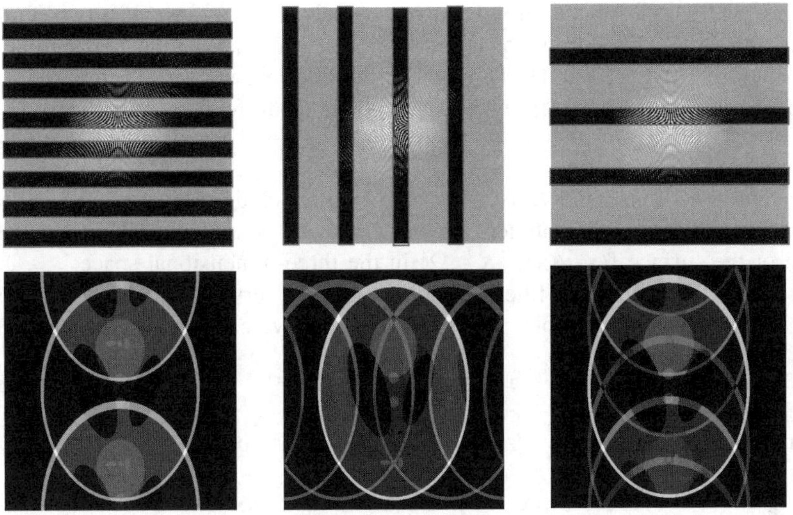

Figure 6.5 The images at the top are subsampled k-space data sets. The images at the bottom are the corresponding inverse Fourier transforms

has received a great deal of attention since the work by Sodickson and Manning (1997), Sodickson *et al.* (1999) and Pruessmann *et al.* (1999). In pMRI, we skip phase encoding lines in k-space during MRI acquisitions in order to reduce the time-consuming phase encoding steps, and in the image reconstruction step, we compensate the skipped k-space data by the use of space-dependent properties of multiple receiver coils. Numerous parallel reconstruction algorithms such as SENSE, SMASH and GRAPPA have been suggested, and those aim to use the least possible data of phase encoding lines, while eliminating aliasing, which is a consequence of violating the Nyquist criterion by skipping the data. We refer to Jakob *et al.* (1998), Kyriakos *et al.* (2000), Heidemann *et al.* (2001), Pruessmann *et al.* (2001), Bydder *et al.* (2002) and Griswold *et al.* (2002) for pMRI algorithms and to Larkman and Nunes (2001) for a review.

6.3 Image Restoration

Numerous algorithms for image restoration and segmentation are based on mathematical models with partial differential equations, level set methods, regularization, energy functionals and others. In these models, we regard the intensity image as a two-dimensional surface in a three-dimensional space with the gray level assigned to the axis. We can use the distribution of its gradient, Laplacian and curvature to interpret the image. For example, Canny (1986) defined edge points as points where the gradient magnitude assumes a local maximum in the gradient direction. The noise is usually characterized as having a local high curvature. Using these properties, one may decompose the image roughly into features, background, texture and noise. Image processing techniques are useful tools for improving the detectability of diagnostic features. We refer the reader to the books by Aubert and Kornprobst (2002) and Osher *et al.* (2002) for PDE-based image processing.

For clear and easy explanation, we restrict ourselves to the case of a two-dimensional grayscale image that is expressed as a function $u \in L^2(\Omega)$ in the rectangular domain $\Omega = \{\mathbf{x} = (x, y) : 0 < x, y < 1\}$. Throughout this section, we assume that the measured (or observed) image $f \in L^2(\Omega)$ and the true image u are related by

$$f = Hu + \eta, \tag{6.34}$$

where η represents noise and H is a linear operator including blurring and shifting. For image restoration and segmentation, we use *a priori* knowledge about the geometric structure of the surface $\{(\mathbf{x}, u(\mathbf{x})) : \mathbf{x} \in \Omega\}$ in the three-dimensional space.

The goal of denoising is to filter out noise while preserving important features. If the noise η is Gaussian random noise satisfying $\int \eta = 0$, we can apply the Gaussian kernel

$$G_\sigma(\mathbf{x}) = \frac{1}{4\pi\epsilon} \mathrm{e}^{-|\mathbf{x}|^2/(4\sigma)}$$

to get an approximation

$$G_\sigma * \eta(\mathbf{x}) = \int_\Omega G_\sigma(\mathbf{x} - \mathbf{x}')\eta(\mathbf{x}')\,\mathrm{d}\mathbf{x}' \approx 0$$

with an appropriate choice of variance $\sigma > 0$. Note that $G_t * f(\mathbf{x})$ is a solution of the heat equation with the initial data $f(\mathbf{x})$:

$$\begin{cases} \left[\dfrac{\partial}{\partial t} - \nabla^2\right] G_t * f(\mathbf{x}) = 0 \quad (\mathbf{x} \in \Omega, t > 0), \\ \lim\limits_{t \to 0+} G_t * f(\mathbf{x}) = f(\mathbf{x}). \end{cases}$$

Hence, we expect that this smoothing method eliminates highly oscillatory noise that results in $G_\sigma * (Hu + \eta) \approx G_\sigma * (Hu)$, while edges having high frequencies are also smeared at the same time. Several schemes have been proposed to deal with this blurring problem.

A typical way of denoising (or image restoration) is to find the best function by minimizing the functional:

$$\Phi(u) = \frac{1}{2}\int_\Omega |Hu - f|^2\,\mathrm{d}\mathbf{x} + \frac{\lambda}{p}\int_\Omega |\nabla u|^p\,\mathrm{d}\mathbf{x} \quad (p = 1, 2), \tag{6.35}$$

where the first term $\|Hu - f\|_{L^2(\Omega)}$, called the fidelity term, forces the residual $Hu - f$ to be small, the second term $\|\nabla u\|_{L^2(\Omega)}$, called the regularization term, enforces the regularity of u, and λ, called the regularization parameter, controls the tradeoff between the residual norm and the regularity. We can associate (6.35) with the Euler–Lagrange equation:

$$-\nabla \cdot (|\nabla u|^{p-2}\nabla u) + \frac{1}{\lambda}H^*(Hu - f) = 0 \quad \text{in } \Omega, \tag{6.36}$$

where H^* is the dual of H. The above equation is nonlinear except for $p = 2$. When $p = 2$, the minimizer u satisfies the linear PDE,

$$-\nabla^2 u + \frac{1}{\lambda}H^*(Hu - f) = 0 \quad \text{in } \Omega,$$

which has computational advantages in computing u. But the Laplace operator has very strong isotropic smoothing properties and does not preserve edges because it penalizes the strong gradients along edges. This is the major reason why many researchers use $p = 1$ since it does not penalize the gradients along edges with alleviating noise:

$$-\nabla \cdot \left(\frac{1}{|\nabla u|} \nabla u \right) + \frac{1}{\lambda} H^*(Hu - f) = 0 \quad \text{in } \Omega.$$

We can solve (6.35) using the gradient descent method by introducing the time variable t:

$$\frac{\partial}{\partial t} u(x, y, t) = -\left[-\nabla \cdot (|\nabla u|^{p-2} \nabla u) + \frac{1}{\lambda} H^*(Hu - f) \right]. \tag{6.37}$$

To understand the above derivation, we present the following simple example.

Example 6.3.1 (Gradient descent) *Let $A \in \mathbb{R}^{n \times n}$ be a symmetric and positive definite matrix. For a given vector $\mathbf{f} \in \mathbb{R}^n$, consider the minimization problem*

$$\textit{minimize} \quad \Phi(\mathbf{u}) = \frac{1}{2} \|A\mathbf{u} - \mathbf{f}\|^2 + \frac{\lambda}{2} \|\mathbf{u}\|^2,$$

where $\|\mathbf{u}\|^2 = \langle \mathbf{u}, \mathbf{u} \rangle$ and $\langle \cdot, \cdot \rangle$ is the usual vector inner product. Then

$$\nabla \Phi(\mathbf{u}) = A^*[A\mathbf{u} - \mathbf{f}] + \lambda \mathbf{u}.$$

If $\mathbf{u}$ is a minimizer, then $\mathbf{u}$ satisfies the Euler–Lagrange equation

$$\nabla \Phi(\mathbf{u}) = A^*[A\mathbf{u} - \mathbf{f}] + \lambda \mathbf{u} = 0,$$

where I is the identity matrix in $\mathbb{R}^n$. Since $-\nabla \Phi(\mathbf{u})$ is the direction of steepest descent *at a vector $\mathbf{u} \in \mathbb{R}^n$, we have the following iteration scheme:*

$$\mathbf{u}^{k+1} = \mathbf{u}^k - \alpha \nabla \Phi(\mathbf{u}^k),$$

where α is the step size. Introducing the time variable t and taking

$$\frac{\mathrm{d}}{\mathrm{d}t} \mathbf{u}(n) \approx \frac{\mathbf{u}^{n+1} - \mathbf{u}^n}{\alpha} \quad (\Delta t = \alpha),$$

we derive

$$\frac{\partial \mathbf{u}}{\partial t} = -\nabla \Phi(\mathbf{u}).$$

Expecting $\lim_{t \to \infty} \mathrm{d}\mathbf{u}(t)/\mathrm{d}t = 0$, $\mathbf{u}(\infty)$ is the minimizer of $\Phi(\mathbf{u})$.

6.3.1 Role of p in (6.35)

Let us investigate the role of the diffusion term $\nabla(|\nabla u|^{p-2} \nabla u)$ in (6.35). The image $u(\mathbf{x})$ can be viewed as a two-dimensional surface $\{(\mathbf{r}, u(\mathbf{r})) : (x, y) \in \Omega\}$. To understand the image structure, we look at the level curve $\mathcal{L}_c := \{\mathbf{x} \in \Omega : u(\mathbf{x}) = c\}$, the curve along which the intensity $u(\mathbf{x})$ is a constant c. Let $\mathbf{x}(s)$ represent a parameterization of the level curve $\mathcal{L}_c$, that is, $\mathcal{L}_c = \{\mathbf{x}(s) = (x(s), y(s)) : 0 < s < 1\}$. The normal vector and tangent vector to the level curve $\mathcal{L}_c$, respectively, are

$$\mathbf{n} = \frac{\nabla u}{|\nabla u|} \quad \text{and} \quad T = \frac{\mathbf{x}'}{|\mathbf{x}'|} = \frac{(-u_y, u_x)}{|\nabla u|},$$

where $u_x = \partial u/\partial x$ and $u_y = \partial u/\partial y$. Let us investigate the double derivative of u in the tangent and normal directions:

$$u_{TT} = \langle T, (D^2u)T\rangle \quad \text{and} \quad u_{\mathbf{nn}} = \langle \mathbf{n}, (D^2u)\mathbf{n}\rangle,$$

where (D^2u) is the Hessian matrix of u. Direct computation shows that

$$\begin{aligned} 0 = \frac{\mathrm{d}^2}{\mathrm{d}s^2}u(\mathbf{x}(s)) &= \frac{\mathrm{d}}{\mathrm{d}s}\left(u_x x' + u_y y'\right) \\ &= (u_{xx}(x')^2 + 2u_{xy}x'y' + u_{yy}(y')^2 + u_x x'' + u_y y''). \end{aligned}$$

This leads to

$$u_{TT} = \frac{-1}{|\mathbf{x}'|^2}[u_x x'' + u_y y''].$$

Substituting $u_x = -y'|\nabla u|/|\mathbf{x}'|$ and $u_y = x'|\nabla u|/|\mathbf{x}'|$ into the above identity and with $0 = \nabla u \cdot \mathbf{x}'$, we obtain

$$u_{TT} = \underbrace{\frac{1}{|\mathbf{x}'|^3}\left[y'x'' - x'y''\right]}_{\kappa} |\nabla u| = \nabla \cdot \mathbf{n}|\nabla u|,$$

where $\nabla \cdot \mathbf{n} = \kappa$ is the curvature. Hence,

$$u_{TT} = \kappa|\nabla u| = \nabla \cdot \mathbf{n}|\nabla u|. \tag{6.38}$$

Similarly, we have

$$\begin{aligned} u_{\mathbf{nn}} &= \frac{1}{|\mathbf{x}'|^2}\left(u_{xx}(y')^2 - 2u_{xy}x'y' + u_{yy}(x')^2\right) \\ &= \frac{1}{|\nabla u|^2}\left(u_{xx}(u_x)^2 + 2u_{xy}u_x u_y + u_{yy}(u_y)^2\right) \\ &= \nabla(|\nabla u|) \cdot \mathbf{n}. \end{aligned} \tag{6.39}$$

- In the case $p = 2$, we have the following isotropic diffusion:

$$\nabla^2 u = \nabla \cdot (|\nabla u|\mathbf{n}) = \nabla(|\nabla u|) \cdot \mathbf{n} + |\nabla u|\nabla \cdot \mathbf{n} = u_{\mathbf{nn}} + u_{TT}. \tag{6.40}$$

  Here, we use the fact that $\nabla u = |\nabla u|\mathbf{n}$. Indeed, the rotation invariance property of the Laplace operator yields (6.40).
- In the case of $p \neq 2$, it follows from (6.38), (6.39) and (6.40) that

$$\begin{aligned} \nabla \cdot \left(|\nabla u|^{p-2}\nabla u\right) &= |\nabla u|^{p-2}\nabla^2 u + (p-2)|\nabla u|^{p-3}\nabla(|\nabla u|) \cdot \nabla u \\ &= |\nabla u|^{p-2}\nabla^2 u + (p-2)|\nabla u|^{p-2}u_{\mathbf{nn}}. \end{aligned}$$

Hence, the diffusion term in (6.35) can be expressed as

$$\nabla \cdot (|\nabla u|^{p-2}\nabla u) = |\nabla u|^{p-2}(u_{TT} + (p-1)u_{\mathbf{nn}}). \tag{6.41}$$

The diffusion rate in the T direction is $|\nabla u|^{p-2}$, while the diffusion rate in the $\mathbf{n}$ direction is $(p-1)|\nabla u|^{p-2}$. Since $\mathbf{n}$ is normal to the edges, it would be preferable to smooth more in the tangential direction than in the normal direction. In order to preserve edges, $p = 1$ would be best, since it annihilates the coefficient of $u_{\mathbf{nn}}$.

6.3.2 *Total Variation Restoration*

In one dimension, the total variation (TV) of a signal f defined in an interval $[0, 1]$ means the total amplitude of signal oscillations in $[0, 1]$. The classical definition of the total variation of f is

$$\|f\|_{BV([0,1])} := \sup\left\{\sum_{j=1}^{N} |f(x_j) - f(x_{j+1})| : x_0 = 0 < x_1 < \cdots < x_N = 1\right\}, \tag{6.42}$$

where the supremum is taken over all partitions in the interval $[0, 1]$. The space of bounded variation functions in $[0, 1]$, denoted by $BV([0, 1])$, is the space of all real-valued functions $f \in L^1([0, 1])$ such that $\|f\|_{BV([0,1])} < \infty$. In the special case when $f \in C^1([0, 1])$ is differentiable, it can be defined by

$$\|f\|_{BV([0,1])} = \int_0^1 |f'(x)|\, \mathrm{d}x. \tag{6.43}$$

Example 6.3.2 *If f is the well-known* Cantor function, *then $\|f\|_{BV(\Omega)} = 1 = f(1) - f(0)$ and it is in $BV(\Omega)$. However, the Cantor function f has the special property that $f' = 0$ almost everywhere. Since $\int_0^1 |f'(x)|\, \mathrm{d}x \neq 1 = \|f\|_{BV(\Omega)}$, the definition (6.43) is not appropriate for general functions in $BV([0, 1])$.*

The classical definition (6.42) of BV in one dimension is not suitable in the two- or three-dimensional cases. Hence, we use the following definition that is closely related to (6.43):

$$\|f\|_{BV([0,1])} := \sup\left\{\int_0^1 f(x)\phi'(x)\, \mathrm{d}x : \phi \in C_0^1([0, 1]),\ \sup_{x\in[0,1]} |\phi(x)| \le 1\right\}. \tag{6.44}$$

Now, we come back to the two-dimensional grayscale image $u(\mathbf{r})$ defined in the rectangular domain $\Omega = \{\mathbf{r} = (x, y) : 0 < x, y < 1\}$.

Definition 6.3.3 *The total variation of a function u on Ω is defined by*

$$\|u\|_{BV(\Omega)} := \sup\left\{\int_\Omega u(\mathbf{x})\nabla \cdot \mathbf{F}(\mathbf{x})\, \mathrm{d}\mathbf{x} : \mathbf{F} \in [C_0^1(\Omega)]^2,\ \sup_{\mathbf{x}\in\Omega} |\mathbf{F}(\mathbf{x})| \le 1\right\}. \tag{6.45}$$

A function u in Ω is a bounded variation on Ω if $\|u\|_{BV(\Omega)} < \infty$:

$$BV(\Omega) := \{u \in L^1(\Omega) : \|u\|_{BV(\Omega)} < \infty\}. \tag{6.46}$$

Theorem 6.3.4 *If $u \in C^1(\bar{\Omega})$, then*

$$\|u\|_{BV} = \int_\Omega |\nabla u(\mathbf{x})|\, d\mathbf{x}.$$

Proof. Let $\mathcal{A} := \{\mathbf{F} \in [C_0^1(\Omega)]^2 : \sup_{\mathbf{x}\in\Omega} |\mathbf{F}(\mathbf{x})| \le 1\}$. From the divergence theorem,

$$\int_\Omega u\nabla\cdot\mathbf{F}\, d\mathbf{r} = \int_\Omega [\nabla\cdot(u\mathbf{F}) - \nabla u\cdot\mathbf{F}]\, d\mathbf{x} = -\int_\Omega \nabla u\cdot\mathbf{F}\, d\mathbf{x} \quad (\forall\, \mathbf{F}\in\mathcal{A}).$$

Hence, for all $\mathbf{F}\in\mathcal{A}$,

$$\int_\Omega u\nabla\cdot\mathbf{F}\, d\mathbf{x} \le \int_\Omega |\nabla u|\,|\mathbf{F}| \le \int_\Omega |\nabla u(\mathbf{x})|\, d\mathbf{x}\left(\sup_{\mathbf{x}\in\Omega}|\mathbf{F}(\mathbf{x})|\right) \le \int_\Omega |\nabla u(\mathbf{x})|\, d\mathbf{x}.$$

This leads to

$$\|u\|_{BV(\Omega)} = \sup_{\mathbf{F}\in\mathcal{A}} \int_\Omega u\nabla\cdot\mathbf{F}\, d\mathbf{x} \le \|\nabla u\|_{L^1(\Omega)}.$$

It remains to prove $\|u\|_{BV(\Omega)} \ge \|\nabla u\|_{L^1(\Omega)}$. We can make a sequence $\mathbf{F}_n \in \mathcal{A}$ such that

$$\mathbf{F}_n(\mathbf{x}) \to -\frac{\nabla u(\mathbf{x})}{|\nabla u(\mathbf{x})|}.$$

For example, we may choose

$$\mathbf{F}_n(\mathbf{x}) = \phi_n G_{1/n} * \left(\frac{\nabla u}{(1/n) + |\nabla u|}\right),$$

where G_σ is the Gaussian kernel and ϕ_n is a positive function in $C_0^1(\Omega)$ such that $\phi(\mathbf{x}) = 1$ if $\mathrm{dist}(\mathbf{x}, \partial\Omega) > 1/n$. Hence, we have

$$\lim_{n\to\infty} \int_\Omega u\nabla\cdot\mathbf{F}_n\, d\mathbf{x} = -\lim_{n\to\infty}\int_\Omega \nabla u\cdot\mathbf{F}_n\, d\mathbf{x} = \|\nabla u\|_{L^1(\Omega)}.$$

This leads to $\|u\|_{BV(\Omega)} \ge \|\nabla u\|_{L^1(\Omega)}$. This completes the proof. □

For a smooth domain D, the bounded variation of the characteristic function χ_D is

$$\|\chi_D\|_{BV} = \text{length of } \partial D.$$

In general, according to the co-area formula, the total variation of $u \in C^2(\bar{\Omega})$ can be expressed as

$$\int_\Omega |\nabla u|\, d\mathbf{x} = \int_{\mathbb{R}} |\partial\{\mathbf{x}\in\Omega : u(\mathbf{x}) > \lambda\}|\, d\lambda.$$

Now, we are ready to explain the TV restoration that is the case for $p = 1$ in (6.35):

$$\Phi(u) := \frac{1}{2}\int_{\Omega} |Hu - f|^2 \, d\mathbf{x} + \lambda \int_{\Omega} |\nabla u| \, d\mathbf{x}. \tag{6.47}$$

Assuming that H is a linear operator with its dual H^*, this is associated with the nonlinear Euler–Lagrange equation:

$$-\nabla\Phi(u) = \lambda\nabla \cdot \left(\frac{\nabla u}{|\nabla u|}\right) + H^*(Hu - f) = 0. \tag{6.48}$$

Hence, the minimizer u can be obtained from the standard method of solving the corresponding time marching algorithm:

$$\begin{cases} u_t = \lambda\nabla \cdot \left(\dfrac{\nabla u}{|\nabla u|}\right) - H^*(Hu - f), & \text{in } \Omega \times (0, T], \\ u(\mathbf{x}, t) = f(\mathbf{x}), & \mathbf{x} \in \Omega, \\ \dfrac{\partial u}{\partial \mathbf{n}} = 0, & \text{on } \partial\Omega \times (0, T]. \end{cases} \tag{6.49}$$

To understand TV denoising effects, consider a smooth function u satisfying

$$u_t(\mathbf{x}, t) = \nabla \cdot \left(\frac{\nabla u(\mathbf{x}, t)}{|\nabla u(\mathbf{x}, t)|}\right) \quad (\mathbf{x} \in \Omega, 0 < t < \infty). \tag{6.50}$$

For a fixed value $\alpha \in \mathbb{R}$, let ∂D_t^α be a simply closed curve in Ω such that $u(\mathbf{x}, t) = \alpha$ on ∂D_t^α and $u(\mathbf{x}, t) > \alpha$ inside D_t^α. Since u satisfies the nonlinear diffusion equation (6.50),

$$\frac{\partial}{\partial t}\int_{D_t^\alpha} (u - \alpha) \, d\mathbf{x} = \int_{D_t^\alpha} \frac{\partial u}{\partial t} \, d\mathbf{x} = \int_{D_t^\alpha} \nabla \cdot \left(\frac{\nabla u}{|\nabla u|}\right) d\mathbf{x} = \int_{\partial D_t^\alpha} \frac{\nabla u}{|\nabla u|} \cdot \mathbf{n} \, d\ell.$$

Note that the outer normal vector $\mathbf{n}$ on ∂D_t^α can be represented as

$$\mathbf{n} = -\text{sign}(u - \alpha)|_{D_t^\alpha} \frac{\nabla u}{|\nabla u|},$$

where $\text{sign}(u - \alpha)|_{D_t^\alpha}$ denotes the sign of $u(\mathbf{x}, t) - \alpha$ in D_t^α. Hence, we have

$$\int_{D_t^\alpha} \nabla \cdot \left(\frac{\nabla u}{|\nabla u|}\right) d\mathbf{x} = -|\partial D_t^\alpha| \tag{6.51}$$

and therefore

$$\frac{\partial}{\partial t}\int_{D_t^\alpha} |u - \alpha| \, d\mathbf{x} = -|\partial D_t^\alpha|$$

or

$$\frac{\partial}{\partial t}\int_{D_t^\alpha} |u - \alpha| \, d\mathbf{x} = -\int_{D_t^\alpha} \frac{|\partial D_t^\alpha|}{|D_t^\alpha|} \, d\mathbf{x}.$$

Hence, the TV regularization quickly removes smaller-scale noise where the level set ratio $|\partial D_t^\alpha|/|D_t^\alpha|$ is very large.

According to (6.48), the noisy image f can be decomposed into

$$f = u + \frac{1}{2\lambda}\nabla\cdot\left(\frac{\nabla u}{|\nabla u|}\right), \tag{6.52}$$

where u is a minimizer of the TV-based ROF (Rudin–Osher–Fatemi) model (Rudin *et al* 1992)

$$\Phi(u) := \lambda\int_\Omega |u-f|^2\,d\mathbf{x} + \int_\Omega |\nabla u|\,d\mathbf{x}.$$

With the use of the G-norm introduced by Meyer (2002), we can understand the role of the parameter λ more precisely (see Figure 6.6):

- $\|f\|_* \le 1/(2\lambda) \Rightarrow u = 0$ and $v = f$, where $\|v\|_*$ is the G-norm of v given by

$$\|v\|_* = \inf\left\{\big\|\, |\mathbf{g}|\,\big\|_{L^\infty} \;\middle|\; v = \nabla\cdot\mathbf{g},\ \mathbf{g} = (g_1, g_2) \in G\right\}.$$

  Here, G, called the G-space, is the distributional closure of the set

$$\left\{ v = \frac{\partial}{\partial x} g_1 + \frac{\partial}{\partial y} g_2 = \nabla\cdot\mathbf{g} \;\middle|\; \mathbf{g}\in C_c^1(\Omega)^2 \right\}.$$

- $\|f\|_* > 1/(2\lambda) \Rightarrow \|v\|_* = 1/(2\lambda)$ and $\int_\Omega uv\,d\mathbf{x} = [1/(2\lambda)]\,\|u\|_{BV}$.

The G-space is somehow very close to the dual of the BV space because of the following inequality:

$$\left|\int_\Omega fv\,d\mathbf{x}\right| \le \|f\|_{BV}\|v\|_*.$$

Example 6.3.5 *Numerous variations of the ROF model have been proposed. We will present some of them.*

1. *Meyer (2002) suggested*

$$u = \arg\min_{u\in BV(\Omega)}\{\|u\|_{BV} + \lambda\|f-u\|_*^2\}.$$

2. *Vese and Osher (2002) proposed*

$$u = \arg\min_{u\in BV(\Omega),\mathbf{g}}\left\{\int_\Omega |\nabla u|\,d\mathbf{x} + \lambda\int_\Omega |f-u-\nabla\cdot\mathbf{g}|^2\,d\mathbf{x} + \mu\left[\int_\Omega |\mathbf{g}|^p\,d\mathbf{x}\right]^{1/p}\right\}$$

   with $p \ge 1$.

3. *Osher et al. (2005) used the Bregman distance function for making an iterative regularization procedure:*

$$u_k = \arg\min_{u\in BV(\Omega)}\{D(u_{k-1}, u) + \lambda\|f-u\|_{L^2(\Omega)}^2\},$$

   where $D(u, v)$ is the Bregman distance between u and v associated with the functional Φ.

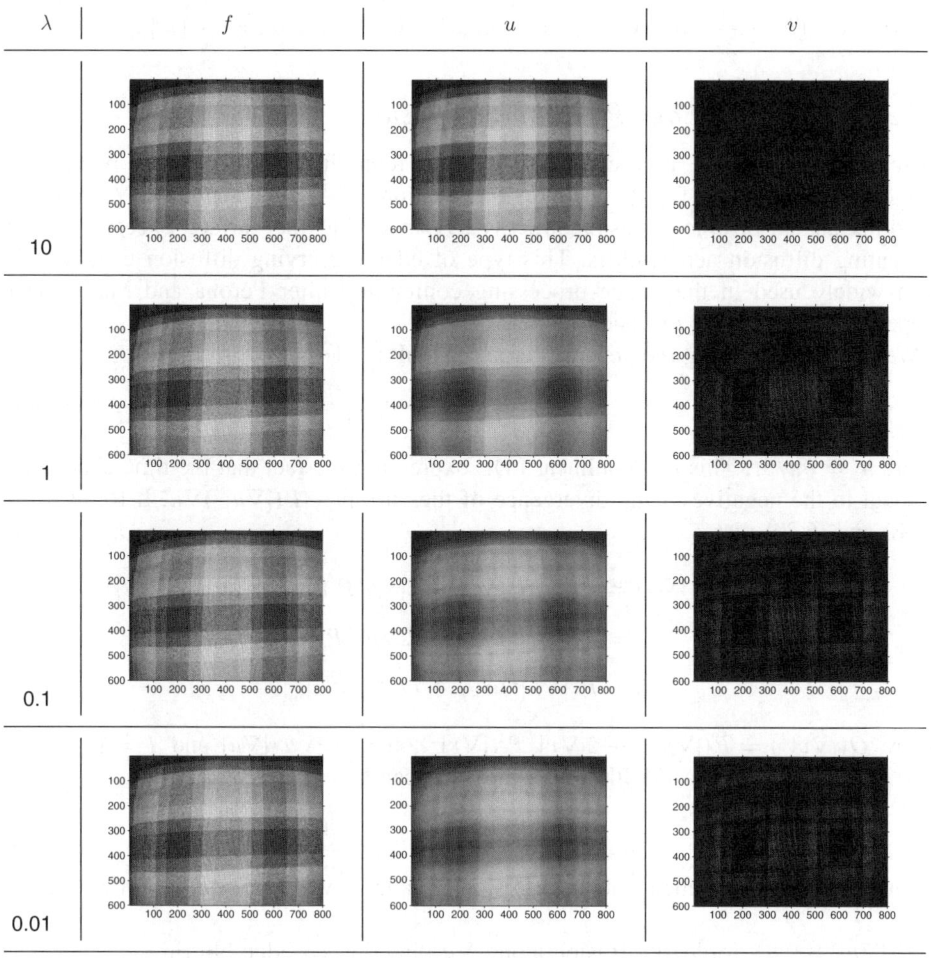

Figure 6.6 Role of the parameter λ: f is the noisy image; u is the image reconstructed by the TV-based ROF model; and $v = f - u$

4. Osher et al. (2002) suggested

$$u = \arg \min_{u \in BV(\Omega)} \{\|u\|_{BV} + \lambda \|\nabla \Delta^{-1}(f - u)\|_{L^2}^2\}.$$

Its Euler–Lagrange equation is

$$-\frac{1}{2\lambda}\nabla \cdot \left(\frac{\nabla u}{|\nabla u|}\right) + \Delta^{-1}(f - u) = 0 \quad \Longleftrightarrow \quad f - u = \frac{1}{2\lambda}\Delta\left(\nabla \cdot \left(\frac{\nabla u}{|\nabla u|}\right)\right).$$

The L^2-fitting in ROF is replaced by the H^{-1}-fitting. In this case, we have

(a) $\|\Delta^{-1} f\|_* \leq \dfrac{1}{2\lambda} \Leftrightarrow u = 0, v = f.$

$$(b)\ \|\Delta^{-1}f\|_* > \frac{1}{2\lambda} \Leftrightarrow \|\Delta^{-1}v\|_* = \frac{1}{2\lambda} \text{ and } -\int_\Omega (\Delta^{-1}v)u\, d\mathbf{x} = \frac{1}{2\lambda}\|u\|_{BV}.$$

6.3.3 Anisotropic Edge-Preserving Diffusion

Diffusion PDE can act as a smoothing tool for removing high-frequency noise in the surface $\{(\mathbf{x}, u(\mathbf{x})) : \mathbf{x} \in \Omega\}$. In the previous section, we learnt that the TV restoration performs an anisotropic diffusion process by smoothing the surface except cliffs and preventing diffusion across cliffs. This type of edge-preserving diffusion technique has been widely used in the image processing community after Perona and Malik (1990) proposed an anisotropic diffusion model.

We begin by studying the general Perona–Malik model:

$$u_t = \nabla \cdot [P(|\nabla u|^2)\nabla u], \tag{6.53}$$

where $u_t = \partial u/\partial t$. This is a nonlinear PDE expressing the fact that the time change of u is equal to the negative of the divergence of the current $-P(|\nabla u|^2)\nabla u$. It follows from (6.38) and (6.39) that

$$\begin{aligned}\nabla \cdot [P(|\nabla u|^2)\nabla u] &= P(|\nabla u|^2)\nabla^2 u + 2|\nabla u|P'(|\nabla u|^2)\nabla u \cdot \nabla(|\nabla u|)\\ &= P(|\nabla u|^2)\nabla^2 u + 2|\nabla u|^2 P'(|\nabla u|^2)u_{\mathbf{nn}}\\ &= P(|\nabla u|^2)u_{TT} + Q(|\nabla v|^2)u_{\mathbf{nn}},\end{aligned}$$

where $Q(|\nabla v|^2) = P(|\nabla v|^2) + 2|\nabla v|^2 P'(|\nabla v|^2)$, $\mathbf{n}(\mathbf{x}) = \nabla u/|\nabla u|$ and $T = (-u_y, u_x)/|\nabla u|$. Hence, the Perona–Malik model in (6.53) can be expressed as

$$\frac{\partial}{\partial t}u = P(|\nabla u|^2)u_+ Q(|\nabla u|)\frac{\partial^2}{\partial \mathbf{n}^2}u. \tag{6.54}$$

From the standard arguments about diffusion equations, we conclude the following:

- $Q(|\nabla u|) > 0 \Rightarrow$ forward diffusion along ∇u direction $\Rightarrow$ edge blurring.
- $Q(|\nabla u|) < 0 \Rightarrow$ backward diffusion along ∇u direction $\Rightarrow$ edge sharpening.

Based on this observation, Perona and Malik (1990) proposed the diffusion coefficient

$$P(|\nabla u|^2) = \frac{1}{1 + |\nabla u|^2/K^2},$$

which corresponds to the *concave* energy functional

$$\Phi(u) = \int_\Omega \ln\left(1 + \frac{|\nabla u(\mathbf{x})|^2}{K^2}\right) d\mathbf{x}.$$

Note that the corresponding Perona–Malik PDE (forward–backward diffusion equation) has the following property:

- $|\nabla u| < K \Rightarrow$ forward diffusion along ∇u direction $\Rightarrow$ blurring.
- $|\nabla u| > K \Rightarrow$ backward diffusion along ∇u direction $\Rightarrow$ edge sharpening.

Hence, the Perona–Malik PDE permits forward diffusion along the edges while sharpening across the edge (Perona and Malik 1990). However, this Perona–Malik PDE raises issues on the discrepancy between numerical and theoretical results in that it is formally ill-posed in the sense of Hadamard, whereas its discrete PDE is nevertheless numerically found to be stable by Calder and Mansouri (2011). Kichenassamy (1997) investigated this phenomenon, termed the Perona–Malik paradox, in which the Perona–Malik PDE admits stable schemes for the initial value problem without a weak solution.

Exercise 6.3.6 *The behavior of solutions of the forward heat equation $u_t = \nabla^2 u$ as t increases is very different from that of the backward heat equation $u_t = -\nabla^2 u$. Consider the one-dimensional forward–backward heat equation:*

$$u_t(x,t) = (2H(x) - 1)u_{xx}(x,t)(x \in \mathbb{R}), \quad u(x,0) = \frac{1}{1+x^2}\sin x,$$

where $H(x)$ is the Heaviside function. Explain why this PDE process causes blurring for $x > 0$ but sharpening for $x < 0$.

Remark 6.3.7 *The Perona–Malik PDE can be unstable with respect to small perturbations of the initial condition (You et al. 1996). The Perona–Malik process may produce a piecewise constant image (staircasing effect) even with a very smooth initial image (Weickert 1997).*

6.3.4 Sparse Sensing

Fast imaging in MRI by accelerating the acquisition is a very important issue since it has a wide range of clinical applications such as cardiac MRI, functional MRI (fMRI), MRE, MREIT and so on. To reduce the MR data acquisition time, we need to skip as many phase encoding lines as possible (violating the Nyquist criterion) in k-space during MRI data acquisitions to minimize the time-consuming phase encoding step. In this case, we need to deal with the under-determined linear problem such as

$$\underbrace{\begin{pmatrix} a_{11} & a_{12} & \cdots & \cdots & a_{1N} \\ \vdots & \vdots & & & \vdots \\ a_1 & a_{m2} & \cdots & \cdots & a_{mN} \end{pmatrix}}_{A} \underbrace{\begin{pmatrix} x_1 \\ x_2 \\ \vdots \\ x_N \end{pmatrix}}_{\mathbf{x}} = \underbrace{\begin{pmatrix} b_1 \\ \vdots \\ b_m \end{pmatrix}}_{\mathbf{b}} \quad \text{and} \quad m < N. \tag{6.55}$$

We know that this under-determined linear system (6.55) has an infinite number of solutions since the null space has $\dim(N(A)) \geq N - m$. Without having some knowledge about the true solution, such as sparsity (having a few non-zero entries of the solution), there is no hope of solving the under-determined linear system (6.55).

Imagine that the true solution, denoted by $\mathbf{x}_{\text{true}}$, has sparsity. Is it possible to reconstruct $\mathbf{x}_{\text{true}}$ by enforcing the sparsity constraint in the under-determined linear system in (6.55)? If so, can the solution be computed reliably? Surprisingly, this very basic linear algebra problem was not studied in depth until 1990.

Donoho and Elad (2003) found the following uniqueness result by introducing the concept of the *spark* of A:

$$\text{spark}(A) = \underbrace{\min\{\|\mathbf{x}\|_0 : A\mathbf{x} = 0 \text{ and } \|\mathbf{x}\|_2 = 1\}}_{\substack{\text{smallest number of linearly} \\ \text{dependent columns of } A}},$$

where $\|\mathbf{x}\|_0 = \sharp\{j : x_j \neq 0\}$ indicates the number of non-zero entries of $\mathbf{x}$.

Theorem 6.3.8 (Donoho and Elad 2003) *If the under-determined linear system (6.55) has a solution* $\mathbf{x}$ *obeying* $\|\mathbf{x}\|_0 < \frac{1}{2}\,\text{spark}(A)$, *this solution is necessarily the sparsest possible.*

Proof. Assume that $\mathbf{x}'$ satisfies $A\mathbf{x} = A\mathbf{x}'$ and $\|\mathbf{x}'\|_0 < \frac{1}{2}\,\text{spark}(A)$. Since $A(\mathbf{x} - \mathbf{x}') = 0$, it follows from the definition of spark that

$$\text{either} \quad \mathbf{x} = \mathbf{x}' \quad \text{or} \quad \|\mathbf{x} - \mathbf{x}'\|_0 \geq \text{spark}(A).$$

Noting that

$$\text{spark}(A) > \|\mathbf{x}\|_0 + \|\mathbf{x}'\|_0 \geq \|\mathbf{x} - \mathbf{x}'\|_0,$$

we must have $\mathbf{x} = \mathbf{x}'$. □

For $S = 1, 2, 3, \ldots$, define the set

$$W_S := \{\mathbf{x} \in \mathbb{R}^N : \|\mathbf{x}\|_0 \leq S\}.$$

Exercise 6.3.9 *Let* $S \leq \frac{1}{2}\,\text{spark}(A)$. *Show that, for* $\mathbf{x}, \mathbf{x}' \in W_S$, $\mathbf{x} \neq \mathbf{x}'$ *if and only if* $A\mathbf{x} \neq A\mathbf{x}'$.

Although the under-determined linear system (6.55) has infinitely many solutions in $\mathbb{R}^N$, according to Theorem 6.3.8, it has at most one solution within the restricted set W_S for $S < \frac{1}{2}\,\text{spark}(A)$. Hence, one may consider the following sparse optimization problem:

$$(\text{P0}) : \qquad \min \|\mathbf{x}\|_0 \quad \text{subject to} \quad A\mathbf{x} = \mathbf{b}. \tag{6.56}$$

Let $\mathbf{x}_0$ be a solution of the ℓ_0-minimization problem (P0). Unfortunately, finding $\mathbf{x}_0$ via ℓ_0-minimization is extremely difficult (NP-hard) due to lack of convexity; we cannot use Newton's iteration.

Admitting fundamental difficulties in handling the ℓ_0-minimization problem (P0), it would be desirable to find a feasible approach for solving the problem (P0). One can consider the relaxed ℓ_1-minimization problem that is the closest convex minimization problem to (P0):

$$(\text{P1}) : \qquad \min \|\mathbf{x}\|_1 \quad \text{subject to} \quad A\mathbf{x} = \mathbf{b}. \tag{6.57}$$

Let $\mathbf{x}^1$ be a solution of the ℓ_1-minimization problem (P1). Then, can the sparsest solution of (P0) be the solution of (P1)? If yes, when? Donoho and Elad (2003) observed that it could be $\mathbf{x}^0 = \mathbf{x}^1$ when $\|\mathbf{x}^0\|_0$ is sufficiently small and A has incoherent columns. Here, the mutual coherence of A measures the largest correlation between different columns from A. For more details see Donoho and Huo (1999).

For robustness of compressed sensing, Candès and Tao (2005) used the notion of the *restricted isometry property* (RIP) condition: A is said to have *RIP of order* S if there exists an isometry constant $\delta_S \in (0, 1)$ such that

$$(1-\delta_S)\|\mathbf{x}\|_2^2 \le \|A\mathbf{x}\|_2^2 \le (1+\delta_S)\|\mathbf{x}\|_2^2, \quad \forall\, \mathbf{x} \in W_S. \tag{6.58}$$

If A has RIP of order $2S$, then the under-determined linear system (6.55) is *well-distinguishable* within the S-sparse set W_S:

$$(1-\delta_{2S}) \le \frac{\|A(\mathbf{x}-\mathbf{x}')\|_2^2}{\|\mathbf{x}-\mathbf{x}\prime\|_2^2} \le (1+\delta_{2S}), \quad \forall\, \mathbf{x}, \mathbf{x}\prime \in W_S.$$

If $\delta_{2S} < 1$, then the map A is injective within the set W_S. If δ_{2S} is close to 0, the transformation A roughly preserves the distance between any two different points.

Candès *et al.* (2006b) observed that the sparse solution of the problem (P0) can be found by solving (P1) under the assumption that A obeys the $2S$*-restricted isometry property* (RIP) condition with δ_{2S} being not close to one (Candès and Tao 2005; Candès *et al.* 2006a).

Theorem 6.3.10 (Candès *et al.* 2006b) *Let $\mathbf{x}^{\text{exact}}$ be a solution of the under-determined linear system (6.55). Assume that A has RIP of order $2S$ with the isometry constant $\delta_{2S} < \sqrt{2}-1$. Then there exists a constant C_0 such that*

$$\|\mathbf{x}^{\text{exact}} - \mathbf{x}^1\|_1 + \sqrt{S}\,\|\mathbf{x}^{\text{exact}} - \mathbf{x}^1\|_2 \le C_0 \min_{\mathbf{x}\in W_S} \|\mathbf{x}^{\text{exact}} - \mathbf{x}\|_1. \tag{6.59}$$

Proof. For ease of explanation, we only prove (6.59) in the case where $N = 3$ and $S = 1$. Denote

$$\mathbf{x}^{\text{exact}} = (a, b, c) \quad \text{and} \quad \mathbf{h} = \mathbf{x}^{\text{exact}} - \mathbf{x}^1 = (h_1, h_2, h_3).$$

We may assume $|a| \ge |b| \ge |c|$. Then

$$\begin{aligned}|a| + |b| + |c| = \|\mathbf{x}^{\text{exact}}\|_1 \ge \|\mathbf{x}^1\|_1 &= \|\mathbf{x}^{\text{exact}} - \mathbf{h}\|_1 \\ &\ge |a| - |h_1| + |h_2| - |b| + |h_3| - |c|,\end{aligned}$$

which leads to

$$|h_2| + |h_3| \le |h_1| + 2(|b| + |c|) = |h_1| + 2 \min_{\mathbf{x}\in W_S} \|\mathbf{x}^{\text{exact}} - \mathbf{x}\|_1.$$

Hence, it is enough to prove that

$$|h_1| \le \alpha(|h_2| + |h_3|) \quad \text{for some} 0 < \alpha < 1 \tag{6.60}$$

since the above estimate implies that

$$\|\mathbf{h}\|_1 \le \underbrace{\frac{2\alpha}{1-\alpha}}_{C_0} \min_{\mathbf{x}\in W_S} \|\mathbf{x}^{\text{exact}} - \mathbf{x}\|_1.$$

Without loss of generality, we may assume that $|h_2| \geq |h_3|$. Denoting $\mathbf{h}_1 = (h_1, 0, 0)$, $\mathbf{h}_2 = (0, h_2, 0)$ and $\mathbf{h}_3 = (0, 0, h_3)$, we have

$$\begin{aligned}
|h_1|^2 = \|\mathbf{h}_1\|_2^2 &= \|\mathbf{h}_1 + \mathbf{h}_2\|_2^2 - \|\mathbf{h}_2\|^2 \leq \frac{1}{1-\delta_2}\|A(\mathbf{h}_1 + \mathbf{h}_2)\|_2^2 - \|\mathbf{h}_2\|^2 \\
&\leq \frac{1}{1-\delta_2}\Bigg(\underbrace{\|A(\mathbf{h})\|_2^2}_{0} + \|A(\underbrace{\mathbf{h} - \mathbf{h}_1 - \mathbf{h}_2}_{\mathbf{h}_3})\|_2^2 \Bigg) - \|\mathbf{h}_2\|_2^2 \\
&= \frac{1+\delta_2}{1-\delta_2}\|\mathbf{h}_3\|_2^2 - \|\mathbf{h}_2\|_2^2 \leq \frac{1}{2}\frac{1+\delta_2}{1-\delta_2}\|\mathbf{h}_2 + \mathbf{h}_3\|_2^2 - \frac{1}{2}\|\mathbf{h}_2 + \mathbf{h}_3\|_2^2 \\
&\leq \underbrace{\frac{\delta_2}{1-\delta_2}}_{<1}\|\mathbf{h}_2 + \mathbf{h}_3\|_2^2 \leq \frac{\delta_2}{1-\delta_2}(|h_2| + |h_3|)^2.
\end{aligned}$$

This completes the proof of (6.60) since $\delta_2/(1-\delta_2) < 1$ for $\delta_2 < \sqrt{2} - 1$. □

From the above theorem, if the true solution $\mathbf{x}^{\text{exact}}$ is S-sparse, then $\mathbf{x}_1 = \mathbf{x}^{\text{exact}} = \mathbf{x}_0$, where $\mathbf{x}_0$ is a solution of (P0). Theorem 6.3.10 also provides robustness of compressed sensing. Consider the case where the measurements $\mathbf{b}$ are corrupted by bounded noise in such a way that

$$\mathbf{b} = A\mathbf{x} + \mathbf{n}_{\text{noise}} \quad \text{with} \quad \|\mathbf{n}_{\text{noise}}\| \leq \epsilon.$$

Corollary 6.3.11 *Let $\tilde{\mathbf{x}}^1$ be a solution of the modified ℓ_1-problem:*

$$\min \|\mathbf{x}\|_1 \quad \textit{subject to} \quad \|\mathbf{b} - A\mathbf{x}\|_2 \leq \epsilon. \tag{6.61}$$

Under the assumption of Theorem 6.3.10, if $\mathbf{x}_S$ is the best S-sparse approximation of $\mathbf{x}^{\text{exact}}$, then

$$\|\mathbf{x}^{\text{exact}} - \tilde{\mathbf{x}}^1\|_2 \leq \frac{C_0}{\sqrt{S}}\|\mathbf{x}^{\text{exact}} - \mathbf{x}_S\|_1 + C_1\epsilon$$

for some constants C_0 and $C_1 > 0$.

Lustig *et al.* (2007) applied these sparse sensing techniques for fast MR imaging. They demonstrated high acceleration in *in vivo* experiments and showed that the sparsity of MR images can be exploited to reduce scan time significantly, or alternatively to improve the resolution of MR images.

6.4 Segmentation

Image segmentation of a target object in the form of a closed curve has numerous medical applications, such as anomaly detection, quantification of tissue volume, planning of surgical interventions, motion tracking for functional imaging and others. With advances in medical imaging technologies, many innovative methods of performing segmentation have been proposed over the past few decades, and these segmentation techniques are based on the basic recipes using thresholding and edge-based detection. In this section, we only consider edge-based methods, which use the strength of the image gradient along the boundary between the target object and the background.

6.4.1 Active Contour Method

The most commonly used method of segmentation in the form of a closed curve would be (explicit or implicit) active contour methods that use an application-dependent energy functional to evolve an active contour toward the boundary of the target region; the direction of the velocity of the active contour is the negative direction of the gradient of the energy functional. To be precise, let u be a given image. We begin by considering the minimization problem of finding a closed curve $\mathcal{C}$ that minimizes the energy functional

$$\Phi(\mathcal{C}) := \int_{\mathcal{C}} g(|\nabla u|)\,\mathrm{d}s, \tag{6.62}$$

where $g(\alpha)$ is a decreasing function, for example, $g(\alpha) = 1/(1+\alpha)$. For computation of a local minimum $\mathcal{C}$ of the functional, we may start from an initial contour $\mathcal{C}_0$ and consider a sequence $\mathcal{C}^1, \mathcal{C}^2, \ldots$ that converges to the local minima $\mathcal{C}$:

$$\Phi(\mathcal{C}^n) \searrow \Phi(\mathcal{C}).$$

For computation of $\mathcal{C}^n$, imagine that the sequence of curves $\{\mathcal{C}^n\}$ is parameterized by $\mathbf{r}(s,n) = x(s,n)\hat{\mathbf{x}} + y(s,n)\hat{\mathbf{y}}$, $0 < s < 1$:

$$\mathcal{C}^n = \{\mathbf{r}(s,n) = x(s,n)\hat{\mathbf{x}} + y(s,n)\hat{\mathbf{y}} \mid 0 < s < 1\}.$$

Then, the energy functional $\phi(\mathcal{C}^n)$ at $\mathcal{C}^n$ can be expressed as

$$\Phi(\mathcal{C}^n) := \int_0^1 g(\nabla u(\mathbf{r}(s,n)))|\mathbf{r}'(s,n)|\,\mathrm{d}s.$$

To calculate the next curve $\mathcal{C}^{n+1}$ from $\mathcal{C}^n$, the gradient descent method based on the Fréchet gradient $-\nabla\Phi(\mathcal{C}^n)$ is widely used. To determine the Fréchet gradient $-\nabla\Phi(\mathcal{C}^n)$, it is convenient to consider a time-varying contour $\mathcal{C}^t$ instead of the sequence $\{\mathcal{C}^n\}$. See Figure 6.7 for the time-varying contour.

Setting $\Phi(t) = \Phi(\mathcal{C}^t)$, we have

$$\Phi(t) := \int_0^1 \tilde{g}(\mathbf{r}(s,t))|\mathbf{r}_s(s,t)|\,\mathrm{d}s, \quad \tilde{g}(\mathbf{r}) = g(\nabla u(\mathbf{r})).$$

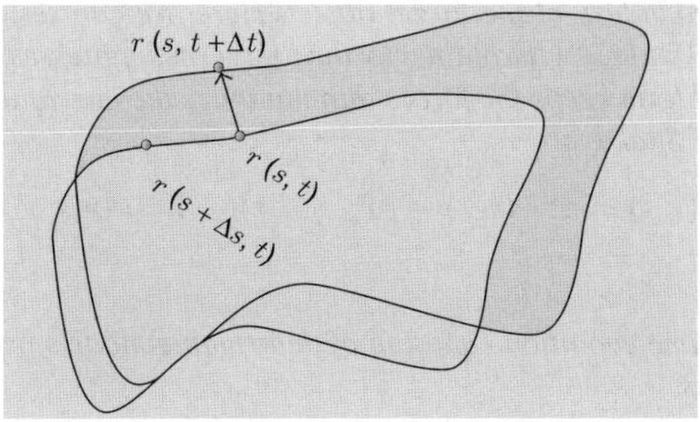

Figure 6.7 Time-varying contour

The variation of the energy functional Φ is

$$\begin{aligned}\Phi'(t) &= \int_0^1 |\mathbf{r}_s|[\nabla\tilde{g}\cdot\mathbf{r}_t]\,ds + \int_0^1 \tilde{g}\left[\frac{\mathbf{r}_s}{|\mathbf{r}_s|}\cdot\mathbf{r}_{ts}\right]ds\\ &= \int_0^1 |\mathbf{r}_s|[\nabla\tilde{g}\cdot\mathbf{r}_t]\,ds - \int_0^1 \left\langle \tilde{g}\left[\frac{\mathbf{r}_s}{|\mathbf{r}_s|}\right]_s + [\nabla\tilde{g}\cdot\mathbf{r}_s]\frac{\mathbf{r}_s}{|\mathbf{r}_s|}, \mathbf{r}_t\right\rangle ds\\ &= \int_0^1 |\mathbf{r}_s|\mathbf{r}_t\cdot[\nabla\tilde{g} - \kappa\tilde{g}\mathbf{n} - \langle T, \nabla\tilde{g}\rangle T]\,ds,\end{aligned}$$

where $\mathbf{r}_t = \partial\mathbf{r}/\partial t$, $\mathbf{r}_s = \partial\mathbf{r}/\partial s$, $\mathbf{n} = \mathbf{n}(s,t)$ is the unit normal to the curve C^t, and $T = T(s,t)$ is the unit tangent vector. Hence, the direction for which $\Phi(t)$ decreases most rapidly is given by

$$\mathbf{r}_t = -[\nabla\tilde{g} - \kappa\tilde{g}\mathbf{n} - \langle T, \nabla\tilde{g}\rangle T]. \tag{6.63}$$

Decomposing $\nabla\tilde{g} = \langle\nabla\tilde{g}, \mathbf{n}\rangle\mathbf{n} + \langle\nabla\tilde{g}, T\rangle T$, equation (6.63) becomes

$$\mathbf{r}_t = \underbrace{(\kappa\tilde{g} - \langle\nabla\tilde{g}, \mathbf{n}\rangle)}_{\pm\text{ speed } |\mathbf{r}_t|}\,\mathbf{n} \quad\text{or}\quad \frac{\partial}{\partial t}C^t = (\kappa g(|\nabla u|) - \nabla g(|\nabla u|)\cdot\mathbf{n})\mathbf{n}, \tag{6.64}$$

which means that the curve $\mathbf{r}(s,t)$ moves along its normal with speed

$$F = \kappa\tilde{g} - \langle\nabla\tilde{g}, \mathbf{n}\rangle. \tag{6.65}$$

Since

$$\frac{\mathbf{r}(s, t+\Delta t) - \mathbf{r}(s,t)}{\Delta t} \approx F(\mathbf{r}(s,t))\mathbf{n}(s,t),$$

we can determine the update C^{n+1} by

$$\mathbf{r}(s, n+1) = \mathbf{r}(s,n) + \Delta t F(\mathbf{r}(s,n))\mathbf{n}(s,n). \tag{6.66}$$

Exercise 6.4.1 *The energy functional Snake (Kass et al. 1987) is defined as*

$$\Phi(\mathbf{r}) = \frac{\alpha}{2}\int_0^1 |\mathbf{r}_s|^2\,ds + \frac{\beta}{2}\int_0^1 |\mathbf{r}_{ss}|^2\,ds + \int_0^1 g(u(\mathbf{r}))\,ds,$$

where the first two terms are regularization terms and $g(u(\mathbf{r}))$ is used for attracting the contour toward the boundary of the target object. Here, we can set $g(u(\mathbf{r})) = -|\nabla u(\mathbf{r})|^2$ or $g(u(\mathbf{r})) = 1/[1 + |\nabla u(\mathbf{r})|^p]$. Explain why the first term suppresses the forces that shrink Snake and the second term keeps the forces that minimize the curvature. Show that the time evolution equation of Snake is

$$\mathbf{r}_t(s,t) = \alpha\mathbf{r}_{ss}(s,t) - \beta\mathbf{r}_{ssss}(s,t) - \nabla g(u(\mathbf{r}(s,t))).$$

Exercise 6.4.2 *The time evolution equation of deformable models by gradient vector flow (GVF) is expressed as*

$$\mathbf{r}_t(s,t) = \alpha\mathbf{r}_{ss}(s,t) - \beta\mathbf{r}_{ssss}(s,t) + \mathbf{v}, \tag{6.67}$$

where $\mathbf{v}$ *is the minimizer of the energy functional:*

$$\int \mu|\nabla \mathbf{v}|^2 + |\nabla u|^2|\mathbf{v} - \nabla u|^2 \,\mathrm{d}x\,\mathrm{d}y, \tag{6.68}$$

where μ *is a regularization parameter and the gradient operator* ∇ *is applied to each component of* $\mathbf{v}$ *separately. Show that* $\mathbf{v}$ *can be achieved by solving the following equation:*

$$\mathbf{v}_t = \mu \underbrace{\nabla^2 \mathbf{v}}_{\text{diffusion}} \underbrace{-(\mathbf{v} - \nabla u)|\nabla u|^2}_{\text{data attraction}}.$$

6.4.2 Level Set Method

The active contour scheme using the explicit expression $\mathcal{C}^t = \{\mathbf{r}(s,t) : 0 \leq s \leq 1\}$ is not appropriate for segmenting multiple targets whose locations are unknown. Using an auxiliary function $\phi(\mathbf{r}, t)$, the propagating contour $\mathcal{C}^t$ changing its topology (splitting multiple closed curves) can be expressed effectively by the implicit expression of the zero level set

$$\mathcal{C}^t = \{\mathbf{r} : \phi(\mathbf{r}, t) = 0\}.$$

With the level set method (Osher *et al.* 2002), the motion of the active contour with the explicit expression (6.64)) is replaced by the motion of level set. Using the property that

$$\phi(\mathbf{r}(s,t), t) = 0,$$

we have the equation of ϕ containing the embedded motion of $\mathcal{C}^t$; for a fixed s,

$$0 = \frac{\mathrm{d}}{\mathrm{d}t}\phi(\mathbf{r}(s,t), t) = \frac{\partial}{\partial t}\phi(\mathbf{r}, t) + \frac{\partial}{\partial t}\mathbf{r}(s,t) \cdot \nabla\phi(\mathbf{r}(s,t), t).$$

Since

$$\frac{\partial}{\partial t}\mathbf{r}(s,t) = F(\mathbf{r}(s,t), t)\mathbf{n}(s,t) \quad \text{and} \quad \mathbf{n}(s,t) = \frac{\nabla\phi(\mathbf{r}(s,t), t)}{|\nabla\phi(\mathbf{r}(s,t), t)|},$$

the above identity leads to

$$\frac{\partial}{\partial t}\phi(\mathbf{r}, t) + F(\mathbf{r}, t)|\nabla\phi(\mathbf{r}, t)| = 0. \tag{6.69}$$

Consider the special model:

$$\frac{\partial}{\partial t}\phi(\mathbf{r}, t) = \underbrace{\nabla \cdot \left(g(|\nabla u|)\frac{\nabla u}{|\nabla u|}\right)}_{-F} |\nabla\phi(\mathbf{r}, t)|.$$

This can be expressed as

$$\phi_t(\mathbf{r}, t) = \left(g(|\nabla u|)\left[\nabla \cdot \frac{\nabla\phi}{|\nabla\phi|}\right] + \frac{\nabla\phi}{|\nabla\phi|} \cdot \nabla g(|\nabla u|)\right)|\nabla\phi|. \tag{6.70}$$

The convection term $\langle\nabla g, \nabla\phi\rangle$ increases the attraction of the deforming contour toward the boundary of an object. Note that (6.70) is related to the geodesic active contour model (6.64).

The level set method is one of the most widely used segmentation techniques; it can be combined with the problem of minimizing the energy functional of a level set function φ:

$$E(\varphi) = \mathrm{Fit}_\varphi + \mu\,\mathrm{Reg}_\varphi, \tag{6.71}$$

where Fit_φ is a fitting term for attracting the zero-level contour $C_\varphi := \{\varphi = 0\}$ toward the target object in the image, Reg_φ is a regularization term of φ for penalizing non-smoothness of the contour, and μ is the regularization parameter. There exist a variety of fitting models, such as edge-based methods (Caselles *et al.* 1993, 1997; Goldenberg *et al.* 2001; Kichenassamy *et al.* 1995; Malladi *et al.* 1995), region-based methods (Chan and Vese 2001; Paragios and Deriche 1998; Yezzi *et al.* 1999), methods based on prior information (Chen *et al.* 2002) and so on. These fitting models are mostly combined with the standard regularization term penalizing the arc length of the contour C_φ. Among the variety of fitting models, the best-fitting model has to be selected depending on characteristics of the image. The selected fitting model is usually combined with an appropriate regularization term.

For example, Chan and Vese (2001) used the following energy functional (Chan–Vese model):

$$\Phi(\phi) = \int |\nabla H(\phi(\mathbf{r}))|\,d\mathbf{r} + \lambda_1 \int H(\phi(\mathbf{r}))\,|u(\mathbf{r}) - ave_{\{\phi\geq 0\}}|^2\,d\mathbf{r}$$
$$+ \lambda_2 \int H(-\phi(\mathbf{r}))\,|u(\mathbf{r}) - ave_{\{\phi<0\}}|^2\,d\mathbf{r}, \tag{6.72}$$

where λ_1 and λ_2 are non-negative parameters, ϕ is the level set function, and $ave_{\{\phi\geq 0\}}$ and $ave_{\{\phi<0\}}$ are the average values of $u(\mathbf{r})$ in the two-dimensional regions $\{\phi < 0\}$ and $\{\phi > 0\}$, respectively. Here, $H(\phi)$ is the one-dimensional Heaviside function, with $H(s) = 1$ if $s \geq 0$, and $H(s) = 0$ if $s < 0$. To compute a minimizer ϕ for the minimization problem (6.72), the following parabolic equation is solved to the steady state:

$$\frac{\partial\phi}{\partial t} = |H\prime(\phi)|\left[\nabla\cdot\left(\frac{\nabla\phi}{|\nabla\phi|}\right) - \lambda_1(u(\mathbf{r}) - ave_{\{\phi\geq 0\}})^2 + \lambda_2(u(\mathbf{r}) - ave_{\{\phi<0\}})^2\right], \tag{6.73}$$

with an appropriate initial level set function. After the evolution comes to a converged state, the zero level set of ϕ becomes the contour that separates the object from the background, as shown in Figure 6.8.

Example 6.4.3 *Sarti et al. (2005) proposed the energy functional using the maximum likelihood estimator in ultrasound images with the Rayleigh distribution as the intensity distribution. Using the level set formulation, the energy functional can be expressed as*

$$\Phi(\varphi) = \mu\int |\nabla H(\varphi)|\,d\mathbf{r} - n_i \ln\left(\frac{1}{n_i}\int_\Omega u^2 H(\varphi)\,d\mathbf{r}\right)$$
$$- n_e \ln\left(\frac{1}{n_e}\int_\Omega u^2(1 - H(\varphi))\,d\mathbf{r}\right). \tag{6.74}$$

Then, the associated flow equation is

$$\frac{\partial\varphi}{\partial t} = \delta(\varphi)\left\{\mu\kappa + \ln\left(\frac{\sigma_i^2}{\sigma_e^2}\right) + \frac{u^2}{2\sigma_i^2} - \frac{u^2}{2\sigma_e^2}\right\}, \tag{6.75}$$

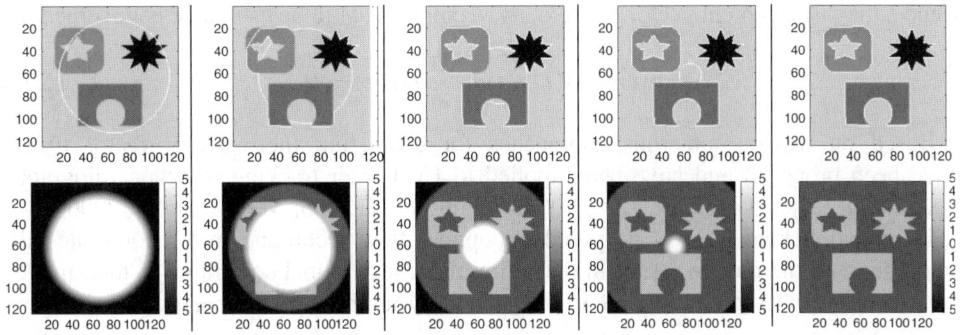

Figure 6.8 Segmentation of different objects with different intensities, with ϕ initialized as a signed distance function. The images at the top show the evolution of the zero level set. The images at the bottom show the evolution of the corresponding level set function. These are the converged results after 323 iterations

where

$$\sigma_i^2 = \frac{\int u^2 H(\varphi)\, d\mathbf{r}}{2\int H(\varphi)\, d\mathbf{r}} \quad \textit{and} \quad \sigma_e^2 = \frac{\int u^2(1-H(\varphi))\, d\mathbf{r}}{2\int (1-H(\varphi))\, d\mathbf{r}}.$$

6.4.3 Motion Tracking for Echocardiography

Ultrasound imaging is widely used because of its non-invasiveness, real-time monitoring, cost-effectiveness and portability compared with other medical imaging modalities such as CT or MRI. In particular, owing to its high temporal resolution, cardiac ultrasound (echocardiogram) has been very successful in providing a quick assessment of the overall health of the heart. For quantitative assessment of cardiac functions, wall motion tracking and left ventricle (LV) volume quantification at each time are needed. Since manual delineation from echocardiography is extremely labor-intensive and time-consuming, the demands for automated analysis methods are rapidly increasing.

The motion tracking of LV is carried out by observing the speckle pattern associated with deforming tissue. Speckle pattern is inherent appearance in ultrasound imaging and its local brightness reflects the local echogeneity of the underlying scatterers. Since it is a difficult task to automatically track the motion of the endocardial border in ultrasound images because of the poor image quality, low contrast and edge dropout by weak signals, some user intervention is required for stable and successful tracking of the endocardial border.

The most commonly used method for motion tracking is optical flow methods, which use the assumption that the intensity of a moving object is constant over time. If $I(\mathbf{r}, t)$ represents the intensity of echocardiography at the location $\mathbf{r} = (x, y)$ and the time t, a voxel at $(\mathbf{r}, t)$ will move by $\Delta\mathbf{r}$ between two image frames that are taken at times t and $t + \Delta t$:

$$I(\mathbf{r}, t) = I(\mathbf{r} + \Delta\mathbf{r}, t + \Delta t).$$

From Taylor's expansion, the time-varying images $I(\mathbf{r}, t)$ approximately satisfy

$$0 = I(\mathbf{r} + \Delta\mathbf{r}, t + \Delta t) - I(\mathbf{r}, t) \approx \Delta\mathbf{r} \cdot \nabla I(\mathbf{r}, t) + \frac{\partial}{\partial t} I(\mathbf{r}, t).$$

Hence, the displacement vector (or velocity vector) $\mathbf{u}(\mathbf{r}, t)$ to be estimated is governed by

$$\mathbf{u}(\mathbf{r}, t) \cdot \nabla I(\mathbf{r}, t) + \frac{\partial}{\partial t} I(\mathbf{r}, t) = 0. \tag{6.76}$$

Based on (6.76), numerous approaches for estimating the velocity vector $\mathbf{u}(\mathbf{r}, t)$ have been proposed and have been applied to LV border tracking in echocardiography (Duan *et al.* 2009; Linguraru *et al.* 2008; Suhling *et al.* 2005; Veronesi *et al.* 2006).

Horn and Schunk (1981) proposed an optical flow technique that incorporates the smoothness of the motion vector in the entire image as a global constraint. In their model, the velocity $\mathbf{u}(\mathbf{r}, t)$ at each time t is determined by minimizing the energy functional:

$$E_t(\mathbf{u}) := \int_\Omega \left(\mathbf{u}(\mathbf{r}) \cdot \nabla I(\mathbf{r}, t) + \frac{\partial}{\partial t} I(\mathbf{r}, t) \right)^2 + \lambda |\nabla \mathbf{u}(\mathbf{r})|^2 \, \mathrm{d}\mathbf{r}, \tag{6.77}$$

where Ω is the image domain and λ is a regularization parameter that controls the balance between the optical flow term and the smoothness on $\mathbf{u}$. The velocity $\mathbf{u}(\mathbf{r}, t)$ at each time t can be computed by solving the corresponding Euler–Lagrange equation, which is a reaction–diffusion equation. Barron *et al.* (1994) observed that this global method with the global smoothness constraint is significantly more sensitive to noise than the local method used by Lucas and Kanade (1981).

Lucas and Kanade (1981) used the assumption of locally constant motion to compute the velocity $\mathbf{u}(\mathbf{r}_0, t)$ at a target location $\mathbf{r}_0 = (x_0, y_0)$ and time t by forcing a constant velocity in a local neighborhood of a point $\mathbf{r}_0 = (x_0, y_0)$, denoted by $\mathcal{N}(\mathbf{r}_0)$. They estimated the velocity $\mathbf{u}(\mathbf{r}_0, t)$ by minimizing the weighted least-squares criterion in the neighborhood $\mathcal{N}(\mathbf{r}_0)$:

$$\mathbf{u}(\mathbf{r}_0, t) := \arg\min_{\mathbf{u}} \int_{\mathcal{N}(\mathbf{r}_0)} \left[w(\mathbf{r} - \mathbf{r}_0) \left(\mathbf{u} \cdot \nabla I(\mathbf{r}, t) + \frac{\partial}{\partial t} I(\mathbf{r}, t) \right)^2 \right] \mathrm{d}\mathbf{r}, \tag{6.78}$$

where w is a weight function that enables more relevance to be given to central terms rather than the ones in the periphery. Since this method determines $\mathbf{u}(\mathbf{r}_0, t)$ at each location $\mathbf{r}_0$ by combining information from all pixels in the neighborhood of $\mathbf{r}_0$, it is reasonably robust against image noise.

Let us explain the numerical algorithm for the Lucas–Kanade method. We denote the endocardial border traced at an initially selected frame (for example, end-systole or end-diastole frame) by a parametric contour $\mathcal{C}^* = \{\mathbf{r}^*(s) = (x^*(s), y^*(s)) \mid 0 \le s \le 1\}$ that can be identified as its n control points $\mathbf{r}_1^* = \mathbf{r}^*(s_0), \ldots, \mathbf{r}_n^* = \mathbf{r}^*(s_n)$. Here, we have $0 = s_1 < s_2 < \cdots < s_n = 1$. Let $\mathcal{C}(t) = \{\mathbf{r}(s, t) = (x(s, t), y(s, t)) \mid 0 \le s \le 1\}$ be the contour deformed from $\mathcal{C}(0) = \mathcal{C}^*$ at time t. The motion of the contour $\mathcal{C}(t)$ will be determined by an appropriately chosen velocity $\mathbf{U}_t$ indicating a time change of control points $(\mathbf{r}_1(t), \ldots, \mathbf{r}_n(t))$:

$$\mathbf{U}(t) := \begin{bmatrix} \mathbf{u}_1(t) \\ \vdots \\ \mathbf{u}_n(t) \end{bmatrix} = \frac{\mathrm{d}}{\mathrm{d}t} \begin{bmatrix} \mathbf{r}_1(t) \\ \vdots \\ \mathbf{r}_n(t) \end{bmatrix} \quad \text{with} \quad \begin{bmatrix} \mathbf{r}_1(0) \\ \vdots \\ \mathbf{r}_n(0) \end{bmatrix} = \begin{bmatrix} \mathbf{r}_1^* \\ \vdots \\ \mathbf{r}_n^* \end{bmatrix}.$$

Here, we identify the contour $\mathcal{C}(t)$ with control points $(\mathbf{r}_1(t), \ldots, \mathbf{r}_n(t))$.

In the Lucas–Kanade method, $\mathbf{U}(t)$ for each time t is a minimizer of the energy functional:

$$\mathbf{U}(t) = \arg\min_{\mathbf{U}} \mathcal{E}_t(\mathbf{U})$$
$$:= \frac{1}{2}\sum_{i=1}^{n}\left[\int_{\mathcal{N}(\mathbf{r}_i(t))} w(\mathbf{r}' - \mathbf{r}_i(t)) \left\{\mathbf{u}_i \cdot \nabla I(\mathbf{r}', t) + \frac{\partial}{\partial t} I(\mathbf{r}', t)\right\}^2 \mathrm{d}\mathbf{r}'\right]. \tag{6.79}$$

To derive the Euler–Lagrange equation from (6.79), we take the partial derivative of $\mathcal{E}_t$ with respect to each $\mathbf{u}_j$:

$$\mathbf{0} = \frac{\partial \mathcal{E}_t}{\partial \mathbf{u}_j} = \int_{\mathcal{N}(\mathbf{r}_j(t))} w(\mathbf{r})\nabla I(\mathbf{r}, t)\{\nabla I(\mathbf{r}, t) \cdot \mathbf{u}_j + I_t(\mathbf{r}, t)\}\,\mathrm{d}\mathbf{r} \quad \text{for}\, j = 1, \ldots, n. \tag{6.80}$$

For a numerical algorithm, we replace the integral over $\mathcal{N}(\mathbf{r}_j(t))$ by summation over pixels around $\mathbf{r}_j(t)$. Assuming that the neighborhood $\mathcal{N}(\mathbf{r}_j(t))$ consists of m pixels $\mathbf{r}_{j1}, \ldots, \mathbf{r}_{jm}$, equation (6.80) becomes

$$\mathbf{0} = A_j^{\mathrm{T}} W_j A_j \mathbf{u}_j + A_j^{\mathrm{T}} W_j \mathbf{b}_j, \tag{6.81}$$

where $A_j = [\nabla I(\mathbf{r}_{j1}, t), \ldots, \nabla I(\mathbf{r}_{jm}, t)]^{\mathrm{T}}$, $W_j = \mathrm{diag}(w(\mathbf{r}_{j1}), \ldots, w(\mathbf{r}_{jm}))$ and $\mathbf{b}_j = [I_t(\mathbf{r}_{j1}, t), \ldots, I_t(\mathbf{r}_{jm}, t)]^{\mathrm{T}}$. As we mentioned before, there often exist some incorrectly tracked points due to weak signals on the cardiac wall, since echocardiography data are acquired through transmitting and receiving ultrasound signals between the ribs, causing considerable shadowing of the cardiac wall (Leung *et al.* 2011). Owing to these incorrectly tracked points, the Lucas–Kanade method may produce a distorted LV shape.

Example 6.4.4 *Suhling et al. (2005) improved the Lucas–Kanade method (6.78) by introducing a linear model for the velocity along the time direction, and the displacement* $\mathbf{u}(\mathbf{r}_0, t)$ *is obtained by evaluating* $\mathbf{u}$ *such that* $\mathbf{u}, \mathbf{b} \in \mathbb{R}^2$ *and the* 2×2 *matrix* A *minimizes the following energy functional:*

$$E_t(\mathbf{u}, A, \mathbf{b}) := \int_{-\Delta t}^{\Delta t}\int_{\mathcal{N}(\mathbf{r}_0)} w(\mathbf{r} - \mathbf{r}_0, s)$$
$$\times \left(\left(\mathbf{u} + A(\mathbf{r} - \mathbf{r}_0) + s\mathbf{b}\right) \cdot \nabla I(\mathbf{r}, t + s) + \frac{\partial}{\partial t} I(\mathbf{r}, t + s)\right)^2 \mathrm{d}\mathbf{r}\,\mathrm{d}s. \tag{6.82}$$

Since this method uses multiple frames between $t - \Delta t$ *and* $t + \Delta t$*, it is more robust than the Lucas–Kanade method (6.78) using a single frame at* t*.*

Example 6.4.5 *Compared with the approaches based on the Lucas–Kanade method, Duan et al. (2009) used the region-based tracking method (also known as the block matching or pattern matching method) with the cross-correlation coefficients as a similarity* *measure. Given two consecutive images I at times t and $t + \Delta t$, the velocity vector $\mathbf{u} = (u, v)$ for each pixel $\mathbf{r} = (x, y) \in \Omega$ is estimated by maximizing the cross-correlation coefficients:*

$$\mathbf{u}(\mathbf{r}_0, t) := \arg\max_{\mathbf{u}} \left\{ \frac{\int_{\mathcal{N}(\mathbf{r}_0)} [I(\mathbf{r}, t)\, I(\mathbf{r} + \mathbf{u}, t + \Delta t)]\, d\mathbf{r}}{\sqrt{\int_{\mathcal{N}(\mathbf{r}_0)} [I(\mathbf{r}, t)]^2\, d\mathbf{r}}\sqrt{\int_{\mathcal{N}(\mathbf{r}_0)} [I(\mathbf{r} + \mathbf{u}, t + \Delta t)]^2\, d\mathbf{r}}} \right\}. \tag{6.83}$$

The block matching method uses similarity measures that are less sensitive to noise, fast motion and potential occlusions and discontinuities (Duan et al. 2009).

References

Aubert G and Kornprobst P 2002 *Mathematical Problems in Image Processing: Partial Differential Equations and the Calculus of Variations*, 2nd edn. Applied Mathematical Sciences Series, vol. 147. Springer, New York.

Barron JL, Fleet DJ and Beauchemin SS 1994 Performance of optical flow techniques. *Int. J. Comput. Vision* **12**(1), 43–77.

Beer 1852 Bestimmung der Absorption des rothen Lichts in farbigen Flussigkeiten. *Ann. Phys. Chem.* **86**, 78–88.

Bracewell RN 1956 Strip integration in radioastronomy. *J. Phys.* **9**, 198–217.

Bracewell RN and Riddle AC 1967 Inversion of fan-beam scans in radio-astronomy. *Astrophys. J.* **150**, 427–434.

Brooks RA and Chiro GD 1976 Beam hardening in x-ray reconstructive tomography. *Phys. Med. Biol.* **21**, 390–398.

Bydder M, Larkman DJ and Hajnal JV 2002 Generalized SMASH imaging. *Magn. Reson. Med.* **47**, 160–170.

Calder J and Mansouri A 2011 Anisotropic image sharpening via well-posed Sobolev gradient flows. *SIAM J. Math. Anal.* **43**, 1536–1556.

Candès EJ and Tao T 2005 Decoding by linear programming. *IEEE Trans. Inform. Theory* **51**(12), 4203–4215.

Candès EJ and Tao T 2006 Near-optimal signal recovery from random projections: universal encoding strategies. *IEEE Trans. Inform. Theory* **52**, 5406–5425.

Candès EJ, Romberg J and Tao T 2006a Robust uncertainty principles: exact signal reconstruction from highly incomplete frequency information. *IEEE Trans. Inform. Theory* **52**(2), 489–509.

Candès EJ, Romberg JK and Tao T 2006b Stable signal recovery from incomplete and inaccurate measurements. *Commun. Pure Appl. Math.* **59**, 1207–1223.

Canny JA 1986 Computational approach to edge detection. *IEEE Trans. Pattern Anal. Machine Intell.* **8**(6), 679–698.

Caselles V, Catte F, Coll T and Dibos F 1993 A geometric model for active contours in image processing. *Numer. Math.* **66**(1), 1–31.

Caselles V, Kimmel R and Sapiro G 1997 Geodesic active contours. *Int. J. Comput. Vision* **22**(1), 61–79.

Chan TF and Vese LA 2001 Active contours without edges. *IEEE Trans. Image Process.* **10**(2), 266–277.

Chen Y, Tagare HD, Thiruvenkadam S, Huang F, Wilson D, Gopinath KS, Briggs RW and Geiser EA 2002 Using prior shapes in geometric active contours in a variational framework. *Int. J. Comput. Vision* **50**(3), 315–328.

Cormack AM 1963 Representation of a function by its line integrals, with some radiological applications. *J. Appl. Phys.* **34**, 2722–2727.

Cormack AM 1992 75 years of radon transform. *J. Comput. Assist. Tomogr.* **16**, 673.

Damadian RV 1971 Tumor detection by nuclear magnetic resonance. *Science* **171**, 1151–1153.

Donoho DL and Elad M 2003 optimally sparse representation in general (non-orthogonal) dictionaries via ℓ_1 minimization. *Proc. Natl Acad. Sci. USA* **100**, 2197–2202.

Donoho DL and Huo X 1999 Uncertainty principles and ideal atomic decomposition. *IEEE Trans. Inform. Theory* **47**, 2845–2862.

Duan Q, Angelini ED, Herz SL, Ingrassia CM, Costa KD, Holmes JW, Homma S and Laine AF 2009 Region-based endocardium tracking on real-time three-dimensional ultrasound. *Ultrasound Med. Biol*. **35**(2), 256–265.

Faridani A 2003 Introduction to the mathematics of computed tomography. In *Inside Out: Inverse Problems and Applications* (ed. G Uhlmann). Math. Sci. Res. Inst. Publications, vol. 47, pp. 1–46. Cambridge University Press, Cambridge.

Feldkamp LA, Davis LC and Kress JW 1984 Practical cone-beam algorithm. *J. Opt. Soc. Am. A* **1**(6), 612–619.

Filler AG 2010 The history, development, and impact of computed imaging in neurological diagnosis and neurosurgery: CT, MRI, DTI *Internet J. Neurosurg*. **7**(1).

Goldenberg R, Kimmel R, Rivlin E and Rudzsky M 2001 Fast geodesic active contours. *IEEE Trans. Image Process*. **10**(10), 1467–1475.

Grangeat P 1991 Mathematical framework of cone beam 3D reconstruction via the first derivative of the Radon transform. In *Mathematical Methods in Tomography (Oberwolfach, 1990)*. Lecture Notes in Mathematics, no. 1497, pp. 66–97. Springer, Berlin.

Griswold MA, Jakob PM, Heidemann RM, Nittka M, Jellus V, Wang J, Kiefer B and Haase A 2002 Generalized autocalibrating partially parallel acquisitions (GRAPPA). *Magn. Reson. Med*. **47**, 1202–1210.

Haacke E, Brown R, Thompson M and Venkatesan R 1999 *Magnetic Resonance Imaging Physical Principles and Sequence Design*. John Wiley & Sons, Inc., New York.

Heidemann RM, Griswold MA, Haase A and Jakob PM 2001 VD-AUTO-SMASH imaging. *Magn. Reson. Med*. **45**, 1066–1074.

Horn B and Schunk B 1981 Determining optical flow. *Artif. Intell*. **17**(2), 185–203.

Hounsfield GN 1973 Computerized transverse axial scanning (tomography): I. Description of system. *Br. J. Radiol*. **46**, 1016–1022.

Jakob PM, Griswold MA, Edelman RR and Sodickson DK 1998 AUTO-SMASH: a self-calibrating technique for SMASH imaging. *Magn. Reson. Mater. Phys., Biol. Med*. **7**, 42–54.

Kak AC and Slaney M 1988 *Principles of Computerized Tomographic Imaging*. IEEE Press, New York.

Kalender WA 2006 X-ray computed tomography. *Phys. Med. Biol*. **51**, R29–R43.

Kalender WA, Hebel R and Ebersberger J 1987 Reduction of CT artifacts caused by metallic implants. *Radiology* **164**(2), 576–577.

Kass M, Witkin A and Terzopoulos D 1987 Snake: active contour models. *Int. J. Comput. Vision* **1**, 321–331.

Kichenassamy S 1997 The Perona–Malik paradox. *SIAM J. Appl. Math*. **57**(5), 1328–1342.

Kichenassamy S, Kumar A, Olver P, Tannenbaum A and Yezzi A 1995 Gradient flows and geometric active contour models. In *Proc. 5th Int. Conf. on Computer Vision*, pp. 810–815. IEEE Press, New York.

Kyriakos WE, Panych LP, Kacher DF, Westin CF, Bao SM, Mulkern RV and Jolesz FA 2000 Sensitivity profiles from an array of coils for encoding and reconstruction in parallel (SPACE RIP). *Magn. Reson. Med*. **44**, 301–308.

Lambert JH 1760 *Photometria, sive de Mensura et gradibus luminis,colorum et umbrae*. Eberhardt Klett, Augsburg.

Larkman DJ and Nunes RG 2001 Parallel magnetic resonance imaging. *Phys. Med. Biol*. **52**, R15.

Lauterbur PC 1973 Image formation by induced local interactions: examples of employing nuclear magnetic resonance. *Nature* **242**(5394), 190–191.

Lauterbur PC 1974 Magnetic resonance zeugmatography. *Pure Appl. Chem*. **40**, 149–157.

Leung KY, Danilouchkine MG, Stralen MV, Jong NE, Steen AFVD and Bosch JG 2011 Left ventricular border tracking using cardiac motion models and optical flow. *Ultrasound Med. Biol*. **37**(4), 605–616.

Linguraru MG, Vasilyev NV, Marx GR, Tworetzky W, Nido PJD and Howe RD 2008 Fast block flow tracking of atrial septal defects in 4D echocardiography. *Med. Image Anal*. **12**(4), 397–412.

Lucas B and Kanade T 1981 An iterative image restoration technique with an application to stereo vision. *Proc. DARPA Image Understanding Workshop*, pp. 121–130.

Lustig M, Donoho DL and Pauly JM 2007 Sparse MRI: the application of compressed sensing for rapid MR imaging. *Magn. Reson. Med*. **58**, 1182–1195.

Malladi R, Sethian JA and Vemuri BC 1995 Shape modeling with front propagation: a level set approach. *IEEE Trans. Pattern Anal. Machine Intell*. **17**(2), 158–175.

Meyer Y *Oscillating Patterns in Image Processing and Nonlinear Evolution Equations*. University Lecture Series, vol. 22. American Mathematical Society, Providence, RI.

Meyer E, Raupach R, Lell M, Schmidt B and Kachelrie M 2010 Normalized metal artifact reduction (NMAR) in computed tomography. *Med. Phys*. **37**, 5482–5493.

Natterer F 2008 *X-Ray Tomography, Inverse Problems and Imaging*. Lecture Notes in Mathematics, no. 1943, pp. 17–34. Springer, Berlin.

Osher S, Solé A and Vese L 2002 Image decomposition and restoration using total variation minimization and the H^{-1} norm. *SIAM Multiscale Model. Simul*. **1**(3), 349–370.

Osher S, Burger M, Goldfarb D, Xu J and Yin W 2005 An iterative regularization method for total variation-based image restoration. *SIAM Multiscale Model. Simul*. **4**(2), 460–489.

Paragios N and Deriche R 1998 A PDE-based level set approach for detection and tracking of moving objects. In *Proc. 6th Int. Conf. on Computer Vision*, pp. 1139–1145. IEEE, New York.

Perona P and Malik J 1990 Scale space and edge detection using anisotropic diffusion. *IEEE Trans. Pattern Anal. Machine Intell*. **12**, 629–639.

Pruessmann KP, Weiger M, Scheidegger MB and Boesiger P 1999 SENSE: sensitivity encoding for fast MRI. *Magn. Reson. Med*. **42**, 952–962.

Pruessmann KP, Weiger M, Bornert P and Boesiger P 2001 Advances in sensitivity encoding with arbitrary *k*-space trajectories. *Magn. Reson. Med*. **46**, 638–651.

Radon JH 1917 Über die Bestimmung von Funktionen durch ihre Integralwerte längs gewisser Mannigfaltigkeiten. *Ber. Sächs. Akad. Wiss. (Leipzig)* **69**, 262–277.

Rudin L, Osher SJ and Fatemi E 1992 Nonlinear total variation based noise removal algorithms. *Physica D* **60**, 259–268.

Sarti A, Corsi C, Mazzini E and Lamberti C 2005 Maximum likelihood segmentation of ultrasound images with Rayleigh distribution. *IEEE Trans. Ultrason. Ferroelectr. Freq. Control* **52**(6), 947–960.

Sodickson DK and Manning WJ 1997 Simultaneous acquisition of spatial harmonics (SMASH): fast imaging withradiofrequency coil arrays. *Magn. Reson. Med*. **38**, 591–603.

Sodickson DK, Griswold MA, Jakob PM, Edelman RR and Manning WJ 1999 Signal-to-noise ratio and signal-to-noise efficiency in SMASH imaging. *Magn. Reson. Med*. **41**, 1009–1022.

Suhling M, Arigovindan M, Jansen C, Hunziker P and Unser M 2005 Myocardial motion analysis from B-mode echocardiograms. *IEEE Trans. Image Process*. **14**(4), 525–536.

Tuy HK 1983 An inversion formula for cone-beam reconstruction. *SIAM J. Appl. Math*. **43**, 546–552.

Veronesi F, Corsi C, Caiani EG, Sarti A and Lamberti C 2006 Tracking of left ventricular long axis from real-time three-dimensional echocardiography using optical flow techniques. *IEEE Trans. Inform. Technol. Biomed*. **10**(1), 174–181.

Vese L and Osher S 2002 Modeling textures with total variation minimization and oscillating patterns in image processing. *J. Sci. Comput*. **19**, 553–572.

Wang G, Snyder DL, O'Sullivan JA and Vannier MW 1996 Iterative deblurring for CT metal artifact reduction. *IEEE Trans. Med. Imag*. **15**, 657–664.

Weickert J 1997 A review of nonlinear diffusion filtering. In *Scale-Space Theory in Computer Vision* (eds BtH Romeny, L Florack, J Koenderink and M Viergever). Lecture Notes in Computer Science, no. 1252, pp. 3–28. Springer, Berlin.

Yezzi A, Tsai A and Willsky A 1999 Binary and ternary flows for image segmentation. In *Proc. Int. Conf. on Image Processing*, vol. 2, pp. 1–5. IEEE, New York.

You Y, Xu W, Tannenbaum A and Kaveh M 1996 Behavioral analysis of anisotropic diffusion in image processing. *IEEE Trans. Image Process*. **5**, 15–53.

Further Reading

Aubert G and Vese L 1997 A variational method in image recovery. *SIAM J. Numer. Anal*. **34**(5), 1948–1979.

Candès EJ and Tao T 2008 Reflections on compressed sensing. *IEEE Inform. Theory Soc. Newslett*. **58**, 20–23.

Carr HY 2004 Field gradients in early MRI. *Physics Today* **57**(7), 83.

Donoho DL 2006 Compressed sensing. *IEEE Trans. Inform. Theory* **52**(4), 1289–1306.

McLachlan GJ and Peel D *Finite Mixture Models*. Wiley Series in Probability and Statistics, no. 84. John Wiley & Sons, Inc., New York.

Mumford D and Shah J 1989 Optimal approximation by piecewise smooth functions and associated variational problems. *Commun. Pure Appl. Math*. **42**, 577–685.

Osher S and Fedkiw R 2002 *Level Set Methods and Dynamic Implicit Surfaces*. Springer, New York.

Zhao HK, Chan T, Merriman B and Osher S 1996 A variational level set approach to multiphase motion. *J. Comput. Phys*. **127**, 179–95.

7

Electrical Impedance Tomography

Electrical impedance tomography (EIT) produces cross-sectional images of an admittivity distribution inside an electrically conducting object. It has a wide range of applications in biomedicine, geophysics, non-destructive testing and so on. Considering the fact that structural imaging modalities such as X-ray CT and MRI provide images with a superior spatial resolution to EIT, the primary goal of biomedical EIT is to supply functional diagnostic information of organs with a high temporal resolution. It may provide diagnostic information on functional and pathological conditions of biological tissues and organs. Following a brief introduction to EIT, we summarize bioimpedance measurement methods, on which an EIT system is based, to acquire data for image reconstruction. Its forward problem is introduced in the context of a practically feasible measurement setting. Modeling of the forward problem and sensitivity analysis will be the key to understanding and designing an inversion method. Three kinds of EIT inverse problems, including static imaging, time-difference imaging and frequency-difference imaging, will be described.

7.1 Introduction

The material properties of electrical conductivity and permittivity may produce image contrast in EIT. The conductivity (σ) and permittivity (ϵ) values of a biological tissue are determined by its ion concentrations in extra- and intracellular fluids, cellular structure and density, molecular composition, membrane characteristics and other factors. In the frequency range of a few hertz to megahertz, numerous experimental findings indicate that different biological tissues have different electrical properties, and their values are influenced by physiological and pathological conditions (Gabriel *et al.* 1996a,b; Geddes and Baker 1967; Grimnes and Martinsen 2008). In biomedical applications of EIT, we deal with the admittivity $\gamma = \sigma + i\omega\epsilon$, where the angular frequency $\omega = 2\pi f$ is in rad s^{-1} with the frequency f in Hz. For most biological tissues, we may assume that $\gamma \approx \sigma$ at low frequencies below 10 kHz. With abundant membraneous structures in an organism, the $\omega\epsilon$ term is not negligible beyond 10 kHz and we should deal with the admittivity $\gamma = \sigma + i\omega\epsilon$ in general at high frequencies.

We consider an electrically conducting object such as the human body with its internal admittivity distribution $\gamma(\mathbf{r})$ as a function of position $\mathbf{r} = (x, y, z)$. To probe the

Nonlinear Inverse Problems in Imaging, First Edition. Jin Keun Seo and Eung Je Woo.

object with the intention of non-invasively sensing γ, we inject current through electrodes attached on its surface. This induces internal current density and voltage distributions that are determined by the admittivity distribution, object geometry and electrode configuration. In the frequency range up to a few MHz, we may adopt the elliptic partial differential equation (PDE) introduced in Chapter 3 to describe the interrelations among the injection current, current density and voltage. By measuring induced voltages on the surface subject to multiple injection currents, an EIT system produces images of the internal admittivity distribution using an inversion method.

Mathematical theory has been developed to support such an EIT system especially for the unique identification of the conductivity σ from knowledge of all possible boundary current-to-voltage data at low frequencies where we can assume $\gamma \approx \sigma$ (Astala and Paivarinta 2006a,b; Brown and Uhlmann 1997; Calderón 1980; Kenig *et al.* 2007; Kohn and Vogelius 1984; Nachman 1988; Nachman 1996; Sylvester and Uhlmann 1986, 1987, 1988). After the early attempt to build an EIT system (Barber and Brown 1984), numerous studies have accumulated knowledge and experience, summarized in the fairly recently published book on EIT (Holder 2005). The nonlinear inverse problem in EIT suffers from its ill-posedness, related to lack of enough measurable information and insensitivity of measured data to a local change of an internal admittivity value. Though there exist numerous image reconstruction algorithms (Barber and Brown 1984; Berenstein *et al.* 1991; Brown *et al.* 1985; Cheney *et al.* 1990; Fuks *et al.* 1991; Gisser *et al.* 1988, 1990; Isaacson and Cheney 1991; Isaacson *et al.* 1989, 1996; Lionheart *et al.* 2005; Newell *et al.* 1988; Santosa and Vogelius 1990; Somersalo *et al.* 1992; Wexler *et al.* 1985; Yorkey 1987), it is difficult to reconstruct accurate admittivity images with a high spatial resolution in a practical setting, where modeling and measurement errors are unavoidable. In this chapter, we focus on robust image reconstructions that may overcome the technical difficulties of the ill-posedness.

7.2 Measurement Method and Data

7.2.1 *Conductivity and Resistance*

We consider a cylinder filled with saline. The saline contains mobile charged ions, and their migration under an external electric field characterizes its conductivity σ (siemens per meter, S m^{-1}). Attaching two electrodes on the top and bottom surfaces, we measure its resistance R (ohms, Ω). Neglecting interfacial phenomena between each electrode and the saline, the resistance R is denoted as

$$R = \frac{1}{\sigma}\frac{L}{A}, \tag{7.1}$$

where L and A are the length (m) and cross-sectional area (m^2) of the cylinder, respectively. If we inject DC current I (amperes, A), the induced DC voltage V (volts, V) follows Ohm's law as

$$V = RI. \tag{7.2}$$

Injecting a known DC current I and measuring the induced DC voltage, we may find the resistance R, as is done in an electrical multimeter. If we have geometrical information about L and A, we can find the conductivity σ. For materials such as biological tissues, we denote the conductivity as σ_ω to emphasize its frequency dependence. We may measure σ_ω by injecting a sinusoidal current $i(t) = I \cos \omega t$ to measure the induced AC voltage $v(t) = V \cos \omega t$, where t is the time (seconds, s). Assuming a linear component, the resistance R at ω also follows Ohm's law as

$$v(t) = Ri(t) = RI \cos \omega t = \frac{1}{\sigma_\omega} \frac{L}{A} I \cos \omega t. \tag{7.3}$$

Note that the current and voltage are in phase. Repeating this measurement for multiple frequencies, we may get a conductivity spectrum, which plots conductivity σ_ω as a function of frequency ω.

7.2.2 Permittivity and Capacitance

We consider a dielectric sandwiched between two parallel conducting plates. When we apply a DC voltage V between the plates, it induces an electric field inside the dielectric. The dielectric contains immobile charges, and their polarization or rotation in the electric field produces surface charges Q and $-Q$ (coulombs, C). The induced charge is proportional to the applied voltage as

$$Q = CV, \tag{7.4}$$

where the proportionality constant C is called the capacitance (coulombs per volt, $\mathrm{C\,V^{-1}}$; or farad, F). The capacitance C between the two plates is given by

$$C = \epsilon \frac{A}{d}, \tag{7.5}$$

where ϵ is the permittivity ($\mathrm{F\,m^{-1}}$), A the surface area and d the gap between the plates. The permittivity is a material property determined by the polarization of the dielectric under an external electric field. For most dielectrics, including biological tissues, the permittivity changes with frequency, and we denote it as ϵ_ω.

If we assume a perfect dielectric, there is no mobile charge and its conductivity σ is zero. Applying DC voltage V to the dielectric, we get zero DC current through it. If we apply a sinusoidal voltage $v(t) = V \sin \omega t$, there occurs an AC displacement current through the dielectric due to time-varying polarizations with frequency ω:

$$i(t) = C \frac{\mathrm{d}v(t)}{\mathrm{d}t} = \omega C V \cos \omega t = I \cos \omega t. \tag{7.6}$$

Note that the current and voltage are out of phase by 90° or the voltage is in quadrature with the current. Assuming that there is no polarization initially, we can express the induced voltage $v(t)$ subject to an injection current $i(t)$ as

$$v(t) = \frac{1}{C} \int_0^t i(\tau)\,\mathrm{d}\tau = \frac{I}{\omega C} \sin \omega t = \frac{Id}{\omega \epsilon_\omega A} \cos \left(\omega t - \frac{\pi}{2} \right). \tag{7.7}$$

With known ω and I, we may find the capacitance C in farads (F), which equals $\mathrm{A\,s\,V^{-1}}$ or $\mathrm{s\,\Omega^{-1}}$. If we have geometrical information on A and d, we can find the permittivity ϵ_ω ($\mathrm{F\,m^{-1}}$). Repeating this measurement for multiple frequencies, we may get a permittivity spectrum, which plots permittivity ϵ_ω as a function of frequency ω.

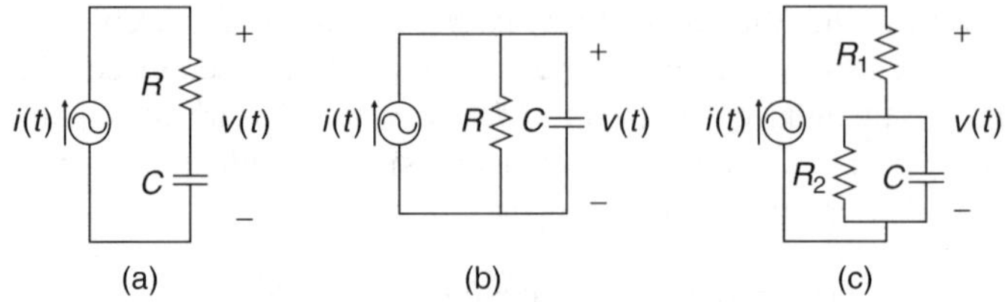

Figure 7.1 (a) Series, (b) parallel and (c) series–parallel RC circuits

7.2.3 *Phasor and Impedance*

Given an electrically conducting object with both mobile and immobile charges, we may view it as a mixture of resistors and capacitors. In this section, we adopt a circuit model using lumped elements since this provides intuitive understanding about the continuum model. Let us consider the series RC circuit in Figure 7.1(a). Injecting a sinusoidal current $i(t) = I\cos\omega t$, we can express the induced voltage $v(t)$ across the series connection of R and C as

$$v(t) = Ri(t) + \frac{1}{C}\int_0^t i(\tau)\,\mathrm{d}\tau = RI\cos\omega t + \frac{I}{\omega C}\sin\omega t = V\cos(\omega t + \theta), \tag{7.8}$$

where

$$V = I\sqrt{R^2 + \frac{1}{\omega^2 C^2}} \quad \text{and} \quad \theta = -\arctan\frac{1}{\omega RC}.$$

Figure 7.2 shows current $i(t)$ and voltage $v(t)$. Noting that there is no change in frequency between current and voltage for all linear components, we adopt the phasor notation. The current and voltage phasors are defined as complex numbers $\mathbf{I} = I\angle 0$ and $\mathbf{V} = V\angle\theta$, so that we can recover time functions $i(t)$ and $v(t)$ from

$$i(t) = \Re\left\{\mathbf{I}\,\mathrm{e}^{\mathrm{i}\omega t}\right\} \quad \text{and} \quad v(t) = \Re\left\{\mathbf{V}\mathrm{e}^{\mathrm{i}\omega t}\right\}, \tag{7.9}$$

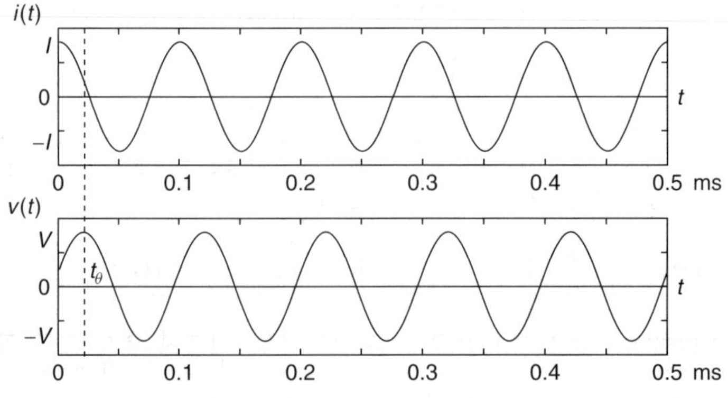

Figure 7.2 Current and voltage waveforms

respectively. Using the phasor notation, we can handle the relation between time functions $i(t)$ and $v(t)$ as an algebraic equation instead of an integrodifferential equation of time.

We define the impedance $\mathbf{Z}$ (Ω) as the ratio of the voltage phasor $\mathbf{V}$ to the current phasor $\mathbf{I}$, and it is a measure of the total opposition to current flow through a component or a collection of components. For the case of the series RC circuit, the total impedance is

$$\mathbf{Z} = Z\angle\theta = \frac{\mathbf{V}}{\mathbf{I}} = \frac{V}{I}\angle\theta = \sqrt{R^2 + \frac{1}{\omega^2 C^2}}\angle - \arctan\frac{1}{\omega RC} = R + \frac{1}{\mathrm{i}\omega C}. \tag{7.10}$$

The real part of $\mathbf{Z}$ is the resistance (Ω) and its imaginary part is the reactance (also Ω). Note that, for a single resistor R, its impedance $\mathbf{Z}_R = R$. For a single capacitor C, $\mathbf{Z}_C = 1/(\mathrm{i}\omega C)$.

Example 7.2.1 *For the series RC circuit in Figure 7.1(a) with* $R = 10\ \mathrm{k\Omega}$ *and* $C = 0.1\ \mathrm{\mu F}$, *compute its impedance* $\mathbf{Z}$ *at a frequency of* $10\ \mathrm{kHz}$. *For* $i(t) = 10\cos(2\pi \times 10^4 t)\ \mathrm{mA}$, *find* $v(t)$ *and plot both* $i(t)$ *and* $v(t)$ *for* $0 \le t \le 0.5\ \mathrm{ms}$.

Example 7.2.2 *For the parallel RC circuit in Figure 7.1(b) with* $R = 10\ \mathrm{k\Omega}$ *and* $C = 0.1\ \mathrm{\mu F}$, *compute its impedance* $\mathbf{Z}$ *at a frequency of* $10\ \mathrm{kHz}$. *For* $i(t) = 10\cos(2\pi \times 10^4 t)\ \mathrm{mA}$, *find* $v(t)$ *and plot both* $i(t)$ *and* $v(t)$ *for* $0 \le t \le 0.5\ \mathrm{ms}$.

Example 7.2.3 *For the series–parallel RC circuit in Figure 7.1(c) with* $R_1 = 500\ \Omega$, $R_2 = 30\ \mathrm{k\Omega}$ *and* $C = 50\ \mathrm{nF}$, *compute its impedance* $\mathbf{Z}$ *at a frequency of* $10\ \mathrm{kHz}$. *For* $i(t) = 10\cos(2\pi \times 10^4 t)\ \mathrm{mA}$, *find* $v(t)$ *and plot both* $i(t)$ *and* $v(t)$ *for* $0 \le t \le 0.5\ \mathrm{ms}$.

Example 7.2.4 *For the series RC circuit in Figure 7.1(a) with* $R = 10\ \mathrm{k\Omega}$ *and* $C = 0.1\ \mathrm{\mu F}$, *plot the magnitude and phase of its impedance* $\mathbf{Z} = Z\angle\theta$ *in the frequency range of 1* Hz *to 1* MHz.

Example 7.2.5 *For the parallel RC circuit in Figure 7.1(b) with* $R = 10\ \mathrm{k\Omega}$ *and* $C = 0.1\ \mathrm{\mu F}$, *plot the magnitude and phase of its impedance* $\mathbf{Z} = Z\angle\theta$ *in the frequency range of 1* Hz *to 1* MHz.

Example 7.2.6 *For the series–parallel RC circuit in Figure 7.1(c) with* $R_1 = 500\ \Omega$, $R_2 = 30\ \mathrm{k\Omega}$ *and* $C = 50\ \mathrm{nF}$, *plot the magnitude and phase of its impedance* $\mathbf{Z} = Z\angle\theta$ *in the frequency range of 1* Hz *to 1* MHz.

7.2.4 Admittivity and Trans-Impedance

When we consider a material including both mobile and immobile charges, its electrical property is expressed as the admittivity γ ($\mathrm{S\,m^{-1}}$). To express its frequency dependence, we denote it as $\gamma_\omega = \sigma_\omega + \mathrm{i}\omega\epsilon_\omega$. Note that σ_ω and $\omega\epsilon_\omega$ have the same unit ($\mathrm{S\,m^{-1}}$). We now assume a cylinder filled with a biological tissue whose admittivity is γ_ω. The impedance $\mathbf{Z}$ between the top and bottom surfaces is

$$\mathbf{Z} = \frac{1}{\sigma_\omega + \mathrm{i}\omega\epsilon_\omega}\frac{L}{A} = \frac{L}{\sigma_\omega A}\frac{1 - \mathrm{i}\omega\epsilon_\omega/\sigma_\omega}{1 + (\omega\epsilon_\omega/\sigma_\omega)^2}, \tag{7.11}$$

where L and A are the length and cross-sectional area of the cylinder, respectively. If $\omega\epsilon_\omega/\sigma_\omega \ll 1$, that is, $\sigma_\omega \gg \omega\epsilon_\omega$, then $\mathbf{Z} \approx L/(\sigma_\omega A) = R$ and the material is resistive. If $\omega\epsilon_\omega/\sigma_\omega \gg 1$, that is, $\sigma_\omega \ll \omega\epsilon_\omega$, then $\mathbf{Z} \approx -\mathrm{i}L/(\omega\epsilon_\omega A) = 1/(\mathrm{i}\omega C)$ and the material is reactive or capacitive.

Most biological tissues are resistive at low frequencies of less than 10 kHz, for example. Since the capacitive term is not negligible beyond 10 kHz, we will denote the admittivity of a biological tissue at position $\mathbf{r}$ as $\gamma_\omega(\mathbf{r})$. We assume an electrically conducting domain Ω with its admittivity distribution $\gamma_\omega(\mathbf{r})$, as illustrated in Figure 7.3. Attaching E electrodes $\mathcal{E}_1, \mathcal{E}_2, \ldots, \mathcal{E}_E$, we inject current $i^j(t) = I^j \cos \omega t$ through a pair of electrodes $\mathcal{E}_j$ and $\mathcal{E}_{j+1}$. Between another pair of electrodes $\mathcal{E}_k$ and $\mathcal{E}_{k+1}$, we measure the induced voltage $v^k(t) = V^k \cos(\omega t + \theta^k)$. We define the trans-impedance from the jth port to the kth port as

$$\mathbf{Z}^{j,k} = \frac{\mathbf{V}^k}{\mathbf{I}^j} = \frac{V^k}{I^j}\angle\theta^k. \tag{7.12}$$

In section 7.4, we will show that the admittivity distribution $\gamma_\omega(\mathbf{r})$, domain geometry and electrode configuration affect the trans-impedance $\mathbf{Z}^{j,k}$. The reciprocity principle explained in section 7.4 indicates that $\mathbf{Z}^{j,k} = \mathbf{Z}^{k,j}$.

7.2.5 Electrode Contact Impedance

To inject current and measure voltage, we use electrodes. An electrode is made of a highly conductive material such as copper, silver, platinum and others. Carbon is also used to make a flexible electrode, though its conductivity is not as large as for metallic conductors. When the electrode makes contact with an electrolyte or the skin of an organic object, the interface can be modeled as a contact impedance and a contact potential in series. The contact impedance includes both resistive and reactive terms, and its typical circuit model is the series–parallel RC circuit in Figure 7.1(c). As long as the interface is mechanically stable, the contact potential is stable and less than 1 V for most electrode materials.

We consider a method to measure the impedance $\mathbf{Z}$ of a cylinder with homogeneous admittivity $\gamma_\omega(\mathbf{r}) = \gamma_\omega$. Attaching a pair of electrodes at the top and bottom surfaces, we

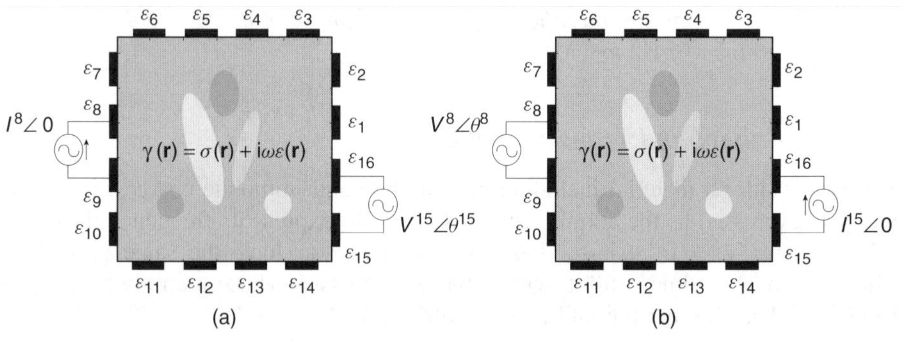

Figure 7.3 Measurement of trans-impedance: (a) $\mathbf{Z}^{8,15}$ and (b) $\mathbf{Z}^{15,8}$. From the reciprocity principle described in section 7.4, we have $\mathbf{Z}^{8,15} = \mathbf{Z}^{15,8}$

inject current $\mathbf{I}$ at ω from the top to the bottom electrode. Denoting the contact impedances of the top and bottom electrodes as $\mathbf{Z}_c^1$ and $\mathbf{Z}_c^2$, respectively, the induced voltage will be expressed as

$$\mathbf{V} = \mathbf{I}(\mathbf{Z}_c^1 + \mathbf{Z} + \mathbf{Z}_c^2), \tag{7.13}$$

assuming that no current flows into the ideal voltmeter. We can ignore the DC contact potential since we measure only the induced voltage at frequency ω. Using this two-electrode or bipolar method shown in Figure 7.4(a), it is not possible to extract only $\mathbf{Z}$ since two contact impedances are in series with $\mathbf{Z}$.

By attaching another pair of electrodes around the cylinder near its top and bottom, we inject current through the first pair and measure the induced voltage between the second pair as shown in Figure 7.4(b). Using a well-designed voltmeter, we may safely assume that there is no current flowing through the second pair of voltage-sensing electrodes. This means that the voltmeter sees only the voltage drop across the impedance of the cylinder $\mathbf{Z}$ between the second pair of electrodes as

$$\mathbf{V} = \mathbf{IZ}. \tag{7.14}$$

This four-electrode or tetrapolar method allows us remove the effects of contact impedances in bioimpedance measurements.

7.2.6 EIT System

We consider an imaging domain Ω with its admittivity distribution $\gamma_\omega(\mathbf{r}) = \sigma_\omega(\mathbf{r}) + i\omega\epsilon_\omega(\mathbf{r})$. We attach E electrodes $\mathcal{E}_1, \mathcal{E}_2, \ldots, \mathcal{E}_E$ on its boundary $\partial\Omega$. We use an EIT system equipped with current sources and voltmeters to measure trans-impedances or equivalent current–voltage data sets. We may do this for multiple frequencies at different times. A typical EIT system comprises one or multiple current sources, one or multiple voltmeters, optional switching networks, a computer system and a DC power supply. The computer controls current sources, voltmeters and switches to acquire current–voltage data sets. It produces images of $\sigma_\omega(\mathbf{r})$ and/or $\omega\epsilon_\omega(\mathbf{r})$ by applying an image reconstruction algorithm to the data sets.

There are several EIT systems with different design concepts and technical details in their implementations. The number of electrodes used in available EIT systems ranges

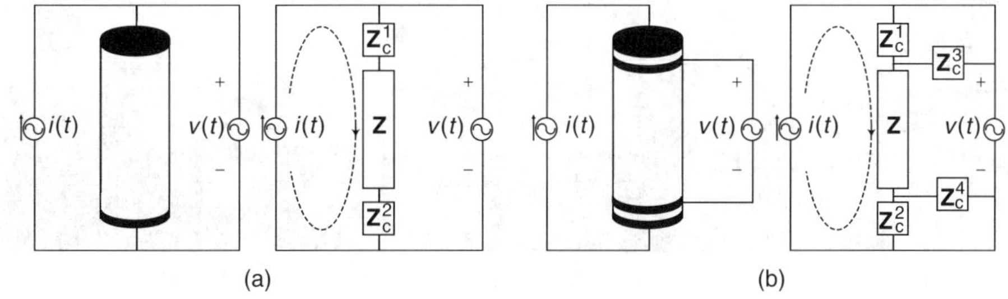

Figure 7.4 Impedance measurements using (a) two-electrode or bipolar method and (b) four-electrode or tetrapolar method. No current flows through the ideal voltmeter

from eight to 256. The human interface gets complicated with a large number of electrodes and lead wires. With a large number of electrodes, the induced voltage between a pair of electrodes tends to become small, since the gap between them gets smaller. In chest imaging, eight or 16 electrodes are commonly used, while more electrodes are used in head or breast imaging.

We may classify recent EIT systems into two types. The first is characterized as one current source with switching networks. In this case, current is sequentially injected between a chosen pair of electrodes and there always exists only one active current source. The second type uses multiple current sources without any switching for current injection. With this type, one may inject a pattern of current through multiple electrodes using multiple active current sources. The sum of currents from all active current sources must be zero. In most EIT systems belonging to both types, voltages between many electrode pairs are simultaneously measured using multiple voltmeters. Typical examples of the first and second types are Mk3.5 from Sheffield (Wilson *et al.* 2001) and ACT3 from RPI (Cook *et al.* 1994), respectively. Boone *et al.* (1997) and Saulnier (2005) summarized numerous techniques in the development of EIT systems. Figure 7.5 shows examples of EIT systems and their use for chest imaging (Oh *et al.* 2007a,b, 2008).

The range of the trans-impedance is from a few milliohms (mΩ) to tens of ohms depending on the imaging object, number of electrodes and their configuration. Assuming injection currents of 1 $\mathrm{mA_{rms}}$, for example, induced voltages are in the range of a few microvolts to tens of millivolts. Allowing a noise level of 1% of the smallest voltage, we should restrict the level below 0.1 μV and this requires state-of-the-art electronic instrumentation technology. Modern EIT systems usually acquire a complete set of current–voltage data within 10 ms for frequencies higher than 10 kHz. Temporal resolutions could be higher than 20 frames per second using a fast image reconstruction algorithm.

7.2.7 Data Collection Protocol and Data Set

A data collection protocol defines a series of injection currents and corresponding voltage measurements. In this section, we introduce only the neighboring protocol. One may

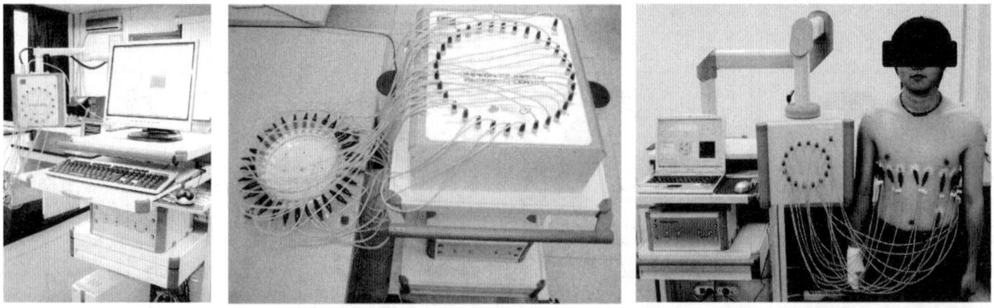

(a) (b) (c)

Figure 7.5 EIT systems: (a) and (b) are KHU Mark 1 16- and 32-channel multi-frequency EIT systems, respectively, and (c) is a set-up for chest imaging

find numerous data collection protocols in Holder (2005) and other literature on EIT. We assume an EIT system with E electrodes. Injecting the jth current between an adjacent pair of electrodes $\mathcal{E}_j$ and $\mathcal{E}_{j+1}$, we measure induced boundary voltages between all neighboring pairs of electrodes $\mathcal{E}_k$ and $\mathcal{E}_{k+1}$ for $k = 1, 2, \ldots, E$. Any index number must be understood as a modulus of the maximal value of the index number. We define this data set as a projection, a term that has its origin in the X-ray CT area. Repeating this for all pairs of current injection electrodes with $j = 1, 2, \ldots, E$, we can obtain a full set of data from E projections. The kth boundary voltage phasor in the jth injection current or the jth projection is denoted as

$$\mathbf{V}^{j,k} = \mathbf{V}^j|_{\mathcal{E}_k} - \mathbf{V}^j|_{\mathcal{E}_{k+1}}$$

for $j, k = 1, 2, \ldots, E$. Since the number of injection currents or projections is E and the number of boundary voltage phasors per projection is also E, the full data set includes E^2 boundary voltage phasors.

From the reciprocity theorem introduced in section 7.4 and Kirchhoff's voltage law, only $E \times (E - 1)/2$ boundary voltage data are independent. This is the maximal amount of measurable information using E electrodes regardless of the adopted data collection protocol. This imposes a fundamental limit on the achievable spatial resolution in EIT using E electrodes regardless of the inversion method.

For each injection current between a chosen pair of neighboring electrodes, boundary voltage data between three adjacent pairs of electrodes are involved with at least one current injection electrode. These three voltage data contain the effects of unknown contact impedances between the electrodes and the skin. We may discard or include these data depending on the way in which contact impedances are treated in the chosen inversion method and electrode model, as discussed in section 7.4.

Figure 7.6 shows examples of the neighboring protocol assuming a 16-channel EIT system. For each projection, 13 boundary voltage phasors between adjacent pairs of electrodes are measured to adopt the four-electrode method. In this example, the number of projections is 16 and the total number of measured boundary voltage phasors is $16 \times 13 = 208$. Among them, only 104 boundary voltage phasors carry independent information. This indicates that the best spatial resolution of a reconstructed admittivity image will be about 10% of the size of the imaging object using a 16-channel EIT system with the neighboring protocol. Using a 32-channel system, we may improve it to 5%.

We can collect boundary voltage data at multiple frequencies for a certain period of time. Assuming that we collected E^2 number of boundary voltage data at each sampling time t and frequency ω, we can express the boundary voltage data set in matrix form as

$$\mathbb{F}_{t,\omega} = \begin{bmatrix} \mathbf{V}_{t,\omega}^{1,1} & \mathbf{V}_{t,\omega}^{1,2} & \cdots\cdots & \mathbf{V}_{t,\omega}^{1,E} \\ \mathbf{V}_{t,\omega}^{2,1} & \mathbf{V}_{t,\omega}^{2,2} & \cdots\cdots & \mathbf{V}_{t,\omega}^{2,E} \\ \vdots & \vdots & & \vdots \\ \vdots & \vdots & & \vdots \\ \mathbf{V}_{t,\omega}^{E,1} & \mathbf{V}_{t,\omega}^{E,2} & \cdots\cdots & \mathbf{V}_{t,\omega}^{E,E} \end{bmatrix} \begin{matrix} \leftarrow \text{1st projection} \\ \leftarrow \text{2nd projection} \\ \\ \\ \leftarrow E\text{th projection} \end{matrix} . \tag{7.15}$$

Alternatively, we may adopt a column vector representation as

$$\mathbb{F}_{t,\omega} = [\mathbf{V}_{t,\omega}^{1,1} \cdots \mathbf{V}_{t,\omega}^{1,E} \quad \mathbf{V}_{t,\omega}^{2,1} \cdots \mathbf{V}_{t,\omega}^{2,E} \quad \cdots \quad \mathbf{V}_{t,\omega}^{E,1} \cdots \mathbf{V}_{t,\omega}^{E,E}]^{\mathrm{T}}, \tag{7.16}$$

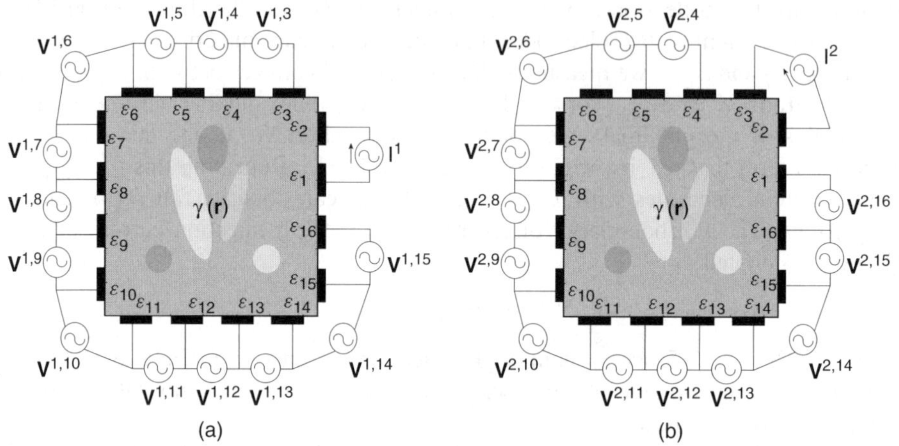

Figure 7.6 Neighboring data collection protocol of a 16-channel EIT system: (a) first projection with the injection current $\mathbf{I}^1$ between $\mathcal{E}_1$ and $\mathcal{E}_2$; and (b) second projection with the injection current $\mathbf{I}^2$ between $\mathcal{E}_2$ and $\mathcal{E}_3$

where the superscript T means the transpose. This column vector representation will be used in sections where we study image reconstruction algorithms. For $\omega = \omega_1, \omega_2, \ldots, \omega_F$, we may collect F data vectors or matrices for each sampling time $t = t_1, t_2, \ldots, t_N$ of total N times.

7.2.8 Linearity between Current and Voltage

Before we move on to mathematical topics in EIT, we note the linear relationship between injection currents and boundary voltages. We assume that the internal admittivity distribution $\gamma_\omega(\mathbf{r})$, domain geometry and electrode configuration are all fixed. For an injection current $\mathbf{I}^j$ or the jth projection, we measure E boundary voltage phasors $\mathbf{V}^{j,k}$ for $k = 1, 2, \ldots, E$ to form the jth projection data vector $\mathbb{V}^j$ as

$$\mathbb{V}^j = [\mathbf{V}^{j,1}\mathbf{V}^{j,1}\ldots\mathbf{V}^{j,E}]^{\mathrm{T}}. \tag{7.17}$$

We now inject current $\mathbf{I}$ as

$$\mathbf{I} = \sum_{j=1}^{E} \alpha^j \mathbf{I}^j \tag{7.18}$$

with some real constants α^j for $j = 1, 2, \ldots, E$. The corresponding projection data vector $\mathbb{V}$ is expressed as

$$\mathbb{V} = \sum_{j=1}^{E} \alpha^j \mathbb{V}^j. \tag{7.19}$$

This stems from the linearity between injection currents and induced voltages when we view the imaging object as a mixture of linear resistors and capacitors.

7.3 Representation of Physical Phenomena

We assume an imaging object occupying a domain Ω with its boundary $\partial\Omega$ and an internal admittivity distribution $\gamma(\mathbf{r})$. Using an E-channel EIT system, we attach E surface electrodes $\mathcal{E}_j$ for $j = 1, \ldots, E$ on $\partial\Omega$ and inject current $i(t) = I\cos\omega t$ through an adjacent pair of electrodes as shown in Figure 7.6. We assume that the current source and sink are connected to electrodes $\mathcal{E}_j$ and $\mathcal{E}_{j+1}$, respectively. The injection current produces internal current density and magnetic flux density distributions, which are dictated by Maxwell's equations, as in Table 7.1. Table 7.2 summarizes the variables used in Maxwell's equations.

In the frequency range of a few hertz to megahertz, we adopt the elliptic PDE studied in Chapter 3 to describe the forward problem in EIT. From Maxwell's equations, we derive the elliptic PDE and its boundary conditions. After analyzing the PDE in terms of its min–max property, we formulate the EIT forward problem and its model.

7.3.1 Derivation of Elliptic PDE

To simplify mathematical derivations, we assume that the admittivity $\gamma(\mathbf{r}) = \sigma(\mathbf{r}) + \mathrm{i}\omega\epsilon(\mathbf{r})$ in Ω is isotropic, $\sigma > 0$ and $\epsilon < \infty$. For some biological tissues, such as muscles and neural tissues, the isotropy assumption is not valid, especially at low frequencies. We assume that the magnetic permeability μ of the imaging object is μ_0, the magnetic permeability of free space.

Table 7.1 Maxwell's equations for time-varying and time-harmonic fields

Name	Time-varying field	Time-harmonic field
Gauss's law	$\nabla\cdot\widetilde{\mathbf{E}} = \dfrac{\rho}{\epsilon}$	$\nabla\cdot\mathbf{E} = \dfrac{\rho}{\epsilon}$
Gauss's law for magnetism	$\nabla\cdot\widetilde{\mathbf{H}} = 0$	$\nabla\cdot\mathbf{H} = 0$
Faraday' law of induction	$\nabla\times\widetilde{\mathbf{E}} = -\dfrac{\partial\widetilde{\mathbf{B}}}{\partial t}$	$\nabla\times\mathbf{E} = -\mathrm{i}\omega\mathbf{B}$
Ampère's circuit law	$\nabla\times\widetilde{\mathbf{H}} = \widetilde{\mathbf{J}} + \dfrac{\partial\widetilde{\mathbf{D}}}{\partial t}$	$\nabla\times\mathbf{H} = \mathbf{J} + \mathrm{i}\omega\mathbf{D}$

Table 7.2 Variables to describe time-harmonic and time-varying electromagnetic fields

Variable	Meaning	Unit	Relation
$\mathbf{E}(\mathbf{r})$ and $\widetilde{\mathbf{E}}(\mathbf{r},t) = \Re\{\mathbf{E}(\mathbf{r})\,\mathrm{e}^{\mathrm{i}\omega t}\}$	electric field intensity	$\mathrm{V\,m^{-1}}$	
$\mathbf{H}(\mathbf{r})$ and $\widetilde{\mathbf{H}}(\mathbf{r},t) = \Re\{\mathbf{H}(\mathbf{r})\,\mathrm{e}^{\mathrm{i}\omega t}\}$	magnetic field intensity	$\mathrm{A\,m^{-1}}$	
$\mathbf{D}(\mathbf{r})$ and $\widetilde{\mathbf{D}}(\mathbf{r},t) = \Re\{\mathbf{D}(\mathbf{r})\,\mathrm{e}^{\mathrm{i}\omega t}\}$	electric flux density	$\mathrm{C\,m^{-2}}$	$\mathbf{D} = \epsilon\mathbf{E}$
$\mathbf{B}(\mathbf{r})$ and $\widetilde{\mathbf{B}}(\mathbf{r},t) = \Re\{\mathbf{B}(\mathbf{r})\,\mathrm{e}^{\mathrm{i}\omega t}\}$	magnetic flux density	T	$\mathbf{B} = \mu\mathbf{H}$
$\mathbf{J}(\mathbf{r})$ and $\widetilde{\mathbf{J}}(\mathbf{r},t) = \Re\{\mathbf{J}(\mathbf{r})\,\mathrm{e}^{\mathrm{i}\omega t}\}$	current density	$\mathrm{A\,m^{-2}}$	$\mathbf{J} = \sigma\mathbf{E}$
ϵ and $\epsilon_0 = 8.85\times10^{-12}$ (free space)	permittivity	$\mathrm{F\,m^{-1}}$	
μ and $\mu_0 = 4\pi\times10^{-7}$ (free space)	permeability	$\mathrm{H\,m^{-1}}$	
σ	conductivity	$\mathrm{S\,m^{-1}}$	
ρ	charge density	$\mathrm{C\,m^{-3}}$	

In the frequency range of a few hertz to megahertz, we neglect the Faraday induction to get

$$\nabla \times \mathbf{E} = -\mathrm{i}\omega\mathbf{B} \approx 0.$$

Since $\mathbf{E}$ is approximately irrotational, it follows from Stokes's theorem that we can define a potential u between any two points $\mathbf{r}_1$ and $\mathbf{r}_2$ as

$$u(\mathbf{r}_2) - u(\mathbf{r}_1) = -\int_{C_{\mathbf{r}_1 \to \mathbf{r}_2}} \mathbf{E} \cdot \mathrm{d}\mathbf{l},$$

where $C_{\mathbf{r}_1 \to \mathbf{r}_2}$ is a curve in Ω joining the starting point $\mathbf{r}_1$ to the ending point $\mathbf{r}_2$. The complex potential u satisfies

$$-\nabla u(\mathbf{r}) = \mathbf{E}(\mathbf{r}) \quad \text{in } \Omega.$$

From $\nabla \times \mathbf{H} = \mathbf{J} + \mathrm{i}\omega\epsilon\mathbf{E} = (\sigma + \mathrm{i}\omega\epsilon)\mathbf{E}$, we have the following relation:

$$\nabla \times \mathbf{H}(\mathbf{r}) = (\sigma(\mathbf{r}) + \mathrm{i}\omega\epsilon(\mathbf{r}))\,\mathbf{E}(\mathbf{r}) = -\gamma(\mathbf{r})\nabla u(\mathbf{r}).$$

Since $\nabla \cdot (\nabla \times \mathbf{H}) = 0$, the complex potential u satisfies the following elliptic PDE with a complex parameter γ:

$$-\nabla \cdot (\gamma(\mathbf{r})\nabla u(\mathbf{r})) = 0 \quad \text{in } \Omega. \tag{7.20}$$

Note that the complex potential u is equivalent to the voltage phasor introduced in section 7.2. In the rest of this chapter, we denote u as the voltage phasor or time-harmonic voltage.

7.3.2 Elliptic PDE for Four-Electrode Method

Using the four-electrode method, we can neglect the contact impedance introduced in section 7.2. Investigating the boundary $\partial\Omega$ of the imaging object Ω with attached electrodes $\mathcal{E}_k$ with $k = 1, 2, \ldots, E$, we can observe the following.

Current injection electrodes. Since the total injection current spreads over each current injection electrode,

$$\int_{\mathcal{E}_j} \widetilde{\mathbf{J}}(\mathbf{r}, t) \cdot \mathbf{n}\,\mathrm{d}s = -\int_{\mathcal{E}_{j+1}} \widetilde{\mathbf{J}}(\mathbf{r}, t) \cdot \mathbf{n}\,\mathrm{d}s = I\cos\omega t,$$

where $\mathbf{n}$ is the unit outward normal vector and $\mathrm{d}s$ the surface element on $\partial\Omega$.

Boundary without any electrode. Since the air is an insulator,

$$\widetilde{\mathbf{J}}(\mathbf{r}, t) \cdot \mathbf{n} = 0 \quad \text{on } \partial\Omega \setminus \left(\bigcup_{k=1}^{E} \mathcal{E}_k\right).$$

Voltage-sensing electrodes. Since there is no current flowing into a voltmeter,

$$\int_{\mathcal{E}_k} \widetilde{\mathbf{J}}(\mathbf{r}, t) \cdot \mathbf{n}\,\mathrm{d}s = 0 \quad \text{for } k \in \{1, 2, \ldots, E\} \setminus \{j, j+1\}.$$

All electrodes. Since u is approximately constant on each electrode with a very high conductivity,

$$\nabla \times \widetilde{\mathbf{J}} \approx 0 \quad \text{on } \mathcal{E}_k \text{ for } k = 1, 2, \ldots, E.$$

From these observations, we can derive the following boundary conditions for the time-harmonic potential u in (7.20).

BC 1: $\int_{\mathcal{E}_j} (\gamma \nabla u) \cdot \mathbf{n}\, ds = -\int_{\mathcal{E}_{j+1}} (\gamma \nabla u) \cdot \mathbf{n}\, ds = I.$

BC 2: $(\gamma \nabla u) \cdot \mathbf{n} = 0 \quad \text{on } \partial\Omega \setminus \left(\bigcup_{k=1}^{E} \mathcal{E}_k \right).$

BC 3: $\int_{\mathcal{E}_k} (\gamma \nabla u) \cdot \mathbf{n}\, ds = 0 \quad \text{for } k \in \{1, 2, \ldots, E\} \setminus \{j, j+1\}.$

BC 4: $\nabla u \times \mathbf{n} \approx 0 \quad \text{on } \mathcal{E}_k$ for $k = 1, 2, \ldots, E.$

We define g as

$$g := \gamma \frac{\partial u}{\partial \mathbf{n}}\bigg|_{\partial\Omega} \tag{7.21}$$

and call it the Neumann data of u. In practice, it is difficult to specify the Neumann data g in a pointwise sense because only the total injection current I is known. Note that the Neumann boundary data g have a singularity along the edge of each electrode and $g \notin L^2(\partial\Omega)$. Fortunately, we can prove that $g \in H^{-1/2}(\partial\Omega)$ by the standard regularity theory in PDE. The total injection current through the electrode $\mathcal{E}_j$ is $I = \int_{\mathcal{E}_j} g\, ds$. The condition $\nabla u \times \mathbf{n}|_{\mathcal{E}_k} \approx 0$ ensures that $u|_{\mathcal{E}_k}$ is approximately a constant for each electrode since ∇u is normal to its level surface.

Expressing the boundary conditions by g, the time-harmonic voltage u is governed by

$$\begin{cases} \nabla \cdot \big(\gamma(\mathbf{r}) \nabla u(\mathbf{r})\big) = 0 & \text{in } \Omega, \\ \gamma \nabla u \cdot \mathbf{n} = g & \text{on } \partial\Omega. \end{cases} \tag{7.22}$$

Since g is the magnitude of the current density on $\partial\Omega$ due to the injection current, $g = 0$ on $\partial\Omega \setminus (\mathcal{E}_j \cup \mathcal{E}_{j+1})$ and $\int_{\mathcal{E}_j} g\, ds = I = -\int_{\mathcal{E}_{j+1}} g\, ds$. Setting a reference voltage $u(\mathbf{r}_0) = 0$ for a fixed point $\mathbf{r}_0 \in \Omega$, we can obtain a unique solution u of (7.22) from γ and g. Note that u depends on γ, g and the geometry of Ω. When γ changes with ω, so does u.

Example 7.3.1 *Assume that the electrodes are perfect conductors and $\omega\epsilon/\sigma = 0$. The potential $u \in \mathbb{R}$ satisfies*

$$\begin{cases} \nabla \cdot (\sigma \nabla u) = 0 \quad \textit{in } \Omega, \\ I = \int_{\mathcal{E}_j} \sigma \frac{\partial u}{\partial \mathbf{n}}\, ds = -\int_{\mathcal{E}_{j+1}} \sigma \frac{\partial u}{\partial \mathbf{n}}\, ds, \\ \int_{\mathcal{E}_k} \mathbf{n} \cdot (\gamma \nabla u)\, ds = 0 \quad \textit{for } k \in \{1, 2, \ldots, E\} \setminus \{j, j+1\}, \\ \nabla u \times \mathbf{n} = 0 \quad \textit{on } \bigcup_{k=1}^{E} \mathcal{E}_k, \\ \sigma \frac{\partial u}{\partial \mathbf{n}} = 0 \quad \textit{on } \partial\Omega \setminus \overline{\bigcup_{k=1}^{E} \mathcal{E}_k}. \end{cases} \tag{7.23}$$

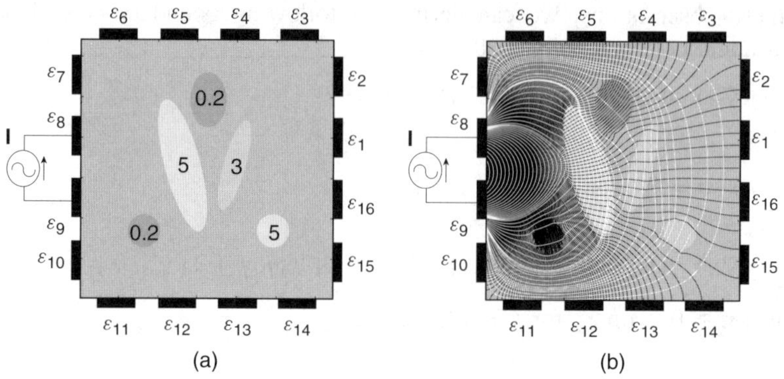

Figure 7.7 (a) An example of an electrically conducting domain with a given conductivity distribution. Numbers inside ellipsoids are conductivity values (S m^{-1}). (b) Voltage and current density distributions induced by the injection current. Black and white lines are equipotential and current density streamlines, respectively

The above non-standard boundary value problem is well-posed and has a unique solution within $H^1(\Omega)$ up to a constant. Figure 7.7 illustrates a numerical example.

Example 7.3.2 *Assume that u is a solution of (7.23). Then, ∇u is singular at the edge of a current injection electrode. To estimate this singularity, we consider the simplified model $\Omega = \mathbb{R}^3_- := \{\mathbf{r} : z < 0\}$ and $\mathcal{E} = \{(x, y, 0) : \sqrt{x^2 + y^2} < 1\}$. Let w be the $H^1(\Omega)$-solution of the following mixed boundary value problem:*

$$\begin{cases} -\nabla^2 w = 0 & \text{in } \Omega, \\ \dfrac{\partial w}{\partial \mathbf{n}} = 0 & \text{on } \partial\Omega \setminus \mathcal{E}, \\ w|_{\mathcal{E}} = 1. \end{cases} \tag{7.24}$$

Let $g = \nabla w \cdot \mathbf{n}|_{\partial\Omega}$. Prove that g satisfies the integral equation:

$$1 = \frac{1}{2\pi} \int_\Gamma \frac{g(x', y')}{\sqrt{(x - x')^2 + (y - y')^2}} \, dx' \, dy' \quad \text{if } \sqrt{x^2 + y^2} < 1. \tag{7.25}$$

Find a representation formula for w and find the behavior of g near the circular edge of the electrode $\mathcal{E}$.

Solution. *Assume that g satisfies (7.25). Define*

$$U(\mathbf{r}) = \int_\Gamma \mathcal{N}(\mathbf{r}, \mathbf{r}') g(\mathbf{r}') \, ds \quad (\mathbf{r} \in \Omega),$$

where $\mathcal{N}$ is the Neumann function

$$\mathcal{N}(\mathbf{r}, \mathbf{r}') := \frac{1}{4\pi} \left(\frac{1}{\sqrt{(x - x')^2 + (y - y')^2 + (z - z')^2}} + \frac{1}{\sqrt{(x - x')^2 + (y - y')^2 + (z + z')^2}} \right).$$

It is easy to see that $U \in H^1(\Omega)$ *satisfies*

$$\nabla^2 U = 0 \text{ in } \Omega, \quad U|_{\mathcal{E}} = 1 \quad \text{and} \quad \left.\frac{\partial U}{\partial \mathbf{n}}\right|_{\partial\Omega\setminus\mathcal{E}} = 0.$$

It follows from the uniqueness theory that the solution w *of (7.24) must be* $U = w$ *in* Ω*. Since* g *is radial,* $g(r) = g(\sqrt{x^2 + y^2})$ *satisfies*

$$1 = \int_0^1 \phi(t, r) g(r)\,\mathrm{d}r \quad (0 \le t < 1),$$

where

$$\phi(t, r) = \frac{1}{2\pi} \int_0^{2\pi} \frac{r}{\sqrt{t^2 - 2tr\cos\theta + r^2}}\,\mathrm{d}\theta.$$

Noting that

$$0 = \frac{\mathrm{d}}{\mathrm{d}t} \int_0^1 \phi(t, r) g(r)\,\mathrm{d}r$$

for $0 < t < 1$*, one can show that* $\lim_{r\to 1^-} |g(r)| = \infty$.

7.3.3 Elliptic PDE for Two-Electrode Method

When we adopt the two-electrode method where we measure voltages on current injection electrodes, we must take into account of the contact impedance. We introduce the complete electrode model (Cheng *et al.* 1989; Somersalo *et al.* 1992; Vauhkonen *et al.* 1996), where the complex potential u satisfies

$$\begin{cases} \nabla \cdot (\gamma \nabla u) = 0 \quad \text{in } \Omega, \\ (u + z_k \gamma \frac{\partial u}{\partial \mathbf{n}})|_{\mathcal{E}_k} = U_k, \quad k = 1, \ldots, E, \\ \gamma \frac{\partial u}{\partial \mathbf{n}} = 0 \quad \text{on } \partial\Omega \setminus \bigcup_{k=1}^{E} \mathcal{E}_k, \\ \int_{\mathcal{E}_k} \gamma \frac{\partial u}{\partial \mathbf{n}} = 0 \quad \text{if } k \in \{1, 2, \ldots, E\} \setminus \{j, j+1\}, \\ \int_{\mathcal{E}_j} \gamma \frac{\partial u}{\partial \mathbf{n}}\,\mathrm{d}s = I = -\int_{\mathcal{E}_{j+1}} \gamma \frac{\partial u}{\partial \mathbf{n}}\,\mathrm{d}s, \end{cases} \tag{7.26}$$

where z_k is the contact impedance of the kth electrode $\mathcal{E}_k$ and U_k is the voltage on $\mathcal{E}_k$. Setting a reference voltage having $\sum_{k=1}^{E} U_k = 0$, we can obtain a unique solution u of (7.26).

In this case, measured boundary voltages are

$$V^{j,1} := U_1^j - U_E^j, \quad V^{j,2} := U_2^j - U_1^j, \quad \ldots, \quad V^{j,E} := U_E^j - U_{E-1}^j.$$

Using an E-channel EIT system, we may inject E number of currents through adjacent pairs of electrodes and measure the following voltage data set:

$$\mathbb{F} = [V^{1,1}, \ldots, V^{1,E}, \quad \ldots, \quad V^{E,1}, \ldots, V^{E,E}]^{\mathrm{T}} \in \mathbb{C}^{E^2}.$$

The voltage data are influenced by contact impedances whose values are unknown. Since the reciprocity principle $V^{k,j} = V^{j,k}$ in section 7.4 still holds, $\mathbb{F}$ contains at most $E(E-1)/2$ number of independent data.

7.3.4 Min–Max Property of Complex Potential

The variational form of the problem (7.22) with the Neumann boundary condition is

$$\int_{\Omega} \gamma \nabla u \cdot \nabla \phi \, d\mathbf{r} = \int_{\partial\Omega} g\phi \, ds \quad \text{for all } \phi \in H^1(\Omega). \tag{7.27}$$

According to the Lax–Milgram theorem in Chapter 4, for a given $g \in H^{-1/2}(\partial\Omega)$ with $\int_{\partial\Omega} g \, ds = 0$, there exists a unique solution $u \in H^1(\Omega)$ with $\int_{\partial\Omega} u \, ds = 0$ satisfying (7.27). When $\omega = 0$, we can figure out the global structure of $u \in \mathbb{R}$ using its weighted mean value property, maximum principle and minimization property of the corresponding energy functional:

$$\int_{\partial\Omega} gu \, ds = \min_{w \in \mathcal{A}_g} \int_{\Omega} \gamma |\nabla w|^2 \, d\mathbf{r}, \quad \mathcal{A}_g = \left\{ w \in H^1(\Omega) : \gamma \frac{\partial w}{\partial \mathbf{n}}\bigg|_{\partial\Omega} = g \right\}. \tag{7.28}$$

When $\omega > 0$, the potential $u \in \mathbb{C}$ does not have the minimization property (7.28), mean value property and maximum principle. Denoting $v = \Re\{u\}$ and $h = \Im\{u\}$, $u = v + ih$ satisfies the following coupled system:

$$\begin{cases} \nabla \cdot (\sigma \nabla v) - \nabla \cdot (\omega\epsilon \nabla h) = 0, & \text{in } \Omega, \\ \nabla \cdot (\omega\epsilon \nabla v) + \nabla \cdot (\sigma \nabla h) = 0, & \text{in } \Omega, \\ \mathbf{n} \cdot (-\sigma \nabla v(x) + \omega\epsilon \nabla h(x)) = g, & \text{on } \partial\Omega, \\ \mathbf{n} \cdot (-\sigma \nabla h(x) - \omega\epsilon \nabla v(x)) = 0, & \text{on } \partial\Omega. \end{cases} \tag{7.29}$$

The complex potential u has the min-max property (Cherkaeva and Cherkaev 1995) in the sense that

$$\int_{\partial\Omega} g\Re\{u\} \, ds = \min_{\substack{v|_{\partial\Omega} = \Re\{u\} \\ v \in H^1(\Omega)}} \max_{\substack{h|_{\partial\Omega} = \Im\{u\} \\ h \in H^1(\Omega)}} \int_{\Omega} [\Re\{\gamma\}(|\nabla v|^2 - |\nabla h|^2) - 2\Im\{\gamma\}\nabla v \cdot \nabla h] \, d\mathbf{r} \tag{7.30}$$

and

$$\int_{\partial\Omega} g\Im\{u\} \, ds = \min_{\substack{v|_{\partial\Omega} = \Re\{u\} \\ v \in H^1(\Omega)}} \max_{\substack{h|_{\partial\Omega} = \Im\{u\} \\ h \in H^1(\Omega)}} \int_{\Omega} [\Im\{\gamma\}(|\nabla v|^2 - |\nabla h|^2) + 2\Re\{\gamma\}\nabla v \cdot \nabla h] \, d\mathbf{r}. \tag{7.31}$$

7.4 Forward Problem and Model

We describe the forward problem of EIT using the Neumann-to-Dirichlet (NtD) data, which depend on the admittivity γ. After introducing the continuous NtD data and some theoretical issues, we formulate the discrete NtD data of an E-channel EIT system.

7.4.1 Continuous Neumann-to-Dirichlet Data

We define the continuous NtD data set Λ_γ as

$$\Lambda_\gamma : H_\diamond^{-1/2}(\partial\Omega) \to H_\diamond^{1/2}(\partial\Omega), \tag{7.32}$$

$$g \rightsquigarrow u_\gamma^g|_{\partial\Omega}, \tag{7.33}$$

where u_γ^g is the unique solution of the Neumann boundary value problem

$$\begin{cases} \nabla\cdot\big(\gamma(\mathbf{r})\nabla u_\gamma^g(\mathbf{r})\big) = 0 \quad \text{in } \Omega, \\ -\gamma\nabla u_\gamma^g\cdot\mathbf{n}|_{\partial\Omega} = g, \quad \displaystyle\int_{\partial\Omega} u\,\mathrm{d}s = 0. \end{cases} \tag{7.34}$$

This NtD data Λ_γ include all possible Cauchy data. With this full data set, the forward problem of EIT is modeled as the map

$$\gamma \to \Lambda_\gamma \tag{7.35}$$

and the inverse problem is to invert the map in (7.35).

There are two major theoretical questions regarding the map.

Uniqueness: Is the map $\gamma \to \Lambda_\gamma$ injective?
Stability: Find the estimate of the form:

$$\|\log\gamma^1 - \log\gamma^2\|_* \le \Psi(\|\Lambda_{\gamma^1} - \Lambda_{\gamma^2}\|_{\mathcal{L}}),$$

where $\|\cdot\|_*$ is an appropriate norm for the admittivity, $\Psi : \mathbb{R}_+ \to \mathbb{R}_+$ is a continuously increasing function with $\Psi(0) = 0$, and $\|\cdot\|_{\mathcal{L}}$ is the operator norm on $\mathcal{L}(H_\diamond^{-1/2}(\partial\Omega), H_\diamond^{1/2}(\partial\Omega))$.

The NtD data Λ_γ are closely related with the Neumann function restricted on $\partial\Omega$. The Neumann function $\mathcal{N}_\gamma(\mathbf{r},\mathbf{r}')$ is the solution of the following Neumann problem: for each $\mathbf{r}$,

$$\begin{cases} \nabla\cdot(\gamma\nabla\mathcal{N}_\gamma(\mathbf{r},\cdot)) = \delta(\mathbf{r}-\cdot) \quad \text{in } \Omega, \\ \gamma\nabla\mathcal{N}_\gamma(\mathbf{r},\cdot)\cdot\mathbf{n} = 0 \quad \text{on } \partial\Omega, \end{cases}$$

where δ is the Dirac delta function. With the use of the Neumann function $\mathcal{N}_\gamma(\mathbf{r},\mathbf{r}')$, we can represent $u_\gamma^g(\mathbf{r})$ in terms of the singular integral:

$$\begin{aligned} u_\gamma^g(\mathbf{r}) &= \int_\Omega \delta(\mathbf{r}-\mathbf{r}')u_\gamma^g(\mathbf{r}')\,\mathrm{d}\mathbf{r}' \\ &= \int_\Omega \nabla\cdot(\gamma(\mathbf{r}')\nabla\mathcal{N}_\gamma(\mathbf{r},\mathbf{r}'))u_\gamma^g(\mathbf{r}')\,\mathrm{d}\mathbf{r}' \\ &= -\int_\Omega \gamma(\mathbf{r}')\nabla\mathcal{N}_\gamma(\mathbf{r},\mathbf{r}')\cdot\nabla u_\gamma^g(\mathbf{r}')\,\mathrm{d}\mathbf{r}' \\ &= \int_{\partial\Omega} \mathcal{N}_\gamma(\mathbf{r},\mathbf{r}')g(\mathbf{r}')\,\mathrm{d}s_{\mathbf{r}'}. \end{aligned}$$

Since Λ_γ is the restriction of u_γ^g to the boundary $\partial\Omega$, we can represent it as

$$\Lambda_\gamma[g](\mathbf{r}) = \int_{\partial\Omega} \mathcal{N}_\gamma(\mathbf{r}, \mathbf{r}')g(\mathbf{r}')\,\mathrm{d}s_{\mathbf{r}'}, \quad \mathbf{r} \in \partial\Omega. \tag{7.36}$$

The kernel $\mathcal{N}_\gamma(\mathbf{r}, \mathbf{r}')$ with $\mathbf{r}, \mathbf{r}' \in \partial\Omega$ can be viewed as an expression of the NtD data Λ_γ. Note that Λ_γ is sensitive to a change in the geometry of the surface $\partial\Omega$ since $\mathcal{N}_\gamma(\mathbf{r}, \mathbf{r}')$ is singular at $\mathbf{r} = \mathbf{r}'$.

For the uniqueness in a three-dimensional problem, Kohn and Vogelius (1985) showed the injectivity of $\gamma \to \Lambda_\gamma$ if γ is piecewise analytic. Sylvester and Uhlmann (1987) showed the injectivity if $\gamma \in C^\infty(\bar{\Omega})$. The smoothness condition on γ and $\partial\Omega$ has been relaxed by several researchers (Astala and Paivarinta 2006b; Brown and Uhlmann 1997; Isakov 1991; Nachman 1988, 1996).

For a two-dimensional problem, Nachman (1996) proved the uniqueness under some smoothness conditions on γ and provided a constructive way of recovering γ. Based on Nachman's proof on two-dimensional global uniqueness, Siltanen *et al.* (2000) developed the *d-bar algorithm*, which solves the full nonlinear EIT problem without iteration.

To reconstruct γ by inverting the map (7.35), it would be ideal if the full continuous NtD data Λ_γ are available. In practice, it is not possible to get them due to a limited number of electrodes with a finite size. It is also difficult to capture the correct geometry of $\partial\Omega$ at a reasonable cost. The map in (7.35) is highly nonlinear and insensitive to a local change of γ, as explained in section 7.4.3. All of these hinder a stable reconstruction of γ with a high spatial resolution.

7.4.2 Discrete Neumann-to-Dirichlet Data

We assume an EIT system using E electrodes $\mathcal{E}_j$ for $j = 1, 2, \ldots, E$. The isotropic admittivity distribution in Ω is denoted as γ. The complex potential u in (7.22) subject to the jth injection current between $\mathcal{E}_j$ and $\mathcal{E}_{j+1}$ is denoted as u^j and it approximately satisfies the following Neumann boundary value problem:

$$\begin{cases} \nabla \cdot \big(\gamma(\mathbf{r})\nabla u^j(\mathbf{r})\big) = 0 & \text{in } \Omega, \\ -\gamma\nabla u^j \cdot \mathbf{n} = g^j & \text{on } \partial\Omega, \end{cases} \tag{7.37}$$

where $\int_{\mathcal{E}_j} g^j\,\mathrm{d}s = I = -\int_{\mathcal{E}_{j+1}} g^j\,\mathrm{d}s$ and the Neumann data g^j are zero on the boundary regions not contacting with the current injection electrodes. Setting a reference voltage at $\mathbf{r}_0 \in \Omega$ as $u^j(\mathbf{r}_0) = 0$, we can obtain a unique solution u^j from γ and g^j.

We assume the neighboring data collection protocol in section 7.2 to measure boundary voltages between adjacent pairs of electrodes, $\mathcal{E}_k$ and $\mathcal{E}_{k+1}$ for $k = 1, 2, \ldots, E$. The kth boundary voltage difference subject to the jth injection current is denoted as

$$V^{j,k} = \frac{1}{|\mathcal{E}_k|}\int_{\mathcal{E}_k} u^j\,\mathrm{d}s - \frac{1}{|\mathcal{E}_{k+1}|}\int_{\mathcal{E}_{k+1}} u^j\,\mathrm{d}s \quad \text{for } j, k = 1, 2, \ldots, E, \tag{7.38}$$

where $(1/|\mathcal{E}_k|)\int_{\mathcal{E}_k} u^j\,\mathrm{d}s$ can be understood as the average of u^j over $\mathcal{E}_k$.

Lemma 7.4.1 *The kth boundary voltage difference subject to the jth injection current satisfies*

$$\int_{\Omega} \gamma \nabla u^j \cdot \nabla u^k \, \mathrm{d}\mathbf{r} = V^{j,k}[\gamma]. \tag{7.39}$$

Proof. Integration by parts yields

$$\int_{\Omega} \gamma \nabla u^j \cdot \nabla u^k \, \mathrm{d}\mathbf{r} = \int_{\partial\Omega} u^j g^k \, \mathrm{d}s = V^{j,k}[\gamma],$$

where the last identity comes from the boundary condition (7.37). □

Since $V^{j,k}[\gamma]$ is uniquely determined by the distribution of γ, it can be viewed as a function of γ. With E projections and E complex boundary voltage data for each projection, we are provided with E^2 complex boundary voltage data, which are expressed in matrix form as

$$\mathbb{F}[\gamma] := \begin{bmatrix} V^{1,1} & V^{1,2} & \cdots\cdots & V^{1,E} \\ V^{2,1} & V^{2,2} & \cdots\cdots & V^{2,E} \\ \vdots & \vdots & & \vdots \\ \vdots & \vdots & & \vdots \\ V^{E,1} & V^{E,2} & \cdots\cdots & V^{E,E} \end{bmatrix}, \quad \begin{matrix} \leftarrow \text{1st projection} \\ \leftarrow \text{2nd projection} \\ \\ \\ \leftarrow E\text{th projection} \end{matrix} \tag{7.40}$$

where $V^{j,k} = V^{j,k}[\gamma]$ for a given γ.

Theorem 7.4.2 (Reciprocity of NtD data) *For a given γ, $V^{j,k}$ in (7.38) satisfies the reciprocity property:*

$$V^{j,k} = V^{k,j} \quad \textit{for all } k, j = 1, 2, \ldots, E. \tag{7.41}$$

Proof. The reciprocity follows from the identity:

$$V^{j,k} = \int_{\partial\Omega} \gamma \frac{\partial u^j}{\partial \mathbf{n}} u^k \, \mathrm{d}s = \int_{\Omega} \gamma \nabla u_j \cdot \nabla u_k \, \mathrm{d}\mathbf{r} = \int_{\partial\Omega} \gamma \frac{\partial u_k}{\partial \mathbf{n}} u_j \, \mathrm{d}s = V^{j,k}. \quad \square$$

Observation 7.4.3 *Assume that γ is constant or homogeneous in Ω. Then,*

$$\gamma = \frac{\int_{\Omega} \nabla w^j \cdot \nabla w^k \, \mathrm{d}\mathbf{r}}{V^{j,k}}, \tag{7.42}$$

where w^j is the solution of (7.37) with $\gamma = 1$.

Proof. Since $w^j = \gamma u^j$, we have

$$\gamma = \frac{\int_{\Omega} (\gamma \nabla u^j) \cdot (\gamma \nabla u^k) \, \mathrm{d}\mathbf{r}}{\int_{\Omega} \gamma \nabla u^j \cdot \nabla u^k \, \mathrm{d}\mathbf{r}} = \frac{\int_{\Omega} \nabla w^j \cdot \nabla w^k \, \mathrm{d}\mathbf{r}}{\int_{\partial\Omega} g^j u^k \, \mathrm{d}S} = \frac{\int_{\Omega} \nabla w^j \cdot \nabla w^k \, \mathrm{d}\mathbf{r}}{V^{j,k}}. \quad \square$$

The data matrix $\mathbb{F}[\gamma]$ in (7.40) can be viewed as a discrete version of the NtD data since it provides all the measurable current-to-voltage relations using the E-channel EIT system. With this discrete NtD data set, the forward problem of the E-channel EIT is modeled as the map

$$\gamma \to \mathbb{F}[\gamma] \tag{7.43}$$

and the inverse problem is to invert the map in (7.43).

The smoothness condition on γ should not be a major issue in a practical EIT image reconstruction. For any discontinuous admittivity γ and an E-channel EIT system, we always find $\tilde{\gamma} \in C^\infty(\bar{\Omega})$, which approximates γ in such a way that

$$\sum_{j=1}^{E}\sum_{k=1}^{E} |V^{j,k}[\gamma] - V^{j,k}[\tilde{\gamma}]| < \text{arbitrary small positive quantity.}$$

Taking account of inevitable measurement noise in the discrete NtD data and the ill-posedness of its inversion process, we conclude that γ and $\tilde{\gamma}$ are not distinguishable in practice.

7.4.3 Nonlinearity between Admittivity and Voltage

As defined in (7.43), the forward model is a map from the admittivity to a set of boundary voltage data. From (7.37), we can see that any change in the admittivity influences all voltage values. Unlike the linear relation between currents and voltages, the map in (7.43) is nonlinear. Understanding the map should precede designing a method to invert it.

A voltage value at a point inside the domain can be expressed as a weighted average of its neighboring voltages, where the weights are determined by the admittivity distribution. In this weighted averaging method, information on the admittivity distribution is conveyed to the boundary voltage, as shown in Figure 7.8. The boundary voltage is entangled with the global structure of the admittivity distribution in a highly nonlinear way, and we investigate the relation in this section.

We assume that the domain Ω is a square in $\mathbb{R}^2$ with its conductivity distribution σ, that is, $\gamma = \sigma$. We divide Ω uniformly into an $N \times N$ square mesh. Each square element

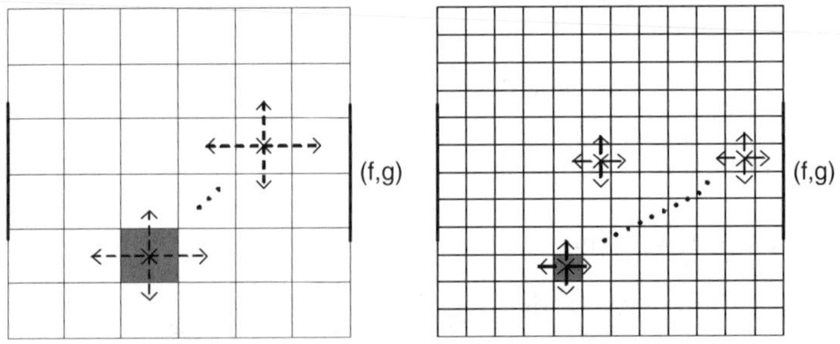

Figure 7.8 Nonlinearity and insensitivity grow exponentially as the matrix size increases

is denoted as $\Omega_{i,j}$ with its center at (x_i, y_j) for $i, j = 1, 2, \ldots, N$. We assume that the conductivity σ is constant in each element $\Omega_{i,j}$, say $\sigma_{i,j}$. Let

$$\Sigma = \left\{\sigma \;\middle|\; \sigma|_{\Omega_{i,j}} = \sigma_{i,j} = \text{constant for all } i, j = 1, 2, \ldots, N\right\}.$$

For a given $\sigma \in \Sigma$, we can express σ as

$$\boldsymbol{\sigma} = \left[\sigma_1, \sigma_2, \ldots, \sigma_{N^2}\right]^{\mathrm{T}}.$$

The solution u of the elliptic PDE in (7.37) with σ in place of γ can be approximated by a vector

$$\mathbf{u} = \left[u_1, u_2, \ldots, u_{N^2}\right]^{\mathrm{T}}$$

such that each voltage u_k for $k = i + jN$ is determined by the weighted average of four neighboring voltages. To be precise, the conductivity equation

$$\nabla \cdot (\sigma(\mathbf{r})\nabla u(\mathbf{r})) = 0$$

can be written as the following discretized form

$$u_k = \frac{1}{a_{k,k}}\left(a_{k,k_T}u_{k_T} + a_{k,k_D}u_{k_D} + a_{k,k_R}u_{k_R} + a_{k,k_L}u_{k_L}\right) \tag{7.44}$$

with

$$a_{k,k} = -\sum_d a_{k,k_d} \quad \text{and} \quad a_{k,k_d} = \frac{\sigma_k \sigma_{k_d}}{\sigma_k + \sigma_{k_d}} \quad \text{for } d = \mathrm{T, D, R, L}, \tag{7.45}$$

where k_{T}, k_{D}, k_{R} and k_{L} denote top, down, right and left neighboring points of the kth point, respectively.

The discretized conductivity equation (7.44) with the Neumann boundary condition can be rewritten as a linear system of equations:

$$\mathbb{A}(\boldsymbol{\sigma})\mathbf{u} = \mathbf{g},$$

where $\mathbf{g}$ is the injection current vector associated with the Neumann boundary data g. Any change in σ_k for $k = 1, 2, \ldots, N^2$ spreads its influence to all u_k for $k = 1, 2, \ldots, N^2$ through the matrix $\mathbb{A}(\boldsymbol{\sigma})$. We should note the following implications of the entanglement among σ_k and u_k.

Geometry. The recursive averaging process in (7.44) with (7.45) makes the influence of a change in σ_k upon u_l smaller and smaller as the distance between positions of σ_k and u_l is further increased.

Nonlinearity. The recursive averaging process in (7.44) with (7.45) causes a nonlinearity between σ_k and u_l for all $k, l = 1, 2, \ldots, N^2$.

Interdependence. The recursive averaging process in (7.44) with (7.45) makes the influence of a change in σ_k upon u_l affected by all other σ_m with $m \in \{1, 2, \ldots, N^2\} \setminus \{k\}$.

7.5 Uniqueness Theory and Direct Reconstruction Method

Before we study practical inversion methods to invert the map in (7.43), we review mathematical theories of uniqueness and a direct reconstruction technique called the d-bar method.

7.5.1 Calderón's Approach

In this section, we will assume a full NtD map Λ_γ as EIT data. Calderón (1980) made the following observation, which plays a key role in achieving the theoretical development of EIT, especially uniqueness theory. For a quick and easy explanation, we assume the following throughout this section:

- $\Omega \subset \mathbb{R}^n$ with its C^2 boundary $\partial\Omega$;
- γ is real and $\gamma - 1 \in C_0^2(\Omega)$ with $\gamma = 1$ in $\mathbb{R}^n \setminus \Omega$;
- $q = (\nabla^2\sqrt{\gamma})/\sqrt{\gamma}$ in Ω and $q = 0$ in $\mathbb{R}^n \setminus \Omega$;
- $\gamma^0 = 1$ is the background conductivity.

To prove his observation, Caldéron used a set of special pairs of harmonic functions that is dense in $L^1(\Omega)$.

Lemma 7.5.1 *If* $\boldsymbol{\xi}, \boldsymbol{\eta} \in \mathbb{R}^3$ *(or* $\mathbb{R}^2$*) satisfy*

$$|\boldsymbol{\xi}| = |\boldsymbol{\eta}| \quad \textit{and} \quad \boldsymbol{\xi} \cdot \boldsymbol{\eta} = 0, \tag{7.46}$$

then both $v = \mathrm{e}^{\mathbf{r}\cdot(\boldsymbol{\eta}+\mathrm{i}\boldsymbol{\xi})}$ *and* $w = \mathrm{e}^{-\mathbf{r}\cdot(\boldsymbol{\eta}+\mathrm{i}\boldsymbol{\xi})}$ *are harmonic in the entire space* $\mathbb{R}^3$ *(or* $\mathbb{R}^2$*). Moreover,*

$$\{\nabla v_{\boldsymbol{\xi}} \cdot \nabla w_{\boldsymbol{\xi}} : \boldsymbol{\xi} \in \mathbb{R}^3\} \quad \textit{is dense in } L^1(\Omega), \tag{7.47}$$

where $v_{\boldsymbol{\xi}} = \mathrm{e}^{\mathbf{r}\cdot(\boldsymbol{\xi}^*+\mathrm{i}\boldsymbol{\xi})}$, $w_{\boldsymbol{\xi}} = \mathrm{e}^{-\mathbf{r}\cdot(\boldsymbol{\xi}^*-\mathrm{i}\boldsymbol{\xi})}$ *and*

$$\boldsymbol{\xi}^* = \frac{|\boldsymbol{\xi}|}{\sqrt{\xi_1^2 + \xi_2^2}}(-\xi_2, \xi_1, 0).$$

Proof. Both v and w are harmonic because

$$\nabla^2 v = [(|\boldsymbol{\eta}|^2 - |\boldsymbol{\xi}|^2) + \mathrm{i}2\,\boldsymbol{\xi}\cdot\boldsymbol{\eta}]v = 0 = \nabla^2 w.$$

Since $\nabla v_{\boldsymbol{\xi}} \cdot \nabla w_{\boldsymbol{\xi}} = -2|\boldsymbol{\xi}|^2 \mathrm{e}^{2\mathrm{i}\boldsymbol{\xi}\cdot\mathbf{r}}$, we have

$$\{\nabla v_{\boldsymbol{\xi}} \cdot \nabla w_{\boldsymbol{\xi}} : \boldsymbol{\xi} \in \mathbb{R}^3\} = \{\mathrm{e}^{2\mathrm{i}\boldsymbol{\xi}\cdot\mathbf{r}} : \boldsymbol{\xi} \in \mathbb{R}^3\},$$

which is clearly dense in $L^1(\Omega)$ due to the Fourier representation formula. □

Theorem 7.5.2 (Calderón's approach) *Let* $\gamma^0 \in \mathbb{C}$ *and* $\delta\sigma, \widetilde{\delta\gamma} \in C_0^2(\Omega)$. *Denote*

$$\gamma_t(\mathbf{r}) = \gamma^0 + t\delta\gamma(\mathbf{r}) \quad \textit{and} \quad \tilde{\gamma}_t(\mathbf{r}) = \gamma^0 + t\,\widetilde{\delta\gamma}(\mathbf{r}) \quad \textit{for } \mathbf{r} \in \mathbb{R}^3.$$

If, for any $g \in H_\diamond^{-1/2}(\partial\Omega)$,

$$\frac{\mathrm{d}}{\mathrm{d}t}\Lambda_{\gamma_t}(g)|_{t=0} = \frac{\mathrm{d}}{\mathrm{d}t}\Lambda_{\tilde{\gamma}_t}(g)|_{t=0} \quad \textit{on } \partial\Omega,$$

then

$$\delta\gamma = \widetilde{\delta\gamma} \quad \textit{in } \Omega.$$

Proof. For $g \in H_\diamond^{-1/2}(\partial\Omega)$, let u_t^g be a solution of

$$\begin{cases} \nabla\cdot\left(\gamma_t \nabla u_t^g\right) = 0 \quad \text{in } \Omega, \\ -\gamma_t \nabla u_t^g \cdot \mathbf{n}|_{\partial\Omega} = 0, \quad \displaystyle\int_{\partial\Omega} u_t^g \, \mathrm{d}s = 0. \end{cases} \tag{7.48}$$

Taking the derivative of the problem (7.48) with respect to t, we have

$$\begin{cases} \nabla\cdot\left(\dfrac{\partial}{\partial t}\gamma_t(\nabla u_t^g)\right) = -\nabla\cdot\left(\gamma_t\nabla\dfrac{\partial}{\partial t}u_t^g\right) \quad \text{in } \Omega, \\ -\gamma_t\nabla(u_t^g)\cdot\mathbf{n}|_{\partial\Omega} = 0. \end{cases} \tag{7.49}$$

Here, we use the assumption that $\delta\sigma|_{\partial\Omega} = 0$. By multiplying (7.49) by u_t^ϕ and integrating over Ω, we have

$$\begin{aligned} \int_\Omega \delta\gamma\nabla u_t^g\cdot\nabla u_t^\phi \, \mathbf{dr} &= -\int_\Omega \gamma_t\nabla\frac{\partial}{\partial t}u_t^g\cdot\nabla u_t^\phi \, \mathbf{dr} \\ &= -\int_{\partial\Omega}\frac{\mathrm{d}}{\mathrm{d}t}\Lambda_{\gamma_t}(g)\phi \, \mathrm{d}s. \end{aligned}$$

At $t = 0$, this becomes

$$\int_\Omega \delta\gamma\nabla u_0^g\cdot\nabla u_0^\phi \, \mathbf{dr} = -\int_{\partial\Omega}\frac{\mathrm{d}}{\mathrm{d}t}\Lambda_{\gamma_t}(g)|_{t=0}\phi \, \mathrm{d}s.$$

We also have the same identity for $\widetilde{\delta\gamma}$:

$$\int_\Omega \widetilde{\delta\gamma}\nabla u_0^g\cdot\nabla u_0^\phi \, \mathbf{dr} = -\int_{\partial\Omega}\frac{\mathrm{d}}{\mathrm{d}t}\Lambda_{\tilde{\gamma}_t}(g)|_{t=0}\phi \, \mathrm{d}s.$$

It follows from the assumption $\Lambda_{\tilde{\gamma}_t}(g)|_{t=0} = \Lambda_{\gamma_t}(g)|_{t=0}$ that

$$\int_\Omega (\delta\gamma - \widetilde{\delta\gamma})\nabla u_0^g\cdot\nabla u_0^\phi \, \mathbf{dr} = 0, \quad \forall g,\phi \in H_\diamond^{-1/2}(\partial\Omega). \tag{7.50}$$

Hence, $\delta\gamma = \widetilde{\delta\gamma}$ because

$$\{\nabla v_{\boldsymbol{\xi}}\cdot\nabla w_{\boldsymbol{\xi}} : \boldsymbol{\xi}\in\mathbb{R}^3\} \subset \{\nabla u_0^g\cdot\nabla u_0^\phi : g,\phi\in H_\diamond^{-1/2}(\partial\Omega)\}$$

and $\{\nabla v_{\boldsymbol{\xi}}\cdot\nabla w_{\boldsymbol{\xi}} : \boldsymbol{\xi}\in\mathbb{R}^3\}$ is dense in $L^1(\Omega)$ from Lemma 7.5.1. □

Let us begin by explaining the scattering transform that transforms the conductivity equation $\nabla\cdot(\gamma\nabla u) = 0$ into the Schrödinger equation $(-\nabla^2 + q)\psi = 0$. This transform was first used to prove the uniqueness of EIT for $\gamma \in C^{1,1}(\bar{\Omega})$ by Sylvester and Uhlmann (1987). The following lemma explains this scattering transform.

Lemma 7.5.3 *Let $\gamma \in C^2(\bar{\Omega})$ and u satisfy*

$$\nabla \cdot (\gamma \nabla u) = 0 \quad \textit{in } \Omega.$$

Then, $\psi := \sqrt{\gamma}\, u$ satisfies

$$-\nabla^2 \psi + \frac{\nabla^2 \sqrt{\gamma}}{\sqrt{\gamma}} \psi = 0. \tag{7.51}$$

Proof. The proof follows from the direct computation:

$$\begin{aligned} 0 &= \nabla \cdot (\gamma \nabla u) \\ &= \nabla \cdot (\sqrt{\gamma}\, \nabla(\sqrt{\gamma}\, u)) - \nabla \cdot (\sqrt{\gamma}\, u \nabla \sqrt{\gamma}) \\ &= \nabla \cdot (\sqrt{\gamma}\, \nabla \psi) - \nabla \cdot (\psi \nabla \sqrt{\gamma}) \\ &= \sqrt{\gamma}\, \nabla^2 \psi - \psi \nabla^2 \sqrt{\gamma}. \end{aligned}$$

□

Remark 7.5.4 *From Lemma 7.5.3, $\psi = \sqrt{\gamma} u$ is the solution of*

$$\begin{cases} -\nabla^2 \psi + q\psi = 0 & \textit{in } \Omega \\ \psi|_{\partial\Omega} = 1 \end{cases} \qquad \left(q = \frac{\nabla^2 \sqrt{\gamma}}{\sqrt{\gamma}} \right). \tag{7.52}$$

This fact has been used to develop a two-dimensional constructive identification method of γ named the $\bar{\partial}$ (or d-bar) method (Nachman 1988).

7.5.2 Uniqueness and Three-Dimensional Reconstruction: Infinite Measurements

In this section, we briefly explain some impressive results on the uniqueness question and three-dimensional reconstruction in EIT mainly by Sylvester and Uhlmann (1987) and Nachman (1988). We, however, note that the reconstruction formula suggested in this section may not be appropriate for practical cases.

We define the DtN map $\Gamma_q : H_\diamond^{-1/2}(\partial\Omega) \to H_\diamond^{1/2}(\partial\Omega)$ by

$$\Gamma_{q_j}(g) = u^j|_{\partial\Omega},$$

where u_j satisfies

$$\begin{cases} L_{q_j} u_j := \nabla^2 u_j - q_j u_j = 0 & \text{in } \Omega, \\ \left.\dfrac{\partial u^j}{\partial \mathbf{n}}\right|_{\partial\Omega} = g. \end{cases} \tag{7.53}$$

The goal is to prove that

$$\Gamma_{q_1} = \Gamma_{q_2} \quad \Longrightarrow \quad q_1 = q_2.$$

Lemma 7.5.5 *Assume that* $\Gamma_{q_1} = \Gamma_{q_2}$. *For any* u_1 *and* u_2 *satisfying* $\nabla^2 u_j - q_j u_j = 0$ *in* Ω, *we have*

$$\int_\Omega (q_2 - q_1) u_1 u_2 = 0.$$

Proof. By the definition, we have

$$\int_\Omega \nabla u_j \cdot \nabla \phi - q_j u_j \phi \, dx = \int_{\partial\Omega} \frac{\partial u_j}{\partial \mathbf{n}} \phi \, ds$$

for any $\phi \in H^1(\Omega)$. Hence,

$$\int_\Omega \nabla u_1 \cdot \nabla u_2 - q_1 u_1 u_2 \, dx = \int_{\partial\Omega} \frac{\partial u_1}{\partial \mathbf{n}} u_2 \, ds,$$

$$\int_\Omega \nabla u_1 \cdot \nabla u_2 - q_2 u_1 u_2 \, dx = \int_{\partial\Omega} u_1 \frac{\partial u_2}{\partial \mathbf{n}} \, ds.$$

Subtracting the above two equations yields

$$\int_\Omega (q_2 - q_1) u_1 u_2 d\mathbf{x} = \int_{\partial\Omega} g(\Gamma_{q_2}(g) - \Gamma_{q_1}(g)) \, ds.$$

It then follows that if $\Gamma_{q_2} = \Gamma_{q_1}$, then

$$\int_\Omega (q_2 - q_1) u_1 u_2 \, d\mathbf{x} = 0.$$

□

Lemma 7.5.6 *For* $\boldsymbol{\zeta} \in \mathbb{C}^3$ *satisfying* $\boldsymbol{\zeta} \cdot \boldsymbol{\zeta} = 0$, *there exists a solution* u *of the equation*

$$\nabla^2 u - qu = 0 \quad in\ \Omega$$

in the form

$$u(x) = e^{ix \cdot \zeta}[1 + \psi_\zeta(x)]$$

and $\psi_\zeta \to 0$ *in* $L^2(\Omega)$ *as* $|\boldsymbol{\zeta}| \to \infty$.

Proof. If $u(x) = e^{ix \cdot \zeta}[1 + \psi_\zeta(\mathbf{x})]$ is a solution of $\nabla^2 u - qu = 0$, then ψ_ζ satisfies

$$(\nabla^2 + 2i\boldsymbol{\zeta} \cdot \nabla)\psi_\zeta = q(1 + \psi_\zeta).$$

Since the symbol of the Fourier transform of the operator $\nabla^2 + 2i\boldsymbol{\zeta} \cdot \nabla$ is $-|\boldsymbol{\xi}|^2 - 2\boldsymbol{\zeta} \cdot \boldsymbol{\xi}$, Green's function for $\nabla^2 + 2i\boldsymbol{\zeta} \cdot \nabla$ can be expressed as

$$g_\zeta(\mathbf{x}) = -\frac{1}{(2\pi)^3} \int_{\mathbb{R}^3} \frac{e^{i\mathbf{x} \cdot \boldsymbol{\xi}}}{|\boldsymbol{\xi}|^2 + 2\boldsymbol{\zeta} \cdot \boldsymbol{\xi}} d\boldsymbol{\xi} \quad (\nabla_\zeta^2 g_\zeta = \delta).$$

This Green's function was first introduced by Faddeev (1965) and the integral should be understood as an oscillatory integral. Hence, ψ_ζ must satisfy

$$\psi_\zeta = g_\zeta * (q\psi_\zeta) + g_\zeta * q.$$

The decay condition $\psi_\zeta \to 0$ in $L^2(\Omega)$ as $|\boldsymbol{\zeta}| \to \infty$ follows from the fact that

$$\|g_\zeta * f\|_{L^2_\delta} \leq \frac{C}{|\boldsymbol{\zeta}|}\|f\|_{L^2_{\delta+1}}, \tag{7.54}$$

where $-1 < \delta < 0$ and

$$\|f\|^2_{L^2_\delta} = \int_{\mathbb{R}^3} (1 + |\mathbf{x}|^2)^\delta |f(\mathbf{x})|^2 \, d\mathbf{x}.$$

□

Theorem 7.5.7 *The set* $\{u_1 u_2 : u_j \in C^2(\bar{\Omega})$ *satisfying* $\nabla^2 u_j - q_j u_j = 0\}$ *is dense in* $L^1(\Omega)$. *Hence,*

$$\Gamma_{q_1} = \Gamma_{q_2} \quad \Longrightarrow \quad q_1 = q_2 \text{ in } \Omega.$$

Proof. For each $\mathbf{k} \in \mathbb{R}^3$ and $r \in \mathbb{R}_+$, we can choose $\boldsymbol{\eta}, \boldsymbol{\xi} \in \mathbb{R}^3$ (six unknowns) satisfying four equations:

$$0 = \boldsymbol{\eta} \cdot \boldsymbol{\xi} = \boldsymbol{\eta} \cdot \mathbf{k} = \mathbf{k} \cdot \boldsymbol{\xi}, \quad |\boldsymbol{\eta}|^2 = r^2 |\boldsymbol{\xi}|^2 + |\mathbf{k}|^2.$$

Denoting

$$\boldsymbol{\zeta}_1 := \mathrm{i}\boldsymbol{\eta} - (r\boldsymbol{\xi} + \mathbf{k}) \quad \text{and} \quad \boldsymbol{\zeta}_2 := -\mathrm{i}\boldsymbol{\eta} - (-r\boldsymbol{\xi} + \mathbf{k}), \quad \boldsymbol{\eta}, \boldsymbol{\xi}, \mathbf{k} \in \mathbb{R}^3,$$

we have $\boldsymbol{\zeta}_1 + \boldsymbol{\zeta}_2 = -2\mathbf{k}$ and $\boldsymbol{\zeta}_j \cdot \boldsymbol{\zeta}_j = 0$ $(j = 1, 2)$. Let u_j be the solution given in Lemma 7.5.6 corresponding to q_j, that is,

$$u_j = \mathrm{e}^{\mathrm{i}\mathbf{x}\cdot\boldsymbol{\zeta}_j}[1 + \psi_{\zeta_j}(x)].$$

Then,

$$u_1 u_2 = \mathrm{e}^{-2\mathrm{i}\mathbf{k}\cdot\mathbf{x}}[1 + \psi_{\zeta_1} + \psi_{\zeta_2} + \psi_{\zeta_1}\psi_{\zeta_2}] \to \mathrm{e}^{-2\mathrm{i}\mathbf{k}\cdot\mathbf{x}} \quad \text{as } r \to \infty$$

in $L^1(\Omega)$-sense. Therefore, we have

$$0 = \int_\Omega (q_2 - q_1) u_1 u_2 \to \int_\Omega (q_2 - q_1)\, \mathrm{e}^{-2\mathrm{i}\mathbf{k}\cdot\mathbf{x}} \quad \text{as } r \to \infty.$$

Since $k \in \mathbb{R}^3$ is arbitrary, we finally have $q_1 = q_2$. □

The next observation provides an explicit representation formula for q from the knowledge of the NtD map.

Observation 7.5.8 (Nachman's reconstruction) *Let* $u_\zeta = \mathrm{e}^{\mathrm{i}\mathbf{x}\cdot\boldsymbol{\zeta}}[1 + \psi_\zeta(\mathbf{x})]$ *in Lemma 7.5.6. Then,*

$$\hat{q}(\boldsymbol{\xi}) = \lim_{|\boldsymbol{\zeta}|\to\infty} \int_{\partial\Omega} \mathrm{e}^{-\mathrm{i}\mathbf{x}\cdot(\boldsymbol{\xi}+\boldsymbol{\zeta})} \left[\frac{\partial u_\zeta}{\partial \mathbf{n}}(\mathbf{x}) + \mathrm{i}(\boldsymbol{\xi} + \boldsymbol{\zeta}) \cdot \mathbf{n}\Lambda_q \left(\left.\frac{\partial u_\zeta}{\partial \mathbf{n}}\right|_{\partial\Omega} \right)(\mathbf{x}) \right] ds. \tag{7.55}$$

Proof. Since $\mathrm{e}^{-\mathrm{i}\mathbf{x}\cdot\boldsymbol{\zeta}} u_\zeta(\mathbf{x}) \to 1$ as $|\boldsymbol{\zeta}| \to \infty$, we have

$$\lim_{|\boldsymbol{\zeta}|\to\infty} \int_{\mathbb{R}^n} \mathrm{e}^{-\mathrm{i}\mathbf{x}\cdot(\boldsymbol{\xi}+\boldsymbol{\zeta})} q(x) u_\zeta(\mathbf{x})\, d\mathbf{x} = \int_{\mathbb{R}^3} \mathrm{e}^{-\mathrm{i}\mathbf{x}\cdot\boldsymbol{\xi}} q(x)\, dx = \hat{q}(\boldsymbol{\xi}). \tag{7.56}$$

On the other hand, since $qu_\zeta = \nabla^2 u_\zeta$, it follows from the divergence theorem that

$$\begin{aligned}\int_\Omega \mathrm{e}^{-\mathrm{i}\mathbf{x}\cdot(\boldsymbol{\xi}+\boldsymbol{\zeta})}\nabla^2 u_\zeta(\mathbf{x})\,\mathrm{d}x &= \int_{\partial\Omega} \mathrm{e}^{-\mathrm{i}\mathbf{x}\cdot(\boldsymbol{\xi}+\boldsymbol{\zeta})}\left[\frac{\partial u_\zeta}{\partial \mathbf{n}}(\mathbf{x}) + \mathrm{i}(\boldsymbol{\xi}+\boldsymbol{\zeta})\cdot\mathbf{n}u_\zeta(\mathbf{x})\right]\mathrm{d}s \\ &= \int_{\partial\Omega} \mathrm{e}^{-\mathrm{i}\mathbf{x}\cdot(\boldsymbol{\xi}+\boldsymbol{\zeta})}\left[\frac{\partial u_\zeta}{\partial \mathbf{n}}(\mathbf{x}) + \mathrm{i}(\boldsymbol{\xi}+\boldsymbol{\zeta})\cdot\mathbf{n}\Lambda_q\left(\left.\frac{\partial u_\zeta}{\partial \mathbf{n}}\right|_{\partial\Omega}\right)(\mathbf{x})\right]\mathrm{d}s.\end{aligned}$$

Thus, we have

$$\hat{q}(\boldsymbol{\xi}) = \lim_{|\zeta|\to\infty}\int_{\partial\Omega} \mathrm{e}^{-\mathrm{i}\mathbf{x}\cdot(\boldsymbol{\xi}+\boldsymbol{\zeta})}\left[\frac{\partial u_\zeta}{\partial \mathbf{n}}(\mathbf{x}) + \mathrm{i}(\boldsymbol{\xi}+\boldsymbol{\zeta})\cdot\mathbf{n}\Lambda_q\left(\left.\frac{\partial u_\zeta}{\partial \mathbf{n}}\right|_{\partial\Omega}\right)(\mathbf{x})\right]\mathrm{d}s. \tag{7.57}$$

□

7.5.3 Nachmann's D-bar Method in Two Dimensions

Siltanen *et al.* (2000) first implemented the d-bar algorithm based on Nachmann's two-dimensional global uniqueness proof of EIT. This d-bar method solves the full nonlinear EIT problem without iteration (Mueller and Siltanen 2003; Murphy and Mueller 2009).

The d-bar method is based on the fact (Lemma 7.5.3) that: $\psi = \sqrt{\gamma}$ is a solution of

$$\begin{cases} -\nabla^2\psi + q\psi = 0 \text{ in } \Omega \\ \psi|_{\partial\Omega} = 1 \end{cases} \quad \left(q = \frac{\nabla^2\sqrt{\gamma}}{\sqrt{\gamma}}\right), \tag{7.58}$$

where $u = \psi/\sqrt{\gamma}$ is the standard solution of the conductivity equation. We know that, for each $k = k_1 + \mathrm{i}k_2$, there exists a unique solution $\psi(\cdot, k)$ of

$$-\nabla^2\psi + q\psi = 0 \quad \text{in } \mathbb{R}^2, \quad \mathrm{e}^{\mathrm{i}k(x+\mathrm{i}y)}(1-\psi) \in W^{1,p}(\mathbb{R}^2).$$

The scattering transform of $q \in C_0(\Omega)$ can be expressed as

$$\begin{aligned}\mathbf{t}(k) &= \int_{\mathbb{R}^2} \mathrm{e}^{\mathrm{i}\bar{k}(x-\mathrm{i}y)} q(x,y)\psi(x,y;k)\,\mathrm{d}z \quad (z=(x,y), k = k_1+\mathrm{i}k_2) \\ &= \int_\Omega \mathrm{e}^{\mathrm{i}\bar{k}(x-\mathrm{i}y)}\nabla^2\psi(z,k)\,\mathrm{d}z \\ &= \int_{\partial\Omega} \mathrm{e}^{\mathrm{i}\bar{k}(x-\mathrm{i}y)}(\Lambda_\gamma^D - \Lambda_1^D)\psi(z,k)\,\mathrm{d}s \quad (\because \nabla^2\mathrm{e}^{\mathrm{i}\bar{k}(x-\mathrm{i}y)} = 0),\end{aligned}$$

where $\Lambda_\gamma^D : H^{1/2}(\partial\Omega) \to H^{-1/2}(\partial\Omega)$ is a Dirichlet-to-Neumann (DtN) map given by

$$\Lambda_\gamma^D f = \gamma \left.\frac{\partial u^f}{\partial \mathbf{n}}\right|_{\partial\Omega},$$

where u^f is a solution of $\nabla\cdot(\gamma\nabla u^f) = 0$ in Ω with the Dirichlet boundary data $u^f|_{\partial\Omega} = f$.

Using the fact that $-\nabla^2\psi + q\psi = 0$ and the above property of $\mathbf{t}(x,k)$, it is easy to prove that

$$\mu(z,k) := \mathrm{e}^{-\mathrm{i}k(x+\mathrm{i}y)}\psi(z,k) \quad (k \notin \mathbb{C}\setminus\{0\})$$

satisfies the d-bar equation:

$$\frac{\partial}{\partial \bar{k}}\mu(z,k) = \frac{\mathbf{t}(k)}{4\pi\bar{k}}\,\mathrm{e}^{-2\mathrm{i}(k_1 x - k_2 y)}\overline{\mu(z,k)}, \quad k \in \mathbb{C}\setminus\{0\}. \tag{7.59}$$

From (7.58), solving the d-bar equation (7.59) for $\mu(z,k)$ leads to the reconstruction algorithm for γ:

$$\sqrt{\gamma(z)} = \lim_{k\to 0}\mu(z,k), \quad z=(x,y)\in\Omega.$$

For the reconstruction algorithm, we need the following steps:

Step 1. Compute $\psi(\cdot,k)|_{\partial\Omega}$ for each $k = k_1 + \mathrm{i}k_2$.
Step 2. Compute $\mathbf{t}(k)$ using step 1.
Step 3. Solve the d-bar equation (7.59) for $\mu(z,k)$.
Step 4. Visualize $\sqrt{\gamma(z)} = \lim_{k\to 0}\mu(z,k)$, $z=(x,y)\in\Omega$.

For a precise explanation of the reconstruction algorithm, let us fix notation and definitions:

- For a complex variable $z = x+\mathrm{i}y$ at a point $z=(x,y)$, define the d-bar operator $\bar{\partial}$ by

$$\bar{\partial}_z = \frac{1}{2}\left(\frac{\partial}{\partial x} + \mathrm{i}\frac{\partial}{\partial x}\right).$$

- For $k = k_1 + \mathrm{i}k_2 \in \mathbb{C}\setminus\{0\}$,

$$g_k(x,y) = \frac{1}{(2\pi)^2}\int_{\mathbb{R}^2}\frac{\mathrm{e}^{(x\xi_1+y\xi_2)}}{|\xi|^2 + 2k(\xi_1+\mathrm{i}\xi_2)}\,\mathrm{d}\xi.$$

  Note that g_k satisfies $(-\nabla^2 - 4\mathrm{i}k\bar{\partial}_z)g_k(x,y) = \delta(x,y)$.
- Define a single-layer operator $\mathcal{S}_k$ for $k = k_1+\mathrm{i}k_2 \in \mathbb{C}\setminus\{0\}$ by

$$\mathcal{S}_k\phi(z) = \int_{\partial\Omega} G_k(z-z')\phi(z')\,\mathrm{d}z', \quad z=(x,y),$$

  where $G_k(z) = \mathrm{e}^{\mathrm{i}k(x+\mathrm{i}y)}g_k(z)$. Note that $-\nabla^2 G_k(z) = \delta(z)$.

The direct method for reconstructing γ without iteration is based on the following theorem.

Theorem 7.5.9 *Nachmann's constructive result:*

1. *For each $k = k_1+\mathrm{i}k_2 \in \mathbb{C}\setminus\{0\}$, there exists a unique solution $\psi(\cdot,k)\in H^{1/2}(\partial\Omega)$ satisfying the integral equation*

$$\psi(\cdot,k)|_{\partial\Omega} = \mathrm{e}^{\mathrm{i}kz} - \mathcal{S}_k(\Lambda_\gamma^D - \Lambda_1^D)\psi(\cdot,k),$$

   where $\mathrm{i}kz = \mathrm{i}k(x+\mathrm{i}y)$ and Λ_1^D denotes the DtN map of the homogeneous conductivity $\gamma = 1$.

2. *For each $z = (x, y)$, the solution μ of (7.59) satisfies the integral equation*

$$\mu(z, k) = 1 + \frac{1}{4\pi^2} \int_{\mathbb{R}^2} \frac{\mathbf{t}(k')}{(k' - k)\bar{k}'} \mathrm{e}^{-2\mathrm{i}(k_1' x - k_2' y)} \overline{\mu(z, k')} \, \mathrm{d}k_1' \, \mathrm{d}k_2'.$$

3. *We reconstruct γ by*

$$\sqrt{\gamma(z)} = \lim_{k \to 0} \mu(z, k) \quad z = (x, y) \in \Omega.$$

Proof. For the detailed proof, please see Nachman (1996). □

7.6 Back-Projection Algorithm

Barber and Brown (1983) introduced the back-projection algorithm as a fast and practically useful algorithm in EIT. Since it was motivated by the X-ray CT algorithm, we can view it as a generalized Radon transform. However, there exists a clear difference between EIT and CT. In CT, we can obtain projected images in various directions; while, in EIT, we cannot control current pathways since the current flow itself depends on the unknown conductivity distribution to be imaged. Under the assumption that the conductivity is a small perturbation of a constant value, we can approximately apply the back-projection algorithm.

Let us begin by reviewing the well-known Radon transform. In CT, we try to reconstruct a cross-sectional image f from its X-ray projections in several different directions $(\cos\theta, \sin\theta)$. The projection of f in direction θ can be defined by

$$P_\theta f(t) = \int_{L_{\theta,t}} f \, \mathrm{d}l \quad (L_{\theta,t} := \{(x, y) : x\cos\theta + y\sin\theta = t\}).$$

Taking the Fourier transform of $P_\theta f$ leads to

$$\begin{aligned}
\widehat{P_\theta f}(k) &= \int_{-\infty}^{\infty} P_\theta f(t) \, \mathrm{e}^{-\mathrm{i}kt} \, \mathrm{d}t \\
&= \int_{-\infty}^{\infty} \left[\int_{x\cos\theta + y\sin\theta = t} f(x, y) \, \mathrm{d}\ell \right] \mathrm{e}^{-\mathrm{i}kt} \, \mathrm{d}t \\
&= \int_{-\infty}^{\infty} \int_{-\infty}^{\infty} f(x, y) \, \mathrm{e}^{-\mathrm{i}k(x\cos\theta + y\sin\theta)} \, \mathrm{d}x \, \mathrm{d}y \\
&= \sqrt{2\pi} \, \hat{f}(k\cos\theta, k\sin\theta).
\end{aligned}$$

The reconstruction algorithm is based on the following expression of f in terms of its projection:

$$\begin{aligned}
f(x, y) &= \frac{1}{2\pi} \iint \hat{f}(\xi_1, \xi_2) \, \mathrm{e}^{\mathrm{i}(x\xi_1 + y\xi_2)} \, \mathrm{d}\xi_1 \, \mathrm{d}\xi_2 \\
&= \frac{1}{2\pi} \int_0^{\pi} \int_{-\infty}^{\infty} \hat{f}(k\cos\theta, k\sin\theta) \, \mathrm{e}^{\mathrm{i}k(x\cos\theta + y\sin\theta)} |k| \, \mathrm{d}k \, \mathrm{d}\theta \\
&\approx \frac{1}{(2\pi)^{3/2}} \frac{\pi}{N} \sum_{j=1}^{N} \int_{-\infty}^{\infty} [\widehat{P_{\theta_j} f}(k)] \mathrm{e}^{\mathrm{i}k(x\cos\theta + y\sin\theta)} |k| \, \mathrm{d}k,
\end{aligned}$$

where $\theta_j = j\pi/N$. Hence, the image f can be computed from knowledge of its projection $P_{\theta_j} f$, $j = 1, 2, \ldots, N$.

To quickly explain the back-projection algorithm in EIT, we assume the following:

- Ω is the unit disk in $\mathbb{R}^2$;
- $\gamma = \gamma_0 + \delta\gamma$ and $\gamma_0 = 1$;
- $\delta\gamma \in C_0^2(\bar{\Omega})$;
- $P_\theta = (\cos\theta, \sin\theta)$ and $z = (x, y)$ (or $z = x + \mathrm{i}y$).

Let u_0 and u denote the electric potentials corresponding to $\gamma_0 = 1$ and γ with the same Neumann dipole boundary data

$$g_\theta = \frac{2\pi}{\epsilon}(\delta_{P_{\theta+\epsilon}} - \delta_{P_\theta}).$$

Writing $u = u_0 + \delta u$, δu approximately satisfies the equation

$$\begin{cases} -\nabla^2 \delta u \approx \nabla\delta\gamma \cdot \nabla u_0 & \text{in } \Omega, \\ \dfrac{\partial \delta u}{\partial \mathbf{n}} = 0 & \text{on } \partial\Omega, \end{cases} \tag{7.60}$$

where we neglect the term $\nabla\delta\gamma \cdot \nabla\delta u$. When ϵ is very small, u_0 can be computed approximately as

$$u_0(z) \approx \frac{z \cdot P_\theta{}^\perp}{|z - P_\theta|^2}, \qquad z = (x, y),$$

where $P_\theta{}^\perp = (\cos(\theta + \pi/2),\ \sin(\theta + \pi/2))$.

Next, we introduce a holomorphic function in Ω whose real part is $-u_0$:

$$\Psi_\theta(z) = \xi + \mathrm{i}\eta = -u_0(z) + \mathrm{i}u_0^*(z) := -\frac{z \cdot P_\theta{}^\perp}{|z - P_\theta|^2} + \mathrm{i}\frac{1 - P_\theta \cdot z}{|z - P_\theta|^2}.$$

Then, Ψ_θ maps from the unit disk onto the upper half-plane as shown in Figure 7.9:

$$\Psi_\theta : \Omega \quad \longrightarrow \quad \Psi_\theta(\Omega) := \{\xi + \mathrm{i}\eta \mid \eta > \tfrac{1}{2}\}.$$

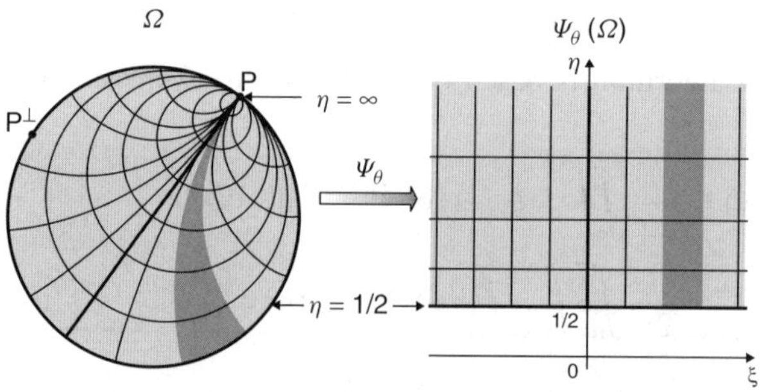

Figure 7.9 The Ψ_θ transformation

Define

$$\widetilde{\delta u}(\xi + \mathrm{i}\eta) := \delta u(\Psi_\theta^{-1}(\xi + \mathrm{i}\eta)) \quad \text{and} \quad \widetilde{\delta\gamma}(\xi + \mathrm{i}\eta) := \delta\gamma(\Psi_\theta^{-1}(\xi + \mathrm{i}\eta)).$$

Viewing $\xi = \xi(x, y)$ and $\eta = \eta(x, y)$, we have

$$\nabla\xi \cdot \nabla\eta = 0 \quad \text{and} \quad |\nabla\xi| = |\nabla\eta|.$$

Hence, the perturbed equation (7.60) implies that $\widetilde{\delta u}$ satisfies

$$-\nabla^2\widetilde{\delta u} = \frac{\partial\widetilde{\delta\gamma}}{\partial\xi} \text{in } \Psi_\theta(\Omega), \qquad \left.\frac{\partial\widetilde{\delta u}}{\partial\eta}\right|_{\eta=1/2} = 0. \tag{7.61}$$

For the moment, we assume that $\widetilde{\delta\gamma}$ is independent of the η variable. With this temporary assumption, $\widetilde{\delta u}$ is independent of the η variable and hence

$$\frac{\partial^2}{\partial\xi^2}\widetilde{\delta u} = -\frac{\partial}{\partial\xi}\widetilde{\delta\gamma} \quad \text{in } \Psi_\theta(\Omega).$$

Therefore,

$$\frac{\partial}{\partial\xi}\widetilde{\delta u} = -\widetilde{\delta\gamma} \quad \text{and} \quad \widetilde{\delta\gamma}(\xi, \eta) = -\frac{\partial\widetilde{\delta u}}{\partial\xi}(\xi + \mathrm{i}\tfrac{1}{2}).$$

For a fixed z, denote $\Psi_\theta(z) = \xi_\theta + \mathrm{i}\eta_\theta$ and $z_\theta^* = \Psi_\theta^{-1}(\xi_\theta + \mathrm{i}\frac{1}{2})$ (see Figure 7.10).

Using the relation among Ψ_θ, z and z^*, Barber and Brown (1983) derived the reconstruction formula

$$\begin{aligned} \delta\gamma(z) = \widetilde{\delta\gamma}(\Psi_\theta(z)) &= \frac{1}{2\pi}\int_0^{2\pi} \frac{\partial}{\partial\xi}\widetilde{\delta u}(\xi_\theta + \mathrm{i}\tfrac{1}{2})\,\mathrm{d}\theta \\ &= \frac{1}{2\pi}\int_0^{2\pi} \frac{\partial}{\partial T}\delta u(z_\theta^*)\,\mathrm{d}\theta, \end{aligned} \tag{7.62}$$

where $\partial/\partial T$ denotes the tangential derivative at $z_\theta^* \in \partial\Omega$.

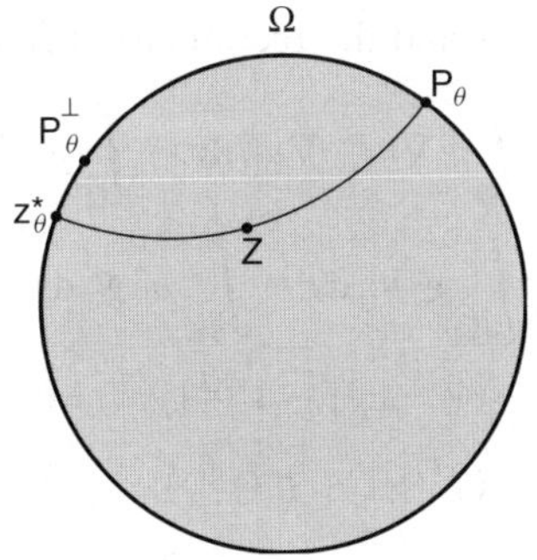

Figure 7.10 Diagram of z_θ^*

7.7 Sensitivity and Sensitivity Matrix

Recently developed image reconstruction algorithms are based on sensitivity analysis. We investigate the sensitivity of a boundary voltage $V^{j,k}[\gamma]$ to a change in γ. We assume that the discrete NtD data $\mathbb{F}$ in (7.40) are available. Since $\mathbb{F}$ can be viewed as a function of γ, we denote it by $\mathbb{F}(\gamma)$. In order to explain the sensitivity matrix, we use the vector form $\mathbb{F}(\gamma)$ as

$$\mathbb{F}(\gamma) = [V^{1,1}[\gamma] \ldots V^{1,E}[\gamma] V^{2,1}[\gamma] \ldots V^{2,E}[\gamma] \quad \ldots \quad V^{E,1}[\gamma] \ldots V^{E,E}[\gamma]]^{\mathrm{T}} \tag{7.63}$$

or

$$l\text{th component of } \mathbb{F}(\gamma) = f^l(\gamma) = V^{j,k}[\gamma] \quad \text{for} \quad l = (j-1) \times E + k, \tag{7.64}$$

for $j, k = 1, 2, \ldots, E$.

We assume a reference admittivity $\gamma^0 = \sigma^0 + i\omega\epsilon^0$, which is a homogeneous admittivity minimizing

$$\|\mathbb{F}(\gamma) - \mathbb{F}(\gamma^0)\| = \sqrt{\sum_{l=1}^{E^2} |f^l(\gamma) - f^l(\gamma^0)|^2}. \tag{7.65}$$

We may assume that $\mathbb{F}(\gamma)$ is a measured data set and $\mathbb{F}(\gamma^0)$ is a computed data set by numerically solving (7.37) with a known γ^0 in place of γ.

7.7.1 *Perturbation and Sensitivity*

We consider γ that is different from a known admittivity γ^0. Assume that we inject the same currents into two imaging domains with γ and γ_0.

Lemma 7.7.1 *The perturbation $\delta\gamma := \gamma - \gamma^0$ satisfies*

$$\int_\Omega \delta\gamma \nabla u^j \cdot \nabla u_0^k \, \mathbf{dr} = f^l(\gamma) - f^l(\gamma^0) \quad \textit{for} \quad l = (j-1) \times E + k, \tag{7.66}$$

where u_0^k is the solution of (7.37) with γ^0 in place of γ.

Proof. Using integration by parts and the reciprocity theorem, we have

$$\begin{aligned}
\int_\Omega \delta\gamma \nabla u^j \cdot \nabla u_0^k \, \mathbf{dr} &= \int_\Omega \gamma \nabla u^j \cdot \nabla u_0^k \, \mathbf{dr} - \int_\Omega \gamma^0 \nabla u^j \cdot \nabla u_0^k \, \mathbf{dr} \\
&= \int_{\partial\Omega} g^j u_0^k \, \mathrm{d}s - \int_{\partial\Omega} u^j g^k \, \mathrm{d}s \\
&= -(V^{k,j}[\gamma] - V^{j,k}[\gamma^0]) = -(V^{j,k}[\gamma] - V^{j,k}[\gamma^0]) \\
&= -(f^l(\gamma) - f^l(\gamma^0)).
\end{aligned}$$

□

The sensitivity expression in (7.66) provides information about how much boundary voltage changes by the admittivity perturbation $\delta\gamma$.

7.7.2 Sensitivity Matrix

The effects of a perturbation $\delta\gamma$ depend on the position $\mathbf{r}$ of the perturbation. In order to construct an explicit expression, we divide the domain Ω into small subregions and assume that γ, γ^0 and $\delta\gamma$ are constant in each subregion. With this kind of discretization, we can transform (7.66) into matrix form.

Observation 7.7.2 *We discretize the domain Ω into N subregions as $\Omega = \bigcup_{n=1}^{N} T_n$. We assume that γ, γ^0 and $\delta\gamma$ are constants in each T_n. We can express (7.66) as*

$$\mathbb{S}_{\gamma,\gamma^0}\boldsymbol{\delta\gamma} = -(\mathbb{F}(\gamma) - \mathbb{F}(\gamma^0)), \tag{7.67}$$

where

$$\boldsymbol{\delta\gamma} = \begin{bmatrix} \delta\gamma_1 \\ \vdots \\ \delta\gamma_N \end{bmatrix} \in \mathbb{C}^N$$

and $\delta\gamma_n = \delta\gamma|_{T_n}$ is the value of $\delta\gamma$ in T_n. The $E^2 \times N$ sensitivity matrix $\mathbb{S}_{\gamma,\gamma^0}$ is given by

$$\mathbb{S}_{\gamma,\gamma^0} = \begin{bmatrix} \vdots \\ \boxed{\begin{array}{c} ((E-1)\times j+1)\text{th row} \\ \vdots \\ ((E-1)\times j+E)\text{th row} \end{array}} \\ \vdots \end{bmatrix}$$

$$= \begin{bmatrix} \vdots \\ \boxed{\begin{array}{ccc} \int_{T_1} \nabla u^j \cdot \nabla u_0^1 & \cdots & \int_{T_N} \nabla u^j \cdot \nabla u_0^1 \\ \vdots & \vdots & \vdots \\ \int_{T_1} \nabla u^j \cdot \nabla u_0^L & \cdots & \int_{T_N} \nabla u^j \cdot \nabla u_0^L \end{array}} \\ \vdots \end{bmatrix}.$$

Note that the sensitivity matrix depends nonlinearly on the admittivity distributions γ and γ^0.

7.7.3 Linearization

We let γ^0 be a variable and make a link between changes in boundary voltages and a small admittivity perturbation $\delta\gamma$ around γ^0.

Observation 7.7.3 *Assuming the same discretization of the domain Ω as explained in the previous section, the admittivity is an N-dimensional variable. When the perturbation $\boldsymbol{\delta\gamma}$*

is small,

$$\boldsymbol{\delta}\mathbb{F} = \mathbb{F}(\gamma^0 + \delta\gamma) - \mathbb{F}(\gamma^0) \approx -\nabla_\gamma \mathbb{F}(\gamma^0)\boldsymbol{\delta\gamma} = -\mathbb{S}_{\gamma_0,\gamma_0}\boldsymbol{\delta\gamma}, \tag{7.68}$$

where $\nabla_\gamma \mathbb{F}(\gamma^0)$ *can be viewed as a Fréchet derivative of* $\mathbb{F}$ *with respect to* γ *at* $\gamma = \gamma_0$.

Proof. Let $p = (j-1) \times E + k$. From (7.66),

$$\begin{aligned}
f^p(\gamma^0 + \delta\gamma) - f^p(\gamma^0) &= \int_\Omega \delta\gamma \nabla u^j \cdot \nabla u_0^k \, \mathbf{dr} \\
&= \int_\Omega \delta\gamma \nabla u_0^j \cdot \nabla u_0^k \, \mathbf{dr} - \int_\Omega \delta\gamma \nabla (u^j - u_0^j) \cdot \nabla u_0^k \, \mathbf{dr} \\
&\approx \int_\Omega \delta\gamma \nabla u_0^j \cdot \nabla u_0^k \, \mathbf{dr} + O(\|\delta\gamma\|^2_{L^\infty(\Omega)}),
\end{aligned}$$

since $\int_\Omega |\nabla(u^j - u_0^j)|^2 \, \mathbf{dr} = O(\|\delta\gamma\|^2_{L^\infty(\Omega)})$. Hence,

$$\mathbf{e}_n \cdot \nabla_\gamma f^p(\gamma_0) = \lim_{h\to 0} \frac{f^p(\gamma^0 + h\chi_{T_n}) - f^p(\gamma^0)}{h} = \int_{T_n} \nabla u_0^j \cdot \nabla u_0^k \, \mathbf{dr},$$

where χ_{T_n} is the characteristic function of T_n. The proof follows from the fact that

$$\int_{T_n} \nabla u_0^j \cdot \nabla u_0^k \, \mathbf{dr} = (n, p)\text{th component of } \mathbb{S}_{\gamma_0,\gamma_0}. \qquad \square$$

Observation 7.7.4 *We let* $\mathbb{S}_{\gamma_0} = \mathbb{S}_{\gamma_0,\gamma_0}$. *The linearized EIT problem is expressed by*

$$\mathbb{S}_{\gamma_0}\boldsymbol{\delta\gamma} = -\boldsymbol{\delta}\mathbb{F} \tag{7.69}$$

or

$$\begin{bmatrix}
\boxed{\begin{matrix}
\int_{T_1} \nabla u_0^1 \cdot \nabla u_0^1 & \cdots & \int_{T_N} \nabla u_0^1 \cdot \nabla u_0^1 \\
\vdots & \vdots & \vdots \\
\int_{T_1} \nabla u_0^1 \cdot \nabla u_0^E & \cdots & \int_{T_N} \nabla u_0^E \cdot \nabla u_0^E
\end{matrix}} \\
\vdots \\
\boxed{\begin{matrix}
\int_{T_1} \nabla u_0^j \cdot \nabla u_0^1 & \cdots & \int_{T_N} \nabla u_0^j \cdot \nabla u_0^1 \\
\vdots & \vdots & \vdots \\
\int_{T_1} \nabla u_0^j \cdot \nabla u_0^E & \cdots & \int_{T_N} \nabla u_0^j \cdot \nabla u_0^E
\end{matrix}} \\
\vdots
\end{bmatrix}
\begin{bmatrix} \delta\gamma_1 \\ \vdots \\ \delta\gamma_N \end{bmatrix}$$

$$
= -\begin{bmatrix} \boxed{\begin{matrix} f^1(\gamma^0+\delta\gamma)-f^1(\gamma_0) \\ \vdots \\ f^E(\gamma^0+\delta\gamma)-f^E(\gamma_0) \end{matrix}} \\ \vdots \\ \boxed{\begin{matrix} f^{(E-1)j+1}(\gamma^0+\delta\gamma)-f^{(E-1)j+1}(\gamma_0) \\ \vdots \\ f^{(E-1)j+E}(\gamma^0+\delta\gamma)-f^{(E-1)j+E}(\gamma_0) \end{matrix}} \\ \vdots \end{bmatrix}.
$$

The matrix $\mathbb{S}_{\gamma_0}$ is called the sensitivity matrix or Jacobian of the linearized EIT problem.

7.7.4 Quality of Sensitivity Matrix

Each data collection protocol is associated with its own sensitivity matrix. We may apply the singular value decomposition explained in Chapter 2 to the sensitivity matrix. Performance of the data collection protocol is closely related with the distribution of singular values. Evaluating several sensitivity matrices from chosen data collection protocols, we may choose a best one. One may also adopt the point spreading function and analyze performance indices of a chosen data collection method, including the spatial resolution, amount of artifacts, uniformity of image contrast and others. This may suggest an optimal data collection method for a specific application.

7.8 Inverse Problem of EIT

Providing intuitive understanding about the inverse problem in EIT using RC circuits as examples, we will formulate three EIT inverse problems including static imaging, time-difference imaging and frequency-difference imaging. Based on the observations in section 7.4.3, we study the ill-posedness in those inverse problems.

7.8.1 Inverse Problem of RC Circuit

We consider two simple examples of elementary inverse problems in RC circuits.

Example 7.8.1 *Consider the series RC circuit. The injection current and measured voltage are* $\mathbf{I} = I\angle 0$ *and* $\mathbf{V} = V\angle\theta$*, respectively, in their phasor forms. The inverse problem is to find the resistance* R *and the capacitance* C *from the relation between* $\mathbf{I}$ *and* $\mathbf{V}$*.*

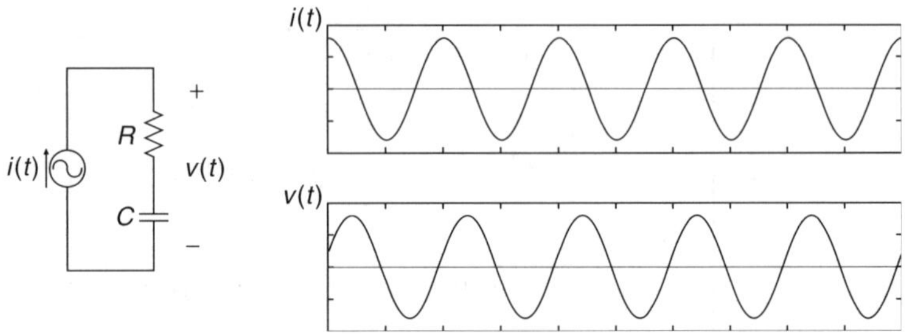

Solution. *From*

$$\mathbf{Z} = R + \frac{1}{\mathrm{i}\omega C} = \frac{\mathbf{V}}{\mathbf{I}} = \frac{V}{I}\angle\theta,$$

we find

$$R = \frac{V\cos\theta}{I} \quad and \quad C = \frac{I}{\omega V\sin(-\theta)}.$$

The number of unknowns is two and the number of measurements is also two, including the real and imaginary parts of the impedance $\mathbf{Z}$.

Example 7.8.2 *Repeat the above example for the parallel RC circuit.*

Example 7.8.3 *Consider the series RC circuit with two resistors and two capacitors. The inverse problem is to find* R_1, R_2, C_1 *and* C_2 *from the data* $\mathbf{I} = I\angle 0$ *and* $\mathbf{V} = V\angle\theta$.

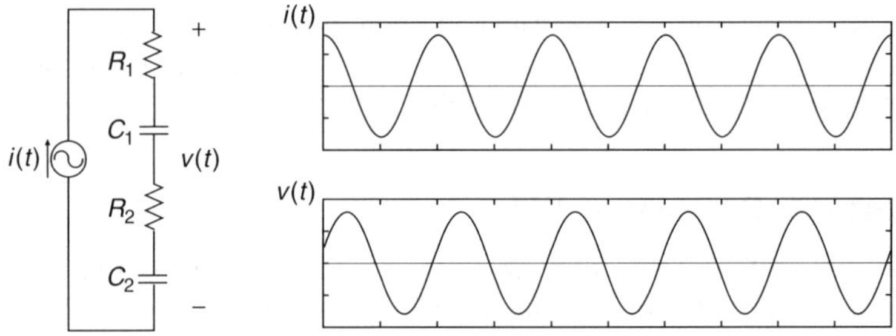

Solution. *From*

$$(R_1 + R_2) + \frac{1}{\mathrm{i}\omega}\left(\frac{1}{C_1} + \frac{1}{C_2}\right) = \frac{\mathbf{V}}{\mathbf{I}} = \frac{V}{I}\angle\theta,$$

we find

$$R_1 + R_2 = \frac{V\cos\theta}{I} \quad and \quad \frac{C_1 + C_2}{C_1 C_2} = \frac{I}{\omega V\sin(-\theta)}.$$

The number of unknowns is four and the number of measurements is two, including the real and imaginary parts of the impedance $\mathbf{Z}$. *This results in infinitely many solutions.*

The inverse problem in Example 7.8.3 has no unique solution and is ill-posed in the sense of Hadamard. Note that we may increase the number of measurements by separately measuring two voltages across R_1C_1 and R_2C_2 to uniquely determine R_1, C_1, R_2 and C_2. One may think of numerous RC circuits with multiple measurements that are either well-posed or ill-posed.

7.8.2 *Formulation of EIT Inverse Problem*

We assume an EIT system using E electrodes $\mathcal{E}_j$ for $j = 1, 2, \ldots, E$. The admittivity inside an imaging domain Ω at time t, angular frequency ω and position $\mathbf{r}$ is denoted as $\gamma_{t,\omega}(\mathbf{r}) = \sigma_{t,\omega}(\mathbf{r}) + \mathrm{i}\omega\epsilon_{t,\omega}(\mathbf{r})$.

7.8.2.1 Static Imaging

Static imaging in EIT is to produce an image of the admittivity $\gamma_{t,\omega}$ from the NtD data $\mathbb{F}[\gamma_{t,\omega}]$ in (7.40). The image reconstruction requires inversion of the map

$$\gamma_{t,\omega} \to \mathbb{F}[\gamma_{t,\omega}]$$

for a fixed time t and frequency ω. We may display images of $\sigma_{t,\omega}$ and $\omega\epsilon_{t,\omega}$ separately. In each image, a pixel value is either $\sigma_{t,\omega}$ or $\omega\epsilon_{t,\omega}$ ($\mathrm{S\,m^{-1}}$). This kind of image is ideal for all applications since it provides absolute quantitative information. One may conduct multi-frequency static imaging by obtaining multiple NtD data sets at the same time at multiple frequencies. We may call this "spectroscopic imaging". We may perform a series of static image reconstructions consecutively at multiple times to provide a time series of admittivity images. Since static EIT imaging is technically difficult in practice, we consider difference imaging methods.

7.8.2.2 Time-Difference Imaging

Time-difference imaging produces an image of any difference, $\gamma_{t_2,\omega} - \gamma_{t_1,\omega}$, between two times t_1 and t_2 from the difference of two NtD data sets, $\mathbb{F}[\gamma_{t_2,\omega}] - \mathbb{F}[\gamma_{t_1,\omega}]$. For single-frequency time-difference imaging, ω is fixed. One may also perform multi-frequency time-difference imaging. Time-difference imaging is desirable for functional imaging to monitor physiological events over time. Though it does not provide absolute values of $\sigma_{t,\omega}$ and $\omega\epsilon_{t,\omega}$, it is more feasible in practice for applications where reference NtD data at some time are available.

7.8.2.3 Frequency-Difference Imaging

For applications where a time-referenced NtD data set is not available, we may consider frequency-difference imaging. It produces an image of any difference between γ_{t,ω_2} and γ_{t,ω_1} using two NtD data sets $\mathbb{F}[\gamma_{t,\omega_2}]$ and $\mathbb{F}[\gamma_{t,\omega_1}]$, which are acquired at the same time. One may perform frequency-difference imaging at multiple frequencies using $\mathbb{F}[\gamma_{t,\omega_n}]$, $n = 1, \ldots, F$. Frequency-difference imaging may classify pathological conditions of tissues without relying on any previous data. Consecutive reconstructions of frequency-difference images at multiple times may provide functional information related to changes over time.

7.8.3 *Ill-Posedness of EIT Inverse Problem*

Before we study these three inverse problems in detail, we investigate their ill-posed characteristics based on the description in section 7.4.3, where we assumed that $\gamma = \sigma$

for simplicity. For an injection current $\mathbf{g}$, we are provided with a limited number of voltage data using a finite number of electrodes. The voltage data vector $\mathbf{f}$ corresponds to measured boundary voltages on portions of $\partial\Omega$ where voltage-sensing electrodes are attached. The inverse problem is to determine the conductivity vector $\boldsymbol{\sigma}$ or equivalently the matrix $\mathbb{A}(\boldsymbol{\sigma})$ from several measurements of current–voltage pairs $(\mathbf{g}^m, \mathbf{f}^m)$ for $m = 1, \ldots, P$, where P is the number of projections.

The ill-posedness of the EIT inverse problem is related to the fact that the difficulty in reconstructing $\mathbb{A}(\boldsymbol{\sigma})$ from $(\mathbf{g}^m, \mathbf{f}^m)$ with $m = 1, \ldots, P$ increases exponentially as the size of $\mathbb{A}(\boldsymbol{\sigma})$ increases. This means that the ill-posedness gets worse as we increase the number of pixels for better spatial resolution. According to (7.44), the voltage at each pixel inside the imaging domain can be expressed as the weighted average of its neighboring voltages, where weights are determined by the conductivity distribution. As explained in section 7.4.3, the measured voltage data vector $\mathbf{f}$ is nonlinearly entangled in the global structure of the conductivity distribution. Any internal conductivity value σ_k has little influence on the boundary measurements $\mathbf{f}$, especially when the position of σ_k is away from the positions of voltage-sensing electrodes. Figure 7.8 depicts these phenomena, from which the ill-posedness originates.

EIT reveals technical difficulties in producing high-resolution images owing to the inherent insensitivity and nonlinearity. For a given finite number of electrodes, the amount of measurable information is limited. Increasing the size of $\mathbb{A}(\boldsymbol{\sigma})$ for better spatial resolution makes the problem more ill-posed. To supply more measurements, we have to increase the number of electrodes. With reduced gaps among a larger number of electrodes, measured voltage differences will become smaller to deteriorate signal-to-noise ratios. Beyond a certain spatial resolution or the pixel size, all efforts to reduce the pixel size using a larger $\mathbb{A}(\boldsymbol{\sigma})$ result in poorer images, since the severe ill-posedness takes over the benefit of additional information from the increased number of electrodes.

Therefore, we should not expect EIT images to have a high spatial resolution needed for structural imaging. EIT cannot compete with X-ray CT or MRI in terms of spatial resolution. One should find clinical significance of biomedical EIT from the fact that it provides unique new contrast information with a high temporal resolution using a portable machine.

7.9 Static Imaging

7.9.1 Iterative Data Fitting Method

Most static image reconstruction algorithms for an E-channel EIT system can be viewed as a data fitting method, as illustrated in Figure 7.11. We first construct a computer model of an imaging object based on (7.37). With the discretization of the imaging domain into N pixels as explained in section 7.7, we can express γ as an admittivity vector

$$\boldsymbol{\gamma} = \left[\gamma_1, \gamma_2, \ldots, \gamma_N\right]^{\mathrm{T}}.$$

Since we do not know the true admittivity $\boldsymbol{\gamma}^*$ of the imaging object, we assume an initial admittivity distribution $\boldsymbol{\gamma}^m$ with $m = 0$ for the model. When we inject currents into both the object and the model, the corresponding measured and computed boundary voltages are different, since $\boldsymbol{\gamma}^m \neq \boldsymbol{\gamma}^*$ in general. An image reconstruction algorithm iteratively updates $\boldsymbol{\gamma}^m$ until it minimizes the difference between measured and computed boundary voltages.

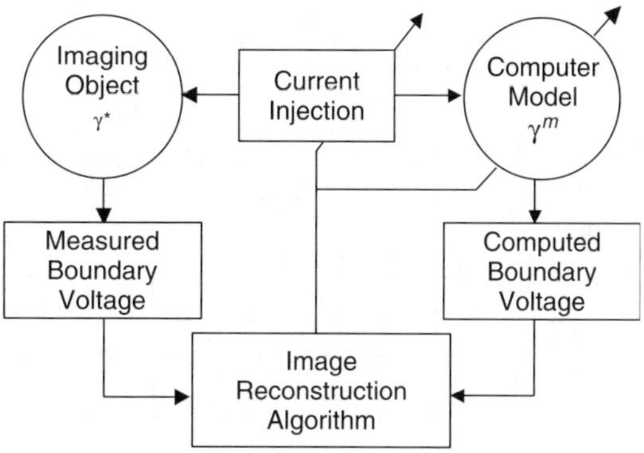

Figure 7.11 Static EIT image reconstruction as a data fitting method

To illustrate this idea, we define the following minimization problem:

$$\boldsymbol{\gamma}^{\natural} = \arg\min_{\boldsymbol{\gamma}\in\mathcal{A}} \quad \Phi(\boldsymbol{\gamma}), \quad \Phi(\boldsymbol{\gamma}) := [\tfrac{1}{2}\|\mathbb{F}(\boldsymbol{\gamma}^*) - \mathbb{F}_C(\boldsymbol{\gamma})\|_2^2] \tag{7.70}$$

where "arg min" is an operator that gives an energy functional minimizer, $\mathbb{F}(\boldsymbol{\gamma}^*)$ is a measured NtD data vector, $\mathbb{F}_C(\boldsymbol{\gamma})$ is the computed NtD data vector and $\mathcal{A}$ is an admissible class for the admittivity. For the solution of (7.70), we may use an iterative nonlinear minimization algorithm such as the Newton–Raphson method (Yorkey and Webster 1987).

In every iteration, we compute the sensitivity matrix or Jacobian $\mathbb{S}_{\gamma^m}$ in (7.69) by solving (7.37) with γ^m in place of γ. Solving the following linear equation

$$\mathbb{S}_{\gamma^m}\boldsymbol{\delta\gamma}^m = \boldsymbol{\delta}\,\mathbb{F}^m = \mathbb{F}(\boldsymbol{\gamma}^*) - \mathbb{F}_C(\boldsymbol{\gamma}^m) \tag{7.71}$$

for $\boldsymbol{\delta\gamma}^m$ by

$$\boldsymbol{\delta\gamma}^m = (\mathbb{S}_{\gamma^m}^{\mathrm{T}}\mathbb{S}_{\gamma^m})^{-1}\mathbb{S}_{\gamma^m}^{\mathrm{T}}\boldsymbol{\delta}\mathbb{F}^m, \tag{7.72}$$

we update $\boldsymbol{\gamma}^m$ as

$$\boldsymbol{\gamma}^{m+1} = \boldsymbol{\gamma}^m + \boldsymbol{\delta\gamma}^m. \tag{7.73}$$

We may stop when

$$\|\boldsymbol{\delta\gamma}^m\| < \delta, \tag{7.74}$$

where δ is a tolerance.

7.9.2 Static Imaging using Four-Channel EIT System

To understand the algorithm in (7.70) clearly, we consider a simple example using a four-channel EIT system. We inject sinusoidal current $i^j(t) = I\cos\omega t$ to each electrode

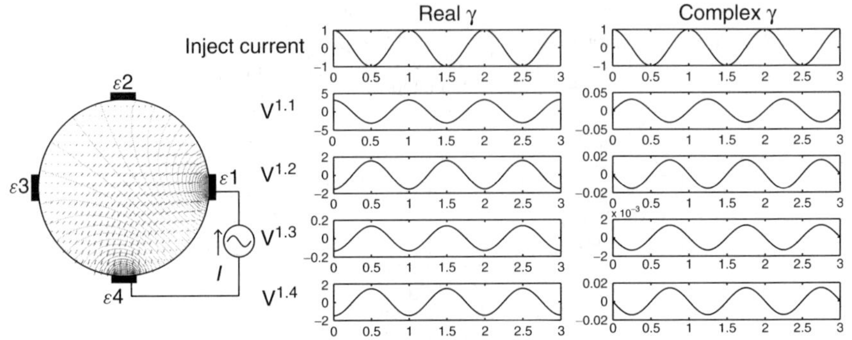

Figure 7.12 Current and voltage signals from a four-channel EIT system

pair $\mathcal{E}_j$ and $\mathcal{E}_{j+1}$ for $j = 1, \ldots, 4$ and $\mathcal{E}_5 = \mathcal{E}_1$. From these four projections, we acquire 16 voltages:

$$\mathbb{F}(\boldsymbol{\gamma}^*) = \begin{pmatrix} V^{1,1} \\ V^{1,2} \\ \vdots \\ V^{4,4} \end{pmatrix} \in \mathbb{C}^{16}.$$

Figure 7.12 shows a circular imaging object Ω, $\Re\{V^{1,1}\,\mathrm{e}^{\mathrm{i}\omega t}\}$, $\Re\{V^{1,2}\,\mathrm{e}^{\mathrm{i}\omega t}\}$, $\Re\{V^{1,3}\,\mathrm{e}^{\mathrm{i}\omega t}\}$ and $\Re\{V^{1,4}\,\mathrm{e}^{\mathrm{i}\omega t}\}$.

We divide the imaging domain as $\Omega = T_1 \cup T_2 \cup T_3 \cup T_4$ in Figure 7.13. Assume that γ is constant on each T_j for $j = 2, 3, 4$ and $\gamma = 1$ on T_1. The goal is to recover γ from the NtD data in Table 7.3 using the following iteration process.

1. Let $\gamma^0 = 1$ be the initial guess.
2. For each $\gamma^m = (\gamma_1^m, \gamma_2^m, \gamma_3^m)^{\mathrm{T}}$ with $m = 1, 2, \ldots,$ solve the forward problem of (7.37) with $\gamma = \gamma^m$ and get $u^j = u^j_{\gamma^m}$. Figure 7.14 shows the distributions of $u^j_{\gamma^0}$ for $j = 1, 2, 3$ and 4.

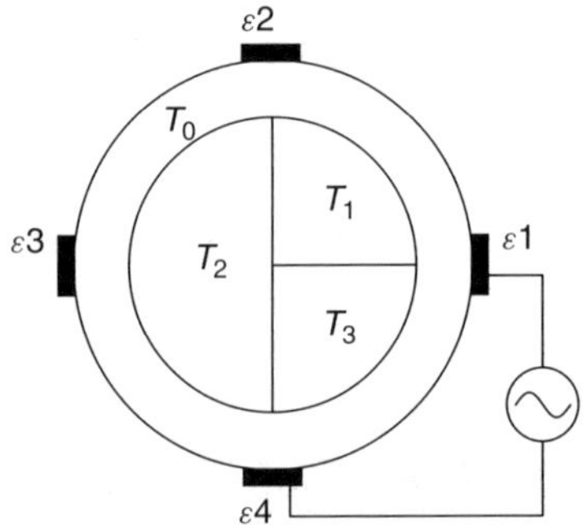

Figure 7.13 Discretized imaging domain for a four-channel EIT system

Table 7.3 NtD data from a four-channel EIT system

$V^{j,k}$	$k=1$	$k=2$	$k=3$	$k=4$
$V^{1,k}$	3.1456	−1.5555	−0.1350	−1.4551
$V^{2,k}$	−1.5555	2.9714	−1.3183	0.0977
$V^{3,k}$	−0.1350	−1.3183	2.7767	−1.3234
$V^{4,k}$	−1.4551	−0.0977	−1.3234	2.8761

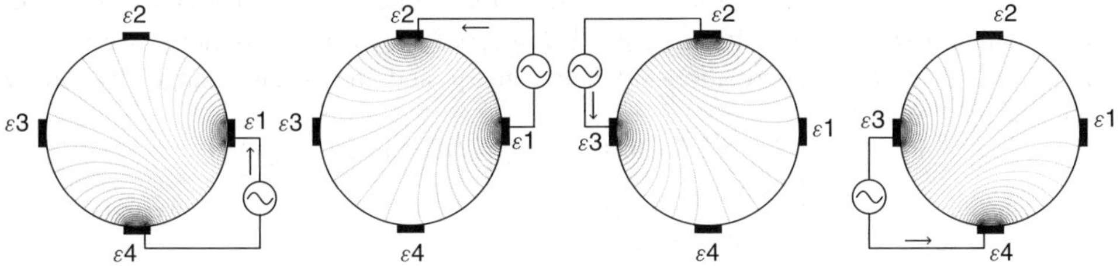

Figure 7.14 Voltage distributions inside the imaging object

3. Compute the sensitivity matrix $\mathbb{S}_{\gamma^m}$ in (7.69) as

$$\mathbb{S}_{\gamma^m} = \begin{pmatrix} \int_{T_1} \nabla u^1 \cdot \nabla u^1 & \int_{T_2} \nabla u^1 \cdot \nabla u^1 & \int_{T_3} \nabla u^1 \cdot \nabla u^1 \\ \int_{T_1} \nabla u^1 \cdot \nabla u^2 & \int_{T_2} \nabla u^1 \cdot \nabla u^2 & \int_{T_3} \nabla u^1 \cdot \nabla u^2 \\ \vdots & \vdots & \vdots \\ \int_{T_1} \nabla u^4 \cdot \nabla u^4 & \int_{T_2} \nabla u^4 \cdot \nabla u^4 & \int_{T_3} \nabla u^4 \cdot \nabla u^4 \end{pmatrix}$$

and compute

$$\mathbb{F}(\boldsymbol{\gamma}^*) - \mathbb{F}_C(\boldsymbol{\gamma}^m) = \begin{pmatrix} V^{1,1} \\ V^{1,2} \\ \vdots \\ V^{4,4} \end{pmatrix} - \begin{pmatrix} V^{1,1}(\boldsymbol{\gamma}^m) \\ V^{1,2}(\boldsymbol{\gamma}^m) \\ \vdots \\ V^{4,4}(\boldsymbol{\gamma}^m) \end{pmatrix}.$$

4. Calculate $\boldsymbol{\delta\gamma} = [\delta\gamma_1, \delta\gamma_2, \delta\gamma_3]^{\mathrm{T}}$ by solving

$$\mathbb{S}_{\gamma^m} \boldsymbol{\delta\gamma}^m = \boldsymbol{\delta}\, \mathbb{F}^m = \mathbb{F}(\boldsymbol{\gamma}^*) - \mathbb{F}_C(\boldsymbol{\gamma}^m).$$

5. Update $\boldsymbol{\gamma}^{m+1} = \boldsymbol{\gamma}^m + \boldsymbol{\delta\gamma}^m$.
6. Repeat steps 2, 3, 4 and 5 until $\|\boldsymbol{\delta\gamma}^m\|$ is smaller than a predetermined tolerance.

In step 4, we used $\mathbb{S}_{\gamma^m} \boldsymbol{\delta\gamma}^m = \mathbb{F}(\boldsymbol{\gamma}^*) - \mathbb{F}_C(\boldsymbol{\gamma}^m)$ to update $\boldsymbol{\gamma}^m$. Recall that solving the minimization problem of $\Phi(\boldsymbol{\gamma})$ with the four-channel EIT is to find a minimizing sequence

$\boldsymbol{\gamma}^m$ such that $\Phi(\boldsymbol{\gamma}^m)$ approaches its minimum effectively. The reason for this choice is that $\boldsymbol{\delta\gamma}^m$ in step 4 makes $\Phi(\boldsymbol{\gamma} + \boldsymbol{\delta\gamma}^m) - \Phi(\boldsymbol{\gamma})$ smallest with a given unit norm of $\|\boldsymbol{\delta\gamma}^m\|$.

To see this rigorously, assume that the true conductivity is γ^* and the measured data are exact so that $V^{j,k} = V^{j,k}[\gamma^*]$. According to (7.66),

$$\Phi(\gamma) = \sum_{j,k=1}^{4} \left|V^{j,k} - V^{j,k}[\gamma]\right|^2 = \sum_{j,k=1}^{4} \left| \int_{\partial\Omega} [u^j_{\gamma^*} - u^j_\gamma]\gamma \frac{\partial u^k_\gamma}{\partial \mathbf{n}} \, \mathrm{d}s \right|^2 .$$

Computation of the Frechét derivative of the functional $\Phi(\gamma)$ requires one to investigate the linear change $\delta u := u_{\gamma+\delta\gamma} - u_\gamma$ subject to a small conductivity perturbation $\delta\gamma$. Note that $\Phi(\gamma + \delta\gamma) \approx \Phi(\gamma) + D\Phi(\gamma)(\delta\gamma) + \frac{1}{2}D^2\Phi(\gamma)(\delta\gamma, \delta\gamma)$. For simplicity, we assume that $\delta\gamma = 0$ near $\partial\Omega$. The relationship between $\delta\gamma$ and the linear change δu can be explained by

$$\begin{cases} \nabla \cdot (\delta\gamma \nabla u) \approx -\nabla \cdot (\gamma \nabla \delta u) & \text{in } \Omega, \\ \left.\dfrac{\partial(\delta u)}{\partial \mathbf{n}}\right|_{\partial\Omega} = 0. \end{cases}$$

We have the following approximation:

$$\int_\Omega \delta\gamma \nabla u^j_\gamma \cdot \nabla u^k_\gamma \, \mathrm{d}x \approx V^{j,k}[\gamma + \delta\gamma] - V^{j,k}[\gamma].$$

We want to find the direction $\delta\gamma$ that makes $\Phi(\gamma + \delta\gamma) - \Phi(\gamma)$ smallest with a given unit norm of $\|\delta\gamma\|$. The steepest descent direction $\delta\gamma = (\delta\gamma_1, \delta\gamma_2, \delta\gamma_3)^{\mathrm{T}}$ can be calculated by solving the matrix equation:

$$\mathbb{S}_\gamma \delta\gamma = \mathbf{f}(\gamma) - \mathbf{f}_{\text{meas}}.$$

To understand this, we recall that $\Phi(\gamma) = \frac{1}{2}\sum_{j,k=1}^{4} \left|V^{j,k}[\gamma] - V^{j,k}\right|^2$ and

$$\Phi(\gamma + \delta\gamma) - \Phi(\gamma) \approx \sum_{j,k=1}^{4} \Re\left\{(V^{j,k}[\gamma] - V^{j,k})\,\overline{(V^{j,k}[\gamma + \delta\gamma] - V^{j,k}[\gamma])}\right\}.$$

We choose the direction $\delta\gamma$ that makes $\Phi(\gamma + \delta\gamma) - \Phi(\gamma)$ smallest with a given norm $\|\delta\gamma\|$ as

$$V^{j,k}[\gamma + \delta\gamma] - V^{j,k}[\gamma] = V^{j,k} - V^{j,k}[\gamma].$$

Owing to

$$\int_\Omega \delta\gamma \nabla u^j_\gamma \cdot \nabla u^k_\gamma \, \mathrm{d}\mathbf{r} \approx V^{j,k}[\gamma + \delta\gamma] - V^{j,k}[\gamma],$$

the steepest descent direction $\delta\gamma$ must satisfy

$$\int_\Omega \delta\gamma \nabla u^j_\gamma \cdot \nabla u^k_\gamma \, \mathrm{d}x = V^{j,k} - V^{j,k}[\gamma] \qquad (j, k = 1, 2, 3, 4).$$

7.9.3 Regularization

Since the Jacobian matrix in (7.71) is ill-conditioned, as explained in section 7.7, we often use a regularization method. Using the Tikhonov type regularization, we set

$$\boldsymbol{\gamma}^{\natural} = \arg\min_{\boldsymbol{\gamma}\in\mathcal{A}}[\Phi(\boldsymbol{\gamma}) + \lambda\eta(\boldsymbol{\gamma})], \tag{7.75}$$

where λ is a regularization parameter and $\eta(\boldsymbol{\gamma})$ is a function measuring a regularity of $\boldsymbol{\gamma}$. This results in the following update equation for the mth iteration:

$$\boldsymbol{\delta\gamma}^m = (\mathbb{S}^{\mathrm{T}}_{\gamma^m}\mathbb{S}_{\gamma^m} + \lambda\mathbb{R})^{-1}\mathbb{S}^{\mathrm{T}}_{\gamma^m}\boldsymbol{\delta}\,\mathbb{F}^m, \tag{7.76}$$

where $\mathbb{R}$ is a regularization matrix.

This kind of method was first introduced in EIT by Yorkey and Webster (1987), followed by numerous variations and improvements (Cheney *et al.* 1990, 1999; Cohen-Bacrie *et al.* 1997; Edic *et al.* 1998; Hyaric and Pidcock 2001; Lionheart *et al.* 2005; Vauhkonen *et al.* 1998; Woo *et al.* 1993). These include utilization of *a priori* information, statistical information, various forms of regularity conditions, adaptive mesh refinement and so on. Though this iterative approach is widely adopted for static imaging, it requires a large amount of computation time and produce static images with a low spatial resolution and poor accuracy for the reasons discussed in the next section. Beyond this classical technique in static imaging, new ideas are in demand for better image quality.

7.9.4 Technical Difficulty of Static Imaging

In a static EIT imaging method, we construct a forward model of the imaging object with a presumed admittivity distribution. Injecting the same currents into the model as the ones used in measurements, boundary voltages are computed to numerically simulate measured data. Since the initially guessed admittivity distribution is in general different from the unknown admittivity distribution of the object, there exist some differences between measured and computed voltages. Most static EIT imaging methods are based on a minimization technique, where a sum of these voltage differences is minimized by adjusting the admittivity distribution of the model (Adler and Lionheart 2006; Cheney *et al.* 1990; Lionheart *et al.* 2005; Woo *et al.* 1993; Yorkey and Webster 1987). Other methods may include layer stripping (Somersalo *et al.* 1991) and d-bar (Siltanen *et al.* 2000) algorithms.

For a static EIT image reconstruction algorithm to be reliable, we should be able to construct a forward model that mimics every aspect of the imaging object except the internal admittivity distribution. This requires knowledge of the boundary geometry, electrode positions and other sources of systematic artifacts in measured data. In practice, it is very difficult to obtain such information within a reasonable accuracy and cost, and most static EIT image reconstruction algorithms are very sensitive to these errors.

When we inject current through a pair of electrodes $\mathcal{E}_j$ and $\mathcal{E}_{j+1}$, the induced voltage $u^j_{\gamma,\Omega}$ is dictated by the applied Neumann data g^j of the injection current, the geometry of the domain Ω and γ. That is, $u^j_{\gamma,\Omega}$ satisfies approximately

$$\nabla \cdot (\gamma \nabla u^j_{\gamma,\Omega}) = 0 \text{ in } \Omega, \quad \gamma \nabla u^j_{\gamma,\Omega} \cdot \mathbf{n} = g^j \text{ on } \partial\Omega, \tag{7.77}$$

where g^j represents the Neumann data in (7.37).

Taking account of the nonlinearity and ill-posedness in EIT, most image reconstruction methods for EIT use the assumption that γ is a perturbation of a known reference distribution γ^0 so that we can linearize the nonlinear problem. The inverse problem is to find $\delta\gamma := \gamma - \gamma^0$ from the integral equation

$$\int_\Omega \delta\gamma \nabla u^j_{\gamma^0,\Omega} \cdot \nabla u^k_{\gamma,\Omega}\, \mathrm{d}x = \int_{\partial\Omega} [\Lambda_{\gamma^0,\Omega}(g^j) - \Lambda_{\gamma,\Omega}(g^j)] g^k \, \mathrm{d}S \quad \text{for all } j, k, \tag{7.78}$$

where $\Lambda_{\gamma,\Omega}(g^j) := u^j_{\gamma,\Omega}|_{\partial\Omega}$ and $\mathrm{d}S$ is the surface element (Cheney *et al.* 1990, 1999, Lionheart *et al.* 2005). In practice, the value of the right-hand side of (7.78) is the potential difference u^j between electrodes $\mathcal{E}_k$ and $\mathcal{E}_{k+1}$.

If the change $\delta\gamma$ is small, we can approximate

$$\int_\Omega \delta\gamma \nabla u^j_{\gamma^0,\Omega} \cdot \nabla u^k_{\gamma,\Omega}\, \mathrm{d}x \approx \int_\Omega \delta\gamma \nabla u^j_{\gamma^0,\Omega} \cdot \nabla u^k_{\gamma^0,\Omega}\, \mathrm{d}x$$

and (7.78) becomes

$$\mathcal{S}_{\gamma^0,\Omega}(\delta\gamma) = \mathbf{b}_{\gamma^0,\Omega} - \mathbf{b}_{\gamma,\Omega}, \tag{7.79}$$

where $\mathcal{S}_{\gamma^0,\Omega}(\delta\gamma)$ and $\mathbf{b}_{\gamma,\Omega}$ are $L \times L$ vectors with $(j-1)L + k$ component

$$\int_\Omega \delta\gamma \nabla u^j_{\gamma^0,\Omega} \cdot \nabla u^k_{\gamma^0,\Omega}\, \mathrm{d}x \quad \text{and} \quad \int_{\partial\Omega} \Lambda_{\gamma,\Omega}(g^j) g^k \, \mathrm{d}S,$$

respectively. We may view $\mathcal{S}_{\gamma^0,\Omega}(\cdot)$ as a linear operator acting on $\delta\gamma$ and its discretized version in terms of the admittivity distribution is called the sensitivity matrix.

To solve the inverse problem (7.79), we construct a forward model of the imaging object with a presumed reference admittivity $\tilde{\gamma}^0$:

$$\nabla \cdot (\tilde{\gamma}^0 \nabla u^j_{\tilde{\gamma}^0,\Omega_c}) = 0 \ \text{ in } \Omega_c, \quad \tilde{\gamma}^0 \nabla u^j_{\tilde{\gamma}^0,\Omega_c} \cdot \nu = \tilde{g}^j \ \text{ on } \partial\Omega_c, \tag{7.80}$$

where $\Omega_c \subset \mathbb{R}^3$ is a computational domain mimicking the geometry of the imaging subject, $\tilde{g}^j$ is the Neumann data mimicking the applied current g^j and $u^j_{\tilde{\gamma}^0,\Omega_c}$ is the internal potential induced by the current corresponding to the Neumann data $\hat{g}^j$.

The forward model (7.80) is used to compute the reference boundary voltage $\Lambda_{\tilde{\gamma}^0,\Omega_c}(\tilde{g}^j) = u^j_{\tilde{\gamma}^0,\Omega_c}|_{\partial\Omega_c}$, which is expected to be substituted for $\Lambda_{\gamma^0,\Omega}(g^j)$ in (7.79). If we have the exact forward modeling $\mathcal{S}_{\tilde{\gamma}^0,\Omega_c}(\cdot) = \mathcal{S}_{\gamma^0,\Omega}(\cdot)$ and $\mathbf{b}_{\tilde{\gamma}^0,\Omega_c} - \mathbf{b}_{\gamma,\Omega} = \mathbf{b}_{\gamma^0,\Omega} - \mathbf{b}_{\gamma,\Omega}$, we may obtain reasonably accurate images of $\delta\gamma$ by inverting the discretized version of the linear operator $\mathcal{S}_{\gamma^0,\Omega}(\cdot)$ with the use of regularization. Knowing that we cannot avoid forward modeling errors, a major drawback of static imaging stems from the fact that the reconstruction problem (7.79) is very sensitive to geometric modeling errors in the computed reference data $\Lambda_{\tilde{\gamma}^0,\Omega_c}(\tilde{g}^j)$, including boundary geometry errors on Ω_c and electrode positioning errors on $\tilde{g}^j$ (Barber and Brown 1988; Kolehmainen *et al.* 2005; Nissinen *et al.* 2008). It would be very difficult to get accurate data $\Lambda_{\tilde{\gamma}^0,\Omega_c}(\tilde{g}^j)$ at a reasonable cost in a practical environment.

To deal with undesirable effects of modeling errors, we investigate two difference imaging methods in the following sections. We expect that time or frequency derivatives of the NtD data $\Lambda_{\sigma,\Omega}$ may cancel out the effects of geometry errors on $\partial\Omega$.

7.10 Time-Difference Imaging

In time-difference EIT (tdEIT), measured data at two different times are subtracted to produce images of changes in the admittivity distribution with respect to time. Since the data subtraction can effectively cancel out common errors, tdEIT has shown its potential as a functional imaging modality in several clinical application areas. In this section, we consider multi-frequency time-difference EIT (mftdEIT) imaging. After formulating the mftdEIT imaging problem, we study the mftdEIT image reconstruction algorithm.

7.10.1 Data Sets for Time-Difference Imaging

We assume an imaging object Ω bounded by its surface $\partial\Omega$. The isotropic admittivity in Ω at time t, angular frequency ω and position $\mathbf{r} = (x, y, z)$ is denoted $\gamma_{t,\omega}(\mathbf{r}) = \sigma_{t,\omega}(\mathbf{r}) + i\omega\epsilon_{t,\omega}(\mathbf{r})$. Attaching surface electrodes $\mathcal{E}_j$ for $j = 1, 2, \ldots, E$ on $\partial\Omega$, we inject a sinusoidal current $i(t) = I\cos(\omega t)$ between a chosen pair of electrodes. A distribution of voltage in Ω is produced and we can express it as $V_{t,\omega}(\mathbf{r})\cos(\omega t + \theta_{t,\omega}(\mathbf{r}))$.

Assuming an EIT system using E electrodes, we inject the jth current between an adjacent pair of electrodes denoted as $\mathcal{E}_j$ and $\mathcal{E}_{j+1}$ for $j = 1, 2, \ldots, E$. The time-harmonic voltage subject to the jth injection current is denoted as $u^j_{t,\omega}$, which is a solution of (7.37) with g replaced by g^j. We assume that the EIT system is equipped with E voltmeters and each of them measures a boundary voltage between an adjacent pair of electrodes, $\mathcal{E}_k$ and $\mathcal{E}_{k+1}$ for $k = 1, 2, \ldots, E$.

Using an mftdEIT system, we collect complex boundary voltage data at multiple frequencies for a certain period of time. Assuming that we collected E^2 number of complex boundary voltage data at each sampling time t and frequency ω, we can express a complex boundary voltage data vector as (7.16). We rewrite it using a column vector representation as

$$\mathbb{F}_{t,\omega} = [\mathbf{V}^{1,1}_{t,\omega} \ldots \mathbf{V}^{1,E}_{t,\omega} \mathbf{V}^{2,1}_{t,\omega} \ldots \mathbf{V}^{2,E}_{t,\omega} \quad \cdots \quad \mathbf{V}^{E,1}_{t,\omega} \ldots \mathbf{V}^{E,E}_{t,\omega}]^{\mathrm{T}}. \tag{7.81}$$

For $t = t_1, t_2, \ldots, t_N$ and $\omega = \omega_1, \omega_2, \ldots, \omega_F$, we are provided with N data vectors for each one of F frequencies. To perform tdEIT imaging, we need a complex boundary voltage data vector at a reference time t_0:

$$\mathbb{F}_{t_0,\omega} = [\mathbf{V}^{1,1}_{t_0,\omega} \ \cdots \ \mathbf{V}^{1,E}_{t_0,\omega} \ \mathbf{V}^{2,1}_{t_0,\omega} \ \cdots \ \mathbf{V}^{2,E}_{t_0,\omega} \quad \cdots \quad \mathbf{V}^{E,1}_{t_0,\omega} \ \cdots \ \mathbf{V}^{E,E}_{t_0,\omega}]^{\mathrm{T}} \tag{7.82}$$

for $\omega = \omega_1, \omega_2, \ldots, \omega_F$. The mftdEIT imaging problem is to produce time series of difference images using $\mathbb{F}_{t,\omega} - \mathbb{F}_{t_0,\omega}$ for $t = t_1, t_2, \ldots, t_N$ at each one of $\omega = \omega_1, \omega_2, \ldots, \omega_F$.

7.10.2 Equivalent Homogeneous Admittivity

For a given admittivity distribution $\gamma_{t,\omega}$, we define the equivalent homogeneous admittivity $\hat{\gamma}_{t,\omega}$ as a complex number that minimizes

$$\sum_{j=1}^{E}\int_\Omega \left|\gamma_{t,\omega}(\mathbf{r})\nabla u^j_{t,\omega}(\mathbf{r}) - \hat{\gamma}_{t,\omega}\nabla \hat{u}^j_{t,\omega}(\mathbf{r})\right|^2 \mathbf{dr} + \eta\int_\Omega |\gamma_{t,\omega}(\mathbf{r}) - \hat{\gamma}_{t,\omega}|^2\,\mathbf{dr},$$

where $\hat{u}^j_{t,\omega}$ is the voltage satisfying (7.37) with $\hat{\gamma}_{t,\omega}$ in place of $\gamma_{t,\omega}$ and η is a weighting constant. We assume that $\gamma_{t,\omega}$ is a small perturbation of $\hat{\gamma}_{t,\omega}$.

We set a reference frequency ω_0 as well as the reference time t_0. We assume that the complex boundary voltage vector $\mathbb{F}_{t_0,\omega_0}$ is available at $t = t_0$ and $\omega = \omega_0$. Defining

$$\alpha_{t,\omega} := \frac{1}{E^2}\sum_{j,k=1}^{E}\frac{\mathbf{V}^{j,k}_{t_0,\omega_0}}{\mathbf{V}^{j,k}_{t,\omega}},$$

it measures the quantity $\hat{\gamma}_{t,\omega}/\hat{\gamma}_{t_0,\omega_0}$ roughly because

$$\begin{aligned}
\alpha_{t,\omega} &= \frac{1}{E^2}\sum_{j,k=1}^{E}\frac{\int_\Omega \gamma_{t_0,\omega_0}\nabla u^j_{t_0,\omega_0}\cdot\nabla u^k_{t_0,\omega_0}\,\mathbf{dr}}{\int_\Omega \gamma_{t,\omega}\nabla u^j_{t,\omega}\cdot\nabla u^k_{t,\omega}\,\mathbf{dr}} \\
&\approx \frac{1}{E^2}\sum_{j,k=1}^{E}\frac{\int_\Omega \hat{\gamma}_{t_0,\omega_0}\nabla \hat{u}^j_{t_0,\omega_0}\cdot\nabla \hat{u}^k_{t_0,\omega_0}\,\mathbf{dr}}{\int_\Omega \hat{\gamma}_{t,\omega}\nabla \hat{u}^j_{t,\omega}\cdot\nabla \hat{u}^k_{t,\omega}\,\mathbf{dr}} \\
&= \frac{\hat{\gamma}_{t,\omega}}{\hat{\gamma}_{t_0,\omega_0}}\frac{1}{E^2}\sum_{j,k=1}^{E}\frac{\int_\Omega \hat{\gamma}_{t_0,\omega_0}\nabla \hat{u}^j_{t_0,\omega_0}\cdot\hat{\gamma}_{t_0,\omega_0}\nabla \hat{u}^k_{t_0,\omega_0}\,\mathbf{dr}}{\int_\Omega \hat{\gamma}_{t,\omega}\nabla \hat{u}^j_{t,\omega}\cdot\hat{\gamma}_{t,\omega}\nabla \hat{u}^k_{t,\omega}\,\mathbf{dr}} \\
&= \frac{\hat{\gamma}_{t,\omega}}{\hat{\gamma}_{t_0,\omega_0}}\frac{1}{E^2}\sum_{j,k=1}^{E}\frac{\int_\Omega \nabla v^j\cdot\nabla v^k\,\mathbf{dr}}{\int_\Omega \nabla v^j\cdot\nabla v^k\,\mathbf{dr}} = \frac{\hat{\gamma}_{t,\omega}}{\hat{\gamma}_{t_0,\omega_0}},
\end{aligned} \tag{7.83}$$

where v^j and v^k are solutions of (7.37) with $\gamma_{t,\omega} = 1$ for the jth and kth injection currents, respectively.

We now relate a time change of the complex boundary voltage with a time change of the internal admittivity. For $p = (k-1)\times E + j$ with $j, k = 1, 2, \ldots, E$,

$$\begin{aligned}
I\left(\mathbb{F}_{t,\omega} - \mathbb{F}_{t_0,\omega}\right)\cdot\mathbf{e}_p &= \int_\Omega (\gamma_{t_0,\omega} - \gamma_{t,\omega})\nabla u^k_{t,\omega}\cdot\nabla u^j_{t_0,\omega}\,\mathbf{dr} \\
&= \int_\Omega\left(\frac{1}{\gamma_{t,\omega}} - \frac{1}{\gamma_{t_0,\omega}}\right)(\gamma_{t,\omega}\nabla u^k_{t,\omega})\cdot(\gamma_{t_0,\omega}\nabla u^j_{t_0,\omega})\,\mathbf{dr} \\
&\approx \int_\Omega\left(\frac{1}{\gamma_{t,\omega}} - \frac{1}{\gamma_{t_0,\omega}}\right)(\hat{\gamma}_{t,\omega}\nabla \hat{u}^k_{t,\omega})\cdot(\hat{\gamma}_{t_0,\omega}\nabla \hat{u}^j_{t_0,\omega})\,\mathbf{dr} \\
&= \int_\Omega\left(\frac{1}{\gamma_{t,\omega}} - \frac{1}{\gamma_{t_0,\omega}}\right)\nabla v^k\cdot\nabla v^j\,\mathbf{dr},
\end{aligned} \tag{7.84}$$

where $\mathbf{e}_p$ is the unit vector in the E^2 dimension having 1 at its pth component. Note that we have utilized the reciprocity theorem in section 7.4. Since we assumed that $\gamma_{t,\omega}$ and $\gamma_{t_0,\omega}$ are small perturbations of $\hat{\gamma}_{t,\omega}$ and $\hat{\gamma}_{t_0,\omega}$, respectively, we have the following approximation:

$$\frac{1}{\gamma_{t,\omega}} - \frac{1}{\gamma_{t_0,\omega}} \approx \frac{\gamma_{t_0,\omega} - \gamma_{t,\omega}}{\hat{\gamma}_{t_0,\omega}\hat{\gamma}_{t,\omega}}.$$

Hence, for all $p = (k-1) \times E + j$, we have

$$\begin{aligned} I\left(\mathbb{F}_{t,\omega} - \mathbb{F}_{t_0,\omega}\right) \cdot \mathbf{e}_p &= \int_\Omega \left(\frac{1}{\gamma_{t,\omega}} - \frac{1}{\gamma_{t_0,\omega}}\right) \nabla v^k \cdot \nabla v^j \, \mathbf{dr} \\ &\approx \frac{1}{\hat{\gamma}_{t_0,\omega}\hat{\gamma}_{t,\omega}} \int_\Omega (\gamma_{t_0,\omega} - \gamma_{t,\omega}) \nabla v^k \cdot \nabla v^j \, \mathbf{dr} \\ &= \frac{1}{\hat{\gamma}^2_{t_0,\omega 0}\alpha_{t_0,\omega}\alpha_{t,\omega}} \int_\Omega (\gamma_{t_0,\omega} - \gamma_{t,\omega}) \nabla v^k \cdot \nabla v^j \, \mathbf{dr}. \end{aligned} \tag{7.85}$$

7.10.3 Linear Time-Difference Algorithm using Sensitivity Matrix

We construct a computer model of the imaging object Ω. Assume that the domain of the model is Λ with its boundary $\partial\Lambda$. Discretizing the model into Q elements or pixels as $\Lambda = \bigcup_{q=1}^{Q} \Lambda_q$, we define the time-difference image $\mathbf{g}_{t,\omega}$ at time t and frequency ω as

$$\mathbb{G}_{t,\omega} = \mathbb{H}_{t,\omega} - \mathbb{H}_{t_0,\omega} \tag{7.86}$$

with

$$\mathbb{H}_{t,\omega} = [\gamma^1_{t,\omega}\ \gamma^2_{t,\omega}\ \cdots\ \gamma^Q_{t,\omega}]^{\mathrm{T}} \quad \text{and} \quad \mathbb{H}_{t_0,\omega} = [\gamma^1_{t_0,\omega}\ \gamma^2_{t_0,\omega}\ \cdots\ \gamma^Q_{t_0\omega}]^{\mathrm{T}},$$

where $\gamma^q_{t,\omega}$ and $\gamma^q_{t_0,\omega}$ for $q = 1, 2, \ldots, Q$ are the admittivity values of the imaging object at times t and t_0, respectively, inside a local region corresponding to the qth pixel Λ_q of the model Λ.

The model is assumed to be homogeneous, with $\gamma_{t_0,\omega} = 1 = \gamma^0$ in Λ. Using E electrodes, we inject current between the jth adjacent pair of electrodes to induce voltage v^j in Λ. We numerically solve (7.37) for v^j by using the finite element method. We can formulate the sensitivity matrix $\mathbb{S}_{\gamma^0} = [s_{pq}]$ in section 7.7 as

$$s_{pq} = \int_{\Lambda_q} \nabla v^k \cdot \nabla v^j \, \mathbf{dr} \quad \text{and} \quad p = (k-1) \times E + j \tag{7.87}$$

for $j, k = 1, 2, \ldots, E$ and $q = 1, 2, \ldots, Q$. The maximal size of $\mathbb{S}_{\gamma^0}$ is $E^2 \times Q$ and all of its elements are real numbers. Using the discretization and linearization, the expression (7.85) becomes

$$\mathbb{F}_{t,\omega} - \mathbb{F}_{t_0,\omega} = \frac{1}{I\hat{\gamma}^2_{t_0,\omega 0}\alpha_{t_0,\omega}\alpha_{t,\omega}} \mathbb{S}_{\gamma^0}(\mathbb{H}_{t_0,\omega} - \mathbb{H}_{t,\omega}). \tag{7.88}$$

Computing the truncated singular value decomposition (TSVD) of $\mathbb{S}_{\gamma^0}$, we find $P \le Q$ singular values that are not negligible. We can compute a pseudo-inverse matrix of $\mathbb{S}_{\gamma^0}$ after truncating its $(Q - P)$ negligible singular values. Denoting this inverse matrix as $\mathbb{A}$, we have

$$\mathbb{G}_{t,\omega} = \mathbb{H}_{t,\omega} - \mathbb{H}_{t_0,\omega} = -I\hat{\gamma}^2_{t_0,\omega_0}\alpha_{t_0,\omega}\alpha_{t,\omega}\mathbb{A}(\mathbb{F}_{t,\omega} - \mathbb{F}_{t_0,\omega}). \tag{7.89}$$

Note that $\mathbb{A}$ is a real matrix whose maximal size is $Q \times E^2$. Since we do not know $\hat{\gamma}_{t_0,\omega_0}$ in (7.89), we replace (7.89) by the following equation:

$$\mathbb{I}_{t,\omega} = \frac{\mathbb{G}_{t,\omega}}{\hat{\gamma}^2_{t_0,\omega_0}} = \mathbb{R}_{t,\omega} + \mathrm{i}\mathbb{X}_{t,\omega} = -I\alpha_{t_0,\omega}\alpha_{t,\omega}\mathbb{A}(\mathbb{F}_{t,\omega} - \mathbb{F}_{t_0,\omega}), \tag{7.90}$$

where $\mathbb{R}_{t,\omega}$ and $\mathbb{X}_{t,\omega}$ are the real and imaginary parts of a reconstructed complex tdEIT image $\mathbb{I}_{t,\omega}$, respectively.

We may reconstruct a time series of mftdEIT images $\mathbb{I}_{t_n,\omega_f}$ for $f = 1, 2, \ldots, F$ at $n = 1, 2, \ldots, N$. Choosing ω_0 at a low frequency below 1 kHz, we may assume that $\hat{\gamma}_{t_0,\omega_0} = \hat{\sigma}_{t_0,\omega_0}$ since we can neglect the effects of the permittivity at low frequencies. In such a case, (7.90) becomes

$$\mathbb{I}_{t_n,\omega_f} = \frac{\mathbb{G}_{t_n,\omega_f}}{\hat{\sigma}^2_{t_0,\omega_0}} = \mathbb{R}_{t_n,\omega_f} + \mathrm{i}\mathbb{X}_{t_n,\omega_f} = -I\alpha_{t_0,\omega_f}\alpha_{t_n,\omega_f}\mathbb{A}(\mathbb{F}_{t_n,\omega_f} - \mathbb{F}_{t_0,\omega_f}). \tag{7.91}$$

Note that $\mathbb{I}_{t_n,\omega_f}$ in (7.91) has the same phase angle as $\mathbb{G}_{t_n,\omega_f}$.

7.10.4 *Interpretation of Time-Difference Image*

The mftdEIT image reconstruction algorithm based on (7.91) produces both real- and imaginary-part tdEIT images at multiple frequencies. It provides a theoretical basis for proper interpretation of a reconstructed image using the equivalent homogeneous complex conductivity. From (7.91), we can see that the real- and imaginary-part images represent $(\sigma_{t,\omega} - \sigma_{t_0,\omega})/\hat{\sigma}^2_{t_0,\omega_0}$ and $(\omega\epsilon_{t,\omega} - \omega\epsilon_{t_0,\omega})/\hat{\sigma}^2_{t_0,\omega_0}$, respectively. We can interpret them as fractional changes of σ and $\omega\epsilon$ between times t and t_0 with respect to the square of the equivalent homogeneous conductivity $\hat{\sigma}^2_{t_0,\omega_0}$ at time t_0 at a low frequency ω_0.

We should note several precautions in using the mftdEIT image reconstruction algorithm of (7.91). First, since (7.84) is based on the reciprocity theorem, the EIT system must have a smallest possible reciprocity error. Second, the true admittivity distribution $\gamma_{t,\omega}$ inside the imaging object at time t and ω should be a small perturbation of its equivalent homogeneous admittivity $\hat{\gamma}_{t,\omega}$ in order for the approximations in (7.84) and (7.85) to be valid. This is the inherent limitation of the difference imaging method using the linearization. Third, the computed voltage v in (7.87) may contain modeling errors. It would be desirable for the model Λ of the imaging object Ω to have correct boundary shape and size. We may improve the model by incorporating a more realistic boundary shape in three dimensions. Fourth, the number of non-negligible singular values of the sensitivity matrix should be maximized by optimizing the electrode configuration and data collection protocol.

7.11 Frequency-Difference Imaging

Since tdEIT requires time-referenced data, it is not applicable to cases where such time-referenced data are not available. Examples may include imaging of tumors (Kulkarni *et al.* 2008; Soni *et al.* 2004; Trokhanova *et al.* 2008) and cerebral stroke (McEwan *et al.* 2006; Romsauerova *et al.* 2006a,b). Noting that admittivity spectra of numerous biological tissues show frequency-dependent changes (Gabriel *et al.* 1996b; Geddes and Baker 1967; Grimnes and Martinsen 2008; Oh *et al.* 2008), frequency-difference EIT (fdEIT) has been proposed to produce images of changes in the admittivity distribution with respect to frequency.

In early fdEIT methods, frequency-difference images were formed by back-projecting the logarithm of the ratio of two voltages at two frequencies (Fitzgerald *et al.* 1999; Griffiths 1987; Griffiths and Ahmed 1987a,b; Griffiths and Zhang 1989; Schlappa *et al.* 2000). More recent studies adopted the sensitivity matrix with a voltage difference at two frequencies (Bujnowski and Wtorek 2007; Romsauerova *et al.* 2006a,b; Yerworth *et al.* 2003). All of these methods are basically utilizing a simple voltage difference at two frequencies and a linearized image reconstruction algorithm. Alternatively, we may consider separately producing two static (absolute) images at two frequencies and then subtract one from the other. This approach, however, will suffer from the technical difficulties in static EIT imaging.

In this section, we describe an fdEIT method using a weighted voltage difference at two frequencies (Seo *et al.* 2008). Since the admittivity spectra of most biological tissues change with frequency, we will assume an imaging object with a frequency-dependent background admittivity in the development of fdEIT theory. We may consider two different contrast mechanisms in a reconstructed frequency-difference image. First, there exists a contrast in admittivity values between an anomaly and background. Second, the admittivity distribution itself changes with frequency.

7.11.1 Data Sets for Frequency-Difference Imaging

We assume the same setting as in section 7.10.1. Using an E-channel EIT system, we may inject E number of currents through adjacent pairs of electrodes and measure the following voltage data set:

$$\mathbb{F}_{t,\omega} = [\mathbf{V}_{t,\omega}^{1,1} \; \ldots \; \mathbf{V}_{t,\omega}^{1,E} \; \mathbf{V}_{t,\omega}^{2,1} \; \ldots \; \mathbf{V}_{t,\omega}^{2,E} \quad \ldots \quad \mathbf{V}_{t,\omega}^{E,1} \; \ldots \; \mathbf{V}_{t,\omega}^{E,E}]^{\mathrm{T}}. \tag{7.92}$$

For $t = t_1, t_2, \ldots, t_N$ and $\omega = \omega_1, \omega_2, \ldots, \omega_F$, we are provided with N data vectors for each one of F frequencies. Let us assume that we inject currents at two frequencies of ω_1 and ω_2 to obtain corresponding voltage data sets $\mathbb{F}_{t,\omega_1}$ and $\mathbb{F}_{t,\omega_2}$, respectively. The goal is to visualize changes of the admittivity distribution between ω_1 and ω_2 by using these two voltage data sets.

In tumor imaging or stroke detection using EIT, we are primarily interested in visualizing an anomaly. This implies that we should reconstruct a local admittivity contrast. For a given injection current, however, the boundary voltage $\mathbb{F}_{t,\omega}$ is significantly affected by the background admittivity, boundary geometry and electrode positions, while the influence of a local admittivity contrast due to an anomaly is much smaller. Since we utilize two sets of boundary voltage data, $\mathbb{F}_{t,\omega_1}$ and $\mathbb{F}_{t,\omega_2}$ in fdEIT, we need to evaluate their capability to perceive the local admittivity contrast. As in tdEIT, the rationale is to eliminate numerous common errors by subtracting the background component of $\mathbb{F}_{t,\omega_1}$ from $\mathbb{F}_{t,\omega_2}$, while preserving the local admittivity contrast component.

7.11.2 *Simple Difference* $\mathbb{F}_{t,\omega_2} - \mathbb{F}_{t,\omega_1}$

The simple voltage difference $\mathbb{F}_{t,\omega_2} - \mathbb{F}_{t,\omega_1}$ may work well for an imaging object whose background admittivity does not change with frequency. A typical example is a saline phantom. For realistic cases where background admittivity distributions change with frequency, it will produce artifacts in reconstructed fdEIT images. To understand this, let us consider a very simple case where the imaging object has a homogeneous admittivity distribution, that is, $\gamma_{t,\omega} = \sigma_{t,\omega} + i\omega\epsilon_{t,\omega}$ is independent of position. In such a homogeneous object, induced voltages $\bar{u}^j_{t,\omega_1}$ and $\bar{u}^j_{t,\omega_2}$ satisfy the Laplace equation with the same boundary data, and the two corresponding voltage data vectors $\bar{\mathbb{F}}_{t,\omega_1}$ and $\bar{\mathbb{F}}_{t,\omega_2}$ are parallel in such a way that

$$\bar{\mathbb{F}}_{t,\omega_2} = \frac{\gamma_{t,\omega_1}}{\gamma_{t,\omega_2}} \bar{\mathbb{F}}_{t,\omega_1}.$$

When there exists a small anomaly inside the imaging object, we may assume that the induced voltages are close to the voltages without any anomaly. In other words, the voltage difference $\mathbb{F}_{t,\omega_2} - \mathbb{F}_{t,\omega_1}$ in the presence of a small anomaly can be expressed as

$$\mathbb{F}_{t,\omega_2} - \mathbb{F}_{t,\omega_1} \approx \bar{\mathbb{F}}_{t,\omega_2} - \bar{\mathbb{F}}_{t,\omega_1} = \frac{\gamma_{t,\omega_1}}{\gamma_{t,\omega_2}} \bar{\mathbb{F}}_{t,\omega_1} - \bar{\mathbb{F}}_{t,\omega_1} = \eta \bar{\mathbb{F}}_{t,\omega_1}$$

for a complex constant η. This means that the simple difference $\mathbb{F}_{t,\omega_2} - \mathbb{F}_{t,\omega_1}$ significantly depends on the boundary geometry and electrode positions except for the special case where $\mathbb{F}_{t,\omega_2} - \mathbb{F}_{t,\omega_1} = 0$. This is the main reason why the use of the simple difference $\mathbb{F}_{t,\omega_2} - \mathbb{F}_{t,\omega_1}$ cannot deal with common modeling errors even for a homogeneous imaging object.

7.11.3 *Weighted Difference* $\mathbb{F}_{t,\omega_2} - \alpha\mathbb{F}_{t,\omega_1}$

An imaging object including an anomaly has an inhomogeneous admittivity distribution $\gamma_{t,\omega}$. We define a weighted difference of the admittivity at two different frequencies ω_1 and ω_2 at time t as

$$\delta\gamma^{\omega_2}_{\omega_1} = \alpha\gamma_{t,\omega_2} - \gamma_{t,\omega_1}, \tag{7.93}$$

where α is a complex number. We assume the following two conditions:

1. In the background region, especially near the boundary, $\delta\gamma^{\omega_2}_{\omega_1} \approx 0$.
2. In the anomaly, $\delta\gamma^{\omega_2}_{\omega_1}$ is significantly different from 0.

In order to extract the anomaly from the background, we investigate the relationship between $\mathbb{F}_{t,\omega_2}$ and $\mathbb{F}_{t,\omega_1}$. We should find a way to eliminate the background influence while maintaining the information of the admittivity contrast across the anomaly. We decompose $\mathbb{F}_{t,\omega_2}$ into a projection part onto $\mathbb{F}_{t,\omega_1}$ and the remaining part:

$$\mathbb{F}_{t,\omega_2} = \alpha\mathbb{F}_{t,\omega_1} + \mathbb{E}_{t,\omega_2}, \quad \alpha = \frac{\langle \mathbb{F}_{t,\omega_2}, \mathbb{F}_{t,\omega_1}\rangle}{\langle \mathbb{F}_{t,\omega_1}, \mathbb{F}_{t,\omega_1}\rangle}, \tag{7.94}$$

where $\langle\cdot,\cdot\rangle$ is the standard inner product of two vectors. Note that $\mathbb{E}_{t,\omega_2}$ is orthogonal to $\mathbb{F}_{t,\omega_1}$.

In the absence of the anomaly, we may set $\gamma_{\omega_2} = (1/\alpha)\gamma_{\omega_1}$ and this results in $\mathbb{F}_{t,\omega_2} = \alpha\mathbb{F}_{t,\omega_1}$. The projection term $\alpha\mathbb{F}_{t,\omega_1}$ mostly contains the background information, while the orthogonal term $\mathbb{E}_{t,\omega_2}$ holds the anomaly information. To be precise, $\alpha\mathbb{F}_{t,\omega_1}$ provides the same information as $\mathbb{F}_{t,\omega_1}$, which includes influences of the background admittivity, boundary geometry and electrode positions. The orthogonal term $\mathbb{E}_{t,\omega_2} = \mathbb{F}_{t,\omega_2} - \alpha\mathbb{F}_{t,\omega_1}$ contains the core information about a nonlinear change due to the admittivity contrast across the anomaly. This explains why the weighted difference $\mathbb{F}_{t,\omega_2} - \alpha\mathbb{F}_{t,\omega_1}$ must be used in fdEIT.

7.11.4 Linear Frequency-Difference Algorithm using Sensitivity Matrix

In this section, we drop the time index t to simplify the notation. Applying the linear approximation in section 7.7, we get the following relation:

$$\int_\Omega \delta\gamma_{\omega_1}^{\omega_2}(\mathbf{r})\nabla u_{\omega_1}^j(\mathbf{r})\cdot\nabla u_{\omega_2}^k(\mathbf{r})\,d\mathbf{r} \approx I(\mathbf{V}_{\omega_2}^{j,k} - \alpha\,\mathbf{V}_{\omega_1}^{j,k}), \quad j,k = 1,2,\ldots,E. \tag{7.95}$$

Given α, we can reconstruct an image of $\delta\gamma_{\omega_1}^{\omega_2}$ using the weighted difference $\mathbb{F}_{\omega_2} - \alpha\mathbb{F}_{\omega_1}$. Since α is not known in practice, we need to estimate it from $\mathbb{F}_{\omega_2}$ and $\mathbb{F}_{\omega_1}$ using (7.94).

We discretize the imaging object Ω as $\Omega = \bigcup_{i=1}^N \Omega_i$, where Ω_i is the ith pixel. Let χ_{Ω_i} be the characteristic function of the ith element Ω_i, that is, $\chi_{\Omega_i} = 1$ in Ω_i and zero otherwise. Let $\xi_1,\ldots,\xi_N$ be complex numbers such that $\sum_{i=1}^N \xi_i\chi_{\Omega_i}$ approximates

$$\sum_{i=1}^N \xi_i\chi_{\Omega_i} \approx \frac{\delta\gamma_{\omega_1}^{\omega_2}}{I\gamma_{\omega_1}\gamma_{\omega_2}}.$$

By approximating $\nabla u_\omega^j \approx (1/\gamma_\omega)\nabla U^j$, where U^j is the solution of (7.26) with $\gamma_\omega = 1$, it follows from (7.95) that

$$\begin{aligned}\sum_{i=1}^N\left(\xi_i\int_{\Omega_i}\nabla U^j(\mathbf{r})\cdot\nabla U^k(\mathbf{r})\,d\mathbf{r}\right) &\approx \int_\Omega \frac{\delta\gamma_{\omega_1}^{\omega_2}}{I\gamma_{\omega_1}\gamma_{\omega_2}}\nabla U^j(\mathbf{r})\cdot\nabla U^k(\mathbf{r})\,d\mathbf{r}\\ &\approx (\mathbf{V}_{\omega_2}^{j,k} - \alpha\mathbf{V}_{\omega_1}^{j,k}), \quad j,k = 1,2,\ldots,E.\end{aligned} \tag{7.96}$$

The reconstruction method using the approximation (7.95) is reduced to reconstructing the $\sum_{i=1}^N \xi_i\chi_{\Omega_i}$ that minimizes the following:

$$\sum_{j,k=1}^N\left|\sum_{i=1}^N\left(\xi_i\int_{\Omega_i}\nabla U^j(\mathbf{r})\cdot\nabla U^k(\mathbf{r})\,d\mathbf{r} - (\mathbf{F}_{\omega_2}^{j,k} - \alpha\mathbf{V}_{\omega_1}^{j,k})\right)\right|^2, \tag{7.97}$$

where α is the complex number described in section 7.11.3. In order to find $\xi = (\xi_1, \dots, \xi_N)$, we use the sensitivity matrix $\mathbb{S}_{\gamma^0}$ in (7.87). We can compute $\xi = (\xi_1, \dots, \xi_N)$ by solving the following linear system through the truncated singular value decomposition (TSVD):

$$\mathbb{S}_{\gamma^0}\xi = \mathbb{F}_{\omega_2} - \alpha\mathbb{F}_{\omega_1}.$$

It remains to compute the fdEIT image $\delta\gamma_{\omega_1}^{\omega_2}$ from knowledge of ξ. We need to estimate the equivalent homogeneous (constant) admittivity $\hat{\gamma}_\omega$ corresponding to γ_ω to use the following approximation

$$\delta\gamma_{\omega_1}^{\omega_2} \approx I\hat{\gamma}_{\omega_1}\hat{\gamma}_{\omega_2}\sum_{i=1}^{N}\xi_i\chi_{\Omega_i}.$$

From the divergence theorem, we obtain the following relation:

$$\frac{\hat{\gamma}_\omega}{\hat{\gamma}_{\omega 0}} = \frac{\hat{\gamma}_\omega}{\hat{\gamma}_{\omega 0}}\frac{\int_\Omega \nabla u_\omega^k \cdot \nabla u_{\omega 0}^j}{\int_\Omega \nabla u_\omega^k \cdot \nabla u_{\omega 0}^j} \approx \frac{\mathbf{V}_{\omega 0}^{j,k}}{\mathbf{V}_\omega^{k,j}} \quad \text{for any } j,k \in \{1,2,\dots,E\}.$$

For an E-channel mfEIT system, we may choose

$$\frac{\hat{\gamma}_\omega}{\hat{\gamma}_{\omega 0}} = \frac{1}{2E}\sum_{j=1}^{E}\left(\frac{\mathbf{V}_{\omega 0}^{j,j+3}}{\mathbf{V}_\omega^{j+3,j}} + \frac{\mathbf{V}_{\omega 0}^{j,j-3}}{\mathbf{V}_\omega^{j-3,j}}\right), \tag{7.98}$$

where we identify $E + j = j$ and $-j = E - j$ for $j = 1, 2, 3$. We reconstruct an fdEIT image $\delta\gamma_{\omega_1}^{\omega_2}$ by

$$\delta\gamma_{\omega_1}^{\omega_2} = I\hat{\gamma}_{\omega_1}\hat{\gamma}_{\omega_2}\mathbb{A}(\mathbb{F}_{\omega_2} - \alpha\mathbb{F}_{\omega_1}) = I\hat{\gamma}_{\omega 0}^2\frac{\hat{\gamma}_{\omega_1}}{\hat{\gamma}_{\omega 0}}\frac{\hat{\gamma}_{\omega_2}}{\hat{\gamma}_{\omega 0}}\mathbb{A}(\mathbb{F}_{\omega_2} - \alpha\mathbb{F}_{\omega_1}), \tag{7.99}$$

where $\mathbb{A}$ is a pseudo-inverse of $\mathbb{S}_{\gamma^0}$.

7.11.5 Interpretation of Frequency-Difference Image

In (7.99), $(\hat{\gamma}_{\omega_1}/\hat{\gamma}_{\omega 0})(\hat{\gamma}_{\omega_2}/\hat{\gamma}_{\omega 0})$ can be estimated from (7.98) using another low-frequency measurement $\mathbb{F}_{\omega 0}$. If we choose ω_0 low enough, $\hat{\gamma}_{\omega 0}$ may have a negligibly small imaginary part. In such a case, we may set $\delta\gamma_{\omega_1}^{\omega_2}/\hat{\gamma}_{\omega 0}^2$ as a reconstructed fdEIT image, which is equivalent to the complex image $\delta\gamma_{\omega_1}^{\omega_2}$ divided by an unknown real constant. In practice, it would be desirable to set ω_0 smaller than 1 kHz, for example 100 Hz.

This scaling will be acceptable for applications where we are mainly looking for a contrast change within an fdEIT image. These may include detections of tumors and strokes. In order to interpret absolute pixel values of an fdEIT image quantitatively, we must estimate the value of $\hat{\gamma}_{\omega 0}$, which requires knowledge of the object size, boundary shape and electrode positions. Alternatively, we may estimate values of $\hat{\gamma}_{\omega_1}$ and $\hat{\gamma}_{\omega_2}$ in (7.99) without using the third frequency ω_0. This will again need geometrical information about the imaging object and electrode positions.

References

Adler A and Lionheart W 2006 Uses and abuses of EIDORS: an extensible software base for EIT. *Physiol. Meas*. **27**(5), S25–S42.

Alessandrini G 1988 Stable determination of conductivity by boundary measurements. *Appl. Anal*. **27**, 153–172.

Astala K and Paivarinta L 2006a A boundary integral equation for Calderón's inverse conductivity problem. In *Proc. 7th Int. Conf. on Harmonic Analysis and Partial Differential Equations, El Escorial, Spain, June 2004*; *Collect. Math*., Extra Volume, pp. 127–139.

Astala K and Paivarinta L 2006b Calderón's inverse conductivity problem in the plane. *Ann. Math*. **163**, 265–299.

Barber DC and Brown BH 1983 Imaging spatial distribution of resistivity using applied potential tomography. *Electron. Lett*. **19**, 933–935.

Barber DC and Brown BH 1984 Applied potential tomography. *J. Phys. E: Sci. Instrum*. **17**, 723–733.

Barber DC and Brown BH 1988 Errors in reconstruction of resistivity images using a linear reconstruction technique. *Clin. Phys. Physiol. Meas*. **9**, 101–104.

Berenstein CA, Tarabusi EC, Cohen JM and Picardello MA 1991 Integral geometry on trees. *Am. J. Math*. **113**, 441–470.

Boone K, Barber DC and Brown BH 1997 Imaging with electricity: report of the European Concerted Action on Impedance Tomography. *J. Med. Eng. Technol*. **21**, 201–232.

Brown R and Uhlmann G 1997 Uniqueness in the inverse conductivity problem with less regular conductivities in two dimensions. *Commun. PDE* **22**, 1009–1027.

Brown BH, Barber DC and Seagar AD 1985 Applied potential tomography: possible clinical applications. *Clin. Phys. Physiol. Meas*. **6**, 109–121.

Bujnowski A and Wtorek J 2007 An excitation in differential EIT selection of measurement frequencies. In *Proc. 13th Int. Conf. on Electrical Bioimpedance and 8th Conf. on Electrical Impedance Tomography, Graz, Austria, 29 August–2 September*. IFMBE Proceedings, vol. 17, eds. H. Scharfetter and R. Merwa. Springer, Berlin.

Calderón AP 1980 On an inverse boundary value problem. *Seminar on Numerical Analysis and its Applications to Continuum Physics*, pp. 65–73. Sociedade Brasileira de Matemática, Rio de Janeiro.

Cheney M, Isaacson D and Isaacson EL 1990 Exact solutions to a linearized inverse boundary value problem. *Inv. Prob*. **6**, 923–934.

Cheney M, Isaacson D, Isaacson E, Somersalo E and Coffey EJ 1991 A layer-stripping reconstruction algorithm for impedance imaging. In *Proc. 13th Annu. Int. Conf. IEEE Engineering in Medicine and Biology Society*, pp. 2–4. IEEE, New York.

Cheney M, Isaacson D and Newell JC 1999 Electrical impedance tomography. *SIAM Rev*. **41**, 85–101.

Cheng KS, Isaacson D, Newell JC and Gisser DG 1989 Electrode models for electric current computed tomography. *IEEE Trans. Biomed. Eng*. **36**(9), 918–924.

Cherkaeva E and Cherkaev A 1995 Bounds for detectability of material damage by noisy electrical measurements. In *Structural and Multidisciplinary Optimization*, eds N Olhoff and GIN Rozvany, pp. 543–548. Pergamon, Oxford.

Cohen-Bacrie C, Goussard Y and Guardo R 1997 Regularized reconstruction in electrical impedance tomography using a variance uniformization constraint. *IEEE Trans. Biomed. Imag*. **6**, 562.

Cook RD, Saulnier GJ, Gisser DG, Goble JC, Newell JC and Isaacson D 1994 ACT3: a high-speed, high-precision electrical impedance tomograph. *IEEE Trans Biomed Eng*. **41**(8), 713–722.

Edic PM, Isaacson D, Saulnier GJ, Jain H and Newell JC 1998 An iterative Newton-Raphson method to solve the inverse admittivity problem. *IEEE Trans. Biomed. Eng*. **45**, 899–908.

Faddeev D 1965 Growing solutions of the Schrödinger equation. *Sov. Phys. Dokl*. **10**, 1033 (Engl. transl.).

Fitzgerald A, Holder D and Griffiths H 1999 Experimental assessment of phase magnitude imaging in multi-frequency EIT by simulation and saline tank studies. *Ann. NY Acad. Sci*. **873**, 381–387.

Fuks LF, Cheney M, Isaacson D, Gisser DG and Newell JC 1991 Detection and imaging of electric conductivity and permittivity at low frequency. *IEEE Trans Biomed Eng*. **38**(11), 1106–1110.

Gabriel C, Gabriel S and Corthout E 1996a The dielectric properties of biological tissues: I. Literature survey. *Phys. Med. Biol*. **41**, 2231–2249.

Gabriel S, Lau RW and Gabriel C 1996b The dielectric properties of biological tissues: II. Measurements in the frequency range 10 Hz to 20 GHz. *Phys. Med. Biol*. **41**, 2251–2269.

Geddes LA and Baker LE 1967 The specific resistance of biological material: a compendium of data for the biomedical engineer and physiologist. *Med. Biol. Eng*. **5**, 271–293.
Gisser DG, Isaacson D and Newell JC 1988 Theory and performance of an adaptive current tomograph system. *Clin. Phys. Physiol. Meas*. **9** (Suppl. A), 35–41.
Gisser DG, Isaacson D and Newell JC 1990 Electric current computed tomography and eigenvalues. *SIAM J. Appl. Math*. **50**, 1623–1634.
Griffiths H 1987 The importance of phase measurement in electrical impedance tomography. *Phys. Med. Biol*. **32**, 1435–1444.
Griffiths H and Ahmed A 1987a A dual-frequency applied potential tomography technique: computer simulations. *Clin. Phys. Physiol. Meas*. **8**, 103–107.
Griffiths H and Ahmed A 1987b Electrical impedance tomography for thermal monitoring of hyperthermia treatment: an assessment using in vitro and in vivo measurements. *Clin. Phys. Physiol. Meas*. **8**, 141–146.
Griffiths H and Zhang Z 1989 A dual-frequency electrical impedance tomography system. *Phys. Med. Biol*. **34**, 1465–1476.
Grimnes S and Martinsen OG 2008 *Bioimpedance and Bioelectricity Basics*, 2nd edn. Academic Press, Oxford.
Holder D (ed.) 2005 *Electrical Impedance Tomography: Methods, History and Applications*. IOP Publishing, Bristol.
Hyaric AL and Pidcock MK 2001 An image reconstruction algorithm for three-dimensional electrical impedance tomography. *IEEE Trans. Biomed. Eng*. **48**, 230–235.
Isaacson D and Cheney M 1991 Effects of measurement precision and finite numbers of electrodes on linear impedance imaging algorithms. *SIAM J. Appl. Math*. **15**, 1705–1731.
Isaacson D, Newell JC and Gisser DG 1989 Rapid assessment of electrode characteristics for impedance imaging. In *Proc. 11th Annu. Int. Conf. IEEE Engineering in Medicine and Biology Society*, pp. 474–475. IEEE, New York.
Isaacson D, Blue RS and Newell JC 1996 A 3-D impedance imaging algorithm for multiple layers of electrodes. In *Proc. Association for the Advancement of Medical Instrumentation Conf*.
Isakov V 1991 Completeness of products of solutions and some inverse problems for PDE. *J. Differ. Eqns* **92**, 305–317.
Kenig CE, Sjostrand J and Uhlmann G 2007 The Calderón problem with partial data. *Ann. Math*. **165**, 567–591.
Kohn R and Vogelius M 1984 Identification of an unknown conductivity by means of measurements at the boundary. In *Inverse Problems*. SIAM–AMS Proceedings, vol. 14, pp. 113–123.
Kohn R and Vogelius M 1985 Determining conductivity by boundary measurements: II. Interior results. *Commun. Pure Appl. Math*. **38**, 643–667.
Kolehmainen V, Lassas M and Ola P 2005 The inverse conductivity problem with an imperfectly known boundary. *SIAM J. Appl. Math*. **66**, 365–383.
Kulkarni R, Boverman G, Isaacson D, Saulnier GJ, Kao TJ and Newell JC 2008 An analytical layered forward model for breasts in electrical impedance tomography. *Physiol. Meas*. **29**, S27–S40.
Lionheart W, Polydorides N and Borsic A 2005 The reconstruction problem. In *Electrical Impedance Tomography: Methods, History and Applications*, ed. D Holder. IOP Publishing, Bristol.
McEwan A, Romsauerova A, Yerworth R, Horesh L, Bayford R and Holder D 2006 Design and calibration of a compact multi-frequency EIT system for acute stroke imaging. *Physiol. Meas*. **27**, S199–S210.
Mueller JL and Siltanen S 2003 Direct reconstructions of conductivities from boundary measurements. *SIAM J. Sci. Comput.tion* **24**(4), 1232–1266.
Murphy EK and Mueller JL 2009 Effect of domain-shape modeling and measurement errors on the 2-D D-bar method for electrical impedance tomography. *IEEE Trans. Med. Imag*. **28**(10), 1576–1584.
Nachman A 1988 Reconstructions from boundary measurements. *Ann. Math*. **128**, 531–576.
Nachman A 1996 Global uniqueness for a two-dimensional inverse boundary value problem. *Ann. Math*. **143**, 71–96.
Nachman A, Sylvester J and Uhlmann G 1988 An n-dimensional Borg–Levinson theorem. *Commun. Math. Phys*. **115**, 593–605.
Newell JC, Gisser DG and Isaacson D 1988 An electric current tomograph. *IEEE Trans. Biomed. Eng*. **35**, 828–833.
Nissinen A, Heikkinen LM and Kaipio JP 2008 The Bayesian approximation error approach for electrical impedance tomography - experimental results. *Meas. Sci. Technol*. **19**, 015501.
Oh TI, Woo EJ and Holder D 2007a Multi-frequency EIT system with radially symmetric architecture: KHU Mark1. *Physiol. Meas*. **28**, S183–S196.

Oh TI, Lee KH, Kim SM, Koo H, Woo EJ and Holder D 2007b Calibration methods for a multi-channel multi-frequency EIT system. *Physiol. Meas*. **28**, S1175–S1188.

Oh TI, Koo W, Lee KH, Kim SM, Lee J, Kim SW, Seo JK and Woo EJ 2008 Validation of a multi-frequency electrical impedance tomography (mfEIT) system KHU Mark1: impedance spectroscopy and time-difference imaging. *Physiol. Meas*. **29**, S295–S307.

Romsauerova A, McEwan A, Horesh L, Yerworth R, Bayford RH and Holder D 2006a Multi-frequency electrical impedance tomography (EIT) of the adult human head: initial findings in brain tumours, arteriovenous malformations and chronic stroke, development of an analysis method and calibration. *Physiol. Meas*. **27**, S147–S161.

Romsauerova A, McEwan A and Holder DS 2006b Identification of a suitable current waveform for acute stroke imaging. *Physiol. Meas*. **27**, S211–S219.

Santosa F and Vogelius M 1990 A backprojection algorithm for electrical impedance imaging. *SIAM J. Appl. Math*. **50**, 216–243.

Saulnier GJ 2005 EIT instrumentation. In *Electrical Impedance Tomography: Methods, History and Applications*, ed. DS Holder. IOP Publishing, Bristol.

Schlappa J, Annese E and Griffiths H 2000 Systematic errors in multi-frequency EIT. *Physiol. Meas*. **21**, S111–S118.

Seo JK, Lee J, Kim SW, Zribi H and Woo EJ 2008 Frequency-difference electrical impedance tomography (fdEIT): algorithm development and feasibility study. *Physiol. Meas*. **29**, S929–S944.

Siltanen S, Mueller JL, Isaacson D 2000 An implementation of the reconstruction algorithm of A Nachman for the 2-D inverse conductivity problem. *Inv. Probs* **16**, 681–699.

Somersalo E, Cheney M, Isaacson D and Isaacson E 1991 Layer stripping: a direct numerical method for impedance imaging. *Inv. Prob*. **7**, 899–926.

Somersalo E, Cheney M and Isaacson D 1992 Existence and uniqueness for electrode models for electric current computed tomography. *SIAM J. Appl. Math*. **52**, 1023–1040.

Soni NK, Hartov A, Kogel C, Poplack SP and Paulsen KD 2004 Multi-frequency electrical impedance tomography of the breast: new clinical results. *Physiol. Meas*. **25**, S301–S314.

Sylvester J and Uhlmann G 1986 A uniqueness theorem for an inverse boundary value problem in electrical prospection. *Commun. Pure Appl. Math*. **39**, 92–112.

Sylvester J and Uhlmann G 1987 A global uniqueness theorem for an inverse boundary value problem. *Ann. Math*. **125**, 153–165.

Sylvester J and Uhlmann G 1988 Inverse boundary value problems at the boundary – continuous dependence. *Commun. Pure Appl. Math*. **41**, 197–221.

Trokhanova OV, Okhapkin MB and Korjenevsky AV 2008 Dual-frequency electrical impedance mammography for the diagnosis of non-malignant breast disease. *Physiol. Meas*. **29**, S331–S334.

Vauhkonen PJ, Vauhkonen M, Savolainen T and Kaipio JP 1996 Three-dimensional electrical impedance tomography based on the complete electrode model. *IEEE Trans. Biomed. Eng* **45**, 1150–1160.

Vauhkonen M, Vadasz D, Karjalainen PA, Somersalo E and Kaipio JP 1998 Tikhonov regularization and prior information in electrical impedance tomography. *IEEE Trans. Med. Imag*. **17**, 285–293.

Wexler A, Fry B and Neuman MR 1985 Impedance-computed tomography algorithm and system. *Appl. Opt*. **24**, 3985–3992.

Wilson AJ, Milnes P, Waterworth AR, Smallwood RH and Brown BH 2001 Mk3.5: a modular, multi-frequency successor to the Mk3a EIS/EIT system. *Physiol. Meas*. **22**, S49–S54.

Woo EJ, Hua P, Tompkins WJ and Webster JG 1993 A robust image reconstruction algorithm and its parallel implementation in electrical impedance tomography. *IEEE Trans. Med. Imag*. **12**(2), 137–146.

Yerworth RJ, Bayford RH, Brown B, Milnes P, Conway M and Holder DS 2003 Electrical impedance tomography spectroscopy (EITS) for human head imaging. *Physiol. Meas*. **24**, S477–S489.

Yorkey JT 1987 Personal communication (Lawrence Livermore National Laboratory, Livermore, CA, USA).

Yorkey TJ and Webster JG 1987 A comparison of impedance tomographic reconstruction algorithms. *Clin. Phys. Physiol. Meas*. **8** (Suppl. A), 55–62.

8

Anomaly Estimation and Layer Potential Techniques

Layer potential techniques have been used widely to deal with the inverse problem of recovering anomalies in a homogeneous background. The reason is that the method provides a concrete expression connecting the anomalies with measured data.

For example, consider the inverse problem of detecting an electrical conductivity anomaly, occupying a region D, inside a three-dimensional region Ω bounded by its surface $\partial\Omega$. Assume that the complex conductivity distribution $\gamma(\mathbf{r}) = \sigma(\mathbf{r}) + i\omega\epsilon(\mathbf{r})$ at angular frequency ω changes abruptly across the boundary ∂D and $\nabla\gamma = 0$ in $\Omega \setminus \partial D$. With the aid of the fundamental solution $F(\mathbf{r}) := -1/(4\pi|\mathbf{r}|)$ of the Laplacian, we can provide a rigorous connection between the anomaly D and the boundary voltage–current data via the following integral equation (Kang and Seo 1996): for $\mathbf{r} \in \mathbb{R}^3 \setminus \bar{\Omega}$,

$$\underbrace{\frac{\gamma|_D - \gamma|_{\Omega\setminus D}}{\gamma|_{\Omega\setminus D}}}_{\substack{\text{admittivity difference}\\ \text{along } \partial D}} \int_D \underbrace{\nabla u(\mathbf{r}') \cdot \nabla\left(\frac{-1}{4\pi|\mathbf{r}-\mathbf{r}'|}\right)}_{\text{sensitivity}} \mathrm{d}\mathbf{r}' = \underbrace{\frac{1}{\gamma|_{\Omega\setminus D}} \mathcal{S}_\Omega g(\mathbf{r}) - \mathcal{D}_\Omega f(\mathbf{r}),}_{\substack{\text{determined by}\\ \text{boundary data}}} \tag{8.1}$$

where g represents Neumann data corresponding to the sinusoidal injection current with an angular frequency ω, u is the induced time-harmonic voltage inside Ω, $f = u|_{\partial\Omega}$, $\mathcal{D}_\Omega f$ is the double-layer potential given by

$$\mathcal{D}_\Omega f(\mathbf{r}) = \int_{\partial\Omega} \frac{\partial}{\partial \mathbf{n}} F(\mathbf{r}-\mathbf{r}') f(\mathbf{r}')\, \mathrm{d}S_{\mathbf{r}'} \quad (\mathbf{r} \in \mathbb{R}^3 \setminus \partial\Omega) \tag{8.2}$$

and $\mathcal{S}g$ is the single-layer potential given by

$$\mathcal{S}g(\mathbf{r}) = \int_{\partial\Omega} F(\mathbf{r}-\mathbf{r}') g(\mathbf{r}')\, \mathrm{d}S_{\mathbf{r}'} \quad (\mathbf{r} \in \mathbb{R}^3 \setminus \partial\Omega). \tag{8.3}$$

When the Neumann data g and Dirichlet data f are available along the boundary $\partial\Omega$, the inverse problem is to estimate the anomaly D from knowledge of the right-hand side

Nonlinear Inverse Problems in Imaging, First Edition. Jin Keun Seo and Eung Je Woo.

of the identity (8.1). Owing to the expression in the sensitivity part in (8.1) containing location information of D, the formula provides useful information in estimating the anomaly D.

8.1 Harmonic Analysis and Potential Theory

8.1.1 Layer Potentials and Boundary Value Problems for Laplace Equation

For simplicity, we will restrict ourselves to three-dimensional cases, although all the arguments in this chapter work for general dimensions with minor modifications. We also assume that both D and Ω are Lipschitz domains and $\overline{D} \subset \Omega$. The boundary value problem of the Laplace equation can be solved by single- or double-layer potentials with a surface potential density. The reason is that a solution $u \in H^1(\Omega)$ of the Laplace equation $\nabla^2 u = 0$ in Ω can be expressed as

$$\mathcal{D}_\Omega f(\mathbf{r}) - \mathcal{S}_\Omega g(\mathbf{r}) = \int_\Omega \nabla^2 F(\mathbf{r}-\mathbf{r}')u(\mathbf{r}')\,\mathrm{d}\mathbf{r}' = \begin{cases} u(\mathbf{r}) & \text{if } \mathbf{r} \in \Omega, \\ 0 & \text{if } \mathbf{r} \notin \Omega \cup \partial\Omega, \end{cases} \tag{8.4}$$

where

$$f = u|_{\partial\Omega} \quad \text{and} \quad g = \frac{\partial}{\partial \mathbf{n}} u|_{\partial\Omega}.$$

Here, the relation between the Dirichlet data f and Neumann data g is dictated by

$$\lim_{t\to 0^+} \mathcal{D}_\Omega f(\mathbf{r} + t\mathbf{n}(\mathbf{r})) = \lim_{t\to 0^+} \mathcal{S}_\Omega g(\mathbf{r} + t\mathbf{n}(\mathbf{r})) \quad (\mathbf{r} \in \partial\Omega). \tag{8.5}$$

To see the relation more clearly, define a trace operator

$$\mathcal{K}_\Omega f(\mathbf{r}) = \int_{\partial\Omega} \underbrace{\frac{\langle \mathbf{r}'-\mathbf{r}, \mathbf{n}(\mathbf{r}')\rangle}{4\pi|\mathbf{r}-\mathbf{r}'|^3}}_{\partial F(\mathbf{r}-\mathbf{r}')/\partial \mathbf{n}_{\mathbf{r}'}} f(\mathbf{r}')\,\mathrm{d}S_{\mathbf{r}'} \quad \text{for } \mathbf{r} \in \partial\Omega. \tag{8.6}$$

The operator $\mathcal{K}_\Omega$ in (8.6) appears to be the same as $\mathcal{D}_\Omega$ in (8.2), but there exists a clear difference between them due to the singular kernel $\partial F(\mathbf{r}-\mathbf{r}')/\partial \mathbf{n}_{\mathbf{r}'}$ at $\mathbf{r} = \mathbf{r}'$. The following theorem explains how the double-layer potential jumps across $\partial\Omega$ due to its singular kernel.

Theorem 8.1.1 *The trace operator $\mathcal{K}_\Omega$ is bounded on $L^p(\partial\Omega)$, $1 < p < \infty$. (When $\partial\Omega$ is a C^1 domain, $\mathcal{K}$ is a compact operator.) For $\phi \in L^2(\partial\Omega)$, the following trace formulas hold* almost everywhere *on $\partial\Omega$ (and limit in $L^2(\partial\Omega)$ sense):*

$$\lim_{t\to 0^+} \mathcal{D}_\Omega \phi(\mathbf{r} \pm t\mathbf{n}(\mathbf{r})) = (\mp\tfrac{1}{2}I + \mathcal{K}_\Omega)\phi(\mathbf{r}), \tag{8.7}$$

$$\lim_{t\to 0^+} \mathcal{S}_\Omega \phi(\mathbf{r} \pm t\mathbf{n}(\mathbf{r})) = \mathcal{S}_\Omega \phi(\mathbf{r}), \tag{8.8}$$

$$\lim_{t\to 0^+} \langle \mathbf{n}(\mathbf{r}), \nabla \mathcal{S}_\Omega \phi(\mathbf{r} \pm t\mathbf{n}(\mathbf{r}))\rangle = (\pm\tfrac{1}{2}I + \mathcal{K}_\Omega^*)\phi(\mathbf{r}), \tag{8.9}$$

where I is the identity operator on $L^2(\partial\Omega)$ and $\mathcal{K}_\Omega^*$ is the dual operator of $\mathcal{K}_\Omega$.

Here, the term "almost everywhere" on $\partial\Omega$ means *all points except a set of measure zero in* $\partial\Omega$. For detailed explanations on these issues in measure theory, see section 4.5. The proof of the boundedness of the trace operator $\mathcal{K}_\Omega$ in Theorem 8.1.1 for the Lipschitz domain Ω requires a deep knowledge on the harmonic analysis (Coifman *et al.* 1982, David and Journé 1984), while the proof in the C^2 domain Ω is a lot simpler (Folland 1976). To prove Theorem 8.1.1, we need to use the following lemma.

Lemma 8.1.2 *Denoting*

$$T(\mathbf{r},\mathbf{r}') = \frac{\langle \mathbf{r}'-\mathbf{r}, \mathbf{n}(\mathbf{r}')\rangle}{4\pi|\mathbf{r}-\mathbf{r}'|^3} = \frac{\partial}{\partial \mathbf{n}_{\mathbf{r}'}} F(\mathbf{r}-\mathbf{r}'),$$

we have

$$\int_{\partial\Omega} T(\mathbf{r},\mathbf{r}')\,\mathrm{d}S_{\mathbf{r}'} = \begin{cases} 1 & \textit{if } \mathbf{r}\in\Omega, \\ 1/2 & \textit{if } \mathbf{r}\in\partial\Omega, \\ 0 & \textit{if } \mathbf{r}\in\mathbb{R}^3\setminus(\Omega\cup\partial\Omega). \end{cases}$$

Proof. For ease of explanation, we restrict ourselves to the case of the C^1 domain Ω. For $\mathbf{r}\in\mathbb{R}^3\setminus(\Omega\cup\partial\Omega)$, we have

$$\int_{\partial\Omega} T(\mathbf{r},\mathbf{r}')\,\mathrm{d}S_{\mathbf{r}'} = \int_{\partial\Omega} \frac{\partial F(\mathbf{r}-\mathbf{r}')}{\partial \mathbf{n}_{\mathbf{r}'}}\,\mathrm{d}S_{\mathbf{r}'} = \int_{\Omega} \underbrace{\nabla^2 F(\mathbf{r}-\mathbf{r}')}_{=0}\,\mathrm{d}\mathbf{r}' = 0.$$

If $\mathbf{r}\in\Omega$, there exists a ball $B_\epsilon(\mathbf{r})$ (as shown in Figure 8.1) such that $\overline{B_\epsilon(\mathbf{r})}\subset\Omega$ and application of the divergence theorem over the region $\Omega\setminus B_{\epsilon(\mathbf{r})}$ leads to

$$\int_{\partial\Omega} T(\mathbf{r},\mathbf{r}')\,\mathrm{d}S_{\mathbf{r}'} = \int_{\partial B_{\epsilon(\mathbf{r})}} T(\mathbf{r},\mathbf{r}')\,\mathrm{d}S_{\mathbf{r}'} + \int_{\Omega\setminus B_{\epsilon(\mathbf{r})}} \underbrace{\nabla^2 F(\mathbf{r}-\mathbf{r}')}_{=0}\,\mathrm{d}\mathbf{r}'. \tag{8.10}$$

Direct calculation over the sphere $\partial B_\epsilon(\mathbf{r})$ yields

$$\int_{\partial\Omega} T(\mathbf{r},\mathbf{r}')\,\mathrm{d}S_{\mathbf{r}'} = \int_{\partial B_\epsilon(\mathbf{r})} T(\mathbf{r},\mathbf{r}')\,\mathrm{d}S_{\mathbf{r}'} = \int_{\partial B_\epsilon(\mathbf{r})} \frac{\epsilon}{4\pi\epsilon^3}\,\mathrm{d}S_{\mathbf{r}'} = 1.$$

Finally, let $\mathbf{r}\in\partial\Omega$ (see Figure 8.1b). Denoting $\Omega_\epsilon = \Omega\setminus\overline{B_\epsilon(\mathbf{r})}$, we have

$$\begin{aligned}\int_{\partial\Omega} T(\mathbf{r},\mathbf{r}')\,\mathrm{d}S_{\mathbf{r}'} &= \lim_{\epsilon\to0}\int_{\partial\Omega_\epsilon} T(\mathbf{r},\mathbf{r}')\,\mathrm{d}S'_{\mathbf{r}} + \lim_{\epsilon\to0}\int_{\partial B_\epsilon(\mathbf{r})\cap\Omega} T(\mathbf{r},\mathbf{r}')\,\mathrm{d}S'_{\mathbf{r}} \\ &= \lim_{\epsilon\to0}\int_{\Omega_\epsilon} \underbrace{\nabla^2 F(\mathbf{r}-\mathbf{r}')}_{=0}\,\mathrm{d}\mathbf{r}' + \lim_{\epsilon\to0}\int_{\partial B_\epsilon(\mathbf{r})\cap\Omega} \frac{\epsilon}{4\pi\epsilon^3}\,\mathrm{d}S_{\mathbf{r}'} \\ &= \lim_{\epsilon\to0}\frac{|\partial B_\epsilon(\mathbf{r})\cap\Omega|}{|\partial B_\epsilon(\mathbf{r})|} = \frac{1}{2}.\end{aligned}$$

□

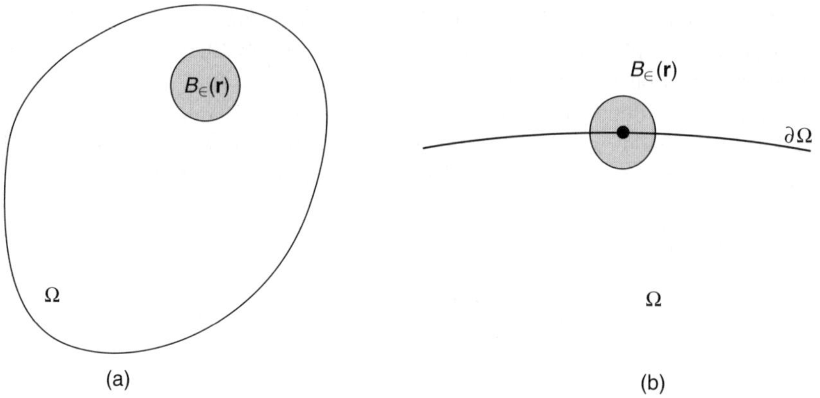

Figure 8.1 Diagrams showing (a) $\mathbf{r} \in \Omega$ and $B_\epsilon(\mathbf{r}) \subset \Omega$ and (b) $\mathbf{r} \in \partial\Omega$. The ratio $|B_\epsilon(\mathbf{r}) \cap \Omega|/|B_\epsilon(\mathbf{r})|$ is about half for small ϵ

Now, we are ready to prove the trace formula in Theorem 8.1.1. We will only prove the trace formula (8.7) under the assumption that $\partial\Omega$ is C^2 and $\phi \in C^1(\partial\Omega)$. For $\mathbf{r} \in \partial\Omega$,

$$\begin{aligned}
&\lim_{t\to 0^+} \mathcal{D}_\Omega\phi(\mathbf{r} \pm t\mathbf{n}(\mathbf{r})) \\
&= \lim_{t\to 0^+} \int_{\partial\Omega} T(\mathbf{r} \pm t\mathbf{n}(\mathbf{r}), \mathbf{r}')\phi(\mathbf{r}')\,\mathrm{d}S_{\mathbf{r}'} \\
&= \lim_{t\to 0^+} \int_{\partial\Omega} \underbrace{T(\mathbf{r} \pm t\mathbf{n}(\mathbf{r}), \mathbf{r}')(\phi(\mathbf{r}') - \phi(\mathbf{r}))}_{\eta_{\pm t,\mathbf{r}}(\mathbf{r}')}\,\mathrm{d}S_{\mathbf{r}'} + \phi(\mathbf{r}) \underbrace{\lim_{t\to 0^+} \int_{\partial\Omega} T(\mathbf{r} \pm t\mathbf{n}(\mathbf{r}), \mathbf{r}')\,\mathrm{d}S_{\mathbf{r}'}}_{=0 \text{ or } 1}.
\end{aligned}$$

From Lemma 8.1.2, $\int_{\partial\Omega} T(\mathbf{r} + t\mathbf{n}(\mathbf{r}), \mathbf{n}_{\mathbf{r}'})\,\mathrm{d}S_{\mathbf{r}'} = 0$ and $\int_{\partial\Omega} T(\mathbf{r} - t\mathbf{n}(\mathbf{r}), \mathbf{n}_{\mathbf{r}'})\,\mathrm{d}S_{\mathbf{r}'} = 1$ and, therefore,

$$\begin{aligned}
\lim_{t\to 0^+} \mathcal{D}_\Omega\phi(\mathbf{r} - t\mathbf{n}(\mathbf{r})) &= \lim_{t\to 0^+} \int_{\partial\Omega} \eta_{\pm t,\mathbf{r}}(\mathbf{r}')\,\mathrm{d}S_{\mathbf{r}'}, \quad \forall\, \mathbf{r} \in \partial\Omega, \\
\lim_{t\to 0^+} \mathcal{D}_\Omega\phi(\mathbf{r} - t\mathbf{n}(\mathbf{r})) &= \lim_{t\to 0^+} \int_{\partial\Omega} \eta_{\pm t,\mathbf{r}}(\mathbf{r}')\,\mathrm{d}S_{\mathbf{r}'} + \phi(\mathbf{r}), \quad \forall\, \mathbf{r} \in \partial\Omega.
\end{aligned} \tag{8.11}$$

Since $\phi \in C^1(\partial\Omega)$ (this property is not necessary for the proof of (8.7), and see Remark 8.1.3), $\phi(\mathbf{r}) - \phi(\mathbf{r}') = O(|\mathbf{r} - \mathbf{r}'|)$ and

$$\sup_{|t|<1} |\eta_{\pm t,\mathbf{r}}(\mathbf{r}')| = \sup_{0<t<1} \left|T(\mathbf{r} \pm t\mathbf{n}(\mathbf{r}), \mathbf{r}')(\phi(\mathbf{r}') - \phi(\mathbf{r}))\right| = O\left(\frac{1}{|\mathbf{r} - \mathbf{r}'|}\right). \tag{8.12}$$

Since $\sup_{|t|<1} |\eta_{t,\mathbf{r}}(\mathbf{r}')|$ as a function of $\mathbf{r}'$ is integrable over $\partial\Omega$, it follows from the Lebesgue dominated convergence theorem and Lemma 8.1.2 that

$$\lim_{t\to 0^+} \int_{\partial\Omega} \eta_{t,\mathbf{r}}(\mathbf{r}')\,\mathrm{d}S_{\mathbf{r}'} = \int_{\partial\Omega} T(\mathbf{r}, \mathbf{r}')(\phi(\mathbf{r}') - \phi(\mathbf{r}))\,\mathrm{d}S_{\mathbf{r}'} = \mathcal{K}_\Omega\phi(\mathbf{r}) - \tfrac{1}{2}\phi(\mathbf{r}). \tag{8.13}$$

From (8.13) and (8.11), we obtain

$$\lim_{t\to 0^+} \mathcal{D}_\Omega \phi(\mathbf{r} \pm t\mathbf{n}(\mathbf{r})) = (\mp\tfrac{1}{2}I + \mathcal{K}_\Omega)\phi(\mathbf{r}).$$

Remark 8.1.3 *The limit (8.13) via the Lebesgue dominated convergence theorem (LDCT) holds true even for the Lipschitz domain Ω and $\phi \in L^2(\partial\Omega)$. To apply the LDCT, we use some estimate using the Hardy–Littlewood maximal function to replace the estimate of (8.12). To understand it simply, we assume that $\Omega = \{\mathbf{r} = (\mathbf{x}, s) \in \mathbb{R}^2 \times \mathbb{R} : s > \eta(\mathbf{x})\}$ is the domain above the graph $\eta \in C^1(\mathbb{R}^2)$. For $\phi \in C(\partial\Omega)$, there is a positive constant C depending only on $\|\eta\|_{C_0^1(\mathbb{R}^2)}$ so that*

$$|\mathcal{D}_\Omega \psi_\mathbf{r}(\mathbf{r}_s) - \mathcal{K}_\Omega \psi_\mathbf{r}(\mathbf{r})| \le C \underbrace{\sup_{t>0} \frac{1}{|\partial\Omega \cap B_t(\mathbf{r})|} \int_{\partial\Omega\cap B_t(\mathbf{r})} |\psi_\mathbf{r}(\mathbf{r}')|\, \mathrm{d}S_{\mathbf{r}'}}_{\text{Hardy–Littlewood maximal function on } \partial\Omega}$$

where $\psi_\mathbf{r}(\mathbf{r}') = \phi(\mathbf{r}') - \phi(\mathbf{r})$ and $\mathbf{r}_s = \mathbf{r} + s\hat{\mathbf{z}}$ for $\mathbf{r} \in \partial\Omega$.

Theorem 8.1.4 *The following operators are all invertible:*

$$\tfrac{1}{2}I + \mathcal{K},\ \tfrac{1}{2}I + \mathcal{K}^* : L^2(\partial\Omega) \to L^2(\partial\Omega), \tag{8.14}$$

$$\tfrac{1}{2}I - \mathcal{K},\ \tfrac{1}{2}I - \mathcal{K}^* : L^2_\diamond(\partial\Omega) \to L^2_\diamond(\partial\Omega), \tag{8.15}$$

$$\mathcal{S} : L^2(\partial\Omega) \to W^{1,2}(\partial\Omega), \tag{8.16}$$

$$\lambda I + \mathcal{K}^* : L^2(\partial\Omega) \to L^2(\partial\Omega) \quad \textit{for } |\lambda| > \tfrac{1}{2} \textit{ and real } \lambda. \tag{8.17}$$

Proof. We only prove that $\frac{1}{2}I + \mathcal{K} : L^2(\partial\Omega) \to L^2(\partial\Omega)$ is invertible. Recall the definition of the null space N and the range R of an operator:

$$N(\tfrac{1}{2}I + \mathcal{K}) := \{\psi \in L^2(\partial\Omega) : (\tfrac{1}{2}I + \mathcal{K})\psi = 0\},$$

$$R(\tfrac{1}{2}I + \mathcal{K}) := \{(\tfrac{1}{2}I + \mathcal{K})\psi : \psi \in L^2(\partial\Omega)\}.$$

It is well known in functional analysis (see Rudin 1973, theorem 4.12) that

$$[R(\tfrac{1}{2}I + \mathcal{K})]^\perp = N(\tfrac{1}{2}I + \mathcal{K}^*). \tag{8.18}$$

We will prove that $(\frac{1}{2}I + \mathcal{K}^*)\phi = 0$ implies $\phi = 0$. From (8.9),

$$\frac{\partial}{\partial\mathbf{n}} \mathcal{S}|_{\partial\Omega^\pm} = (\pm\tfrac{1}{2}I + \mathcal{K}^*)\phi,$$

where $f|_{\partial\Omega^+}$ and $f|_{\partial\Omega^-}$ denote the limits of a function f from outside and inside of Ω, respectively. If $(\frac{1}{2}I + \mathcal{K}^*)\phi = 0$, then $\mathcal{S}\phi$ satisfies

$$\begin{cases} \nabla^2 \mathcal{S}\phi = 0 & \text{in } \mathbb{R}^3 \setminus \bar{\Omega}, \\ \dfrac{\partial}{\partial\mathbf{n}} \mathcal{S}|_{\partial\Omega^+} = (\tfrac{1}{2}I + \mathcal{K}^*)\phi = 0 & \text{on } \partial\Omega. \end{cases}$$

From the uniqueness (up to a constant) of the exterior Neumann boundary value problem, $\nabla\mathcal{S}\phi = 0$ in $\mathbb{R}^3 \setminus \Omega$ and $\mathcal{S}\phi \equiv 0$ in $\mathbb{R}^3 \setminus \Omega$ because $\mathcal{S}\phi(\mathbf{r}) = 0$ for $|\mathbf{r}| = \infty$. Hence $\mathcal{S}\phi$ satisfies the interior Dirichlet boundary value problem

$$\nabla^2\mathcal{S}\phi = 0 \text{ in } \Omega, \quad \mathcal{S}\phi|_{\partial\Omega} = 0.$$

From the maximum principle, $\mathcal{S}\phi = 0$ in Ω and, therefore, $\mathcal{S}\phi = 0$ in $\mathbb{R}^3$. As a result, we have

$$(\pm\tfrac{1}{2}I + \mathcal{K}^*)\phi = \frac{\partial}{\partial\mathbf{n}}\mathcal{S}\phi|_{\partial\Omega^\pm} = 0,$$

and hence,

$$\phi = \underbrace{(\tfrac{1}{2}I + \mathcal{K}^*)\phi}_{=0} - \underbrace{(-\tfrac{1}{2}I + \mathcal{K}^*)\phi}_{=0} = 0.$$

From (8.18), $[R(\frac{1}{2}I + \mathcal{K})]^\perp = \{0\}$ and, therefore, the closure of $R(\frac{1}{2}I + \mathcal{K})$ is $L^2(\partial\Omega)$.

Now we will prove $R(\frac{1}{2}I + \mathcal{K}) = L^2(\partial\Omega)$ by showing that the range of $\frac{1}{2}I + \mathcal{K}$ is closed. According to the Banach closed range theorem (Rudin 1973, p. 96), it suffices to show that $\frac{1}{2}I + \mathcal{K}^*$ has a closed range. We need to prove that, if $g \in L^2(\partial\Omega)$ is the limit of $(\frac{1}{2}I + \mathcal{K}^*)\phi_n$, then there exists $\phi \in L^2(\partial\Omega)$ such that $(\frac{1}{2}I + \mathcal{K}^*)\phi = g$.

Case (i). Consider the case where $\sup_n \|\phi_n\| < \infty$. Owing to the weak compactness (Evans 2010), there exists a subsequence $\{\phi_{n_k}\} \to \phi$ weakly in $L^2(\partial\Omega)$. Then, we have

$$\begin{aligned}\int_{\partial\Omega} g\eta \, dS &= \lim_{k\to\infty}\int_{\partial\Omega}(\tfrac{1}{2}I + \mathcal{K}^*)\phi_{n_k}\eta \, dS \\ &= \lim_{k\to\infty}\int_{\partial\Omega}\phi_{n_k}(\tfrac{1}{2}I + \mathcal{K})\eta \, dS = \textstyle\int_{\partial\Omega}\phi(\tfrac{1}{2}I + \mathcal{K})\eta \, dS \\ &= \int_{\partial\Omega}(\tfrac{1}{2}I + \mathcal{K}^*)\phi\eta \, dS \quad \text{for all } \eta \in L^2(\partial\Omega).\end{aligned}$$

Hence, $g = (\frac{1}{2}I + \mathcal{K}^*)\phi$.

Case (ii). It remains to prove the case where $\sup_n \|\phi_n\| = \infty$. We may assume that $\|\phi_n\| \nearrow \infty$. Denoting $\tilde{\phi}_n = \phi_n/\|\phi_n\|$, we have

$$\lim_{n\to\infty}(\tfrac{1}{2}I + \mathcal{K}^*)\tilde{\phi}_n = \lim_{n\to\infty}\underbrace{\frac{1}{\|\phi_n\|}}_{\to 0}\underbrace{(\tfrac{1}{2}I + \mathcal{K}^*)\phi_n}_{\to g} = 0. \tag{8.19}$$

Since $\|\tilde{\phi}_n\| = 1$, there exists a subsequence $\{\tilde{\phi}_{n_k}\}$ such that $\tilde{\phi}_n \to \tilde{\phi}$ weakly in $L^2(\partial\Omega)$.

We repeat the previous argument to get

$$0 = \lim_{k\to\infty}\int_{\partial\Omega}(\tfrac{1}{2}I + \mathcal{K}^*)\tilde{\phi}_{n_k}\eta \, dS = \textstyle\int_{\partial\Omega}\tilde{\phi}\eta \, dS, \quad \forall\, \eta \in L^2(\Omega).$$

Hence $\tilde{\phi} = 0$ and $\tilde{\phi}_n \to 0$ weakly on $\partial\Omega$. Verchota (1984) obtained the following remarkable contradiction, which indicates that case (ii) is not possible.

$$\begin{aligned} 1 = \|\tilde{\phi}_n\| &= \|(\tfrac{1}{2}I + \mathcal{K}^*)\tilde{\phi}_n - (-\tfrac{1}{2}I + \mathcal{K}^*)\tilde{\phi}_n\| \\ &\leq \underbrace{\|(\tfrac{1}{2}I + \mathcal{K}^*)\tilde{\phi}_n\|}_{\to 0} \quad + \quad \underbrace{\|(-\tfrac{1}{2}I + \mathcal{K}^*)\tilde{\phi}_n\|}_{\leq C[\|(\frac{1}{2}I+\mathcal{K}^*)\tilde{\phi}_n\| + |\int_{\partial\Omega} \mathcal{S}\tilde{\phi}_n|]} \\ &\leq (1+C)\underbrace{\|(\tfrac{1}{2}I + \mathcal{K}^*)\tilde{\phi}_n\|}_{\to 0} + C\underbrace{\left|\int_{\partial\Omega} \mathcal{S}\tilde{\phi}_n\right|}_{\to 0} \to 0. \end{aligned}$$

Here, we have used the estimate in Lemma 8.15 (see below), (8.19) and the fact that, if $\tilde{\phi}_n \to 0$ weakly on $\partial\Omega$, then $\int_{\partial\Omega} \mathcal{S}\tilde{\phi}_n \, \mathrm{d}S \to 0$.

From (i) and (ii), we obtain $R(\frac{1}{2}I + \mathcal{K}) = L^2(\partial\Omega)$. Hence, to prove its invertibility, it remains to prove that $(\frac{1}{2}I + \mathcal{K}) : L^2(\partial\Omega) \to L^2(\partial\Omega)$ is one-to-one. It is sufficient to prove that $(\frac{1}{2}I + \mathcal{K}^*)$ has a dense range on $L^2(\partial\Omega)$. For the Lipschitz domain Ω, we may use a sequence $\{\Omega_j\}$ of C^2 domains such that $\Omega_j \searrow \Omega$. We know that for C^2 boundary $\partial\Omega_j$, $\mathcal{K}^*_{\partial\Omega_j}$ is a compact operator and the Fredholm alternative theorem can be applied to get $R(\frac{1}{2}I + \mathcal{K}^*_{\partial\Omega_j}) = L^2(\partial\Omega_j)$. With the aid of this property, we can obtain that the closure of $R(\frac{1}{2}I + \mathcal{K}^*_{\partial\Omega})$ is $L^2(\partial\Omega)$. (Please refer to Verchota (1984) for the details of the proof.) This completes the proof of the invertibility of $(\frac{1}{2}I + \mathcal{K}) : L^2(\partial\Omega) \to L^2(\partial\Omega)$. Hence, we also get the invertibility of $(\frac{1}{2}I + \mathcal{K}^*) : L^2(\partial\Omega) \to L^2(\partial\Omega)$. To prove the invertibility of $(\frac{1}{2}I - \mathcal{K}) : L^2_\diamond(\partial\Omega) \to L^2_\diamond(\partial\Omega)$, we may repeat similar arguments. □

Lemma 8.1.5 (Verchota 1984) *Let Ω be a bounded Lipschitz domain in $\mathbb{R}^3$ with a connected boundary. Then, for all $\phi \in L^2(\partial\Omega)$, we have*

$$\|(\tfrac{1}{2}I - \mathcal{K}^*)\phi\|_2 \leq C\left\{\|(\tfrac{1}{2}I + \mathcal{K}^*)\phi\|_2 + \left|\int_{\partial\Omega} \mathcal{S}\phi \, \mathrm{d}S\right|\right\}, \tag{8.20}$$

$$\|(\tfrac{1}{2}I + \mathcal{K}^*)\phi\|_2 \leq C\left\{\|(\tfrac{1}{2}I - \mathcal{K}^*)\phi\|_2 + \left|\int_{\partial\Omega} \mathcal{S}\phi \, \mathrm{d}S\right|\right\}, \tag{8.21}$$

where C depends only on the Lipschitz constant for Ω.

Proof. Let $u = \mathcal{S}\phi$ and denote

$$u(\mathbf{r}) = \mathcal{S}\phi(\mathbf{r}) = \begin{cases} u^+(\mathbf{r}) & \text{if } \mathbf{r} \in \Omega^+ = \Omega, \\ u^-(\mathbf{r}) & \text{if } \mathbf{r} \in \Omega^- = \mathbb{R}^3 \setminus \Omega. \end{cases} \tag{8.22}$$

We decompose ∇u on $\partial\Omega$ into

$$\nabla u = \underbrace{\langle \nabla u, \mathbf{n}\rangle \mathbf{n}}_{\nabla_{\mathbf{n}} u} + \underbrace{(\nabla u - \langle \nabla u, \mathbf{n}\rangle \mathbf{n})}_{\nabla_T u}.$$

The proof is based on the following Rellich identity.

(Rellich identity) Let $\vec{\beta} \in [C_0^\infty(\mathbb{R}^3)]^3$ be a vector field satisfying $\langle \vec{\beta}, \mathbf{n}(\mathbf{r}) \rangle \geq c > 0$ on $\partial\Omega$. Then, we have the following identity:

$$\int_{\partial\Omega} \langle \vec{\beta}, \mathbf{n} \rangle \left(|\nabla_T u^\pm|^2 - |\nabla_\mathbf{n} u^\pm|^2 \right) \mathrm{d}S = 2 \int_{\partial\Omega} \langle \vec{\beta}, \nabla_T u^\pm \rangle \frac{\partial u^\pm}{\partial \mathbf{n}} \mathrm{d}S + \Upsilon_\pm(u, \vec{\beta}), \tag{8.23}$$

where $\Upsilon_\pm(u, \vec{\beta})$ is

$$\begin{aligned} \Upsilon_\pm(u, \vec{\beta}) &:= \int_{\Omega^\pm} [\nabla \cdot \vec{\beta} |\nabla u|^2 - 2\langle \nabla \vec{\beta} \nabla u, \nabla u \rangle] \, \mathrm{d}\mathbf{r} \\ &= O(\|\nabla \vec{\beta}\|_{L^\infty(\Omega)} \|\nabla u\|^2_{L^2(\Omega^\pm)}). \end{aligned}$$

The identity (8.23) follows from

$$\int_{\partial\Omega} \langle \vec{\beta}, \mathbf{n} \rangle |\nabla u^\pm|^2 \, \mathrm{d}S = 2 \int_{\partial\Omega} \langle \vec{\beta}, \nabla u \rangle \frac{\partial u^\pm}{\partial \mathbf{n}} \mathrm{d}S + \Upsilon(u, \beta), \tag{8.24}$$

which can be obtained by applying the Gauss divergence to the following identity:

$$\begin{aligned} \nabla \cdot (\vec{\beta} |\nabla u|^2) &= \nabla \cdot \vec{\beta} |\nabla u|^2 + 2\langle \vec{\beta}, (\nabla \nabla u) \nabla u \rangle = \nabla \cdot \vec{\beta} |\nabla u|^2 + 2\langle (\nabla \nabla u) \vec{\beta}, \nabla u \rangle \\ &= \nabla \cdot \vec{\beta} |\nabla u|^2 + 2\langle \nabla(\vec{\beta} \cdot \nabla u), \nabla u \rangle - 2\langle \nabla \vec{\beta} \nabla u, \nabla u \rangle. \end{aligned}$$

The Rellich identity (8.23) provides the following estimates:

$$\int_{\partial\Omega} (|\nabla_T u^\pm|^2 - |\nabla_\mathbf{n} u^\pm|^2) \, \mathrm{d}S \leq C(\|\nabla_T u^\pm\|_{L^2(\partial\Omega)} \|\nabla_\mathbf{n} u^\pm\|_{L^2(\partial\Omega)} + \|\nabla u\|^2_{L^2(\mathbb{R}^3)}),$$

$$\int_{\partial\Omega} (-|\nabla_T u^\pm|^2 + |\nabla_\mathbf{n} u^\pm|^2) \, \mathrm{d}S \leq C(\|\nabla_T u^\pm\|_{L^2(\partial\Omega)} \|\nabla_\mathbf{n} u^\pm\|_{L^2(\partial\Omega)} + \|\nabla u\|^2_{L^2(\mathbb{R}^3)}),$$

where the constant C depends on $\|\nabla \vec{\beta}\|_{L^\infty}$ and $\langle \vec{\beta}, \mathbf{n}(\mathbf{r}) \rangle \geq c > 0$. Since $\|\nabla_T u^+\|_{L^2(\partial\Omega)} = \|\nabla_T u^-\|_{L^2(\partial\Omega)}$, we obtain

$$\|\nabla_\mathbf{n} u^\pm\|^2_{L^2(\partial\Omega)} \leq C(\|\nabla_T u\|^2_{L^2(\partial\Omega)} + \|\nabla u\|^2_{L^2(\mathbb{R}^3)}), \tag{8.25}$$

$$\|\nabla_T u\|^2_{L^2(\partial\Omega)} \leq C(\|\nabla_\mathbf{n} u^\pm\|^2_{L^2(\partial\Omega)} + \|\nabla u\|^2_{L^2(\mathbb{R}^3)}), \tag{8.26}$$

$$\|\nabla_\mathbf{n} u^\pm\|^2_{L^2(\partial\Omega)} \leq C(\|\nabla_\mathbf{n} u^\mp\|^2_{L^2(\partial\Omega)} + \|\nabla u\|^2_{L^2(\mathbb{R}^3)}), \tag{8.27}$$

where the constant C is independent of u and ϕ. Setting

$$a = \frac{1}{|\partial\Omega|} \int_{\partial\Omega} u \, \mathrm{d}S = \frac{1}{|\partial\Omega|} \int_{\partial\Omega} \mathcal{S}\phi \, \mathrm{d}S$$

(the average), the term $\|\nabla u\|^2_{L^2(\mathbb{R}^3)}$ can be estimated by

$$\begin{aligned} \int_{\Omega^\pm} |\nabla u|^2 \, \mathrm{d}\mathbf{r} &= \int_{\partial\Omega} u^\pm \frac{\partial u^\pm}{\partial \mathbf{n}} \mathrm{d}S = \int_{\partial\Omega} (u^\pm - a) \frac{\partial u^\pm}{\partial \mathbf{n}} \mathrm{d}S + a \int_{\partial\Omega} \frac{\partial u^\pm}{\partial \mathbf{n}} \mathrm{d}S \\ &\leq C(\|\nabla_T u\|_{L^2(\partial\Omega)} \|\nabla_\mathbf{n} u^\pm\|_{L^2(\partial\Omega)} + |a| \|\nabla_\mathbf{n} u^\pm\|_{L^2(\partial\Omega)}), \end{aligned} \tag{8.28}$$

where the constant C is independent of u and ϕ. Here, we have used the Poincaré inequality

$$\int_{\partial\Omega} |u^{\pm} - a|^2 \, dS \leq C \|\nabla_T u\|^2_{L^2(\partial\Omega)}.$$

Combining all the estimates (8.25)–(8.28) with the Hölder inequality, we obtain

$$\|\nabla_{\mathbf{n}} u^{\pm}\|^2_{L^2(\partial\Omega)} \leq C(\|\nabla_{\mathbf{n}} u^{\mp}\|^2_{L^2(\partial\Omega)} + |a|), \tag{8.29}$$

where the constant C is independent of u and ϕ. This completes the proof. □

We briefly summarize the layer potential method for the boundary value problems (BVP) of the Laplace equation on the Lipschitz domain Ω (Fabes *et al.* 1978). With the aid of the layer potential method, solving the Neumann BVP can be converted into the invertibility of $-\frac{1}{2}I + \mathcal{K}^* : L^2_\diamond(\partial\Omega) \to L^2_\diamond(\partial\Omega)$. Since it is invertible, the potential

$$u(\mathbf{r}) = \mathcal{S}(-\tfrac{1}{2}I + \mathcal{K}^*)^{-1} g(\mathbf{r}) \quad (\forall\, \mathbf{r} \in \Omega)$$

satisfies

$$\nabla^2 u = 0 \text{ in } \Omega, \quad \left.\frac{\partial u}{\partial u}\right|_{\partial\Omega} = g.$$

8.1.2 *Regularity for Solution of Elliptic Equation along Boundary of Inhomogeneity*

In this section, we study the regularity of solutions of elliptic equations on a Lipschitz interface based on the work by Escauriaza and Seo (1993). Throughout this section, let $\Omega = B = B_1$ be the unit ball and let D be a Lipschitz domain with a connected boundary contained in $B_{1/2}$ where B_r denotes the ball with radius r and center at the origin. We consider a weak solution $u \in H^1(\Omega)$ for the elliptic equation

$$Lu = \nabla \cdot ((A_0 + (A_D - A_0)\chi_D)\nabla u) = 0 \quad \text{in } \Omega, \tag{8.30}$$

where A_0 and A_D are positive constant matrices.

Denote by $F_0(\mathbf{r})$ and $F_D(\mathbf{r})$, respectively, the fundamental solutions of the constant-coefficient elliptic operators

$$L_1 = \nabla \cdot (A_0 \nabla) \quad \text{and} \quad L_2 = \nabla \cdot (A_D \nabla).$$

We define the corresponding single-layer potentials $\mathcal{S}$ and $\widetilde{\mathcal{S}}$ as

$$\mathcal{S} f(\mathbf{r}) = \int_{\partial D} F_0(\mathbf{r} - \mathbf{r}') f(\mathbf{r}') \, dS_{\mathbf{r}'},$$

$$\widetilde{\mathcal{S}} g(\mathbf{r}) = \int_{\partial D} F_D(\mathbf{r} - \mathbf{r}') g(\mathbf{r}') \, dS_{\mathbf{r}'}.$$

The corresponding trace operators are defined as, for $\mathbf{r} \in \partial D$,

$$\mathcal{K}^* f(\mathbf{r}) = \int_{\partial D} \langle A_0 \nabla F_0(\mathbf{r} - \mathbf{r}'), \mathbf{n}(\mathbf{r}) \rangle f(\mathbf{r}') \, dS_{\mathbf{r}'},$$

$$\widetilde{\mathcal{K}}^* f(\mathbf{r}) = \int_{\partial D} \langle A_D \nabla F_D(\mathbf{r}-\mathbf{r}'), \mathbf{n}(\mathbf{r})\rangle g(\mathbf{r}')\,\mathrm{d}S_{\mathbf{r}'}.$$

For $\phi \in H^1(B)$, we define the interior and exterior non-tangential maximal functions of ϕ at $\mathbf{r} \in \partial D$, respectively, as

$$(\phi^+)^\natural(\mathbf{r}) = \sup\{|u(\mathbf{r}')| : |\mathbf{r}-\mathbf{r}'| < \tfrac{3}{2}\,\mathrm{dist}(\mathbf{r}', \partial D), \mathbf{r}' \in D\},$$

$$(\phi^-)^\natural(\mathbf{r}) = \sup\{|u(\mathbf{r}')| : |\mathbf{r}-\mathbf{r}'| < \tfrac{3}{2}\,\mathrm{dist}(\mathbf{r}', \partial D), \mathbf{r}' \in B_{2/3}\}.$$

We write

$$u^+ = u|_D \quad \text{and} \quad u^+ = u|_{\Omega\setminus\bar{D}}.$$

We have the following regularity result for a solution $u \in H^1(\Omega)$ of (8.30).

Theorem 8.1.6 (Escauriaza and Seo 1993) *Suppose $A_D - A_0$ is either a positive or negative definite matrix. If $u \in H^1(\Omega)$ is a weak solution to (8.30), then*

$$\|(\nabla u^\pm)^\natural\|_{L^2(\partial D)} \le C\|u\|_{H^1(\Omega)},$$

where the constant C depends only on the Lipschitz character of D, A_0, A_D and $A_D - A_0$.

Theorem 8.1.7 *Under the assumptions on Theorem 8.1.6, the mapping*

$$\mathcal{T} : L^2(\partial D) \times L^2(\partial D) \to H^1(\partial D) \times L^2(\partial D)$$

defined by

$$\mathcal{T}(f, g) = \big(\widetilde{\mathcal{S}}g - \mathcal{S}f, (-\tfrac{1}{2}I + \widetilde{\mathcal{K}}^*)g - (\tfrac{1}{2}I + \mathcal{K}^*)f\big)$$

is invertible.

We will prove Theorem 8.1.6 assuming that Theorem 8.1.7 has already been proved. Take $\phi \in C_0^\infty(\mathbb{R}^3)$ with $\phi = 1$ on $B_{3/5}$, and $\phi = 0$ outside $B_{4/5}$. Then, $u\phi$ satisfies

$$L_1(u\phi) = 0 \quad \text{in } (B_{3/5} \setminus D) \cup (\mathbb{R}^3 \setminus B_{4/5}).$$

Introducing the Newtonian potential

$$v(\mathbf{r}) := \int_{\Omega\setminus D} F_D(\mathbf{r}-\mathbf{r}')L_1(u\phi)(\mathbf{r}')\,\mathrm{d}\mathbf{r},$$

we have

$$L_1(u\phi - v) = 0 \quad \text{in } \Omega \setminus D.$$

Since $L_1(u\phi) \in C_0^\infty(B_{4/5} \setminus B_{3/5})$, it is easy to see that $\|v\|_{H^1(\partial D)} \le C\|u\|_{H^1(\Omega)}$. Let

$$\phi_1 = v|_{\partial D} \quad \text{and} \quad \phi_2 = \langle A_0\mathbf{n}, \nabla v\rangle|_{\partial D}.$$

According to Theorem 8.1.7,

$$\exists (f, g) \in L^2(\partial D) \times L^2(\partial D) \text{ s.t. } \mathcal{T}(f, g) = (-\phi_1, -\phi_2). \tag{8.31}$$

Define

$$w(\mathbf{r}) := \begin{cases} u(\mathbf{r})\phi(\mathbf{r}) - v(\mathbf{r}) + \mathcal{S}f(\mathbf{r}) & \text{for } \mathbf{r} \in \mathbb{R}^3 \setminus D, \\ u(\mathbf{r}) + \widetilde{\mathcal{S}}g(\mathbf{r}) & \text{for } \mathbf{r} \in D. \end{cases}$$

From (8.31), w satisfies

$$Lw = 0 \quad \text{in the entire domain } \mathbb{R}^3,$$

and we must have $w = 0$ in $\mathbb{R}^3$ from the maximum principle, since $w(\mathbf{r}) = O(|\mathbf{r}|^{-1})$ at infinity. Hence, we have the following representation formula:

$$u(\mathbf{r}) = \begin{cases} v(\mathbf{r}) - \mathcal{S}f(\mathbf{r}) & \text{for } \mathbf{r} \in B_{4/5} \setminus D, \\ -\widetilde{\mathcal{S}}g(\mathbf{r}) & \text{for } \mathbf{r} \in D. \end{cases}$$

The theorem follows from the property of the singular integral on the Lipschitz domain.

The main step in the proof of Theorem 8.1.7 is the following estimate.

Lemma 8.1.8 *Under the assumption of Theorem 8.1.7, there exists a positive constant C depending only on the Lipschitz character of D, A_0, A_D and $A_D - A_0$ such that, for $(f, g) \in L^2(\partial D) \times L^2(\partial D)$,*

$$\|f\|_{L^2(\partial D)} + \|g\|_{L^2(\partial D)} \leq C \left\{ \|\mathcal{T}(f, g)\|_{H^1(\partial D) \times L^2(\partial D)} + \left| \int_{\partial D} \mathcal{S}f \right| + \left| \int_{\partial D} \widetilde{\mathcal{S}}g \right| \right\}.$$

Proof. After a linear change of coordinates, we may assume that $A_0 = I$ and $A_D = A$, where I denotes the identity matrix. From the assumptions, either $A - I \geq \alpha I$ or $I - A \geq \alpha I$ for some $\alpha > 0$. We only consider the case where $A - I \geq \alpha I$ for $\alpha > 0$. Denote

$$u := \begin{cases} u^+ := \widetilde{\mathcal{S}}g & \text{in } D, \\ u^- := \mathcal{S}f & \text{in } \mathbb{R}^3 \setminus \bar{D} \end{cases}$$

and

$$(\phi_1, \phi_2) := \mathcal{T}(f, g) \quad \text{on } \partial D. \tag{8.32}$$

Then, u satisfies

$$\int_D \langle A\nabla u, \nabla \eta \rangle \, \mathrm{d}\mathbf{r} + \int_{\mathbb{R}^3 \setminus \bar{D}} \langle \nabla u, \nabla \eta \rangle \, \mathrm{d}\mathbf{r} = \int_{\partial D} \phi_2 \eta \, \mathrm{d}S \quad (\forall \eta \in C_0^\infty(\mathbb{R}^3)).$$

There exists C depending on the Lipschitz character of ∂D, A and α such that

$$\begin{aligned} &\|\nabla u^+\|_{L^2(\partial D)} + \|\nabla u^-\|_{L^2(\partial D)} \\ &\quad \leq C\{\|\phi_1\|_{H^1(\partial D)} + \|\phi_2\|_{L^2(\partial D)} + \|\nabla u^+\|_{L^2(D)} + \|\nabla u^-\|_{L^2(\Omega \setminus D)}\}. \end{aligned} \tag{8.33}$$

To prove (8.33), we will derive Rellich-type identities as in the proof of Lemma 8.1.5. We choose a vector field $\vec{\beta} \in C_0^\infty(B_{3/4})$ with $\langle \vec{\beta}, \mathbf{n}\rangle \geq C_1$ on ∂D, where C_1 depends only on the Lipschitz character of D. Then,

$$\nabla \cdot (\vec{\beta}\langle A\nabla u^+, \nabla u^+\rangle) = 2\nabla \cdot (\langle \vec{\beta}, \nabla u^+\rangle A\nabla u^+) + O(|\nabla u^+|^2) \quad \text{on } D, \tag{8.34}$$

$$\nabla \cdot (\vec{\beta}\langle \nabla u^-, \nabla u^-\rangle) = 2\nabla \cdot (\langle \vec{\beta}, \nabla u^-\rangle \nabla u^-) + O(|\nabla u^-|^2 \chi_{B\setminus D}) \quad \text{on } \mathbb{R}^3 \setminus D. \tag{8.35}$$

Integrating (8.34) over D and applying the divergence theorem, we get

$$\int_{\partial D} \langle \vec{\beta}, \mathbf{n}\rangle \langle A\nabla u^+, \nabla u^+\rangle = 2\int_{\partial D} \langle \vec{\beta}, \nabla u^+\rangle \langle A\nabla u^+, \mathbf{n}\rangle + O(\|\nabla u^+\|_{L^2(D)}^2). \tag{8.36}$$

Using the orthonormal basis $\{\mathbf{n}, \mathbf{t}_1, \mathbf{t}_2\}$, we can decompose

$$\langle A\nabla u^+, \nabla u^+\rangle = \sum_{l=1}^{2} \langle A\nabla u^+, \mathbf{t}_l\rangle \langle \nabla u^+, \mathbf{t}_l\rangle + \langle A\nabla u^+, \mathbf{n}\rangle \langle \nabla u^+, \mathbf{n}\rangle$$

and

$$\langle \vec{\beta}, \nabla u^+\rangle = \langle \vec{\beta}, \mathbf{n}\rangle \langle \nabla u^+, \mathbf{n}\rangle + \sum_{l=1}^{2} \langle \vec{\beta}, \mathbf{t}_l\rangle \langle \nabla u^+, \mathbf{t}_l\rangle.$$

Applying the above identities to (8.36), we obtain

$$\begin{aligned}
&\int_{\partial D} \langle \vec{\beta}, \mathbf{n}\rangle \left\{ \sum_{l=1}^{2} \langle A\nabla u^+, \mathbf{t}_l\rangle \langle \nabla u^+, \mathbf{t}_l\rangle - \langle A\nabla u^+, \mathbf{n}\rangle \langle \nabla u^+, \mathbf{n}\rangle \right\} \mathrm{d}S \\
&\quad = 2\int_{\partial D} \sum_{l=1}^{2} \langle \vec{\beta}, \mathbf{t}_l\rangle \langle \nabla u^+, \mathbf{t}_l\rangle \langle A\nabla u^+, \mathbf{n}\rangle \,\mathrm{d}S + O(\|\nabla u^+\|_{L^2(D)}^2).
\end{aligned} \tag{8.37}$$

Similarly, (8.35) for u^- leads to

$$\begin{aligned}
&\int_{\partial D} \langle \vec{\beta}, \mathbf{n}\rangle \left\{ \sum_{l=1}^{2} |\langle \nabla u^-, \mathbf{t}_l\rangle|^2 - |\langle \nabla u^-, \mathbf{n}\rangle|^2 \right\} \mathrm{d}S \\
&\quad = 2\int_{\partial D} \sum_{l=1}^{2} \langle \vec{\beta}, \mathbf{t}_l\rangle \langle \nabla u^-, \mathbf{t}_l\rangle \langle \nabla u^-, \mathbf{n}\rangle \,\mathrm{d}S + O(\|\nabla u^-\|_{L^2(\mathbb{R}^3\setminus D)}^2).
\end{aligned} \tag{8.38}$$

We can rewrite the last equality as

$$\begin{aligned}
&\int_{\partial D} \langle \vec{\beta}, \mathbf{n}\rangle \Bigg\{ \sum_{l=1}^{2} [\langle \nabla u^+, \mathbf{t}_l\rangle + \underbrace{(\langle \nabla u^-, \mathbf{t}_l\rangle - \langle \nabla u^+, \mathbf{t}_l\rangle)}_{=-\mathbf{t}_l\cdot\nabla\phi_1}]^2 \\
&\qquad - [\langle A\nabla u^+, \mathbf{n}\rangle + \underbrace{(\langle \nabla u^-, \mathbf{n}\rangle - \langle A\nabla u^+, \mathbf{n}\rangle)}_{-\phi_2}]^2 \Bigg\} \mathrm{d}S
\end{aligned}$$

$$= 2\int_{\partial D}\sum_{l=1}^{2}\langle\vec{\beta},\mathbf{t}_l\rangle[\langle\nabla u^+,\mathbf{t}_l\rangle + \underbrace{(\langle\nabla u^-,\mathbf{t}_l\rangle - \langle\nabla u^+,\mathbf{t}_l\rangle)}_{=-\mathbf{t}_l\cdot\nabla\phi_1}]$$

$$\times[\langle A\nabla u^+,\mathbf{n}\rangle + \underbrace{(\langle\nabla u^-,\mathbf{n}\rangle - \langle A\nabla u^+,\mathbf{n}\rangle)}_{-\phi_2}]\,\mathrm{d}S$$

$$+ O(\|\nabla u^-\|^2_{L^2(B\setminus D)}).$$

From the transmission conditions (8.32), we can rewrite the last equality in terms of the gradient of u^+, ϕ_1 and ϕ_2 on ∂D, obtaining

$$\int_{\partial D}\langle\vec{\beta},\mathbf{n}\rangle\left\{\sum_{l=1}^{2}|\langle\nabla u^+,\mathbf{t}_l\rangle|^2 - |\langle A\nabla u^+,\mathbf{n}\rangle|^2\right\}\mathrm{d}S$$

$$= 2\int_{\partial D}\sum_{l=1}^{2}\langle\vec{\beta},\mathbf{t}_l\rangle\langle\nabla u^+,\mathbf{t}_l\rangle\langle A\nabla u^+,\mathbf{n}\rangle\,\mathrm{d}S + O(\|\nabla u^-\|_{L^2(B\setminus D)}) \tag{8.39}$$

$$+ O(\{\|\phi_1\|^2_{H^1(\partial D)} + \|\phi_2\|^2_{L^2(\partial D)}\}) + \|\nabla u^+\|_{L^2(\partial D)}\{\|\phi_1\|_{H^1(\partial D)} + \|\phi_2\|_{L^2(\partial D)}\}.$$

Subtracting (8.39) from (8.37), we have

$$\int_{\partial D}\langle\vec{\beta},\mathbf{n}\rangle\left\{\sum_{l=1}^{2}\langle(A-I)\nabla u^+,\mathbf{t}_l\rangle\langle\nabla u^+,\mathbf{t}_l\rangle + \langle(A-I)\nabla u^+,\mathbf{n}\rangle\langle A\nabla u^+,\mathbf{n}\rangle\right\}\mathrm{d}S$$

$$= O(\|\nabla u^-\|^2_{L^2(B\setminus D)} + \|\nabla u^+\|^2_{L^2(D)}) + O(\{\|\phi_1\|^2_{H^1(\partial D)} + \|\phi_2\|^2_{L^2(\partial D)}\}) \tag{8.40}$$

$$+ \|\nabla u^+\|_{L^2(\partial D)}\{\|\phi_1\|_{H^1(\partial D)} + \|\phi_2\|_{L^2(\partial D)}\}.$$

From the orthonormality of the linear base $\{\mathbf{n},\mathbf{t}_1,\mathbf{t}_2\}$, we have

$$\sum_{l=1}^{2}\langle(A-I)\nabla u^+,\mathbf{t}_l\rangle\langle\nabla u^+,\mathbf{t}_l\rangle + \langle(A-I)\nabla u^+,\mathbf{n}\rangle\langle A\nabla u^+,\mathbf{n}\rangle \geq C|\nabla u^+|^2$$

for a positive constant C depending only on A and α. From (8.40) with the above estimate, we have

$$\int_{\partial D}|\nabla u^+|^2\,\mathrm{d}S \leq C(\|\phi_1\|^2_{H^1(\partial D)} + \|\phi_2\|^2_{L^2(\partial D)} + \|\nabla u^-\|^2_{L^2(B\setminus D)} + \|\nabla u^+\|^2_{L^2(D)}).$$

The above estimate together with a similar estimate for $\int_{\partial D}|\nabla u^-|^2\,\mathrm{d}S$ give the estimate (8.33). The proof of Lemma 8.1.8 follows from the estimate (8.33) and Lemma 8.1.5. □

Theorem 8.1.9 *Under the assumption of Theorem 8.1.7, there exists a positive constant C depending only on the Lipschitz character of D, A_D and A_0 such that, given $h \in C_0^\infty(\mathbb{R}^3)$, there exists a function u satisfying*

$$\nabla\cdot((A_D\chi_D + A_0\chi_{\mathbb{R}^3\setminus\bar{D}})\nabla u) = 0 \quad \textit{on } \mathbb{R}^3\setminus\partial D, \tag{8.41}$$

$$u^+ - u^- = 0 \quad \textit{on } \partial D, \tag{8.42}$$

$$\langle \mathbf{n}, A_D \nabla u^+ \rangle - \langle \mathbf{n}, A_0 \nabla u^- \rangle = h \quad \text{on } \partial D, \tag{8.43}$$

$$\|u\|_{L^6(\mathbb{R}^3)} + \|\nabla u\|_{L^2(\mathbb{R}^3)} \leq C\|h\|_{L^2(\partial D)}, \tag{8.44}$$

$$|u(\mathbf{r})| + |\mathbf{r}||\nabla u(\mathbf{r})| = O(1/|\mathbf{r}|) \quad \text{at } |\mathbf{r}| \text{ infinity.} \tag{8.45}$$

Moreover, u can be represented as

$$u = \begin{cases} \widetilde{\mathcal{S}} g & \text{in } D, \\ \mathcal{S} f & \text{in } \mathbb{R}^3 \setminus D, \end{cases} \qquad \text{for some } f, g \in L^2(\partial D).$$

Proof. With a linear change of the coordinate system, we may assume that $A_0 = I$. Denote $A_D = A$. From the Lax–Milgram theorem, for each $r > 1$, there is a unique $u_r \in H_0^1(B_r)$ such that

$$\int_{B_r} \langle (A\chi_D + I\chi_{\mathbb{R}^3\setminus\bar{D}})\nabla u_r, \nabla\phi\rangle \, \mathrm{d}\mathbf{r} = \int_{\partial D} \phi h \, \mathrm{d}S \quad (\forall \phi \in H_0^1(B_r)).$$

In particular,

$$\int_{B_r} \langle (A\chi_D + I\chi_{\mathbb{R}^3\setminus\bar{D}})\nabla u_r, \nabla u_r\rangle \, \mathrm{d}\mathbf{r} = \int_{\partial D} u_r h \, \mathrm{d}S.$$

Hence, we have

$$\int_{B_r} |\nabla u_r|^2 \, \mathrm{d}\mathbf{r} \lesssim \frac{1}{\epsilon} \int_{\partial D} |h|^2 \, \mathrm{d}S + \epsilon \int_{\partial D} |u_r|^2 \, \mathrm{d}S.$$

Here, $\Phi \lesssim \Psi$ means that there exists a constant C depending only on the Lipschitz character of D, A_D and A_0 such that $\Phi \leq C\Psi$. From the trace theorem, we obtain

$$\int_{B_r} |\nabla u_r|^2 \, \mathrm{d}\mathbf{r} \lesssim \frac{1}{\epsilon} \int_{\partial D} |h|^2 \, \mathrm{d}\mathbf{r} + \epsilon \int_{D} |u_r|^2 \, \mathrm{d}S.$$

From the Sobolev inequality, we have

$$\left(\int_{B_r} |u_r|^6 \, \mathrm{d}\mathbf{r} \right)^{1/3} \lesssim \frac{1}{\epsilon} \int_{\partial D} |h|^2 \, \mathrm{d}\mathbf{r} + \epsilon \int_{D} |u_r|^2 \, \mathrm{d}S.$$

Using a small constant ϵ and Jenson's inequality, we get

$$\left(\int_{B_r} |u_r|^6 \, \mathrm{d}\mathbf{r} \right)^{1/3} \lesssim \frac{1}{\epsilon} \int_{\partial D} |h|^2 \, \mathrm{d}\mathbf{r}$$

and, therefore,

$$\|u_r\|_{L^q(B_r)} + \|\nabla u_r\|_{L^2(B_r)} \leq C\|h\|_{L^2(\partial D)},$$

where the constant C is independent of r. Therefore, we can choose a sequence $\mathbf{r}_n \nearrow \infty$ such that $u_{r_n j} \to u \in H^1(\Omega)$ weakly. Clearly, this u satisfies (8.42)–(8.44). From the standard Schauder interior estimate, we have

$$|u(\mathbf{r})| + |\mathbf{r}| \, |\nabla u(\mathbf{r})| = O(|\mathbf{r}|^{-1/2}) \quad \text{at infinity.}$$

From Theorem 8.1.4 by Verchota (1984), we can find $f, g \in L^2(\partial D)$ such that

$$\mathcal{S} f = \widetilde{\mathcal{S}} g = u \quad \text{on } \partial D.$$

From the maximum principle, we have

$$u = \begin{cases} \widetilde{\mathcal{S}}g & \text{in } D, \\ \mathcal{S}f & \text{in } \mathbb{R}^3 \setminus D. \end{cases}$$

Hence,

$$|u(\mathbf{r})| + |\mathbf{r}|\,|\nabla u(\mathbf{r})| = O(|\mathbf{r}|^{-1}) \quad \text{at infinity.}$$

This completes the proof of Theorem 8.1.9. □

Proof of Theorem 8.1.7. To prove that the operator $\mathcal{T}$ is one-to-one, let $f, g \in L^2(\partial D)$ satisfy $\widetilde{\mathcal{S}}g - \mathcal{S}f = 0$ and $\langle A_0\mathbf{n}, \nabla\widetilde{\mathcal{S}}f^+\rangle - \langle A_D\mathbf{n}, \nabla\widetilde{\mathcal{S}}g^-\rangle = 0$ on ∂D. Then, the function u defined as $u = \widetilde{\mathcal{S}}g$ in D and $u = \mathcal{S}f$ in $\mathbb{R}^3 \setminus D$ lies in $H^1(\mathbb{R}^3)$ and is a weak solution of $Lu = 0$ in $\mathbb{R}^3$. Thus u must be identically zero by the maximum principle. Hence $\widetilde{\mathcal{S}}g = 0$ in $\mathbb{R}^3$ and $\mathcal{S}f = 0$ in $\mathbb{R}^3$. From the jump relations,

$$\begin{aligned} 0 = \int_{\mathbb{R}^3} \langle \nabla\mathcal{S}f, \nabla\phi\rangle \, d\mathbf{r} &= \int_D \langle \nabla\mathcal{S}f, \nabla\phi\rangle \, d\mathbf{r} + \int_{\mathbb{R}^3\setminus\bar{D}} \langle \nabla\mathcal{S}f, \nabla\phi\rangle \, d\mathbf{r} \\ &= \int_{\partial D} (-\tfrac{1}{2}I + \mathcal{K}_0^*) f\phi \, dS - \textstyle\int_{\partial D} (\tfrac{1}{2}I + \mathcal{K}_0^*) f\phi \, dS \\ &= -\int_{\partial D} f\phi \, dS \quad (\forall\phi \in C_0^\infty(\mathbb{R}^3)). \end{aligned}$$

Therefore, f must be zero on ∂D. Similar arguments give $g = 0$ on ∂D.

Next, we will prove that $\mathcal{T}$ has a closed range. As in Theorem 8.1.4, we first assume that

$$\mathcal{S}(f_j) - \widetilde{\mathcal{S}}(g_j) \to \phi_1 \in H^1(\partial D),$$
$$\langle \mathbf{n}, A_0\nabla\mathcal{S}(f_j)^+\rangle - \langle \mathbf{n}, A_D\nabla\widetilde{\mathcal{S}}(g_j)^-\rangle \to \phi_2 \in L^2(\partial D).$$

If $\|f_j\|_{L^2(\partial D)} + \|g_j\|_{L^2(\partial D)} < c < \infty$, we may then assume that $f_j \to f$ weakly for some $f \in L^2(\partial D)$ and $g_j \to g$ weakly for some $g \in L^2(\partial D)$ and, therefore, we can easily conclude that $\widetilde{\mathcal{S}}g - \mathcal{S}f = \phi_1$ and $\langle A_0\mathbf{n}, \nabla\widetilde{\mathcal{S}}g^+\rangle - \langle A_D\mathbf{n}, \nabla\mathcal{S}g^-\rangle = \phi_2$.

If $\|f_j\|_{L^2(\partial D)} + \|g_j\|_{L^2(\partial D)} \to \infty$, we may assume $\|f_j\|_{L^2(\partial D)} + \|g_j\|_{L^2(\partial D)} = 1$, $\mathcal{S}(f_j) - \widetilde{\mathcal{S}}(g_j) \to 0 \in H^1(\partial D)$ and $\langle A_0\mathbf{n}, \nabla\mathcal{S}(f_j)^+\rangle - \langle A_D\mathbf{n}, \nabla\mathcal{S}(g_j)^-\rangle \to 0 \in L^2(\partial D)$. Since $\mathcal{T}$ is one-to-one, we may assume that $f_j \to 0$ weakly in $L^2(\partial D)$ and $g_j \to 0$ weakly in $L^2(\partial D)$. Since $\mathcal{S}$ and $\widetilde{\mathcal{S}}$ are compact operators from $L^2(\partial D)$ to $L^2(\partial D)$, $\mathcal{S}(f_j)$ and $\widetilde{\mathcal{S}}(g_j)$ converge strongly to zero in $L^2(\partial D)$. But, since

$$\begin{aligned} 1 &= \|f_j\|_{L^2(\partial D)} + \|g_j\|_{L^2(\partial D)} \\ &\le C\left\{ \|\mathcal{S}(f_j) - \widetilde{\mathcal{S}}(g_j)\|_{H^1(\partial D)} + \|\langle \mathbf{n}, A_0\nabla\mathcal{S}(f_j)^+\rangle - \langle \mathbf{n}, A_D\nabla\widetilde{\mathcal{S}}(g_j)^-\rangle\|_{L^2(\partial D)} \right. \\ &\quad \left. + \left|\int_{\partial D} \mathcal{S}(f_j) \, dS\right| + \left|\int_{\partial D} \widetilde{\mathcal{S}}(g_j) \, dS\right| \right\} \\ &\to 0, \end{aligned}$$

there is a contradiction.

Now it remains to prove that $\mathcal{T}$ has a dense range. As in Verchota (1984), approximation of D by smooth domains and the estimate in Lemma 8.1.8 will give that the operator in Theorem 8.1.7 has a dense range, if the range of this operator is dense when D is a smooth domain. Let ϕ_1 and ϕ_2 in $C_0^\infty(\mathbb{R}^3)$ be given. From Theorem 8.1.9, we can find f_1 and g in $L^2(\partial D)$ such that

$$\mathcal{S}f_1(\mathbf{r}) = \widetilde{\mathcal{S}}(g) \quad \text{on } \partial D,$$

$$\langle A_D\mathbf{n}, \nabla\widetilde{\mathcal{S}}g^+\rangle - \langle \mathbf{n}, A_0\nabla\mathcal{S}f_1^-\rangle = h \quad \text{on } \partial D,$$

where $h = \langle \mathbf{n}, A_0\nabla\mathcal{S}(\mathcal{S}^{-1}(\phi_1)^+)\rangle + \phi_2$. From Verchota (1984), we can find $f \in L^2(\partial D)$ such that

$$\mathcal{S}f = \phi_1 + \mathcal{S}f_1 \quad \text{on } \partial D.$$

Clearly, $\mathcal{T}(f, g) = (\phi_1, \phi_2)$ and $\mathcal{T}$ has a dense range when D is smooth. This completes the proof of Theorem 8.1.7. □

8.2 Anomaly Estimation using EIT

In this section, we focus our attention on the estimation of the sizes of anomalies with different conductivity values compared with the background tissues. We describe how to estimate their size using the relationship between injection currents and measured boundary voltages. There are many potential applications where the locations and sizes of anomalies or changes in them with time or space are of primary concern. They include monitoring of impedance-related physiological events, breast cancer detection, bubble detection in two-phase flow and others in medicine and non-destructive testing.

Let $\Omega \subset \mathbb{R}^3$ denote an electrically conducting medium and let the anomalies occupy a region D contained in the homogeneous medium Ω. Then, the conductivity distribution σ can be written as

$$\sigma(\mathbf{r}) = \sigma_0(1 + \mu\chi_D(\mathbf{r})), \quad \mathbf{r} \in \Omega, \tag{8.46}$$

where σ_0 is a positive constant (which will be assumed to be 1 for simplicity) and μ is a constant such that $-1 < \mu \neq 0 < \infty$. Physically, σ_0 is the conductivity of the homogeneous background $\Omega \setminus \bar{D}$ and $\sigma_0\mu := \sigma_D - \sigma_0$, where σ_D is the conductivity of the anomaly D. A high contrast in conductivity occurs at the interface ∂D between the anomaly D and the background $\Omega \setminus \bar{D}$.

The goal is to develop an algorithm for extracting quantitative core information about D from the relationship between the applied Neumann data $g := (\partial u/\partial\mathbf{n})|_{\partial\Omega} \in H_\diamond^{-1/2}(\partial\Omega)$ and the measured Dirichlet data $f := u|_{\partial\Omega} \in H_\diamond^{1/2}(\partial\Omega)$. Here u is the induced potential due to the Neumann data g, and it is determined by solving the Neumann problem:

$$\begin{cases} \nabla\cdot(\sigma_0(1 + \mu\chi_D)\nabla u) = 0 & \text{in } \Omega, \\ \left.\dfrac{\partial u}{\partial\mathbf{n}}\right|_{\partial\Omega} = g \quad \text{and} \quad \displaystyle\int_{\partial\Omega} u \, \mathrm{d}S = 0. \end{cases} \tag{8.47}$$

Throughout this section, $f_0 := u_0|_{\partial\Omega}$, where u_0 is the potential satisfying

$$\begin{cases} \nabla\cdot(\sigma_0\nabla u_0) = 0 \quad \text{in } \Omega, \\ \sigma_0 \dfrac{\partial u_0}{\partial \mathbf{n}} = g \text{ on } \partial\Omega \quad \text{and} \quad \displaystyle\int_{\partial\Omega} u_0 \, \mathrm{d}S = 0. \end{cases} \tag{8.48}$$

Theorem 8.2.1 (Kang and Seo 1996) *The function* $H(\mathbf{r}) := \mathcal{D}_\Omega f(\mathbf{r}) - (1/\sigma_0)\mathcal{S}_\Omega g(\mathbf{r})$ *for* $\mathbf{r} \in \mathbb{R}^3 \setminus \partial D$ *can be expressed as*

$$H(\mathbf{r}) = \begin{cases} u(\mathbf{r}) - \displaystyle\int_D \frac{\mu}{\sigma_0} \nabla_{\mathbf{r}'} F(\mathbf{r}-\mathbf{r}')\cdot\nabla u(\mathbf{r}')\,\mathrm{d}\mathbf{r}' & \textit{for } \mathbf{r}\in\Omega, \\ -\displaystyle\int_D \frac{\mu}{\sigma_0} \nabla_{\mathbf{r}'} F(\mathbf{r}-\mathbf{r}')\cdot\nabla u(\mathbf{r}')\,\mathrm{d}\mathbf{r}' & \textit{for } \mathbf{r}\in\mathbb{R}^n\setminus\bar{\Omega}, \end{cases} \tag{8.49}$$

where $F(\mathbf{r}) = -1/(4\pi|\mathbf{r}|)$. *The difference* $\mathcal{D}_\Omega(f - f_0)$ *can be expressed as*

$$\mathcal{D}_\Omega(f - f_0)(\mathbf{r}) = \begin{cases} (u-u_0)(\mathbf{r}) - \dfrac{\mu}{\sigma_0}\displaystyle\int_D \nabla F(\mathbf{r}-\mathbf{r}')\cdot\nabla u(\mathbf{r}')\,\mathrm{d}\mathbf{r}' & \textit{for } \mathbf{r}'\in\Omega, \\ -\displaystyle\int_D \frac{\mu}{\sigma_0}\nabla F(\mathbf{r}-\mathbf{r}')\cdot\nabla u(\mathbf{r}')\,\mathrm{d}\mathbf{r}' & \textit{for } \mathbf{r}\in\mathbb{R}^3\setminus\bar{\Omega}. \end{cases} \tag{8.50}$$

Proof. Using the fact that $\nabla^2\mathbf{F}(\mathbf{r},\mathbf{r}') = \delta(\mathbf{r}-\mathbf{r}')$ in the distribution sense, we have

$$u(\mathbf{r})\chi_\Omega(\mathbf{r}) = -\int_\Omega \nabla F(\mathbf{r}-\mathbf{r}')\cdot\nabla u(\mathbf{r}')\,\mathrm{d}\mathbf{r}' + \underbrace{\int_{\partial\Omega}\frac{\partial}{\partial \mathbf{n}'_{\mathbf{r}}}F(\mathbf{r}-\mathbf{r}')f(\mathbf{r}')\,\mathrm{d}S}_{=\mathcal{S}f}$$

for all $\mathbf{r}\in\mathbb{R}^3\setminus\partial\Omega$. We now use the refraction condition of u along ∂D to get

$$\begin{aligned} &\int_\Omega \nabla_{\mathbf{r}'}F(\mathbf{r},\mathbf{r}')\cdot\nabla_{\mathbf{r}'}u(\mathbf{r}')\,\mathrm{d}\mathbf{r}' \\ &\quad = -\frac{\mu}{\sigma_0}\int_D \nabla_{\mathbf{r}'}F(\mathbf{r},\mathbf{r}')\cdot\nabla_{\mathbf{r}'}u(\mathbf{r}')\,\mathrm{d}\mathbf{r}' \\ &\qquad + \underbrace{\int_{\Omega\setminus\bar{D}} \nabla_{\mathbf{r}'}F(\mathbf{r},\mathbf{r}')\cdot\nabla_{\mathbf{r}'}u(\mathbf{r}')\,\mathrm{d}\mathbf{r}' + (1+\mu)\int_D \nabla_{\mathbf{r}'}F(\mathbf{r},\mathbf{r}')\cdot\nabla_{\mathbf{r}'}u(\mathbf{r}')\,\mathrm{d}\mathbf{r}'}_{=(1/\sigma_0)\int_{\partial\Omega}F(\mathbf{r},\mathbf{r}')g(\mathbf{r}')\,\mathrm{d}S_{\mathbf{r}'} = (1/\sigma_0)\mathcal{S}g(\mathbf{r})}. \end{aligned} \tag{8.51}$$

These prove (8.49). The expression (8.50) follows from (8.49) and

$$H_0(\mathbf{r}) := \mathcal{D}_\Omega f_0(\mathbf{r}) - \frac{1}{\sigma_0}\mathcal{S}_\Omega g(\mathbf{r}) = \begin{cases} u_0(\mathbf{r}) & \text{for } \mathbf{r}\in\Omega, \\ 0 & \text{for } \mathbf{r}\in\mathbb{R}^3\setminus\bar{\Omega}. \end{cases} \tag{8.52}$$

□

This theorem plays an important role in extracting location information about D. This section considers non-iterative anomaly estimation algorithms for searching its location and estimating its size.

8.2.1 Size Estimation Method

Kang *et al.* (1997) derived that, with the special Neumann data $g = \mathbf{a} \cdot \mathbf{n}$ where $\mathbf{a}$ is a unit constant vector, the volume of D can be estimated by

$$\frac{\min\{1, 1+\mu\}}{|\mu|} \left| \int_{\partial\Omega} (u_0 - u) g \, \mathrm{d}S \right| \leq |D| \leq \frac{\max\{1, 1+\mu\}}{|\mu|} \left| \int_{\partial\Omega} (u_0 - u) g \, \mathrm{d}S \right|, \quad (8.53)$$

where u_0 is the corresponding solution of (8.47) with the homogeneous conductivity distribution. Alessandrini *et al.* (2000) provided a careful analysis on the bound of the size of inclusions for a quite general g and conductivity distribution. In this section, we will explain the results of Alessandrini *et al.* (2000) in detail.

Before presenting an analysis of size estimation, we begin by explaining the algorithm to estimate the *total size of anomalies* proposed by Kwon and Seo (2001). The total size estimation of anomalies $D = \bigcup_{j=1}^{M} D_j$ uses the projection current $g = \mathbf{a} \cdot \mathbf{n}$, where $\mathbf{a}$ is a unit constant vector. We may assume that Ω contains the origin. Define the scaled domain $\Omega_t := \{\mathbf{r} : \mathrm{dist}(\mathbf{r}, \partial\Omega) > t\}$ for a scaling factor $t > 0$. Let v_t be the solution of the problem

$$\begin{cases} \nabla \cdot (1 + \mu\chi_{\Omega_t})\nabla v_t) = 0 & \text{in } \Omega, \\ \left.\dfrac{\partial v_t}{\partial \mathbf{n}}\right|_{\partial\Omega} = g, & \displaystyle\int_{\partial\Omega} v_t = 0. \end{cases}$$

The following lemma provides a way to compute $|D|$.

Lemma 8.2.2 (Kwon and Seo 2001) *There exists a unique* t_0, $0 < t_0 < 1$, *so that*

$$\int_{\partial\Omega} (u - v_{t_0}) g \, \mathrm{d}S = 0.$$

Proof. Let $\eta(t) := \int_{\partial\Omega} v_t g \, \mathrm{d}S$ as a function of t defined in the interval $(0, 1)$. If $t_1 < t_2$, it follows from integration by parts that

$$\begin{aligned} \eta(t_1) - \eta(t_2) &= \int_{\partial\Omega} (v_{t_1} - v_{t_2}) g \, \mathrm{d}S \\ &= \int_{\Omega} (1 + \mu\chi_{\Omega_{t_1}}) |\nabla(v_{t_1} - v_{t_2})|^2 \, \mathrm{d}\mathbf{r} + \mu \int_{\Omega_{t_2} \setminus \Omega_{t_1}} |\nabla v_{t_2}|^2 \, \mathrm{d}\mathbf{r}, \end{aligned}$$

$$\eta(t_1) - \eta(t_2) = -\int_{\Omega} (1 + \mu\chi_{\Omega_{t_2}}) |\nabla(v_{t_1} - v_{t_2})|^2 \, \mathrm{d}\mathbf{r} + \mu \int_{\Omega_{t_2} \setminus \Omega_{t_1}} |\nabla v_{t_1}|^2 \, \mathrm{d}\mathbf{r}.$$

These identities give the monotonicity of $\eta(t)$:

$$\eta(t_1) < \eta(t_2) \text{if } \mu < 0 \quad \text{and} \quad \eta(t_1) > \eta(t_2) \text{ if } \mu > 0.$$

Since $D \subset \Omega$, a similar monotonicity argument leads to the following inequalities;

$$\eta(0) < \int_{\partial\Omega} ug \, \mathrm{d}s < \eta(1) \quad \text{if } \mu < 0,$$

$$\eta(0) > \int_{\partial\Omega} ug\,\mathrm{d}s > \eta(1) \quad \text{if } \mu > 0.$$

Since $\eta(t)$ is continuous, there exists a unique t_0 so that $\eta(t_0) = \int_{\partial\Omega} ug\,\mathrm{d}S$. □

With the aid of the above lemma, Kwon and Seo (2001) developed the following method of finding the total size of multiple anomalies.

- Suppose that $D = \bigcup_{j=1}^{M} D_j \subset B_R(0)$. With an applied current $g = \mathbf{a} \cdot \mathbf{n}$, where $\mathbf{a}$ is a unit constant vector, choose the unique t_0, $0 < t_0 < R$, so that

$$\int_{\partial\Omega} (u - v_{t_0})g\,\mathrm{d}S = 0.$$

- Then, the size of the ball Ω_{t_0} is a good approximation of the total size of $\bigcup_{j=1}^{M} D_j$.

Various numerical experiments indicate that the above algorithm gives a nearly exact estimate for arbitrary multiple anomalies with a quite general conductivity distribution, as shown in Figure 8.2.

Next, we will provide an explanation on the background idea of the size estimation for the case where μ is small. Integrating by parts yields

$$\int_{\partial\Omega} (u - v_t)g\,\mathrm{d}\sigma = \int_{\Omega} \sigma|\nabla(u - v_t)|^2\,\mathrm{d}\mathbf{r} + \mu\int_{\Omega_t} |\nabla v_t|^2\,\mathrm{d}\mathbf{r} - \mu\int_{D} |\nabla v_t|^2\,\mathrm{d}\mathbf{r},$$

$$\int_{\partial\Omega} (u - v_t)g\,\mathrm{d}\sigma = -\int_{\Omega} \sigma_t|\nabla(u - v_t)|^2\,\mathrm{d}\mathbf{r} + \mu\int_{\Omega_t} |\nabla u|^2\,\mathrm{d}\mathbf{r} - \mu\int_{D} |\nabla u|^2\,\mathrm{d}\mathbf{r},$$

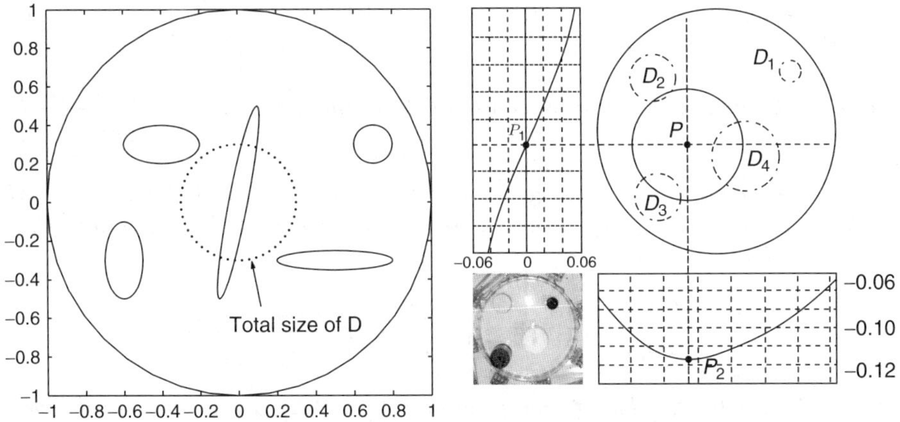

Figure 8.2 Numerical simulation for size estimation using Lemma 8.2.2 and phantom experiments for size estimation and location search. From Kwon *et al.* (2003). Reproduced with permission from IEEE

where $\sigma = 1 + \mu\chi_D$ and $\sigma_t = 1 + \mu\chi_{\Omega_t}$. By adding the above two identities, we obtain

$$2\int_{\partial\Omega}(u - v_t)g\,d\sigma = \mu\left[\int_D |\nabla(u - v_t)|^2\,\mathbf{dr} + \int_{\Omega_t} |\nabla v_t|^2 + |\nabla u|^2\,\mathbf{dr}\right]$$

$$- \mu\left[\int_{\Omega_t} |\nabla(u - v_t)|^2\,\mathbf{dr} + \int_D |\nabla v_t|^2 + |\nabla u|^2\,\mathbf{dr}\right]$$

$$= 2\mu\left[\int_{\Omega_t} \nabla u \cdot \nabla v_t\,\mathbf{dr} - \int_D \nabla u \cdot \nabla v_t\,\mathbf{dr}\right].$$

According to the choice of t_0,

$$\int_{\Omega_{t_0}} \nabla u \cdot \nabla v_{t_0}\,\mathbf{dr} = \int_D \nabla u \cdot \nabla v_{t_0}\,\mathbf{dr}.$$

If $\mu \approx 0$, then ∇u and ∇v_t are approximately constant, $\nabla u_0 = \mathbf{a}$ and the above identity is possible when the volume of Ω_{t_0} is close to the total volume $D = \bigcup_{j=1}^{M} D_j$.

Theorem 8.2.3 (Alessandrini *et al.* 2000) *Let Ω be a bounded domain in $\mathbb{R}^3$ with a connected C^1 boundary such that $\Omega^d := \{\mathbf{r} \in \Omega : \mathrm{dist}(\mathbf{r}, \Omega) > d\}$ is connected for all $d > 0$. Let $D \subset \Omega^{d_0}$ for a fixed $d_0 > 0$. Assume that the conductivity distribution is*

$$\sigma(\mathbf{r}) = A_0\chi_{\Omega\setminus\bar{D}}(\mathbf{r}) + A_D\chi_D(\mathbf{r}) \quad (\mathbf{r} \in \Omega),$$

where $\lambda I \leq \sigma \leq (1/\lambda)I$ for some $\lambda > 0$, and A_D and A_0 are constant matrices such that $\alpha I \leq A_D - A_0$ and $A_D \leq \mu A_0$ for some positive constants α, μ. Then, we have

$$\frac{1}{\mu - 1}C_1\frac{\int_{\partial\Omega}(f_0 - f)g}{\int_{\partial\Omega} f_0 g} \leq |D| \leq \left(\frac{\mu}{\alpha}\right)^{1/p} C_2\left(\frac{\int_{\partial\Omega}(f_0 - f)g}{\int_{\partial\Omega} f_0 g}\right)^{1/p}, \tag{8.54}$$

where $p > 1$, C_1 and C_2 are positive constants depending on d_0, λ, Ω and $\dfrac{\|g\|_{L^2(\partial\Omega)}}{\|g\|_{H^{-1/2}(\partial\Omega)}}$.

Remark 8.2.4 *Definitely, we can get an estimate similar to (8.54) for the case where $\alpha I \leq A_0 - A_D$ and $\mu A_0 \leq A_D$. In this case, we can get the following result, which is similar to (8.54):*

$$\frac{1}{1 - \mu}C_1\frac{\int_{\partial\Omega}(f - f_0)g}{\int_{\partial\Omega} f_0 g} \leq |D| \leq \left(\frac{1}{\alpha}\right)^{1/p} C_2\left(\frac{\int_{\partial\Omega}(f - f_0)g}{\int_{\partial\Omega} f_0 g}\right)^{1/p}. \tag{8.55}$$

Theorem 8.2.3 can be stated in greater generality. A_0 and $\mathbf{A}_D$ are not necessarily constant matrices. We refer to Alessandrini et al. (2000) for a precise statement for conditions on A_0 and $\mathbf{A}_D$. The proof of the estimate (8.54) relies on the results by Garofalo and Lin (1986), in which they developed an elegant theory that connects the unique continuation property for solutions of elliptic partial differential equations with the theory of A_p Muckenhoupt weights.

Lemma 8.2.5 *Under the assumption of Theorem 8.2.3, we have*

$$\frac{\alpha\lambda}{\mu}\int_D |\nabla u_0|^2 \leq \int_{\partial\Omega}(f_0 - f)g \leq \frac{\mu - 1}{\lambda}\int_D |\nabla u_0|^2. \tag{8.56}$$

Proof. Denote $G = A_D - A_0$. The proof is based on the following three identities:

$$\int_D \langle G\nabla u_0, \nabla u_0\rangle - \int_\Omega \langle (A_0 + G\chi_D)\nabla(u - u_0), \nabla(u - u_0)\rangle = \int_{\partial\Omega}(f_0 - f)g, \tag{8.57}$$

$$\int_\Omega \langle A_0\nabla(u - u_0), \nabla(u - u_0)\rangle + \int_D \langle G\nabla u, \nabla u\rangle = \int_{\partial\Omega}(f_0 - f)g, \tag{8.58}$$

$$\int_D \langle G\nabla u, \nabla u_0\rangle = \int_{\partial\Omega}(f_0 - f)g. \tag{8.59}$$

The identity (8.57) yields

$$\int_{\partial\Omega}(f_0 - f)g \leq \int_D \langle G\nabla u_0, \nabla u_0\rangle \leq \frac{\mu - 1}{\lambda}\int_D |\nabla u_0|^2.$$

From (8.57) and (8.58), we have

$$\int_D \langle G\nabla u_0, \nabla u_0\rangle = \int_D \langle G\nabla u, \nabla u\rangle + \underbrace{\int_D \langle G\nabla(u - u_0), \nabla(u - u_0)\rangle}_{\leq (\mu-1)\int_D \langle A_0\nabla(u-u_0), \nabla(u-u_0)\rangle}$$

and, therefore,

$$\mu\int_{\partial\Omega}(f_0 - f)g \geq \int_D G\nabla u_0 \cdot \nabla u_0 \geq \alpha\lambda\int_D |\nabla u_0|^2.$$

This completes the proof. □

Lemma 8.2.6 *Under the assumption of Theorem 8.2.3, we have the following estimate for every $\rho > 0$:*

$$\int_{B_\rho(\mathbf{r})} |\nabla u_0|^2 \geq C_\rho \int_\Omega |\nabla u_0|^2 \quad (\forall\, \mathbf{r} \in \Omega^{4\rho}), \tag{8.60}$$

where C_ρ depends only on ρ, d_0, λ, Ω and $\dfrac{\|g\|_{L^2(\partial\Omega)}}{\|g\|_{H^{-1/2}(\partial\Omega)}}$.

Proof. From the three sphere inequality given by Garofalo and Lin (1986), there exist $C \geq 1$ and $\delta \in (0, 1)$ depending on λ and d_0 such that

$$\|u_0 - c\|_{L^2(B_{7\rho/2}(\mathbf{r}))} \leq C\|u_0 - c\|^{\delta}_{L^2(B_\rho(\mathbf{r}))}\|u_0 - c\|^{1-\delta}_{L^2(B_{4\rho}(\mathbf{r}))} \quad (\forall\, \mathbf{r} \in \Omega^{4\rho}),$$

where

$$c = \left(\frac{1}{|B_\rho(\mathbf{r})|}\right)\int_{B_\rho(\mathbf{r})} u_0.$$

Applying the Caccioppoli and Poincaré inequalities, we have

$$\|\nabla u_0\|_{L^2(B_{3\rho}(\mathbf{r}))} \le C\|\nabla u_0\|^{\delta}_{L^2(B_\rho(\mathbf{r}))}\|\nabla u_0\|^{1-\delta}_{L^2(B_{4\rho}(\mathbf{r}))} \quad (\forall \mathbf{r} \in \Omega^{4\rho}), \tag{8.61}$$

which leads to

$$\frac{\|\nabla u_0\|_{L^2(B_{3\rho}(\mathbf{r}))}}{\|\nabla u_0\|_{L^2(\Omega)}} \le C\left(\frac{\|\nabla u_0\|_{L^2(B_\rho(\mathbf{r}))}}{\|\nabla u_0\|_{L^2(\Omega)}}\right)^{\delta} \quad (\forall \mathbf{r} \in \Omega^{4\rho}). \tag{8.62}$$

Here, the constant C differs on each occurrence but all of the C are independent of u_0 and depend only on ρ, d_0, λ, Ω and $\dfrac{\|g\|_{L^2(\partial\Omega)}}{\|g\|_{H^{-1/2}(\partial\Omega)}}$.

Let $\mathbf{r}, \mathbf{r}' \in \Omega^{4\rho}$ be any two points. Note that the two points $\mathbf{r}$ and $\mathbf{r}'$ can be joined by a polygonal arc with node points $\mathbf{r}' = \mathbf{r}_1, \mathbf{r}_2, \ldots, \mathbf{r}_L = \mathbf{r} \in \Omega^{d_0}$ such that $|\mathbf{r}_j - \mathbf{r}_{j+1}| = 2\rho$ for $j = 1, 2, \ldots, L-1$ and $L \le L^* := |\Omega|/(4\pi\rho^3)$. From the estimate (8.62), we have

$$\frac{\|\nabla u_0\|_{L^2(B_\rho(\mathbf{r}_{i+1}))}}{\|\nabla u_0\|_{L^2(\Omega)}} \le C\left(\frac{\|\nabla u_0\|_{L^2(B_\rho(\mathbf{r}_i))}}{\|\nabla u_0\|_{L^2(\Omega)}}\right)^{\delta} \quad (j = 1, \ldots, L-1) \tag{8.63}$$

and by induction

$$\frac{\|\nabla u_0\|_{L^2(B_\rho(\mathbf{r}'))}}{r\|\nabla u_0\|_{L^2(\Omega)}} \le C^{1/(1-\delta)}\left(\frac{\|\nabla u_0\|_{L^2(B_\rho(\mathbf{r}))}}{\|\nabla u_0\|_{L^2(\Omega)}}\right)^{\delta^L}. \tag{8.64}$$

By covering $\Omega^{5\rho}$ with balls of radius ρ using the estimate (8.64), we have

$$\frac{\|\nabla u_0\|_{L^2(\Omega^{5\rho})}}{\|\nabla u_0\|_{L^2(\Omega)}} \le C\left(\frac{\|\nabla u_0\|_{L^2(B_\rho(\mathbf{r}))}}{\|\nabla u_0\|_{L^2(\Omega)}}\right)^{\delta^{L^*}} \tag{8.65}$$

and, therefore,

$$\int_{B_\rho(\mathbf{r})} |\nabla u_0|^2 \ge \underbrace{\left(\frac{1}{C}\frac{\|\nabla u_0\|_{L^2(\Omega^{5\rho})}}{\|\nabla u_0\|_{L^2(\Omega)}}\right)^{1/\delta^{L^*}}}_{:=W} \int_\Omega |\nabla u_0|^2 \quad (\forall \mathbf{r} \in \Omega^{4\rho}). \tag{8.66}$$

It remains to derive an appropriate lower bound for W to estimate (8.60). To be precise, it suffices to prove that there exists $\rho_* > 0$ depending only on λ, Ω, $|\Omega|$ and $\dfrac{\|g\|_{L^2(\partial\Omega)}}{\|g\|_{H^{-1/2}(\partial\Omega)}}$ such that

$$\frac{\int_{\Omega^{5\rho}} |\nabla u_0|^2}{\int_\Omega |\nabla u_0|^2} \ge \frac{1}{2} \quad (\forall \rho < \rho_*). \tag{8.67}$$

If (8.67) holds true, then $W \ge [1/(\sqrt{2}C)]^{1/\delta^{L^*}}$ for $\rho < \rho_*$, and (8.66) and (8.67) yield

$$\rho < \rho_* \quad \Longrightarrow \quad \int_{B_\rho(\mathbf{r})} |\nabla u_0|^2 \ge \left(\frac{1}{\sqrt{2}C}\right)^{1/\delta^{L^*}} \int_\Omega |\nabla u_0|^2 \quad (\forall\, \mathbf{r} \in \Omega^{4\rho}). \tag{8.68}$$

If (8.68) holds true, it is obvious that the estimate on the right-hand side of (8.68) holds true for $\rho > \rho_*$.

Hence, it remains to prove the estimates (8.67). It can be proven by showing that there exists a ρ_* such that

$$\frac{\int_{\Omega\setminus\Omega^{5\rho_*}} |\nabla u_0|^2}{\int_\Omega |\nabla u_0|^2} < \frac{1}{2}. \tag{8.69}$$

The proof of (8.69) follows from a careful adaptation of results by Kenig *et al.* (2007):

$$\int_{\Omega\setminus\Omega_{4\rho}\}} |\nabla u_0|^2 \le C\rho \|g\|^2_{L^2(\partial\Omega)}, \tag{8.70}$$

where $C > 0$ depends on λ and the Lipschitz character of Ω only. The proof of the estimate (8.70) requires a deep knowledge of harmonic analysis when $\partial\Omega$ is only Lipschitz, while it is a lot simpler when $\partial\Omega$ is C^2. For ease of explanation, we will give the proof under the assumption that $\partial\Omega$ is in C^2. By the Hölder inequality and Sobolev inequality,

$$\|\nabla u_0\|^2_{L^2(\Omega\setminus\Omega_{4\rho})} \le \underbrace{|\Omega \setminus \Omega^{5\rho}|^{1/3}}_{O(\rho^{1/3})} \|\nabla u_0\|^2_{L^3(\Omega\setminus\Omega^{5\rho})} \le C\rho^{1/3}\|\nabla u_0\|^2_{H^{1/2}(\Omega)}, \tag{8.71}$$

where C depends on Ω only. Moreover, we have

$$\|\nabla u_0\|_{H^{1/2}(\Omega)} \le C\|g\|_{L^2(\partial\Omega)}, \tag{8.72}$$

where C depends on λ and Ω only. From (8.71) and (8.72), we have

$$\|\nabla u_0\|^2_{L^2(\Omega\setminus\Omega_{4\rho})} \le C\rho^{1/3}\|g\|^2_{L^2(\partial\Omega)}. \tag{8.73}$$

Using the standard estimate $\|g\|^2_{H^{-1/2}(\partial\Omega)} \le C\int_\Omega |\nabla u_0|^2$ and (8.73), we have

$$\frac{\|\nabla u_0\|^2_{L^2(\Omega\setminus\Omega^{5\rho})}}{\|\nabla u_0\|_{L^2(\Omega)}} \le C\rho^{1/3}\frac{\|g\|^2_{L^2(\partial\Omega)}}{\|g\|^2_{H^{-1/2}(\partial\Omega)}}, \tag{8.74}$$

where C depends on λ and Ω. Hence, we obtain (8.69) by choosing a sufficiently small ρ_*. This completes the proof. □

Now, we will prove Theorem 8.2.3. For $p > 1$, we have

$$|D| = \int_D |\nabla u_0|^{-2/p}|\nabla u_0|^{2/p} \le \left(\int_D |\nabla u_0|^{-2/(p-1)}\right)^{(p-1)/p}\left(\int_D |\nabla u_0|^2\right)^{1/p}. \tag{8.75}$$

We can cover D with internally non-overlapping closed cubes Q_j, $j = 1, \ldots, J$, with diameter ρ and $\rho < d_0/6$. Then, (8.75) yields

$$|D| \le \left(\rho^3 \sum_{j=1}^{J} \frac{1}{|Q_j|}\int_{Q_j} |\nabla u_0|^{-2/(p-1)}\right)^{(p-1)/p}\left(\int_D |\nabla u_0|^2\right)^{1/p}. \tag{8.76}$$

We will take advantage of the fact that $|\nabla u_0|^2$ is a Muckenhoupt weight (Garofalo and Lin 1986):

$$\left(\frac{1}{|Q_j|}\int_{Q_j}|\nabla u_0|^2\right)\left(\frac{1}{|Q_j|}\int_{Q_j}|\nabla u_0|^{-2/(p-1)}\right)^{p-1}\leq M, \tag{8.77}$$

where M and p depend only on ρ, d_0, λ, Ω and $\dfrac{\|g\|_{L^2(\partial\Omega)}}{\|g\|_{H^{-1/2}(\partial\Omega)}}$. The left term in (8.77) can be estimated by

$$\begin{aligned}\left(\rho^3\sum_{j=1}^{J}\frac{1}{|Q_j|}\int_{Q_j}|\nabla u_0|^{-2/(p-1)}\right)^{(p-1)/p} &\leq \left(\rho^3\sum_{j=1}^{J}\left(\frac{M}{(1/|Q_j|)\int_{Q_j}|\nabla u_0|^2}\right)^{1/(p-1)}\right)^{(p-1)/p}\\ &\leq \frac{(\rho^3 J)^{(p-1)/p}M^{1/p}}{\min_j\left((1/|Q_j|)\int_{Q_j}|\nabla u_0|^2\right)^{1/p}}.\end{aligned}$$

From (8.76) and the above estimate, we have

$$|D|\leq|\Omega|^{(p-1)/p}M^{1/p}\left(\frac{\rho^3\int_D|\nabla u_0|^2}{\min_j\int_{Q_j}|\nabla u_0|^2}\right)^{1/p}. \tag{8.78}$$

By (8.60) and (8.78), we have

$$\int_D|\nabla u_0|^2\geq C_1|D|^p\underbrace{\min_j\int_{Q_j}|\nabla u_0|^2}_{\geq C_2\int_\Omega|\nabla u_0|^2}\geq C_\rho|D|^p\int_{\partial\Omega}f_0 g, \tag{8.79}$$

where $C_1=|\Omega|^{1-p}M^{-1}\rho^{-3}$, $C_2=C_1C_\rho\lambda$ and C_ρ is the constant in (8.60). From (8.79) and the left-hand side of (8.56), we have

$$|D|\leq\frac{1}{(C_2)^{1/p}}\left(\frac{\int_D|\nabla u_0|^2}{\int_{\partial\Omega}f_0 g}\right)^{1/p}\leq\left(\frac{\mu}{C_2\lambda\alpha}\right)^{1/p}\left(\frac{\int_{\partial\Omega}(f-f_0)g}{\int_{\partial\Omega}f_0 g}\right)^{1/p}.$$

This completes the proof of the right-hand side of (8.54).

8.2.2 *Location Search Method*

Kwon *et al.* (2002) developed a location search method to detect an anomaly using a pattern injection current g and boundary voltage f. We assume that the object contains a single anomaly D that is small compared with the object itself and is located away from the boundary $\partial\Omega$. The location search algorithm is based on simple aspects of the function $H(\mathbf{r})$ outside the domain Ω, which can be computed directly from the data g and f.

We choose $g(\mathbf{r})=\mathbf{a}\cdot\mathbf{n}(\mathbf{r})$ for some fixed constant vector $\mathbf{a}$.

1. Take two observation regions Σ_1 and Σ_2 contained in $\mathbb{R}^3 \setminus \Omega$ given by

$$\Sigma_1 := \text{a line parallel to } \mathbf{a},$$
$$\Sigma_2 := \text{a plane normal to } \mathbf{a}.$$

2. Find two points $P_i \in \Sigma_i (i = 1, 2)$ so that

$$H(P_1) = 0$$

and

$$H(P_2) = \begin{cases} \min_{\mathbf{r}\in\Sigma_2} H(\mathbf{r}) & \text{if } \mu > 0, \\ \max_{\mathbf{r}\in\Sigma_2} H(\mathbf{r}) & \text{if } \mu < 0. \end{cases}$$

3. Draw the corresponding plane $\Pi_1(P_1)$ and the line $\Pi_2(P_2)$ given by

$$\Pi_1(P_1) := \{\mathbf{r} : \mathbf{a} \cdot (\mathbf{r} - P_1) = 0\},$$
$$\Pi_2(P_2) := \{\mathbf{r} : (\mathbf{r} - P_2) \text{ is parallel to } \mathbf{a}\}.$$

4. Find the intersecting point P of the plane $\Pi_1(P_1)$ and the line $\Pi_2(P_2)$. Then this point P can be viewed as the location of the anomaly D.

Figure 8.3 illustrates how the location search method works. The above location search method is based on the assumption that $\nabla u|_D \approx \alpha \nabla u_0$ for some constant α. Noting that $\nabla u_0|_\Omega = (1/\sigma_0)\mathbf{a}$ with the injection current $g = \mathbf{a} \cdot \mathbf{n}$, it follows from (8.50) that

$$H(\mathbf{r}) = \mathcal{D}(f - f_0)(\mathbf{r}) \approx \frac{\mu\alpha}{\sigma_0} w(\mathbf{r}) \quad \text{for } \mathbf{r} \in \mathbb{R}^3 \setminus \bar{\Omega}, \tag{8.80}$$

where

$$w(\mathbf{r}) := \int_D \frac{(\mathbf{r} - \mathbf{r}') \cdot \mathbf{a}}{4\pi |\mathbf{r} - \mathbf{r}'|^3} \, d\mathbf{r}'.$$

We can express w as a single-layer potential $w = \mathcal{S}(\mathbf{a} \cdot \mathbf{n}|_{\partial D})$ and, therefore, w is harmonic both in D and in $\mathbb{R}^3 \setminus \bar{D}$. Hence, we can get the following observation by the mean value property of harmonic functions in $\mathbb{R}^3 \setminus \bar{D}$, and the uniqueness of the interior Dirichlet problem for the Laplace equation in D by considering the limit value of $w(\mathbf{r})$ to the boundary ∂D.

Observation 8.2.7 (Kwon *et al.* 2002) *The function $w(\mathbf{r})$ in (8.80) is harmonic in $\mathbb{R}^3 \setminus \partial D$. If $D = B_r(\mathbf{r}_*)$ is a ball, $w(\mathbf{r})$ satisfies*

$$w(\mathbf{r}) = \begin{cases} \dfrac{r^3}{3} \dfrac{(\mathbf{r} - \mathbf{r}_*) \cdot \mathbf{a}}{|\mathbf{r} - \mathbf{r}_*|^3} & \text{for } \mathbf{r} \in \mathbb{R}^3 \setminus \bar{D}, \\ \dfrac{1}{3}(\mathbf{r} - \mathbf{r}_*) \cdot \mathbf{a} & \text{for } \mathbf{r} \in D. \end{cases} \tag{8.81}$$

Assuming $H \approx (\mu\alpha/\sigma_0)w$ outside Ω and examining the sign of w in (8.81) as depicted in Figure 8.3, we have the following:

$$\text{both } \Pi_1(P_1) \text{ and } \Pi_2(P_2) \text{ hit the point near the center of } D = B_r(\mathbf{r}_*). \tag{8.82}$$

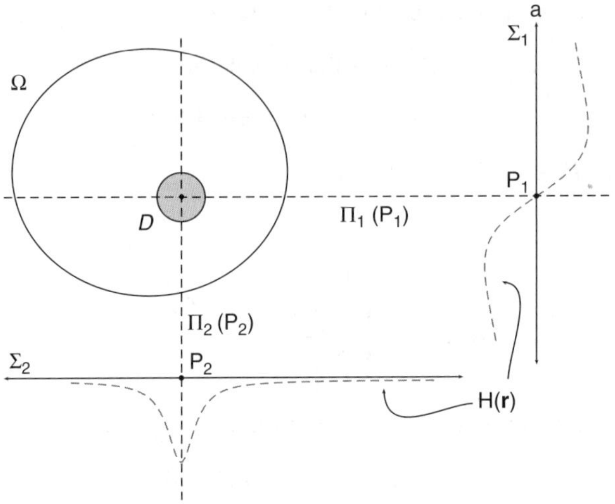

Figure 8.3 Relations between the location of the anomaly and the pattern of $H(\mathbf{r})$ in the case of $\mu > 0$

We should note that this nice observation is made under the assumption that $\nabla u \times \nabla u_0 \approx 0$, that is, ∇u inside the anomaly D is a fixed constant vector $\nabla u_0 = \mathbf{a}$. Hence, we need to check whether the current $g = \mathbf{a} \cdot \mathbf{n}|_{\partial\Omega}$ generates $\nabla u|_D \approx \alpha\mathbf{a}$ for some scalar α. In the special case where Ω and D are concentric balls, we can compute u explicitly via Fourier expansion.

Exercise 8.2.8 *Show that if Ω and D are concentric balls, then*

$$\nabla u = \alpha\mathbf{a} \quad \text{in } D, \tag{8.83}$$

for some scalar α.

Observation 8.2.9 *For ease of explanation, let $\sigma_0 = 1$ and $\mathbf{a} = \hat{\mathbf{y}}$. Let $\Omega = B_R(0) \subset \mathbb{R}^3$ and $D = B_\rho(\mathbf{r}_*)$, the ball centered at $\mathbf{r}_*$ of radius ρ. Assume that $\operatorname{dist}(D, \partial\Omega) \geq d$ for a positive number d and that $\mu \neq 0$ satisfies $-1 + 1/\mu_0 \leq \mu \leq \mu_0$ for a number $\mu_0 \geq 1$. Take an observation line $\Sigma_1 :=$ a line parallel to $\mathbf{a}$ with $\operatorname{dist}(\Sigma_1, \Omega) = L$. Then, there exists $\epsilon(\rho)$ with $\epsilon(\rho) \to 0$ as $\rho \to 0$, so that for the point $P_1 \in \Sigma_1$ satisfying $H(P_1) = 0$, the plane or the line $\Pi_1(P_1) := \{\mathbf{r} : \mathbf{a} \cdot (\mathbf{r} - P_1) = 0\}$ hits the $\epsilon(\rho)$-neighborhood of the center $\mathbf{r}_*$ of D.*

Proof. We will compare u with a new function v (which makes quantitative analysis a lot easier) defined by

$$v(\mathbf{r}) := \mathbf{a} \cdot \mathbf{r} + \frac{3\mu}{3 + \mu} w(\mathbf{r}) - c \quad (\mathbf{r} \in \Omega), \tag{8.84}$$

where the constant c is chosen so that $\int_{\partial\Omega} v = 0$. Here, the term $3\mu/(3 + \mu)$ is chosen so that v satisfies $\nabla \cdot ((1 + \mu\chi_D)\nabla v) = 0$ in $\mathbb{R}^3$, which will be explained below.

From Observation 8.2.7, we have

$$v(\mathbf{r}) = \begin{cases} \mathbf{a}\cdot\mathbf{r} + \dfrac{\mu\rho^3}{3+\mu}\dfrac{(\mathbf{r}_* - \mathbf{r})\cdot\mathbf{a}}{|\mathbf{r}-\mathbf{r}_*|^3} - c & \text{for } \mathbf{r}\in\mathbb{R}^3\setminus\bar{D}, \\ \mathbf{a}\cdot\mathbf{r} + \dfrac{\mu}{3+\mu}(\mathbf{r}_* - \mathbf{r})\cdot\mathbf{a} - c & \text{for } \mathbf{r}\in D. \end{cases} \tag{8.85}$$

A straightforward calculation of (8.85) using parametric spherical coordinates centered at $\mathbf{r}_*$ yields

$$\frac{\partial v^-}{\partial \mathbf{n}}(\mathbf{r}) = \frac{(1+\mu)\,3}{3+\mu}\,\mathbf{a}\cdot\mathbf{n}(\mathbf{r}) \quad\text{and}\quad \frac{\partial v^+}{\partial \mathbf{n}}(\mathbf{r}) = \frac{3}{3+\mu}\,\mathbf{a}\cdot\mathbf{n}(\mathbf{r}) \quad \text{for } \mathbf{r}\in\partial D,$$

where $v^+ = v|_D$ and $v^- = v|_{\mathbb{R}^3\setminus\bar{D}}$. The above identities lead to the transmission condition

$$\frac{\partial v^-}{\partial \mathbf{n}} = (1+\mu)\frac{\partial v^+}{\partial \mathbf{n}} \quad \text{on } \partial D$$

and, therefore, v satisfies the conductivity equation $\nabla\cdot((1+\mu\chi_D)\nabla v) = 0$ in $\mathbb{R}^3$. The Neumann data of v on $\partial\Omega$ is given by

$$\frac{\partial v}{\partial \mathbf{n}}(\mathbf{r}) = g(\mathbf{r}) - \frac{\mu\rho^3}{n+\mu}\frac{E(\mathbf{r},\mathbf{r}_*)}{|\mathbf{r}-\mathbf{r}_*|^3} \quad \text{for } \mathbf{r}\in\partial\Omega,$$

where $E(\mathbf{r},\mathbf{r}_*)$ is defined as

$$E(\mathbf{r},\mathbf{r}_*) = \frac{(\mathbf{r}\cdot\mathbf{a})(|\mathbf{r}_*|^2 - 2|\mathbf{r}|^2 + \mathbf{r}\cdot\mathbf{r}_*) + 3(\mathbf{r}_*\cdot\mathbf{a})(|\mathbf{r}|^2 - \mathbf{r}\cdot\mathbf{r}_*)}{R|\mathbf{r}-\mathbf{r}_*|^2}.$$

Hence, $\eta := u - v$ solves the following Neumann problem:

$$\begin{cases} \nabla\cdot((1+\mu\chi_D)\nabla\eta) = 0 & \text{in } \Omega, \\ \dfrac{\partial\eta}{\partial\mathbf{n}} = \dfrac{\mu\rho^3}{3+\mu}\dfrac{E(\mathbf{r},\mathbf{r}_*)}{|\mathbf{r}-\mathbf{r}_*|^3} := \tilde{g} & \text{on } \partial\Omega, \\ \displaystyle\int_{\partial\Omega}\eta\,d\sigma = 0. \end{cases} \tag{8.86}$$

We will estimate that

$$|E(\mathbf{r},\mathbf{r}_*)| \le 2. \tag{8.87}$$

By simple arithmetic, we can decompose $E(\mathbf{r},\mathbf{r}_*)$ into the following two terms:

$$E(\mathbf{r},\mathbf{r}_*) = \underbrace{\frac{(\mathbf{r}\cdot\mathbf{a})(|\mathbf{r}_*|^2 - |\mathbf{r}|^2) + 2(\mathbf{r}_*\cdot\mathbf{a})(|\mathbf{r}|^2 - \mathbf{r}\cdot\mathbf{r}_*)}{R|\mathbf{r}-\mathbf{r}_*|^2}}_{:=U(\mathbf{r},\mathbf{r}_*)} + \underbrace{\frac{\{(\mathbf{r}_*-\mathbf{r})\cdot\mathbf{a}\}\{(\mathbf{r}-\mathbf{r}_*)\cdot\mathbf{r}\}}{R|\mathbf{r}-\mathbf{r}_*|^2}}_{:=V(\mathbf{r},\mathbf{r}_*)}.$$

Since $R = |\mathbf{r}|$, we get the following by a simple Cauchy–Schwarz inequality:

$$|V(\mathbf{r},\mathbf{r}_*)| \le \left|\frac{(\mathbf{r}_*-\mathbf{r})\cdot\mathbf{a}}{|\mathbf{r}-\mathbf{r}_*|}\right| \left|\frac{\mathbf{r}\cdot(\mathbf{r}-\mathbf{r}_*)}{R\,|\mathbf{r}-\mathbf{r}_*|}\right| \le \frac{|\mathbf{r}_*-\mathbf{r}|}{|\mathbf{r}-\mathbf{r}_*|}\frac{|\mathbf{r}|}{|\mathbf{r}|}\frac{|\mathbf{r}-\mathbf{r}_*|}{|\mathbf{r}-\mathbf{r}_*|} = 1.$$

In order to estimate $U(\mathbf{r}, \mathbf{r}_*)$, consider the spherical coordinates for $\mathbf{r}$ and $\mathbf{r}_*$ with θ and α as latitudes measured from the xz plane to the y axis, respectively:

$$\mathbf{r} = R(\cos\theta\cos\eta,\ \sin\theta,\ \cos\theta\sin\eta) \quad \text{and} \quad \mathbf{r}_* = \tau(\cos\alpha\cos\beta,\ \sin\alpha,\ \cos\alpha\sin\beta)$$

for $-\pi/2 \le \theta, \alpha \le \pi/2$ and $0 \le \eta, \beta \le 2\pi$. Denoting $t := \tau/R$ and $r := \cos(\eta - \beta)$, $U(\mathbf{r}, \mathbf{r}_*)$ is expressed as

$$\begin{aligned} U(\mathbf{r}, \mathbf{r}_*) &= \frac{(t^2 - 1)\sin\theta + 2t\sin\alpha(1 - t\{\cos\theta\cos\alpha\cos(\eta - \beta) + \sin\theta\sin\alpha\})}{1 + t^2 - 2t\{\cos\theta\cos\alpha\cos(\eta - \beta) + \sin\theta\sin\alpha\}} \\ &= \frac{\{\sin\theta(t^2\cos 2\alpha - 1) + 2t\sin\alpha\} - \{t^2\sin 2\alpha\cos\theta\}r}{\{1 + t^2 - 2t\sin\alpha\sin\theta\} - \{2t\cos\alpha\cos\theta\}r}. \end{aligned} \tag{8.88}$$

By a tedious calculation with (8.88), we get

$$\frac{dU}{dr}(\mathbf{r}, \mathbf{r}_*) = \frac{2t(1 - t^2)\cos\alpha\cos\theta(t\sin\alpha - \sin\theta)}{[\{1 + t^2 - 2t\sin\alpha\sin\theta\} - \{2t\cos\alpha\cos\theta\}r]^2}.$$

Thus, $U(r; t, \alpha, \beta, \theta) := U(\mathbf{r}, \mathbf{r}_*)$ is monotonic for $-1 \le r \le 1$ for fixed t, α, β (that is, for fixed $\mathbf{r}_*$) and θ. To check the extremes of $U(\mathbf{r}, \mathbf{r}_*)$, we only need to check

$$\begin{aligned} U(\mathbf{r}, \mathbf{r}_*) &= \frac{\{\sin\theta(t^2\cos 2\alpha - 1) + 2t\sin\alpha\} \pm \{t^2\sin 2\alpha\cos\theta\}}{\{1 + t^2 - 2t\sin\alpha\sin\theta\} \pm \{2t\cos\alpha\cos\theta\}} \\ &= \frac{\{(t^2 - 1)\cos\alpha\}\sin(\theta \pm \alpha) - \{(1 + t^2)\sin\alpha\}\{\mp\cos(\theta \pm \alpha)\} + 2t\sin\alpha}{1 + t^2 - 2t\{\mp\cos(\theta \pm \alpha)\}} \end{aligned}$$

for $0 \le t < 1$ and $-\pi/2 \le \theta, \alpha \le \pi/2$. Letting $s := \mp\cos(\theta \pm \alpha)$, we get

$$U(\mathbf{r}, \mathbf{r}_*) = \frac{\pm\{(t^2 - 1)\cos\alpha\}\sqrt{1 - s^2} - \{(1 + t^2)\sin\alpha\}s + 2t\sin\alpha}{1 + t^2 - 2ts}.$$

Then, it is easy to get $|U(\mathbf{r}, \mathbf{r}_*)| \le 1$. Hence, we obtain the desired assertion (8.87):

$$|E(\mathbf{r}, \mathbf{r}_*)| \le |U(\mathbf{r}, \mathbf{r}_*)| + |V(\mathbf{r}, \mathbf{r}_*)| \le 2.$$

Since $\text{dist}(D, \partial\Omega) \ge d$, we see that $|\mathbf{r} - \mathbf{r}_*| \ge d$ for all $\mathbf{r} \in \partial\Omega$. In addition, the fact that $|\mu|/(3 + \mu) \le 1$ for all $-1 < \mu < \infty$ gives the estimate

$$\|\tilde{g}\|_{L^\infty(\partial\Omega)} \le 2(\rho/d)^3. \tag{8.89}$$

Under the assumption that $-1 < -1 + 1/\mu_0 \le \mu \le \mu_0 < \infty$, we are ready to compare $H(\mathbf{r})$ with

$$\widetilde{H}(\mathbf{r}; v) := \frac{\mu}{12\pi}\int_D \frac{(\mathbf{r} - \mathbf{r}')\cdot\nabla v}{|\mathbf{r} - \mathbf{r}'|^3}\,dy = \frac{\mu}{4\pi(3 + \mu)}\int_D \frac{(\mathbf{r} - \mathbf{r}')\cdot\mathbf{a}}{|\mathbf{r} - \mathbf{r}'|^3}\,dy$$

in the region $\mathbb{R}^3 \setminus \bar{\Omega}$. From (8.49), we have

$$H(\mathbf{r}) = \widetilde{H}(\mathbf{r}; v) + \frac{\mu}{n4\pi}\int_D \frac{(\mathbf{r} - \mathbf{r}')\cdot\nabla\eta}{|\mathbf{r} - \mathbf{r}'|^3}\,dy \quad \text{for } \mathbf{r} \in \mathbb{R}^3 \setminus \bar{\Omega}. \tag{8.90}$$

Therefore, by the observation that

$$\begin{aligned}\operatorname{dist}(\mathbf{r}, D) &= \left|\mathbf{r}-\mathbf{r}_*\right| - \rho \\ &= \frac{\left|\mathbf{r}-\mathbf{r}_*\right| - \rho}{\left|\mathbf{r}-\mathbf{r}_*\right|}\left|\mathbf{r}-\mathbf{r}_*\right| \geq \frac{(d+\rho)-\rho}{d+\rho}\left|\mathbf{r}-\mathbf{r}_*\right| \geq \frac{d}{R}\left|\mathbf{r}-\mathbf{r}_*\right|,\end{aligned}$$

we obtain the following estimate by the standard Hölder estimate of (8.90):

$$\left| H(\mathbf{r}) - \widetilde{H}(\mathbf{r}; v) \right| \leq \left(\frac{R}{d}\right)^2 \frac{|\mu|\rho^3 \|\nabla\eta\|_{L^\infty(D)}}{3\left|\mathbf{r}-\mathbf{r}_*\right|^2} \quad \text{for } \mathbf{r} \in \mathbb{R}^3 \setminus \bar{\Omega}. \tag{8.91}$$

We claim that there exists a positive constant $C_1 = C_1(\mu_0, d, \Omega)$ so that

$$\|\nabla\eta\|_{L^\infty(D)} \leq C_1\rho. \tag{8.92}$$

The proof of (8.92) is a bit technical and lengthy, so we omit its here. Conditions (8.91) and (8.92) lead to the following estimate:

$$|H(\mathbf{r}) - \widetilde{H}(\mathbf{r}; v)| \leq \left(\frac{R}{d}\right)^2 \frac{|\mu| C_1 \rho^4}{3\left|\mathbf{r}-\mathbf{r}_*\right|^2} \quad \text{for } \mathbf{r} \in \mathbb{R}^3 \setminus \bar{\Omega}. \tag{8.93}$$

Note that by using the mean value theorem for harmonic functions, we easily get

$$\widetilde{H}(\mathbf{r}; v) = \frac{\mu}{n+\mu} \frac{\rho^3}{\left|\mathbf{r}-\mathbf{r}_*\right|^{n-1}} \frac{(\mathbf{r}-\mathbf{r}_*)\cdot\mathbf{a}}{|\mathbf{r}-\mathbf{r}_*|} \quad \text{for } \mathbf{r} \in \mathbb{R}^3 \setminus \bar{\Omega}. \tag{8.94}$$

From (8.93) and (8.94), we obtain the final estimate

$$\frac{\rho^3}{\left|\mathbf{r}-\mathbf{r}_*\right|^2}\left(\frac{(\mathbf{r}-\mathbf{r}_*)\cdot\mathbf{a}}{|\mathbf{r}-\mathbf{r}_*|} - C\rho\right) \leq \frac{n+\mu}{\mu} H(\mathbf{r}) \leq \frac{\rho^3}{\left|\mathbf{r}-\mathbf{r}_*\right|^2}\left(\frac{(\mathbf{r}-\mathbf{r}_*)\cdot\mathbf{a}}{|\mathbf{r}-\mathbf{r}_*|} + C\rho\right), \tag{8.95}$$

where $C = C(\mu_0, d, \Omega)$ is given by

$$C = C_1 \frac{3+\mu_0}{3}\left(\frac{R}{d}\right)^2. \tag{8.96}$$

Now let $\epsilon(\rho) \in (0, \infty]$ be defined as follows, which clearly satisfies $\epsilon(\rho) \to 0$ as $\rho \to 0$:

$$\epsilon(\rho) := \frac{(2R+L)C\rho}{1-C\rho} \text{for } \rho < 1/C \quad \text{and} \quad \epsilon(\rho) := \infty \text{ otherwise},$$

where $C = C(\mu_0, d, \Omega)$, independent of ρ, is given by (8.96). Because there is nothing to be proved if $\epsilon(\rho) = \infty$, we may assume that $\rho < 1/C$ from now on.

For $\mathbf{r} \in \Sigma_1$ satisfying $\mathbf{r}\cdot\mathbf{a} > \mathbf{r}_*\cdot\mathbf{a} + \epsilon(\rho)$, we have

$$|\mathbf{r}-\mathbf{r}_*| \leq (\mathbf{r}-\mathbf{r}_*)\cdot\mathbf{a} + (2R+L).$$

Hence, by the positiveness of $(\mathbf{r}-\mathbf{r}_*)\cdot\mathbf{a}$ and the definition of $\epsilon(\rho)$, we obtain

$$\frac{(\mathbf{r}-\mathbf{r}_*)\cdot\mathbf{a}}{|\mathbf{r}-\mathbf{r}_*|}-C\rho \geq \frac{(\mathbf{r}-\mathbf{r}_*)\cdot\mathbf{a}}{(\mathbf{r}-\mathbf{r}_*)\cdot\mathbf{a}+(2R+L)}-C\rho > \frac{\epsilon(\rho)}{\epsilon(\rho)+(2R+L)}-C\rho = 0.$$

On the other hand, for $\mathbf{r}\in\Sigma_1$ satisfying $\mathbf{r}\cdot\mathbf{a} < z\cdot\mathbf{a}-\epsilon(\rho)$, we have

$$|\mathbf{r}-\mathbf{r}_*| \leq (\mathbf{r}_*-\mathbf{r})\cdot\mathbf{a}+(2R+L).$$

Being careful of the negativeness of $(\mathbf{r}-\mathbf{r}_*)\cdot\mathbf{a}$, we similarly obtain

$$\frac{(\mathbf{r}-\mathbf{r}_*)\cdot\mathbf{a}}{|\mathbf{r}-\mathbf{r}_*|}+C\rho \leq -\frac{(\mathbf{r}_*-\mathbf{r})\cdot\mathbf{a}}{(\mathbf{r}_*-\mathbf{r})\cdot\mathbf{a}+(2R+L)}+C\rho < -\frac{\epsilon(\rho)}{\epsilon(\rho)+(2R+L)}+C\rho = 0.$$

In the end, by the estimate (8.95) we have, if $\mu > 0$,

$$\begin{cases} H(\mathbf{r}) > 0 & \text{for } \mathbf{r}\in\Sigma_1 \text{ with } \mathbf{r}\cdot\mathbf{a} > \mathbf{r}_*\cdot\mathbf{a}+\epsilon(\rho), \\ H(\mathbf{r}) < 0 & \text{for } \mathbf{r}\in\Sigma_1 \text{ with } \mathbf{r}\cdot\mathbf{a} < \mathbf{r}_*\cdot\mathbf{a}-\epsilon(\rho). \end{cases}$$

Note that the sign of $H(\mathbf{r})$ is exchanged if $\mu < 0$. Therefore, the zero point $P_1\in\Sigma_1$ of $H(\mathbf{r})$ satisfies $\mathbf{r}_*\cdot\mathbf{a}-\epsilon(\rho) \leq P_1\cdot\mathbf{a} \leq \mathbf{r}_*\cdot\mathbf{a}+\epsilon(\rho)$, which completes the proof. □

Although the basic idea of the algorithm is simple, several technical arguments are needed for its proof. Combining this location search algorithm with the size estimation algorithm proposed in Kwon and Seo (2001), one can select an appropriate initial guess. Figure 8.4 explains this algorithm.

In order to test the feasibility of the location search and size estimation methods, Kwon *et al.* (2003) carried out phantom experiments. They used a circular phantom with 290 mm diameter as a container and filled it with NaCl solution of conductivity 0.69 S m^{-1}. Anomalies with different conductivity values, shapes and sizes were placed inside the phantom. A total of 32 equally spaced electrodes were attached on the surface of the phantom. Using a 32-channel EIT system, they applied the algorithms to

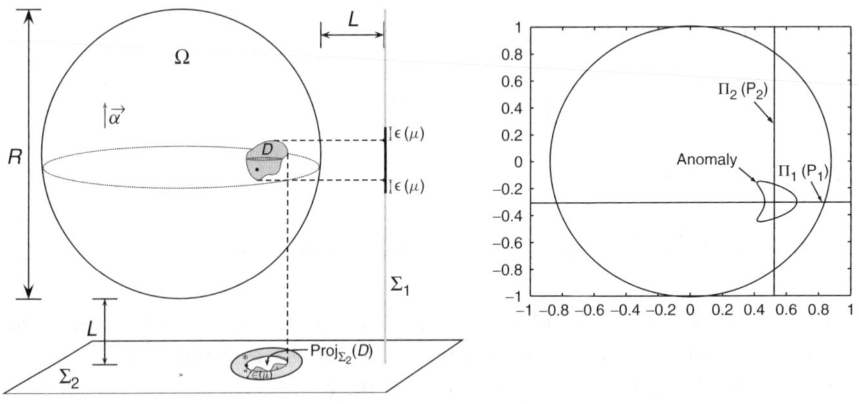

Figure 8.4 Location detection by finding an intersecting point of two lines $\Pi_1(P_1)$ and $\Pi_2(P_2)$

measured boundary voltage data. The circular phantom can be regarded as a unit disk $\Omega := B_1(0,0)$ by normalizing the length scale. To demonstrate how the location search and size estimation algorithm work, they placed four insulators $D = \bigcup_{j=1}^{4} D_j$ into the phantom:

$$D_1 = B_{0.1138}(0.5172, 0.5172), \qquad D_2 = B_{0.1759}(-0.5172, 0.5172),$$

$$D_3 = B_{0.1828}(-0.5172, -0.5172), \qquad D_4 = B_{0.2448}(0.1724, -0.1724).$$

They injected a projection current $g = \vec{a} \cdot \mathbf{n}$ with $\vec{a} = (0, 1)$ and measured the boundary voltage f. For the location search, they chose two observation lines:

$$\Sigma_1 := \{(-1.5, s) \mid s \in \mathbb{R}\} \quad \text{and} \quad \Sigma_2 := \{(s, -1.5) \mid s \in \mathbb{R}\}.$$

They evaluated the two-dimensional version of $H(\mathbf{r})$ with F replaced by $F(\mathbf{r}) = [1/(2\pi)] \log \sqrt{x^2 + y^2}$. In Figure 8.3, the left-hand plot is the graph of $H(\mathbf{r})$ on Σ_1 and the right-hand plot is the graph of $H(\mathbf{r})$ on Σ_2. They found the zero point of $H(\mathbf{r})$ on Σ_1 and the maximum point of $|H(\mathbf{r})|$ on Σ_2 as denoted by the black dots in Figure 8.3. The intersecting points were calculated as $P(-0.1620, -0.0980)$, which was close to the center of mass $P_M(-0.1184, -0.0358)$. For the case of a single anomaly or a cluster of multiple anomalies, the intersecting point furnished meaningful location information.

For the size estimation, the estimated total size was 0.4537 compared with the true total size of 0.4311. In Figure 8.2, the corresponding disk with the size of 0.4537 centered at $P(-0.1620, -0.0980)$ is drawn with a solid line and the corresponding disk with the true size centered at P_M is drawn with a dotted line. The relative error of the estimated size was about 5.24%.

Remark 8.2.10 *Numerous experimental results by Kwon et al. (2003) using a circular saline phantom showed the feasibility of the method for many applications in medicine and non-destructive testing. The algorithm is also fast enough for real-time monitoring of impedance-related events. The performance of the location and size estimation algorithm is not sensitive to anomaly shapes, locations and configurations. In practice, μ defined as the conductivity contrast between the background and anomaly is unknown. Note that the location search method does not depend on μ, whereas the size estimation does depend on μ.*

In the case of multiple anomalies, the location search algorithm produces a point close to the center of mass of multiple anomalies. If they are widely scattered, this information may not be useful or may be misleading in some applications. However, when they form a cluster of small anomalies, this point becomes meaningful. For more detailed explanations on anomaly estimations, see Ammari and Kang (2007).

8.3 Anomaly Estimation using Planar Probe

In this section, we describe anomaly estimation techniques using a local measurement from a planar probe placed on a portion of an electrically conducting object. The detection of breast cancer using a trans-admittance scanner (TAS) is a typical example of this setting.

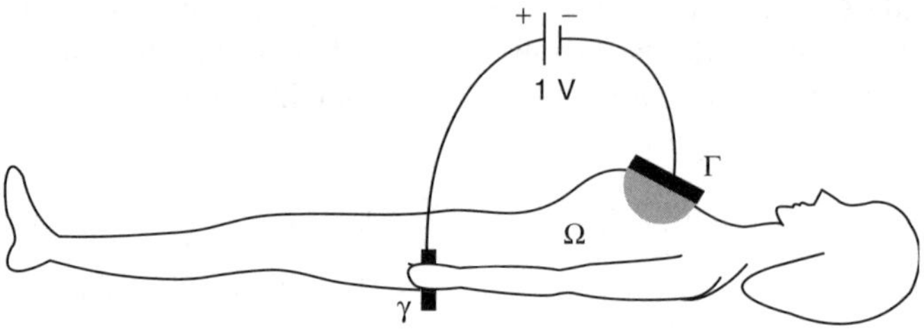

Figure 8.5 Configuration for breast cancer detection using TAS

TAS is based on the experimental findings showing that the complex conductivity values of breast tumors differ significantly from those of surrounding normal tissues (Assenheimer *et al.* 2001; Hartov *et al.* 2005; Jossinet and Schmitt 1999; Silva *et al.* 2000). In TAS, with one hand a patient holds a reference electrode through which a sinusoidal voltage $V_0 \sin \omega t$ is applied, while a scanning probe at the ground potential is placed on the surface of the breast. The voltage difference $V_0 \sin \omega t$ produces an electric current flowing through the breast region (see Figure 8.5). The resulting electric potential at position $\mathbf{r} = (x, y, z)$ and time t can be expressed as the real part of $u(\mathbf{r})\,\mathrm{e}^{\mathrm{i}\omega t}$, where the complex potential $u(\mathbf{r})$ is governed by the equation $\nabla \cdot ((\sigma + \mathrm{i}\omega\epsilon)\nabla u) = 0$ in the object, where σ and ϵ denote the conductivity and permittivity, respectively. The scanning probe is equipped with a planar array of electrodes and we measure exit currents (Neumann data) $g = -(\sigma + \mathrm{i}\omega\epsilon)\partial u/\partial \mathbf{n}$, which reflect the electrical properties of tissues under the scanning probe.

The inverse problem of TAS is to detect a suspicious abnormality in a breast region underneath the probe from measured Neumann data g. One may utilize the difference $g - g_0$, where g_0 is reference Neumann data measured beforehand without any anomaly inside the breast region (Ammari *et al.* 2004; Kim *et al.* 2008; Seo *et al.* 2004). This difference $g - g_0$ can be viewed as a kind of background subtraction, so that it makes the anomaly apparently visible. However, it may not be available in practice, and calculating g_0 is not possible since the inhomogeneous complex conductivity of a specific normal breast is unknown. In such a case, we should use a frequency-difference TAS method.

8.3.1 Mathematical Formulation

Let the human body occupy a three-dimensional domain Ω with a smooth boundary $\partial\Omega$. Let Γ and γ be portions of $\partial\Omega$, denoting the probe plane placed on the breast and the surface of the metallic reference electrode, respectively. Through γ, we apply a sinusoidal voltage of $V_0 \sin \omega t$ with frequency $f = \omega/2\pi$ in the range of 50 Hz to 500 kHz. Then the corresponding complex potential u_ω at ω satisfies the following mixed boundary value problem:

$$\begin{cases} \nabla \cdot ((\sigma + \mathrm{i}\omega\epsilon)\nabla u_\omega(\mathbf{r})) = 0 & \text{in } \Omega, \\ u_\omega(\mathbf{r}) = 0, \quad x \in \Gamma, \\ u_\omega(\mathbf{r}) = V_0, \quad x \in \gamma, \\ (\sigma + \mathrm{i}\omega\epsilon)\nabla u_\omega(\mathbf{r}) \cdot \mathbf{n}(\mathbf{r}) = 0, & x \in \partial\Omega \setminus (\Gamma \cup \gamma), \end{cases} \tag{8.97}$$

where $\mathbf{n}$ is the unit outward normal vector to the boundary $\partial\Omega$. Note that both $\sigma = \sigma(\mathbf{r}, \omega)$ and $\epsilon = \epsilon(\mathbf{r}, \omega)$ depend on ω. The scan probe Γ consists of a planar array of electrodes $\mathcal{E}_1, \ldots, \mathcal{E}_m$ and we measure the exit current $g_\omega(j)$ through each electrode $\mathcal{E}_j$:

$$g_\omega := -(\sigma + \mathrm{i}\omega\epsilon)\nabla u_\omega \cdot \mathbf{n}|_\Gamma.$$

In the frequency-difference TAS, we apply voltage at two different frequencies, $f_1 = \omega_1/2\pi$ and $f_2 = \omega_2/2\pi$, with $50\,\text{Hz} \le f_1 < f_2 \le 500\,\text{kHz}$, and measure two sets of corresponding Neumann data g_{ω_1} and g_{ω_2} through Γ at the same time. We assume that there exists a region of breast tumor D beneath the probe Γ so that $\sigma + \mathrm{i}\omega\epsilon$ changes abruptly across ∂D. The inverse problem of frequency-difference TAS is to detect the anomaly D beneath Γ from the difference between g_{ω_1} and g_{ω_2}.

In order for any detection algorithm to be practicable, we must take account of the following limitations.

(a) Since Ω differs for each subject, the algorithm should be robust against any change in the geometry of Ω and also any change in the complex conductivity distribution outside the breast region.
(b) The Neumann data g_ω are available only on a small surface Γ instead of the whole surface $\partial\Omega$.
(c) Since the inhomogeneous complex conductivity of the normal breast without D is unknown, it is difficult to obtain the reference Neumann data g_ω^* in the absence of D.

These limitations are indispensable to a TAS model in practical situations. In the frequency-difference TAS model, we use a weighted frequency difference of Neumann data $g_{\omega_2} - \alpha g_{\omega_1}$ instead of $g_{\omega_2} - g_{\omega_1}$. The weight α is approximately

$$\alpha \approx \frac{\int_\Gamma g_{\omega_2}\,\mathrm{d}s}{\int_\Gamma g_{\omega_1}\,\mathrm{d}s},$$

and it is a crucial factor in anomaly detection. We should note that the simple difference $g_{\omega_2} - g_{\omega_1}$ may fail to extract the anomaly owing to the complicated structure of the solution of the complex conductivity equation. We need to understand how $g_{\omega_2} - \alpha g_{\omega_1}$ reflects a contrast in complex conductivity values between the anomaly D and surrounding normal tissues.

We assume that σ and ϵ are isotropic, positive and piecewise smooth functions in $\bar{\Omega}$. Let u_ω be the $H^1(\Omega)$ solution of (8.97). Denoting the real and imaginary parts of u_ω by $v_\omega = \Re\, u_\omega$ and $h_\omega = \Im\, u_\omega$, the mixed boundary value problem (8.97) can be expressed as the following coupled system:

$$\begin{cases} \nabla\cdot(\sigma\nabla v_\omega) - \nabla\cdot(\omega\epsilon\nabla h_\omega) = 0 & \text{in } \Omega, \\ \nabla\cdot(\omega\epsilon\nabla v_\omega) + \nabla\cdot(\sigma\nabla h_\omega) = 0 & \text{in } \Omega, \\ v_\omega = 0 \quad \text{and} \quad h_\omega = 0 \quad \text{on } \Gamma, \\ v_\omega = V_0 \quad \text{and} \quad h_\omega = 0 \quad \text{on } \gamma, \\ \mathbf{n}\cdot\nabla v_\omega = 0 \quad \text{and} \quad \mathbf{n}\cdot\nabla h_\omega = 0 \quad \text{on } \partial\Omega\setminus(\Gamma\cup\gamma). \end{cases} \tag{8.98}$$

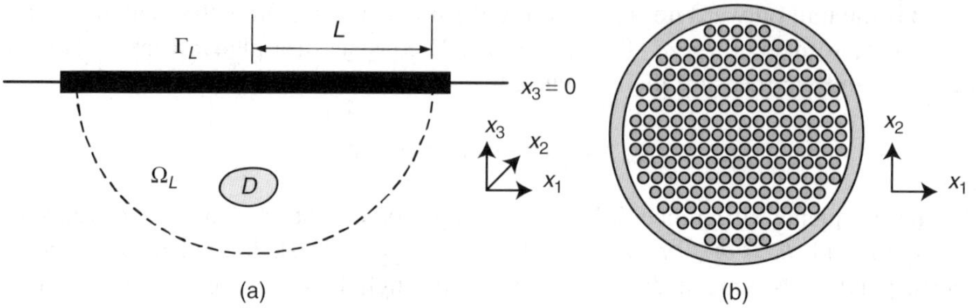

Figure 8.6 (a) Simplified model of the breast region with a cancerous lesion D under the scanning probe. (b) Schematic of the scanning probe in the (x, y) plane

The measured Neumann data g_ω can be decomposed into

$$g_\omega(\mathbf{r}) := \underbrace{\mathbf{n}\cdot(-\sigma\nabla v_\omega(\mathbf{r}) + \omega\epsilon\nabla h_\omega(\mathbf{r}))}_{\text{realpart}} + \mathrm{i}\underbrace{\mathbf{n}\cdot(-\sigma\nabla h_\omega(\mathbf{r}) - \omega\epsilon\nabla v_\omega(\mathbf{r}))}_{\text{imaginarypart}}, \quad x \in \Gamma.$$

The solution of the coupled system (8.98) is a kind of saddle point (Borcea 2002; Cherkaev and Gibiansky 1994), and we have the following relations:

$$V_0 \int_\Gamma \Re(g_\omega)\,\mathrm{d}s = \min_{v\in\mathcal{H}_{\mathrm{re}}} \max_{h\in\mathcal{H}_{\mathrm{im}}} \int_\Omega [\sigma|\nabla v|^2 - 2\omega\epsilon\nabla v\cdot\nabla h - \sigma|\nabla h|^2]\,\mathrm{d}x \tag{8.99}$$

and

$$V_0 \int_\Gamma \Im(g_\omega)\,\mathrm{d}s = \min_{v\in\mathcal{H}_{\mathrm{re}}} \max_{h\in\mathcal{H}_{\mathrm{im}}} \int_\Omega [\omega\epsilon|\nabla v|^2 + 2\sigma\nabla v\cdot\nabla h - \omega\epsilon|\nabla h|^2]\,\mathrm{d}x, \tag{8.100}$$

where we have $\mathcal{H}_{\mathrm{re}} := \{v \in H^1(\Omega) : v|_\Gamma = 0,\ v|_\gamma = V_0,\ (\partial v/\partial\mathbf{n})|_{\partial\Omega\setminus(\Gamma\cup\gamma)} = 0\}$ and $\mathcal{H}_{\mathrm{im}} := \{h \in H^1(\Omega) : h|_{\Gamma\cup\gamma} = 0,\ (\partial h/\partial\mathbf{n})|_{\partial\Omega\setminus(\Gamma\cup\gamma)} = 0\}$.

In order to detect a lesion D underneath the scanning probe Γ, we define a local region of interest under the probe plane Γ as shown in Figure 8.6. For simplicity, we let z be the axis normal to Γ and let the center of Γ be the origin. Hence, the probe region Γ can be approximated as a two-dimensional region $\Gamma = \{(x, y, 0) : \sqrt{x^2 + y^2} < L\}$, where L is the radius of the scan probe. We set the region of interest inside the breast as a half-ball $\Omega_L = \Omega \cap B_L$, as shown in Figure 8.6, where B_L is a ball with radius L and its center at the origin.

Remark 8.3.1 *In Table 8.1 we summarize the conductivity and permittivity values of normal and tumor tissues in the breast. Both σ and $\omega\epsilon$ have a unit of* $\mathrm{S\,m^{-1}}$ *and $\sigma + \mathrm{i}\omega\epsilon = \sigma + \mathrm{i}2\pi f\epsilon_0\epsilon_r$, where $\epsilon_0 \approx 8.854 \times 10^{-12}\,\mathrm{F\,m^{-1}}$ is the permittivity of free space and ϵ_r is the relative permittivity. Note that $\omega\epsilon_n/\sigma_n \geq \frac{1}{50}$ for a frequency $f = \omega/2\pi \geq 50\,\mathrm{kHz}$ (Surowiec et al. 1988).*

For successful anomaly detection, we should carefully choose the two frequencies ω_1 and ω_2. One may choose $f_1 = \omega_1/2\pi$ and $f_2 = \omega_2/2\pi$ such that

$$50\,\mathrm{Hz} \leq f_1 \leq 500\,\mathrm{Hz} \quad \text{and} \quad 50\,\mathrm{kHz} \leq f_2 \leq 500\,\mathrm{kHz}. \tag{8.101}$$

Table 8.1 Conductivity and permittivity values of normal and tumor breast tissues

$f = \omega/2\pi$ (Hz)	σ_n (S m^{-1})	σ_c (S m^{-1})	$\omega\epsilon_n$ (S m^{-1})	$\omega\epsilon_c$ (S m^{-1})
≤ 500	0.03	0.2	$\ll \sigma_n$	$\ll \sigma_c$
50×10^3	0.03	0.2	5.6×10^{-4}	1.7×10^{-2}
100×10^3	0.03	0.2	2.8×10^{-4}	2.2×10^{-2}
500×10^3	0.03	0.2	1.1×10^{-3}	5.6×10^{-2}

We denote by $u_1 = v_1 + ih_1$ and $u_2 = v_2 + ih_2$ the complex potentials satisfying (8.98) at ω_1 and ω_2, respectively, and let $g_1 = g_{\omega_1}$ and $g_2 = g_{\omega_2}$. The frequency-difference TAS aims to detect D from a weighted difference between g_1 and g_2.

Now, let us investigate the connection between u_1 and u_2 and whether the frequency-difference Neumann data $g_2 - \alpha g_1$ contain any information about D. Since both σ and ϵ depend on ω and $\mathbf{r}$, $\sigma(\mathbf{r}, \omega_1) \neq \sigma(\mathbf{r}, \omega_2)$ and $\epsilon(\mathbf{r}, \omega_1) \neq \epsilon(\mathbf{r}, \omega_2)$. For simplicity, we denote

$$\sigma_j(\mathbf{r}) = \sigma(\mathbf{r}, \omega_j) \quad \text{and} \quad \epsilon_j(\mathbf{r}) = \epsilon(\mathbf{r}, \omega_j), \quad j = 1, 2.$$

Suppose there is a cancerous lesion D inside Ω_L and the complex conductivity $\sigma_j + i\omega_j\epsilon_j$ changes abruptly across ∂D as in Table 8.1. To distinguish them, we denote

$$\sigma_j = \begin{cases} \sigma_{j,n} & \text{in } \Omega_L \setminus \bar{D}, \\ \sigma_{j,c} & \text{in } D, \end{cases} \quad \text{and} \quad \epsilon_j = \begin{cases} \epsilon_{j,n} & \text{in } \Omega_L \setminus \bar{D}, \\ \epsilon_{j,c} & \text{in } D. \end{cases} \tag{8.102}$$

With the use of this notation, u_1 and u_2 satisfy

$$\begin{cases} \nabla \cdot ((\sigma_1 + i\omega_1\epsilon_1)\nabla u_1) = 0 \text{ in } \Omega, \\ u_1|_\Gamma = 0, \quad u_1|_\gamma = V_0, \\ (\sigma_1 + i\omega_1\epsilon_1)\dfrac{\partial u_1}{\partial \mathbf{n}}\Big|_{\partial\Omega\setminus(\Gamma\cup\gamma)} = 0, \end{cases} \quad \text{and} \quad \begin{cases} \nabla \cdot ((\sigma_2 + i\omega_2\epsilon_2)\nabla u_2) = 0 \text{ in } \Omega, \\ u_2|_\Gamma = 0, \quad u_2|_\gamma = V_0, \\ (\sigma_2 + i\omega_2\epsilon_2)\dfrac{\partial u_2}{\partial \mathbf{n}}\Big|_{\partial\Omega\setminus(\Gamma\cup\gamma)} = 0. \end{cases} \tag{8.103}$$

Remark 8.3.2 *Owing to the complicated structure of (8.99) and (8.100) for the solution u_ω, it is quite difficult to analyze the interrelation between the complex conductivity contrast $\nabla(\sigma + i\omega\epsilon)$ and the Neumann data g_ω. The simple frequency-difference data $g_2 - g_1$ on Γ may fail to extract the anomaly for more general cases of complex conductivity distributions in Ω because of the complicated structure of the solution of (8.98). To be precise, the use of the weighted difference is essential when the background comprises biological materials with non-negligible frequency-dependent complex conductivity values. To explain this clearly, consider a homogeneous complex conductivity distribution in Ω, where $\sigma(\mathbf{r}, \omega) + i\omega\epsilon(\mathbf{r}, \omega)$ depends only on ω. Owing to the frequency dependence, the simple difference $g_2 - g_1$ is not zero, while $g_2 - \alpha g_1 = 0$. Hence, any reconstruction method using $g_2 - g_1$ always produces artifacts because $g_2 - g_1$ does not eliminate modeling errors. See (8.113) for an approximation of $g_2 - g_1$ in the presence of an anomaly D.*

Remark 8.3.3 *Here, we do not consider effects of contact impedances along electrode–skin interfaces. For details about contact impedances, please see Somersalo et al. (1992), Hyvonen (2004) and other publications cited therein. In TAS, we may adopt a skin preparation procedure and electrode gels to reduce contact impedances. Since we cannot expect complete removal of contact impedances, however, we need to investigate how the exit currents are affected by the contact impedances of a planar array of electrodes that are kept at the ground potential. The contact impedance of each electrode leads to a voltage drop across it and therefore the voltage beneath the electrode–skin interface layer would be slightly different from zero. In other words, when contact impedances are not negligible, the surface area in contact with* Γ *cannot be regarded as an equipotential surface any more, and this will result in some changes in exit currents. Future studies are needed to estimate how the contact impedance affects the weighted difference of the Neumann data. We should also investigate experimental techniques, including choice of frequencies to minimize their effects.*

The next observation explains why we should use a weighted difference $g_2 - \alpha g_1$ instead of $g_2 - g_1$.

Observation 8.3.4 *Denoting*

$$\eta := \frac{\sigma_2 + i\omega_2\epsilon_2}{\sigma_1 + i\omega_1\epsilon_1},$$

it follows from a direct computation that $u_2 - u_1$ *satisfies*

$$\begin{cases} \nabla \cdot ((\sigma_1 + i\omega_1\epsilon_1)\nabla(u_2 - u_1)) = -(\sigma_1 + i\omega_1\epsilon_1)\nabla \log \eta \cdot \nabla u_2 & \text{in } \Omega, \\ (u_2 - u_1)|_{\Gamma\cup\gamma} = 0, \\ (\sigma_1 + i\omega_1\epsilon_1)\dfrac{\partial(u_2 - u_1)}{\partial \mathbf{n}}\Big|_{\partial\Omega\setminus(\Gamma\cup\gamma)} = 0. \end{cases} \tag{8.104}$$

For the detection of D, we use the following weighted difference:

$$g_2 - \alpha g_1 = \eta(\sigma_1 + i\omega_1\epsilon_1)\mathbf{n} \cdot \nabla(u_2 - u_1) \quad \text{on } \Gamma,$$

where $\alpha = \eta|_\Gamma$. *If* $\nabla \log \eta = 0$ *in (8.104), then* $u_1 = u_2$ *in* Ω *and* $g_2 - \alpha g_1 = 0$ *on* Γ. *In other words, if* $\nabla \log \eta = 0$ *in* Ω_L, *it is impossible to detect D from* $g_2 - \alpha g_1 = 0$ *regardless of the contrasts in* σ *and* ϵ *across* ∂D. *Any useful information on D could be found from non-zero* $g_2 - \alpha g_1$ *on* Γ *when* $|\nabla \log \eta|$ *is large along* ∂D.

For chosen frequencies ω_1 *and* ω_2, *we can assume that* σ *and* ϵ *are approximately constant in the normal breast region* $\Omega_L \setminus \bar{D}$ *and also in the cancerous region D. Hence, if* η *changes abruptly across* ∂D, *we roughly have*

$$\nabla \log \eta \approx 0 \text{ in } \Omega_L \setminus \bar{D} \quad \text{and} \quad |\nabla \log \eta| = \infty \quad \text{on } \partial D,$$

and therefore the term $(\sigma_1 + i\omega_1\epsilon_1)\nabla \log \eta \cdot \nabla u_2$ *in (8.104) is supported on* ∂D *in the breast region* Ω_L. *This explains why the difference* $g_2 - \alpha g_1$ *on* Γ *can provide information of* ∂D. *We note that the inner product* $\nabla \log \eta \cdot \nabla u_2$ *is to be interpreted in a suitable distributional sense if the coefficients jump at* ∂D.

8.3.2 Representation Formula

Observation 8.3.4 in the previous section roughly explains how D is related to $g_2 - \alpha g_1$. In this section, the observation will be justified rigorously in a simplified model. We assume that $\sigma_{j,n}$, $\sigma_{j,c}$, $\epsilon_{j,n}$ and $\epsilon_{j,c}$ are constants. According to Table 8.1, the change in conductivity due to the change in frequency is small, so we assume that

$$\sigma_{1,n} = \sigma_{2,n} := \sigma_n \quad \text{and} \quad \sigma_{1,c} = \sigma_{2,c} := \sigma_c. \tag{8.105}$$

Since the breast region of interest is relatively small compared with the entire body Ω, we may assume that Ω is the lower half-space $\Omega = \mathbb{R}^3_- := \{\mathbf{x} = (x, y, z) \mid z < 0\}$ and $\gamma = \infty$.

Suppose that v_j and h_j are H^1-solutions of the following coupled system for $j = 1, 2$:

$$\begin{cases} \nabla \cdot (\sigma \nabla v_j) - \nabla \cdot (\omega_j \epsilon_j \nabla h_j) = 0 & \text{in } \Omega = \mathbb{R}^3_-, \\ \nabla \cdot (\omega_j \epsilon_j \nabla v_j) + \nabla \cdot (\sigma \nabla h_j) = 0 & \text{in } \Omega = \mathbb{R}^3_-, \\ v_j = 1 \quad \text{and} \quad h_j = 0 & \text{on } \Gamma, \\ \mathbf{n} \cdot \nabla v_j = 0 \quad \text{and} \quad \mathbf{n} \cdot \nabla h_j = 0 & \text{on } \Omega \setminus \Gamma. \end{cases} \tag{8.106}$$

Let $u_j = v_j + ih_j$. Then $V_0(1 - u_j)$ can be viewed as a solution of (8.103) with $\Omega = \mathbb{R}^3_-$ and $\gamma = \infty$.

Let us introduce a key representation formula explaining the relationship between D and the weighted difference $g_2 - \alpha g_1$. For each $x \in \mathbb{R}^3 \backslash \Gamma$, we define

$$\Psi(\mathbf{r}, \mathbf{v}) = \Phi(\mathbf{r}, \mathbf{r}') + \Phi(\mathbf{r}, \mathbf{r}'_+) + \varphi(\mathbf{r}, \mathbf{r}'),$$

where $\mathbf{r}'_+ = (x', y', -z')$ is the reflection point of $\mathbf{r}'$ with respect to the plane $\{z = 0\}$ and $\varphi(\mathbf{r}, \cdot)$ is the $H^1(\mathbb{R}^3 \backslash \Gamma)$-solution of the following PDE:

$$\begin{cases} \nabla^2_{\mathbf{r}'} \varphi(\mathbf{r}, \mathbf{r}') = 0, & y \in \mathbb{R}^3 \setminus \Gamma, \\ \varphi(\mathbf{r}, \mathbf{r}') = \dfrac{1}{2\pi |\mathbf{r} - \mathbf{r}'|}, & \mathbf{r}' \in \Gamma, \\ \varphi(\mathbf{r}, \mathbf{r}') = 0 & \text{as } |\mathbf{r}'| \to \infty. \end{cases}$$

The following theorem explains an explicit relation between D and $\Im(g_2 - \alpha g_1)$.

Theorem 8.3.5 (Kim *et al.* 2008) *The imaginary part of the weighted difference $g_2 - \alpha g_1$ satisfies the following (for $\mathbf{r} \in \Gamma$):*

$$\frac{1}{2\sigma_n} \Im(g_2 - \alpha g_1)(\mathbf{r}) = \int_D \nabla_{\mathbf{r}'} \frac{\partial \Phi(\mathbf{r}, \mathbf{r}')}{\partial z} \cdot \Theta(\mathbf{r}') \, \mathrm{d}\mathbf{r}' + \frac{\partial}{\partial z} \int_{\partial\Omega \backslash \Gamma} \frac{\partial \Phi(\mathbf{r}, \mathbf{r}')}{\partial z'} \left[\int_D \nabla_{\tilde{\mathbf{r}}} \Psi(\mathbf{r}', \tilde{\mathbf{r}}) \cdot \Theta(\tilde{\mathbf{r}}) \, \mathrm{d}\tilde{\mathbf{r}} \right] \mathrm{d}S, \tag{8.107}$$

where

$$\Theta(\mathbf{r}') = \frac{\sigma_n - \sigma_c}{\sigma_n} \nabla(h_2 - h_1)(\mathbf{r}') + \frac{\omega_2(\epsilon_{2,n} - \epsilon_{2,c})}{\sigma_n} \nabla(v_2 - v_1)(\mathbf{r}') - \Im(\beta \nabla u_1(\mathbf{r}'))$$

and

$$\beta = \frac{\mathrm{i}}{1+\mathrm{i}\omega_1\epsilon_{1,n}/\sigma_n}\left[\frac{\omega_2\epsilon_{2,n}}{\sigma_n}\left(\frac{\epsilon_{2,c}}{\epsilon_{2,n}}-\frac{\sigma_c}{\sigma_n}\right)-\frac{\omega_1\epsilon_{1,n}}{\sigma_n}\left(\frac{\epsilon_{1,c}}{\epsilon_{1,n}}-\frac{\sigma_c}{\sigma_n}\right)\right.$$
$$\left.-\,\mathrm{i}\frac{\omega_1\omega_2\epsilon_{1,n}\epsilon_{2,n}}{\sigma_n^2}\left(\frac{\epsilon_{1,c}}{\epsilon_{1,n}}-\frac{\epsilon_{2,c}}{\epsilon_{2,n}}\right)\right].$$

Now, let us derive a constructive formula extracting D from the representation formula (8.107) under some reasonable assumptions. We assume that

$$\bar{D}\subset\Omega_{L/2},\quad D=B_\delta(\boldsymbol{\xi})\quad\text{and}\quad\delta\le\mathrm{dist}(D,\Gamma)\le C_1\delta, \tag{8.108}$$

where C_1 is a positive constant, B_δ is a ball with radius δ and center $\boldsymbol{\xi}$, and $\delta/L\le\frac{1}{10}$. Suppose we choose $\omega_1/2\pi\approx 50\,\mathrm{Hz}$ and $\omega_2/2\pi\approx 100\,\mathrm{kHz}$. Then the experimental data in Remark 8.3.1 shows that

$$\frac{\omega_2\epsilon_{2,n}}{\sigma_n}\approx\frac{1}{100}\quad\text{and}\quad\frac{\omega_1\epsilon_{1,n}}{\sigma_n}\le\frac{1}{10\,000}.$$

Hence, in practice, we can assume that

$$\frac{\omega_1\epsilon_{1,n}}{\sigma_n}\approx 0,\quad\left(\frac{\delta}{L}\right)^3\approx 0,\quad\left(\frac{\omega_2\epsilon_{2,n}}{\sigma_n}\right)^2\approx 0. \tag{8.109}$$

Based on the experimental data in Remark 8.3.1, we assume that

$$\max\left\{\frac{\epsilon_{j,n}}{\epsilon_{j,c}},\frac{\sigma_n}{\sigma_c}\right\}\le\kappa_1,\qquad\frac{\omega_2\epsilon_{2,n}}{\sigma_n}\le\kappa_2\frac{\sigma_n}{\sigma_c},\qquad\frac{\sigma_c}{\sigma_n}\le\kappa_3, \tag{8.110}$$

where κ_1 and κ_2 are positive constants less than $\frac{1}{2}$ and κ_3 is a positive constant less than 10. Taking advantage of these, we can simplify the representation formula (8.107).

Theorem 8.3.6 *Under the assumptions (8.108) and (8.110), the imaginary part of the weighted frequency difference* $g_2-\alpha g_1$ *can be expressed as*

$$\frac{1}{2\sigma_n}\Im\left(g_2-\alpha g_1\right)(\mathbf{r})=\int_D\frac{\partial}{\partial z}\frac{(\mathbf{r}-\mathbf{r}')\cdot\widetilde{\Theta}(\mathbf{r}')}{4\pi|\mathbf{r}-\mathbf{r}'|^3}\,\mathrm{d}\mathbf{r}'+\mathrm{Error}(\mathbf{r}),\quad x\in\Gamma_{L/2}, \tag{8.111}$$

where

$$\widetilde{\Theta}=\frac{\sigma_n-\sigma_c}{\sigma_n}\nabla h_2-\frac{\omega_2\epsilon_{2,n}}{\sigma_n}\left(\frac{\epsilon_{2,c}}{\epsilon_{2,n}}-\frac{\sigma_c}{\sigma_n}\right)\nabla v_1,$$

and the error term Error($\mathbf{r}$) *is estimated by*

$$|\mathrm{Error}(\mathbf{r})|\le\left\{\frac{\omega_2\epsilon_{2,n}}{\sigma_n}\mathcal{P}_1\left(\left|\frac{\epsilon_{2,c}}{\epsilon_{2,n}}-\frac{\sigma_c}{\sigma_n}\right|\right)\frac{\delta^3}{L^3}\right. \tag{8.112}$$
$$\left.+\left[\frac{\omega_1\epsilon_{1,n}}{\sigma_n}\mathcal{P}_1\left(\left|\frac{\epsilon_{1,c}}{\epsilon_{1,n}}-\frac{\sigma_c}{\sigma_n}\right|\right)+\left(\frac{\omega_2\epsilon_{2,n}}{\sigma_n}\right)^2\mathcal{P}_2\left(\left|\frac{\epsilon_{2,c}}{\epsilon_{2,n}}-\frac{\sigma_c}{\sigma_n}\right|\right)\right]\frac{\delta^3}{|\mathbf{r}-\boldsymbol{\xi}|^3}\right\}.$$

Here, $\mathcal{P}_n(\lambda)$ is a polynomial function of order n such that $\mathcal{P}_n(0) = 0$ and its coefficients depend only on κ_j, $j = 1, 2, 3$.

Remark 8.3.7 *According to Theorem 8.3.6,*

$$\frac{1}{2\sigma_n}\Im\left(g_2 - \alpha g_1\right) = 0 \quad \textit{when} \quad \left|\frac{\epsilon_{j,c}}{\epsilon_{j,n}} - \frac{\sigma_c}{\sigma_n}\right| = 0, \quad j = 1, 2.$$

Hence, even if $\epsilon_{2,c}$ and $\epsilon_{1,c}$ are quite different, we cannot extract any information on D when

$$4\left|\frac{\epsilon_{j,c}}{\epsilon_{j,n}} - \frac{\sigma_c}{\sigma_n}\right| = 0, \quad j = 1, 2.$$

On the other hand, even if $\epsilon_{2,c} = \epsilon_{1,c}$, we can extract information on D whenever

$$\left|\frac{\epsilon_{j,c}}{\epsilon_{j,n}} - \frac{\sigma_c}{\sigma_n}\right| \neq 0, \quad j = 1, 2.$$

Remark 8.3.8 *Based on (8.111), we can derive the following simple approximate formula for the reconstruction of D:*

$$\begin{aligned}\frac{1}{2\sigma_n}\Im\left(g_2 - \alpha g_1\right)(\mathbf{r}) &\approx \frac{\omega_2\epsilon_{2,n}}{\sigma_n}\frac{(3\sigma_n)^2}{(2\sigma_n + \sigma_c)^2}\left(\frac{\epsilon_{2,c}}{\epsilon_{2,n}} - \frac{\sigma_c}{\sigma_n}\right)\partial_z U(\boldsymbol{\xi}) \\ &\times |D|\frac{2\xi_3^2 - (x - \xi_1)^2 - (y - \xi_2)^2}{4\pi|\mathbf{r} - \boldsymbol{\xi}|^5}, \quad x \in \Gamma_{L/2}, \end{aligned} \tag{8.113}$$

where U is the solution of (8.106) in the absence of any anomaly at $\omega = 0$. Note that the difference $g_2 - g_1$ can be approximated by

$$\begin{aligned}\frac{1}{2\sigma_n}(g_2 - g_1)(\mathbf{r}) &\approx \mathrm{i}\frac{\omega_2\epsilon_{2,n}}{\sigma_n}g_1(\mathbf{r}) + \mathrm{i}\frac{\omega_2\epsilon_{2,n}}{\sigma_n}\frac{(3\sigma_n)^2}{(2\sigma_n + \sigma_c)^2}\left(\frac{\epsilon_{2,c}}{\epsilon_{2,n}} - \frac{\sigma_c}{\sigma_n}\right)\partial_z U(\boldsymbol{\xi}) \\ &\times |D|\frac{2\xi_3^2 - (x - \xi_1)^2 - (y - \xi_2)^2}{4\pi|\mathbf{r} - \boldsymbol{\xi}|^5}, \quad x \in \Gamma_{L/2}, \end{aligned} \tag{8.114}$$

and therefore any detection algorithm using the above approximation will be disturbed by the term $(\omega_2\epsilon_{2,n}/\sigma_n)g_1$.

Remark 8.3.9 *The reconstruction algorithm is based on the approximation formula (8.113). In practice, we may not have* a priori *knowledge of the background conductivities. In that case, α is unknown. But α can be evaluated approximately by the ratio of the measured Neumann data as follows:*

$$\alpha = \frac{\int_\Gamma g_{\omega_2}\,\mathrm{d}s}{\int_\Gamma g_{\omega_1}\,\mathrm{d}s} + \frac{\int_D((1-\alpha)\sigma_c + \mathrm{i}(\omega_2\epsilon_{2,c} - \alpha\omega_1\epsilon_{1,c}))\nabla u_2 \cdot \nabla u_1\,\mathrm{d}x}{\int_\Omega(\sigma + \mathrm{i}\omega_1\epsilon_1)|\nabla u_1|^2\,\mathrm{d}x}. \tag{8.115}$$

Hence, we may choose $\alpha \approx \dfrac{\int_\Gamma g_{\omega_2}\,\mathrm{d}s}{\int_\Gamma g_{\omega_1}\,\mathrm{d}s}$.

We can prove the identity (8.115) for a bounded domain Ω. Using $u_1|_\gamma = u_2|_\gamma = V_0$, we have

$$\begin{aligned}\int_\Gamma (g_2 - \alpha g_1)\,\mathrm{d}s &= -\int_\gamma (g_2 - \alpha g_1)\,\mathrm{d}s = -\frac{1}{V_0}\int_\gamma (g_2 u_1 - \alpha g_1 u_1)\,\mathrm{d}s \\ &= -\frac{1}{V_0}\int_{\partial\Omega} (g_2 - \alpha g_1)u_1\,\mathrm{d}s \\ &= \frac{1}{V_0}\int_\Omega ((\sigma + \mathrm{i}\omega_1\epsilon_2)\nabla u_2 - \alpha(\sigma + \mathrm{i}\omega_1\epsilon_1)\nabla u_1)\cdot\nabla u_1\,\mathrm{d}x \\ &= \frac{1}{V_0}\int_\Omega \big[\alpha(\sigma + \mathrm{i}\omega_1\epsilon_1)(\nabla u_2 - \nabla u_1)\cdot\nabla u_1 \\ &\quad + ((1-\alpha)\sigma + \mathrm{i}(\omega_2\epsilon_2 - \alpha\omega_1\epsilon_1))\nabla u_2\cdot\nabla u_1\big]\,\mathrm{d}x \\ &= \frac{1}{V_0}\int_D ((1-\alpha)\sigma_c + \mathrm{i}(\omega_2\epsilon_{2,c} - \alpha\omega_1\epsilon_{1,c}))\nabla u_2\cdot\nabla u_1\,\mathrm{d}x.\end{aligned}$$

The identity (8.115) follows from the fact that

$$V_0\int_\Gamma g_{\omega_1}\,\mathrm{d}s = \int_\Omega (\sigma + \mathrm{i}\omega_1\epsilon_1)|\nabla u_1|^2\,\mathrm{d}x.$$

References

Alessandrini G, Rosset E and Seo JK 2000 Optimal size estimates for the inverse conductivity problem with one measurement. *Proc. Am. Math. Soc*. **128**, 53–64.

Ammari H and Kang H 2007 *Polarization and Moment Tensors: With Applications to Inverse Problems and Effective Medium Theory*. Applied Mathematical Sciences, no. 162. Springer, New York.

Ammari H, Kwon O, Seo JK and Woo EJ 2004 T-scan electrical impedance imaging system for anomaly detection. *SIAM J. Appl. Math*. **65**, 252–266.

Assenheimer M, Laver-Moskovitz O, Malonek D, Manor D, Nahliel U, Nitzan R and Saad A 2001 The T-scan technology: electrical impedance as a diagnostic tool for breast cancer detection. *Physiol. Meas*. **22**, 1–8.

Borcea L 2002 EIT electrical impedance tomography. *Inv. Prob*., **18**(6), R99–R136.

Cherkaev AV and Gibiansky LV 1994 Variational principles for complex conductivity, viscoelasticity and similar problems in media with complex moduli. *J. Math. Phys*. **35**(1), 127–145.

Coifman RR, McIntosh A and Meyer Y 1982 L'intégrale de Cauchy definit un opérateur bournée sur L^2 pour courbes lipschitziennes. *Ann. Math*. **116**, 361–387.

David G and Journé JL 1984 A boundedness criterion for generalized Calderón–Zygmund operators. *Ann. Math*. **120**, 371–397.

Escauriaza L and Seo JK 1993 Regularity properties of solutions to transmission problems. *Trans. Am. Math. Soc*. **338**, 405–430.

Evans LC 2010 *Partial Differential Equations*. Graduate Studies in Mathematics, vol. 19. American Mathematical Society, Providence, RI.

Fabes E, Jodeit M and Riviére N 1978 Potential techniques for boundary value problems on C^1 domains. *Acta Math*. **141**, 165–186.

Folland G 1976 *Introduction to Partial Differential Equations*. Princeton University Press, Princeton, NJ.

Garofalo N and Lin F 1986 Monotonicity properties of variational integrals, A_p weights and unique continuation. *Indiana Univ. Math. J*. **35**, 245–268.

Hartov A, Soni N and Halter R 2005 Breast cancer screening with electrical impedance tomography. In *Electrical Impedance Tomography: Methods, History and Applications*, ed. DS Holder, pp. 167–185. IOP Publishing, Bristol.

Hyvonen N 2004 Complete electrode model of electric impedance tomography: approximation properties and characterization of inclusions. *SIAM J. Appl. Math*. **64**, 902–931.
Jossinet J and Schmitt M 1999 A review of parameters for the bioelectrical characterization of breast tissue. *Ann. N.Y. Acad. Sci*. **873**, 30–41.
Kang H and Seo JK 1996 Layer potential technique for the inverse conductivity problem. *Inv. Prob*. **12**, 267–278.
Kang H, Seo JK and Sheen D 1997 The inverse conductivity problem with one measurement: stability and estimation of size. *SIAM J. Math. Anal*. **28**, 1389–1405.
Kenig C, Sjostrand J and Uhlmann G 2007 The Calderon problem with partial data. *Ann. Math*. **165**, 567–591.
Kim S, Lee J, Seo JK, Woo EJ and Zribi H 2008 Multi-frequency trans-admittance scanner: mathematical framework and feasibility. *SIAM J. Appl. Math*. **69**, 22–36.
Kwon O and Seo JK 2001 Total size estimation and identification of multiple anomalies in the inverse electrical impedance tomography. *Inv. Prob*. **17**, 59–75.
Kwon O, Seo JK and Yoon JR 2002 A real-time algorithm for the location search of discontinuous conductivities with one measurement. *Commun. Pure Appl. Math*. **55**, 1–29.
Kwon O, Yoon JR, Seo JK, Woo EJ and Cho YG 2003 Estimation of anomaly location and size using electrical impedance tomography. *IEEE Trans. Biomed. Eng*. **50**, 89–96.
Rudin W 1973 *Functional Analysis*. McGraw-Hill, New York.
Seo JK, Kwon O, Ammari H and Woo EJ 2004 Mathematical framework and anomaly estimation algorithm for breast cancer detection: electrical impedance technique using TS2000 configuration. *IEEE Trans. Biomed. Eng*. **51**(11), 1898–1906.
Silva JE, Marques JP and Jossinet J 2000 Classification of breast tissue by electrical impedance spectroscopy. *Med. Biol. Eng. Comput*. **38**, 26–30.
Somersalo E, Cheney M and Isaacson D 1992 Existence and uniqueness for electrode models for electric current computed tomography. *SIAM J. Appl. Math*. **52**, 1023–1040.
Surowiec AJ, Stuchly SS, Barr JR and Swarup A 1988 Dielectric properties of breast carcinoma and the surrounding tissues. *IEEE Trans. Biomed. Eng*. **35**, 257–263.
Verchota G 1984 Layer potentials and boundary value problems for Laplace's equation in Lipschitz domains. *J. Funct. Anal*. **59**, 572–611.

Further Reading

Adler A, Arnold JH, Bayford R, Borsic A, Brown B, Dixon P, Faes TJC, Frerichs I, Gagnon H, Gärber Y, Grychtol B, Hahn G, Lionheart WRB, Malik A, Patterson RP, Stocks J, Tizzard A, Weiler N and Wolf GK 2009 GREIT: a unified approach to 2D linear EIT reconstruction of lung images. *Physiol. Meas*. **30**, S35–S55.
Alessandrini G and Magnanini R 1992 The index of isolated critical points and solutions of elliptic equations in the plane. *Ann. Scu. Norm. Sup. Pisa Cl. Sci*. **19**, 567–589.
Alessandrini G, Isakov V and Powell J 1995 Local uniqueness in the inverse problemwith one measurement. *Trans. Am. Math. Soc*. **347**, 3031–3041.
Ammari H and Seo JK 2003 An accurate formula for the reconstruction of conductivity inhomogeneity. *Adv. Appl. Math*. **30**, 679–705.
Ammari H, Moskow S and Vogelius MS 2003 Boundary integral formulae for the reconstruction of electric and electromagnetic inhomogeneities of small volume. *ESAIM: Control Optim. Calc. Var*. **9**, 49–66.
Astala K and Päivärinta L 2006 Calderon's inverse conductivity problem in the plane. *Ann. Math*. **163**, 265–299.
Barber DC and Brown BH 1984 Applied potential tomography. *J. Phys. Sci. Instrum*. **17** 723–733.
Bellout H and Friedman A 1988 Identification problem in potential theory. *Arch. Rat. Mech. Anal*. **101**, 143–160.
Bellout H, Friedman A and Isakov V 1992 Inverse problem in potential theory. *Trans. Am. Math. Soc*. **332**, 271–296.
Berenstein C and Tarabusi EC 1991 Inversion formulas for the k-dimensional Radon transform in real hyperbolic spaces. *Duke Math. J*. **62**, 1–9.
Boone K, Barber D and Brown B 1997 Imaging with electricity: report of the European Concerted Action on Impedance Tomography. *J. Med. Eng. Technol*. **21**(6), 201–202.

Brown R and Uhlmann G 1997 Uniqueness in the inverse conductivity problem with less regular conductivities in two dimensions. *Commun. Part. Differ. Eqns* **22**, 1009–1027.

Brown BH, Barber DC and Seagar AD 1985 Applied potential tomography: possible clinical applications. *Clin. Phys. Physiol. Meas.* **6**, 109–121.

Brühl M and Hanke M 2000 Numerical implementation of two non-iterative methods for locating inclusions by impedance tomography. *Inv. Prob.* **16**, 1029–1042.

Bryan K 1991 Numerical recovery of certain discontinuous electrical conductivities. *Inv. Prob.* **7**, 827–840.

Calderón AP 1980 On an inverse boundary value problem. In *Seminar on Numerical Analysis and its Applications to Continuum Physics*, pp. 65–73. Sociedade Brasileira de Matemática, Rio de Janeiro.

Cedio-Fengya DJ, Moskow S and Vogelius M 1998 Identification of conductivity imperfections of small parameter by boundary measurements. Continuous dependence and computational reconstruction. *Inv. Prob.* **14**, 553–595.

Cheney M, Isaacson D, Newell J, Goble J and Simske S 1990 NOSER: an algorithm for solving the inverse conductivity problem. *Int. J. Imag. Syst. Technol.* **2**, 66–75.

Cheney M, Isaacson D and Newell JC 1999 Electrical impedance tomography. *SIAM Rev.* **41**, 85–101.

Cherepenin V, Karpov A, Korjenevsky A, Kornienko V, Mazaletskaya A, Mazourov D and Meister J 2001 A 3D electrical impedance tomography (EIT) system for breast cancer detection. *Physiol. Meas.* **22**, 9–18.

Cherepenin V, Karpov A, Korjenevsky A, Kornienko V, Kultiasov Y, Ochapkin M, Trochanova O and Meister J 2002 Three-dimensional EIT imaging of breast tissues: system design and clinical testing. *IEEE Trans. Med. Imag.* **21**, 662–667.

Cohen-Bacrie C and Guardo R 1997 Regularized reconstruction in electrical impedance tomography using a variance uniformization constraint. *IEEE Trans. Med. Imag.* **16**(5), 562–571.

Colton D and Kress R 1998 *Inverse Acoustic and Electromagnetic Scattering Theory*, 2nd edn. Springer, Berlin.

Cook RD, Saulnier GJ, Gisser DG, Goble JG, Newell JC and Isaacson D 1994 ACT3: a high-speed, high-precision electrical impedance tomography. *IEEE Trans. Biomed. Eng.* **41**, 713–722.

Dobson D and Santosa F 1994 An image-enhancement technique for electrical impedance tomography. *Inv. Prob.* **10**, 317–334.

Escauriaza L, Fabes E and Verchota G 1992 On a regularity thoerem for weak solutions to transmission problems with internal Lipschitz boundaries. *Proc. Am. Math. Soc.* **115**, 1069–1076.

Fabes E, Sand M and Seo JK 1992 The spectral radius of the classical layer potentials on convex domains. *IMA Vol. Math. Appl.* **42**, 129–137.

Fabes E, Kang H and Seo JK 1999 Inverse conductivity problem: error estimates and approximate identification for perturbed disks. *SIAM J. Math. Anal.* **30**, 699–720.

Feldman J and Uhlmann G 2003 *Inverse Problems*, Lecture Note. See http://www.math.ubc.ca/~feldman/ibook/.

Franco S 2002 *Design with Operational Amplifiers and Analog Integrated Circuits*, 3rd edn. McGraw-Hill, New York.

Friedman A and Isakov V 1989 On the uniqueness in the inverse conductivity problem with one measurement. *Indiana Univ. Math. J.* **38**, 563–579.

Friedman A and Vogelius MS 1989 Identification of small inhomogeneities of extreme conductivity by boundary measurements: a theorem on continuous dependence. *Arch. Rat. Mech. Anal.* **105**, 299–326.

Fuks LF, Cheney M, Isaacson D, Gisser DG and Newell JC 1991 Detection and imaging of electric conductivity and permittivity at low frequency. *IEEE Trans. Biomed. Eng.* **3**, 1106–1110.

Gabriel C, Gabriel S and Corthout E 1996 The dielectric properties of biological tissues: I. Literature survey. *Phys. Med. Biol.* **41**, 2231–2249.

Gabriel S, Lau RW and Gabriel C 1996 The dielectric properties of biological tissues: II. Measurements in the frequency range 10 Hz to 20 GHz. *Phys. Med. Biol.* **41**, 2251–2269.

Geddes LA and Baker LE 1967 The specific resistance of biological material: a compendium of data for the biomedical engineer and physiologist. *Med. Biol. Eng.* **5**, 271–293.

Giaquinta M 1983 *Multiple Integrals in the Calculus of Variations and Non-Linear Elliptic Systems*. Princeton University Press, Princeton, NJ.

Gilbarg D and Trudinger N 1998 *Elliptic Partial Differential Equations of Second Order*. Springer, Berlin.

Gisser DG, Isaacson D and Newell JC 1988 Theory and performance of an adaptive current tomography system. *Clin. Phys. Physiol. Meas.* **9** (Suppl. A), 35–41.

Gisser DG, Isaacson D and Newell JC 1990 Electric current computed tomography and eigenvalues. *SIAM J. Appl. Math.* **50**, 1623–1634.

Grimnes S and Martinsen OG *Bioimpedance and Bioelectricity Basics*. Academic Press, London.

Grisvard P 1985 *Elliptic Problems in Nonsmooth Domains*. Monographs and Studies in Mathematics, no. 24. Pitman, Boston, MA.

Henderson RP and Webster JG 1978 An impedance camera for spatially specific measurements of the thorax. *IEEE Trans. Biomed. Eng*., **25**, 250–254.

Hettlich F and Rundell W 1998 The determination of a discontinuity in a conductivity from a single boundary measurement. *Inv. Prob*. **14**, 67–82.

Holder D (ed.) 2005 *Electrical Impedance Tomography: Methods, History and Applications*. IOP Publishing, Bristol.

Hua P, Tompkins W and Webster J 1988 A regularized electrical impedance tomography reconstruction algorithm. *Clin. Phys. Physiol. Meas*. **9**, 137–141.

Hyaric AL and Pidcock MK 2001 An image reconstruction algorithm for three-dimensional electrical impedance tomography. *IEEE Trans. Biomed. Eng*. **48**(2), 230–235.

Ikehata M 2000 On reconstruction in the inverse conductivity problem with one measurement. *Inv. Prob*. **16**, 785–793.

Isaacson D 1986 Distinguishability of conductivities by electric current computed tomography. *IEEE Trans. Med. Imag*. **5**(2), 91–95.

Isaacson D and Cheney M 1991 Effects of measurement precision and finite numbers of electrodes on linear impedance imaging algorithms. *SIAM J. Appl. Math*. **51** 1705–1731.

Isaacson D and Cheney M 1996 *Process for producing optimal current patterns for electrical impedance tomography*. US Patent 5588429, 31 December.

Isaacson D and Isaacson E 1989 Comment on Calderon's paper: "On an inverse boundary value problem". *Math. Comput*. **52**, 553–559.

Isakov V 1988 On uniqueness of recovery of a discontinuous conductivity coefficient. *Commun. Pure Appl. Math*. **41**, 856–877.

Isakov V 1998 *Inverse Problems for Partial Differential Equations*. Applied Mathematical Sciences, no. 127. Springer, New York.

Lieb EH and Loss M 2001 *Analysis*, 2nd edn. Graduate Studies in Mathematics, vol. 14. American Mathematical Society, Providence, RI.

Kao T, Newell JC, Saulnier GJ and Isaacson D 2003 Distinguishability of inhomogeneities using planar electrode arrays and different patterns of applied excitation. *Physiol. Meas*. **24**, 403–411.

Kerner TE, Paulsen KD, Hartov A, Soho SK and Poplack SP 2002 Electrical impedance spectroscopy of the breast: clinical imaging results in 26 subjects. *IEEE Trans. Med. Imag*. **21**, 638–645.

Kellogg OD 1953 *Foundations of Potential Theory*. Dover, New York.

Kohn R and Vogelius M 1984 Determining conductivity by boundary measurements. *Commun. Pure Appl. Math*. **37**, 113–123.

Larson-Wiseman JL 1998 Early breast cancer detection utilizing clustered electrode arrays in impedance imaging. PhD Thesis, Rensselaer Polytechnic Institute, Troy, NY.

Lionheart W, Polydorides W and Borsic A 2005 The reconstruction problem. In *Electrical Impedance Tomography: Methods, History and Applications*, ed. DS Holder. IOP Publishing, Bristol.

Liu N, Saulnier GJ, Newell JC, Isaacson D and Kao TJ 2005 ACT4: a high-precision, multi-frequency electrical impedance tomography. In *Proc. Conf. on Biomedical Applications of Electrical Impedance Tomography*, University College London, 22–24 June.

Marsden JE 1974 *Elementary Classical Analysis*. W.H. Freeman, San Francisco.

Mast TD, Nachman A and Waag RC 1997 Focusing and imagining using eigenfunctions of the scattering operator. *J. Acoust. Soc. Am*. **102**, 715–725.

Metherall P, Barber DC, Smallwood RH and Brown BH 1996 Three-dimensional electrical impedance tomography. *Nature* **380**, 509–512.

Mueller JL, Isaacson D and Newell JC 1999 A reconstruction algorithm for electrical impedance tomography data collected on rectangular electrode arrays. *IEEE Trans. Biomed. Eng*. **46**, 1379–1386.

Nachman A 1988 Reconstructions from boundary measurements. *Ann. Math*. **128**, 531–577.

Nachman A 1996 Global uniqueness for a two-dimensional inverse boundary value problem. *Ann. Math*. **142**, 71–96.

Newell JC, Gisser DG and Isaacson D 1988 An electric current tomograph. *IEEE Trans. Biomed. Eng*. **35**, 828–833.

Oh TI, Lee J, Seo JK, Kim SW and Woo EJ 2007 Feasibility of breast cancer lesion detection using multi-frequency trans-admittance scanner (TAS) with 10 Hz to 500 kHz bandwidth. *Physiol. Meas.* **28**, S71–S84.

Reed M and Simon B 1980 *Methods of Modern Mathematical Physics I: Functional Analysis*, revised and enlarged edition. Academic Press, San Diego.

Rudin W 1970 *Real and complex analysis*. McGraw-Hill, New York.

Santosa F and Vogelius MS 1990 A backprojection algorithm for electrical impedance imaging. *SIAM J. Appl. Math.* **50**, 216–243.

Scholz B 2002 Towards virtual electrical breast biopsy: space-frequency MUSIC for trans-admittance data. *IEEE Trans. Med. Imag.* **21**, 588–595.

Seo JK 1996 On the uniqueness in the inverse conductivity problem. *J. Fourier Anal. Appl.* **2**, 227–235.

Seo JK, Lee J, Kim SW, Zribi H and Woo EJ 2008 Frequency-difference electrical impedance tomography (fdEIT): algorithm development and feasibility study. *Physiol. Meas.* **29**, 929–944.

Stein EM and Shakarchi R 2005 *Real Analysis: Measure Theory, Integration and Hilbert Spaces*. Princeton University Press, Princeton, NJ.

Sylvester J and Uhlmann G 1986 A uniqueness theorem for an inverse boundary value problem in electrical prospection. *Commun. Pure Appl. Math.* **39**, 91–112.

Sylvester J and Uhlmann G 1987 A global uniqueness theorem for an inverse boundary value problem. *Ann. Math.* **125**, 153–169.

Sylvester J and Uhlmann G 1988 Inverse boundary value problems at the boundary – continuous dependence. *Commun. Pure Appl. Math.* **21**, 197–221.

Tidswell AT, Gibson A, Liston A, Yerworth RJ, Bagshaw A, Wyatt J, Bayford RH and Holder DS 2001 3D electrical impedance tomography of neonatal brain activity. In *Biomedical Applications of EIT, EPSRC 3rd Engineering Network Meeting*, London.

Vauhkonen M, Vadasz D, Karjalainen PA, Somersalo E and Kaipio JP 1998 Tikhonov regularization and prior information in electrical impedance tomography. *IEEE Trans. Med. Imag.* **17**(2), 285–293.

Webster J 1990 *Electrical Impedance Tomography*. Adam Hilger, Bristol.

Wexler A, Fry B and Neuman MR 1985 Impedance-computed tomography algorithm and system. *Appl. Opt.* **24**, 3985–3992.

Wheeden RL and Zygmund A 1977 *Measure and Integral: An Introduction to Real Analysis*. Monographs and Textbooks in Pure and Applied Mathematics, vol. 43. Marcel Dekker, New York.

Wilson AJ, Milnes P, Waterworth AR, Smallwood RH and Brown BH 2001 Mk3.5: a modular, multi-frequency successor to the Mk3a EIS/EIT system. *Physiol. Meas.* **22**, 49–54.

Woo EJ, Hua P, Webster J and Tompkins W 1993 A robust image reconstruction algorithm and its parallel implementation in electrical impedance tomography. *IEEE Trans. Med. Imag.* **12**, 137–146.

Yorkey T, Webster J and Tompkins W 1987 Comparing reconstruction algorithms for electrical impedance tomography. *IEEE Trans. Biomed. Engr.* **34**, 843–852.

Zhang N 1992 Electrical impedance tomography based on current density imaging. MS Thesis, University of Toronto, Canada.

9

Magnetic Resonance Electrical Impedance Tomography

For high-resolution static imaging of a conductivity distribution inside the human body, there have been strong needs for supplementary data to make the inverse problem well-posed and to overcome the fundamental limitations of the electrical impedance tomography (EIT) imaging methods. To bypass the ill-posed nature of EIT, magnetic resonance electrical impedance tomography (MREIT) was proposed in the early 1990s to take advantage of an MRI scanner as a tool to capture internal magnetic flux density data induced by externally injected currents (Birgul and Ider 1995; Birgul and Ider 1996; Woo *et al.* 1994; Zhang 1992).

MREIT aims to visualize conductivity images of an electrically conducting object using the current injection MRI technique (Joy *et al.* 1989; Scott *et al.* 1991, 1992). To probe the passive material property of the conductivity, low-frequency electrical current is injected into the imaging object through surface electrodes. This induces internal distributions of voltage u, current density $\mathbf{J} = (J_x, J_y, J_z)$ and magnetic flux density $\mathbf{B} = (B_x, B_y, B_z)$ dictated by Maxwell's equations. At a low frequency of less than a few kilohertz, we can ignore the effects of permittivity and consider only conductivity. For conductivity image reconstructions, MREIT relies on a set of internal magnetic flux density data, since it is dictated by the conductivity distribution σ according to Ampère's law

$$-\sigma \nabla u = \mathbf{J} = \frac{1}{\mu_0} \nabla \times \mathbf{B},$$

where μ_0 is the magnetic permeability of free space.

In early MREIT systems, all three components of $\mathbf{B} = (B_x, B_y, B_z)$ have been utilized as measured data, and this requires mechanical rotations of the imaging object within the MRI scanner (Birgul *et al.* 2003; Ider *et al.* 2003; Khang *et al.* 2002; Kwon *et al.* 2002a; Lee *et al.* 2003a). Assuming knowledge of the full components of $\mathbf{B}$, we can directly compute the current density $\mathbf{J} = (1/\mu_0)\nabla \times \mathbf{B}$ and reconstruct σ using an image reconstruction algorithm such as the J-substitution algorithm (Khang *et al.* 2002; Kwon *et al.* 2002b; Lee *et al.* 2003a), current-constrained voltage-scaled reconstruction

Nonlinear Inverse Problems in Imaging, First Edition. Jin Keun Seo and Eung Je Woo.

(CCVSR) algorithm (Birgul *et al.* 2003) and equipotential line methods (Ider *et al.* 2003; Kwon *et al.* 2002a). Recently, a new non-iterative conductivity image reconstruction method called current density impedance imaging (CDII) has been suggested and experimentally verified (Hasanov *et al.* 2008; Nachman *et al.* 2007, 2009). These methods using $\mathbf{B} = (B_x, B_y, B_z)$ suffer from technical difficulties related to object rotations within the main magnet of the MRI scanner.

To make the MREIT technique applicable to clinical situations, it is desirable to use only B_z data to avoid object rotation. In 2001, a constructive B_z-based MREIT algorithm called the harmonic B_z algorithm was developed, and numerical simulations and phantom experiments showed that high-resolution conductivity imaging is possible without rotating the object (Oh *et al.* 2003, 2004, 2005; Seo *et al.* 2003a). This novel algorithm is based on the key observation that the Laplacian of B_z, ΔB_z, probes a change of $\ln \sigma$ along any curve having its tangent direction to the vector field $\mathbf{J} \times (0, 0, 1)$. Since then, imaging techniques in MREIT have advanced rapidly and have now reached the stage of *in vivo* animal and human imaging experiments (Kim *et al.* 2008a,b; Kwon *et al.* 2005; Liu *et al.* 2007; Oh *et al.* 2005; Park *et al.* 2004a,b; Seo *et al.* 2004). In this chapter, we study MREIT techniques following the descriptions in Woo and Seo (2008) and Seo and Woo (2011).

9.1 Data Collection using MRI

An MREIT system comprises an MRI scanner, constant-current source and conductivity image reconstruction software. The current source is interfaced to the spectrometer of the MRI scanner to inject current in a synchronized way with a chosen MR pulse sequence. We will assume that the scanner has its main magnetic field in the z direction. The homogeneity and gradient linearity of the main magnetic field are especially important in MREIT. Conventional RF coils can be adopted as long as there is enough space for electrodes and lead wires. The sensitivity and B_1 field uniformity of an RF coil significantly affect the image quality in MREIT. All possible means must be sought to minimize noise and artifacts in collected k-space data when we construct an MREIT system.

Figure 9.1(a) shows a typical MREIT current source. Oh *et al.* (2006) describe details of its design, including user interface, spectrometer interface and timing control for interleaved current injections. It is controlled by a microprocessor and includes circuits for waveform generation, current output, switching, discharge and auxiliary voltage measurement. The current source is usually located inside the shield room, and coaxial cables are

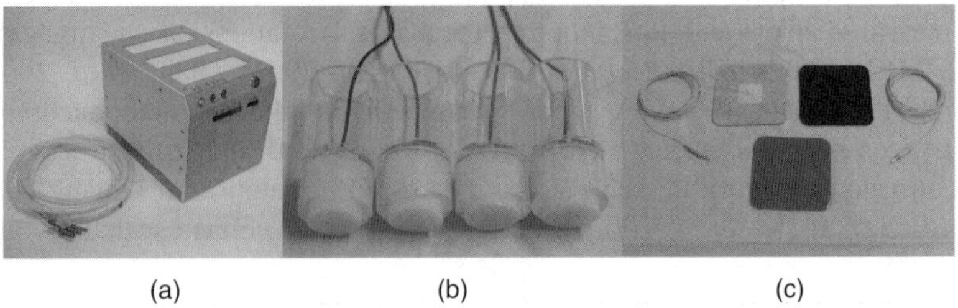

Figure 9.1 (a) MREIT current source, (b) recessed electrode and (c) carbon-hydrogel electrode

mostly used for the connection to surface electrodes on an imaging object placed inside the bore. Carbon cables could be advantageous when a high-field MRI scanner is used. Near the imaging object inside the bore, it would be better to place the cables in the z direction.

For electrodes, we may use non-magnetic conductive materials such as copper, silver, carbon or others. An artifact occurs when a highly conductive electrode is directly attached on the surface of the imaging object, since it shields RF signals. Lee *et al.* (2003a) proposed a recessed electrode that has a gap of moderately conductive gel between the object's surface and a copper electrode (Figure 9.1b). Thin and flexible carbon-hydrogel electrodes with conductive adhesive (Figure 9.1c) are more commonly used in *in vivo* animal and human experiments, replacing bulky and rigid recessed electrodes (Jeong *et al.* 2008; Kim *et al.* 2008b; Minhas *et al.* 2008).

9.1.1 *Measurement of B_z*

Figure 9.2 shows a typical set-up for MREIT imaging experiments. We attach electrodes on the imaging object, which is positioned inside the bore of the MRI scanner. The current source is located outside the bore near the imaging object. The electrodes are connected to the current source by lead wires running in the z direction as much as possible. This is to minimize the amount of B_z inside the object induced by the currents flowing through the external lead wires (Lee *et al.* 2003a). Through a pair of electrodes, current is injected in a form of pulses whose timing is synchronized with an MR pulse sequence. Injected current induces a magnetic flux density $\mathbf{B}$ and it produces extra phase shifts. Phase accumulation is proportional to the z component B_z of $\mathbf{B}$.

We inject current in a form of pulse, whose timing is synchronized with an MRI pulse sequence, as shown in Figure 9.3(a). To eliminate any systematic phase artifact of the MRI scanner, we sequentially inject positive and negative currents denoted as I^+ and I^-, respectively. For the positive injection current I^+, we inject current as a positive pulse

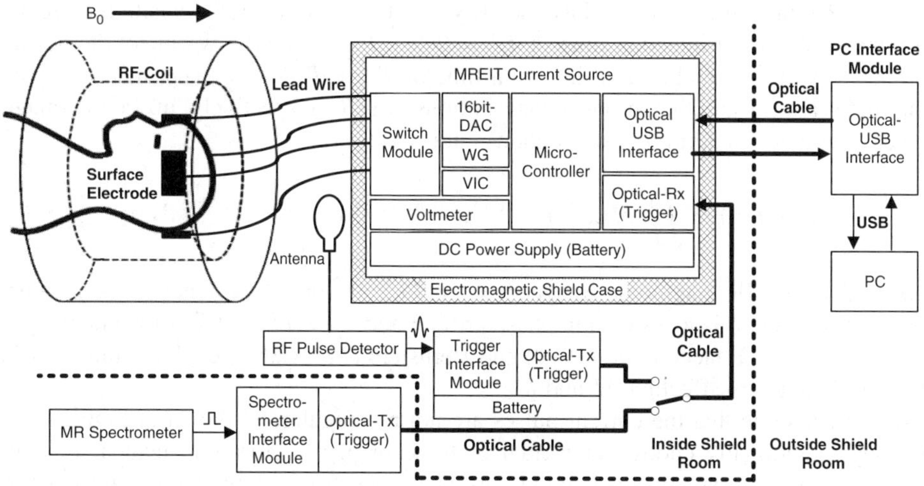

Figure 9.2 Structure of an MREIT current source. WG and VIC are the waveform generator and voltage-to-current converter, respectively

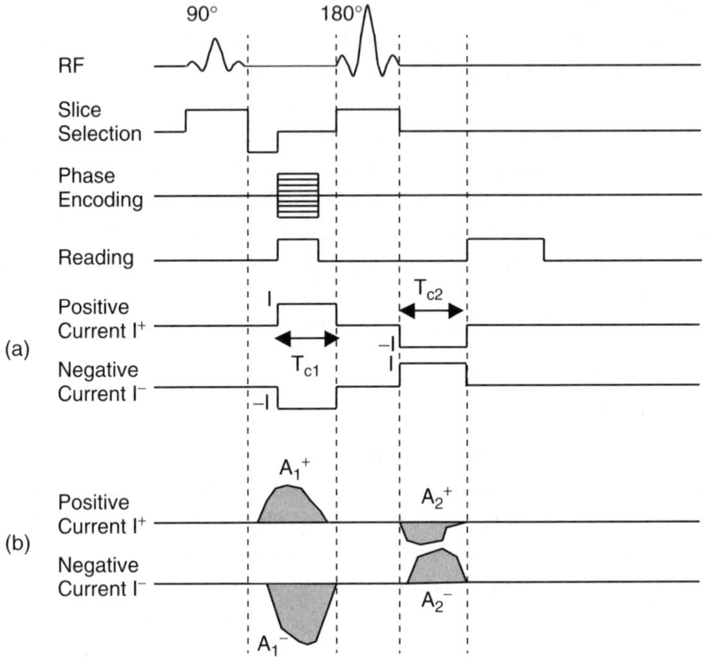

Figure 9.3 Example of a pulse sequence for MREIT. (a) For rectangular pulses, we need to know their amplitudes and widths. (b) For arbitrary waveforms, we need to know their areas. It is desirable to have a net zero DC current for a time duration of a few seconds

between the 90° and 180° radio-frequency (RF) pulses for T_{c1}^+ milliseconds. After the 180° RF pulse, we inject current again as a negative pulse for T_{c2}^+ milliseconds, since the 180° RF pulse reverses the phase polarity. In effect, we inject the positive current I^+ with the total current injection time of $T_c^+ = T_{c1}^+ + T_{c2}^+$. Similarly, we inject the negative current I^- with the total current injection time of $T_c^- = T_{c1}^- + T_{c2}^-$. The injection current amplitudes are $\pm I$ and they produce distributions of $\pm B_z$, respectively, inside the imaging object. The corresponding k-space data $\mathcal{S}^\pm$ are

$$\mathcal{S}^\pm(m,n) = \iint M(x,y)\, \mathrm{e}^{\mathrm{i}\delta(x,y)} \mathrm{e}^{\pm\mathrm{i}\gamma B_z(x,y)T_c} \mathrm{e}^{\mathrm{i}(xm\Delta k_x + yn\Delta k_y)}\, \mathrm{d}x\, \mathrm{d}y, \tag{9.1}$$

where M is an MR magnitude image, $\gamma = 26.75 \times 10^7\,\mathrm{rad\,T^{-1}\,s^{-1}}$ is the gyromagnetic ratio of hydrogen, δ is a systematic phase artifact and $T_c = T_{c1}^\pm + T_{c2}^\pm$ is the current pulse width in seconds. Haacke *et al.* (1999) and Bernstein *et al.* (2004) explain numerous MR imaging parameters affecting M and δ.

We now assume that the current pulses are not rectangular, as shown in Figure 9.3(b). We may intentionally choose sinusoidal or trapezoidal waveforms instead of rectangular pulses. In addition, practical limitations of the electronic components and reactance within the circuits may distort any chosen waveform. We denote the total area of the current pulses during T_{c1}^+ and T_{c2}^+ as $A^+ = A_1^+ + A_2^+$. Similarly, we find the total area of

$A^- = A_1^- + A_2^-$ during T_{c1}^- and T_{c2}^-. Pretending that we have injected positive and negative currents with perfectly rectangular waveforms with amplitudes $\pm I$, we may express (9.1) as

$$\mathcal{S}^{\pm}(m, n) = \iint M(x, y)\, \mathrm{e}^{\mathrm{i}\delta(x,y)} \mathrm{e}^{\pm\mathrm{i}\gamma B_z(x,y)A^{\pm}/I} \mathrm{e}^{\mathrm{i}(xm\Delta k_x + yn\Delta k_y)}\, \mathrm{d}x\, \mathrm{d}y. \tag{9.2}$$

Note that the instantaneous value of $B_z(x, y)$ is proportional to the instantaneous value of the current amplitude at every point (x, y) (Lee *et al.* 2003a). Throughout the entire current injection time, the distributions of B_z are identical except for scaling factors, which are proportional to the instantaneous values of the current amplitude.

We compute complex images $\mathcal{M}^{\pm}$ by two-dimensional discrete Fourier transformations of $\mathcal{S}^{\pm}$ in either (9.1) or (9.2):

$$\mathcal{M}^{\pm}(x, y) = M(x, y)\, \mathrm{e}^{\mathrm{i}\delta(x,y)} \mathrm{e}^{\pm\mathrm{i}\gamma B_z(x,y)T_c} = M(x, y)\, \mathrm{e}^{\mathrm{i}\delta(x,y)} \mathrm{e}^{\pm\mathrm{i}\gamma B_z(x,y)A^{\pm}/I}. \tag{9.3}$$

Dividing the two complex images to reject δ, we get the phase change Ψ due to B_z as

$$\Psi(x, y) = \arg\left(\frac{\mathcal{M}^{+}(x, y)}{\mathcal{M}^{-}(x, y)}\right) = 2\gamma B_z(x, y) T_c = \gamma B_z(x, y) \frac{A^+ + A^-}{I}. \tag{9.4}$$

Finally, we get B_z as

$$B_z(x, y) = \frac{\Psi(x, y)}{C} = \frac{1}{C} \arg\left(\frac{\mathcal{M}^{+}(x, y)}{\mathcal{M}^{-}(x, y)}\right), \tag{9.5}$$

where the constant C is either $2\gamma T_c$ or $\gamma(A^+ + A^-)/I$.

Since B_z is proportional to I, the phase change Ψ in (9.4) is proportional to IT_c or $(A^+ + A^-)/I$. To maximize the phase change, we should maximize the product IT_c or the areas A^+ and A^- within a permissible range (Reilly 1998). There are numerous technical issues to achieve this in terms of MREIT pulse sequence designs (Hamamura and Muftuler 2008; Minhas *et al.* 2009; Nam and Kwon 2010; Park *et al.* 2006).

Figure 9.4(a) shows an MR magnitude image M of a cylindrical saline phantom including an agar object whose conductivity was different from that of the saline. Injection current from the top to the bottom electrodes produced the wrapped phase image in Figure 9.4(b). Such phase wrapping may not occur when the amplitude of the injection current is small. Figure 9.4(c) is the B_z image after applying a phase unwrapping algorithm. We can observe the deflection of B_z across the boundary of the agar object where a conductivity contrast exists.

9.1.2 *Noise in Measured B_z Data*

Noise in measured B_z data is the primary limiting factor in determining the spatial resolution of a reconstructed conductivity image. To distinguish noise due to the MRI scanner itself from noise due to the current source, we first assume that the current source is noise-free and ideal in terms of its current amplitude and timing. Since the MR signals $\mathcal{S}^{\pm}$ are contaminated by random noise originating from the MRI scanner itself, the measured B_z data contain random noise. The noise standard deviation s_{B_z} in the B_z data is

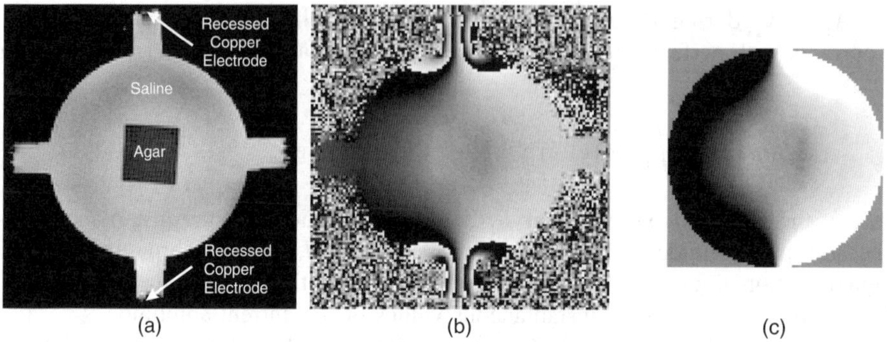

Figure 9.4 (a) MR magnitude image M of a cylindrical saline phantom including an agar object. Conductivity values of the saline and agar were different. (b) Wrapped phase image subject to an injection current from the top to the bottom electrodes. (c) Corresponding image of induced B_z after applying a phase unwrapping algorithm

inversely proportional to the signal-to-noise ratio (SNR) of the MR magnitude image Υ_M and the total current injection time T_c (Sadleir 2005; Scott *et al.* 1992):

$$s_{B_z} = \frac{1}{\sqrt{2}\gamma T_c \Upsilon_M}. \tag{9.6}$$

We now consider a practical current source with noise and errors in its amplitude and timing, which will add extra noise in addition to that in (9.6). Examining (9.3)–(9.5), we should note that the total area $(A^+ + A^-)$ must be identical for all TR cycles. This requires a high amplitude stability of the current source through an entire scan time including many TR cycles. The timing jitter in trigger signals is not critical as long as we control the timing within one microsecond, for example, since both $T_{c1}^{\pm}$ and $T_{c2}^{\pm}$ are in the range of a few milliseconds.

The current source includes analog and high-speed digital electronic circuits with inherent noise sources, which may increase noise in MR images. The MRI scanner itself radiates RF as well as audio-frequency (AF) electromagnetic waves, which are external interferences to the current source. RF interferences are from RF coils at around 128 MHz in a 3 T scanner. AF interferences are mostly from gradient coils with frequencies below 1 kHz (Hedeen and Edelstein *et al.* 1997). Connections among modules increase the vulnerability of the entire system to electromagnetic interference. Under such circumstances, a proper design of the current source is important to reduce measurement noise in MREIT.

In MREIT, the raw data are the incremental phase change, Ψ in (9.4). This phase change is proportional to the product of B_z and T_c. Since B_z is directly proportional to I, we must optimize the MREIT pulse sequence to maximize the product of I and T_c in Figure 9.3. In human imaging experiments, permissible pairs of (I, T_c) must be sought carefully, considering their physiological effects (Reilly 1998).

The spin–echo pulse sequence in Figure 9.3 has been widely used in MREIT, since it is most robust to many kinds of undesirable perturbations to the phase image. As expressed in (9.6), a prolonged current pulse width (i.e. larger T_c) reduces the noise level in measured B_z data. Park *et al.* (2006) proposed a new MREIT pulse sequence called injection current nonlinear encoding (ICNE), where the duration of the injection current pulse is extended until the end of the reading gradient. Since the current injection during the reading gradient disturbs the gradient linearity, they developed an algorithm to extract B_z data from the acquired MR signal using the ICNE pulse sequence. They could reduce the noise level by about 25%. For a chosen pulse sequence, Lee *et al.* (2006) and Kwon *et al.* (2007) analyzed the associated noise level and provided a way to optimize the pulse sequence to minimize it.

9.1.3 Measurement of $\mathbf{B} = (B_x, B_y, B_z)$

Using an MRI scanner, we can measure only one component **B** that is in the direction of the main magnetic field of the MRI scanner. To measure the other two components of **B**, the imaging object must be rotated twice, as shown in Figure 9.5. The electrodes must be kept at the same positions all the time and all the pixels should not be shifted. Though we may prevent pixels from being shifted by carefully designing the object rotation method, soft tissues and fluids inside the human body will move under gravitational forces.

9.2 Forward Problem and Model Construction

We let the object to be imaged occupy a three-dimensional bounded domain $\Omega \subset \mathbb{R}^3$ with a smooth boundary $\partial\Omega$. We attach a pair of surface electrodes $\mathcal{E}^+$ and $\mathcal{E}^-$ on the boundary $\partial\Omega$ through which we inject current I at a fixed low angular frequency ω ranging over $0 < \omega/2\pi < 500\,\text{Hz}$, for example. Then, the time-harmonic current density **J**, electric field intensity **E** and magnetic flux density **B** due to the injection current approximately satisfy the following:

$$\nabla \cdot \mathbf{J} = 0 = \nabla \cdot \mathbf{B}, \quad \mathbf{J} = \frac{1}{\mu_0} \nabla \times \mathbf{B} \quad \text{in } \Omega, \tag{9.7}$$

$$\mathbf{J} = \sigma \mathbf{E}, \quad \nabla \times \mathbf{E} = 0 \quad \text{in } \Omega, \tag{9.8}$$

$$I = -\int_{\mathcal{E}^+} \mathbf{J} \cdot \mathbf{n}\, ds = \int_{\mathcal{E}^-} \mathbf{J} \cdot \mathbf{n}\, ds, \tag{9.9}$$

$$\mathbf{J} \cdot \mathbf{n} = 0 \ \text{ on } \partial\Omega \setminus \overline{\mathcal{E}^+ \cup \mathcal{E}^-}, \quad \mathbf{J} \times \mathbf{n} = 0 \ \text{ on } \mathcal{E}^+ \cup \mathcal{E}^-, \tag{9.10}$$

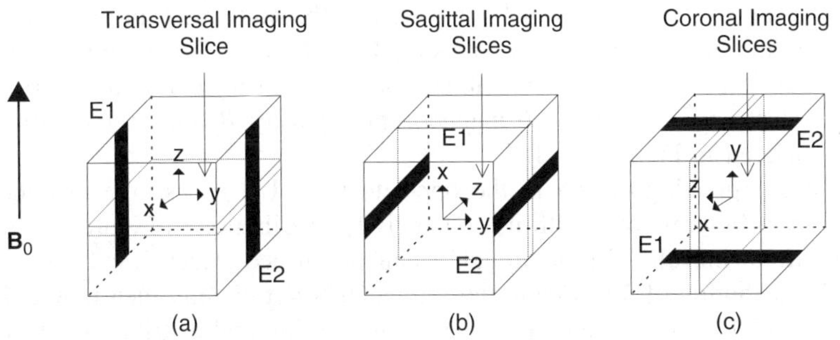

Figure 9.5 (a) Current is injected into xy plane and induced B_z is measured. (b) Current is injected into yz plane and B_x is measured. (c) Current is injected into xz plane and B_y is measured. $\mathbf{B}_0$ is the main magnetic field of the MRI scanner

where $\mathbf{n}$ is the outward unit normal vector on $\partial\Omega$ and $\mathrm{d}s$ is the surface area element. In order to simplify the MREIT problem, we will assume that the conductivity distribution σ in Ω is isotropic, $0 < \sigma < \infty$ and smooth.

9.2.1 *Relation between* $\mathbf{J}$, B_z *and* σ

The induced voltage u in the Sobolev space $H^1(\Omega)$ satisfies the following boundary value problem:

$$\begin{cases} \nabla \cdot (\sigma \nabla u) = 0 \quad \text{in } \Omega, \\ I = \int_{\mathcal{E}^+} \sigma \dfrac{\partial u}{\partial \mathbf{n}}\, \mathrm{d}s = -\int_{\mathcal{E}^-} \sigma \dfrac{\partial u}{\partial \mathbf{n}}\, \mathrm{d}s, \\ \nabla u \times \mathbf{n} = 0 \ \text{ on } \mathcal{E}^+ \cup \mathcal{E}^-, \quad \sigma \dfrac{\partial u}{\partial \mathbf{n}} = 0 \ \text{ on } \partial\Omega \setminus \overline{\mathcal{E}^+ \cup \mathcal{E}^-}, \end{cases} \tag{9.11}$$

where $\partial u / \partial \mathbf{n}$ is the derivative of u normal to the boundary. Setting a reference voltage $u|_{\mathcal{E}^-} = 0$, we can obtain a unique solution u. Among u, $\mathbf{J}$ and $\mathbf{E}$, we have the following relation:

$$\mathbf{J} = \sigma \mathbf{E} = -\sigma \nabla u. \tag{9.12}$$

The relation between the internal B_z data and the conductivity σ can be expressed implicitly by the z component of the Biot–Savart law:

$$B_z(\mathbf{r}) = \frac{\mu_0}{4\pi} \int_\Omega \frac{\langle \mathbf{r} - \mathbf{r}', -\sigma(\mathbf{r}')\nabla u(\mathbf{r}') \times \mathbf{e}_z \rangle}{|\mathbf{r} - \mathbf{r}'|^3}\, \mathrm{d}\mathbf{r}' + \mathcal{H}(\mathbf{r}) \quad \text{for } \mathbf{r} \in \Omega, \tag{9.13}$$

where $\mathbf{r} = (x, y, z)$ is a position vector in $\mathbb{R}^3$, $\mathbf{e}_z = (0, 0, 1)$, $\mathcal{H}(\mathbf{r})$ is a harmonic function in Ω representing a magnetic flux density generated by currents flowing through external lead wires and u is the induced voltage. In practice, the harmonic function $\mathcal{H}$ is unknown, so we should eliminate its effects in any conductivity image reconstruction algorithm.

Figure 9.6(a) is an MR magnitude image of a cylindrical phantom whose background was filled with an agar gel. It contained chunks of three different biological tissues. Its conductivity image is shown in Figure 9.6(b), where we used an MREIT image reconstruction algorithm described later. From multi-slice conductivity images of the three-dimensional phantom, we solved the equation (9.11) for u using the finite element method (FEM) and computed the internal current density $\mathbf{J}$ using $\mathbf{J} = \sigma\mathbf{E} = -\sigma\nabla u$. Figure 9.6(c) is a plot of $|\mathbf{J}|$; the thin lines are current streamlines subject to an injection current from the left to the right electrodes. The induced magnetic flux density B_z due to the current density in (c) is visualized in Figure 9.6(d).

Now, we interpret Figure 9.6 in the opposite way. Let us assume that the imaging object shown in Figure 9.6(a) with its conductivity distribution in Figure 9.6(b) is given. We inject current into the object through a pair of surface electrodes. Then, it produces an internal distribution of $\mathbf{J}$ in Figure 9.6(c), which is not directly measurable. Following the relation in (9.13), the current density generates an internal distribution of the induced magnetic flux density B_z in Figure 9.6(d), which is measurable by using an MRI scanner. The goal in MREIT is to reconstruct an image of the internal conductivity distribution in Figure 9.6(b) by using the measured data of B_z in Figure 9.6(d) or $\mathbf{B}$.

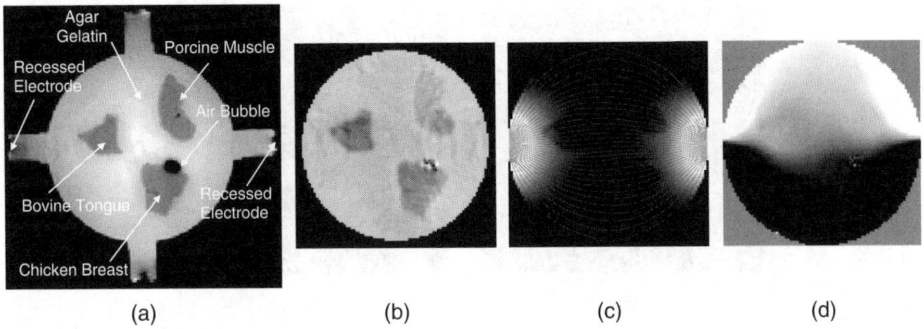

Figure 9.6 (a) MR magnitude image M of a cylindrical phantom including chunks of three different biological tissues. Its background was filled with an agar gel. (b) Reconstructed conductivity image of the same slice using an MREIT conductivity image reconstruction algorithm. (c) Image of the magnitude of the current density $|\mathbf{J}|$, where thin lines are current streamlines. Current was injected from the left to the right electrodes. (d) Induced magnetic flux density B_z image subject to the current density in (c)

9.2.2 Three Key Observations

The right-hand side of (9.13) is a sum of a nonlinear function of σ and the harmonic function $\mathcal{H}$, which is independent of σ. We may consider an inverse problem of recovering the conductivity distribution σ entering the nonlinear problem (9.13) from knowledge of the measured data B_z, the geometry of $\partial\Omega$, the positions of the electrodes $\mathcal{E}^{\pm}$ and the size of the injection current.

First, there is a scaling uncertainty of σ in the nonlinear problem (9.13) owing to the fact that, if σ is a solution of (9.13), so is a scaled conductivity $\alpha\sigma$ for any scaling factor $\alpha > 0$. Hence, we should resolve the scaling uncertainty of σ by measuring the voltage difference at any two fixed boundary points or by including a piece of electrically conducting material with a known conductivity value as part of the imaging object (Ider *et al.* 2003; Kwon *et al.* 2002a).

Second, any change of σ in the direction ∇u normal to the equipotential surface is invisible from B_z data. Assume that a function $\varphi : \mathbb{R} \to \mathbb{R}$ is strictly increasing and continuously differentiable. Then $\varphi(u)$ is a solution of (9.11) with σ replaced by $\sigma/\varphi'(u)$ because

$$\sigma(\mathbf{r})\nabla u(\mathbf{r}) = \frac{\sigma(\mathbf{r})}{\varphi'(u(\mathbf{r}))}\nabla\varphi(u(\mathbf{r})), \quad \mathbf{r} \in \Omega. \tag{9.14}$$

Noting that this is true for any strictly increasing $\varphi \in C^1(\mathbb{R})$, we can see that the data B_z cannot trace a change of σ in the direction ∇u. This means that there are infinitely many conductivity distributions that satisfy (9.11) and (9.13) for given B_z data. Figure 9.7 shows an example of two conductivity distributions producing the same B_z data.

Third, B_z data can trace a change of σ in the tangent direction $\mathbb{L}\nabla u$ to the equipotential surface where

$$\mathbb{L} = \begin{pmatrix} 0 & 1 & 0 \\ -1 & 0 & 0 \\ 0 & 0 & 0 \end{pmatrix}.$$

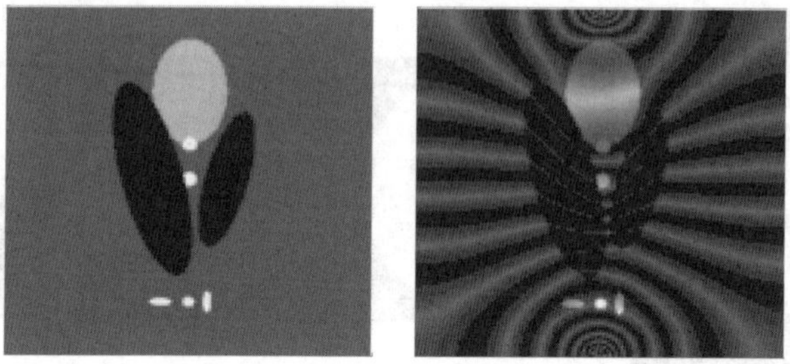

(a) (b)

Figure 9.7 Two different conductivity distributions (a) and (b) that produce the same B_z data subject to Neumann data $g(x, y) = \delta((x, y) - (0, 1)) - \delta((x, y) - (0, -1)), x \in \partial\Omega$, where $\Omega = (-1, 1) \times (-1, 1)$

To see this, we change (9.13) into the following variational form, where the unknown harmonic term $\mathcal{H}$ is eliminated,

$$\int_\Omega \nabla B_z \cdot \nabla\eta \, d\mathbf{r} = \int_\Omega \sigma \left(\nabla u \times \mathbf{e}_z\right) \cdot \nabla\eta \, d\mathbf{r} \quad \text{for all } \eta \in C_0^1(\Omega), \tag{9.15}$$

or using the smoothness assumption of σ and the fact that $\nabla \cdot (\nabla u \times \mathbf{e}_z) = 0$,

$$\Delta B_z = \nabla \ln\sigma \cdot (\sigma \nabla u \times \mathbf{e}_z) \quad \text{in } \Omega. \tag{9.16}$$

The two expressions (9.15) and (9.16) clearly explain that B_z data probes a change of $\ln\sigma$ along the vector field $\sigma\nabla u \times \mathbf{e}_z$.

Remark 9.2.1 (Smoothness assumption of σ) *The identity (9.16) definitely does not make any sense when σ is discontinuous. However, we can still use (9.16) to develop any MREIT image reconstruction algorithm for a non-smooth conductivity distribution. To see this, suppose that $\tilde{\sigma}$ is a C^1-approximation of a non-smooth function σ with a finite bounded variation $\|\sigma\|_{BV(\Omega)} < \infty$. An MRI scanner provides a two-dimensional array of B_z intensities inside voxels of a field of view, and each intensity is affected by the amount of protons in each voxel and an adopted pulse sequence. Hence, any practically available B_z data are always a blurred version, which cannot distinguish σ from $\tilde{\sigma}$. Admitting the obvious fact that an achievable spatial resolution of a reconstructed conductivity image cannot be better than the voxel size, the Laplacian and gradient in the identity (9.16) should be understood as discrete differentials at the voxel size of the MR image.*

9.2.3 Data B_z Traces $\sigma\nabla u \times \mathbf{e}_z$ Directional Change of σ

From the formula $\Delta B_z = \nabla \ln\sigma \cdot \left(\sigma\nabla u \times \mathbf{e}_z\right)$ in (9.16), the distribution of B_z traces a change of σ to the direction $\mathbb{L}\nabla u$ in the following ways:

(i) If B_z is super-harmonic at $\mathbf{r}$, then $\ln\sigma$ is increasing at $\mathbf{r}$ in the direction $\sigma\nabla u(\mathbf{r}) \times \mathbf{e}_z$.
(ii) If B_z is sub-harmonic at $\mathbf{r}$, then $\ln\sigma$ is decreasing at $\mathbf{r}$ in the direction $\sigma\nabla u(\mathbf{r}) \times \mathbf{e}_z$.
(iii) If B_z is harmonic at $\mathbf{r}$, then $\ln\sigma$ is not changing at $\mathbf{r}$ in the direction $\sigma\nabla u(\mathbf{r}) \times \mathbf{e}_z$.

According to the above observations, if we could predict the direction of $\sigma\nabla u \times \mathbf{e}_z$, we may estimate a spatial change of σ in that direction from measured B_z data. However, the vector field $\sigma\nabla u \times \mathbf{e}_z$ is a nonlinear function of the unknown conductivity σ, and hence estimation of the direction of $\sigma\nabla u \times \mathbf{e}_z$ without explicit knowledge of σ appears to be paradoxical.

Assume that the conductivity contrast is reasonably small as $\|\nabla \ln\sigma\|_{L^\infty(\Omega)} \le 1$. The distribution of the current density $\mathbf{J} = -\sigma\nabla u$ is mostly dictated by the given positions of the electrodes $\mathcal{E}^\pm$, the size of the injection current I, and the geometry of the boundary $\partial\Omega$, while the influence of changes in σ on $\mathbf{J}$ is relatively small. This means that $\sigma\nabla u \approx \nabla v$, where v is a solution of the Laplace equation $\Delta v = 0$ with the same boundary data as in (9.11). Hence, under the assumption of low conductivity contrast, the change in $\ln\sigma$ along any characteristic curve having its tangent direction $\mathbf{J} \times \mathbf{e}_z$ can be evaluated by using the following approximation:

$$\nabla \ln\sigma \cdot (\nabla v \times \mathbf{e}_z) \approx \nabla \ln\sigma \cdot (\sigma\nabla u \times \mathbf{e}_z) = \Delta B_z. \tag{9.17}$$

9.2.4 Mathematical Analysis toward MREIT Model

Since our goal is to use MREIT techniques in practical clinical applications, we must set up a mathematical model of MREIT that agrees with a planned medical imaging system. To simplify our study, let us make several assumptions, which should not go astray from the practical model. Let the object to be imaged occupy a three-dimensional bounded domain $\Omega \subset \mathbb{R}^3$ with a smooth connected boundary $\partial\Omega$, and each $\Omega_{z_0} := \Omega \cap \{z = z_0\} \subset \mathbb{R}^2$, the slice of Ω cut by the plane $\{z = z_0\}$, has a smooth connected boundary. We assume that the conductivity distribution σ of the subject Ω is isotropic, $C^1(\bar{\Omega})$ and $0 < \sigma_- < \sigma < \sigma_+$ with two known constants $\sigma_\pm$. Though σ is usually piecewise-smooth in practice, this can be approximated by a $C^1(\bar{\Omega})$-function and so it is a matter of how big $\|\sigma\|_{C^1(\Omega)}$ is. We attach a pair of copper electrodes $\mathcal{E}^+$ and $\mathcal{E}^-$ on $\partial\Omega$ in order to inject current, and let $\mathcal{E}^+ \cup \mathcal{E}^-$ be the portion of the surface $\partial\Omega$ where the electrodes are attached.

The injection current I produces an internal current density $\mathbf{J} = (J_x, J_y, J_z)$ inside the subject Ω satisfying the following problem:

$$\begin{cases} \nabla \cdot \mathbf{J} = 0 \quad \text{in } \Omega, \\ I = -\displaystyle\int_{\mathcal{E}^+} \mathbf{J}\cdot\mathbf{n}\,\mathrm{d}s = \int_{\mathcal{E}^-} \mathbf{J}\cdot\mathbf{n}\,\mathrm{d}s, \quad \mathbf{J}\times\mathbf{n} = 0 \quad \text{on } \mathcal{E}^+ \cup \mathcal{E}^-, \\ \mathbf{J}\cdot\mathbf{n} = 0 \quad \text{on } \partial\Omega \setminus \overline{\mathcal{E}^+ \cup \mathcal{E}^-}, \end{cases} \tag{9.18}$$

where $\mathbf{n}$ is the outward unit normal vector on $\partial\Omega$ and $\mathrm{d}s$ is the surface area element. The condition $\mathbf{J} \times \mathbf{n} = 0$ on $\mathcal{E}^+ \cup \mathcal{E}^-$ comes from the fact that copper electrodes are highly conductive. Since $\mathbf{J}$ is expressed as $\mathbf{J} = -\sigma\nabla u$, where u is the corresponding electrical potential, (9.18) can be converted to

$$\begin{cases} \nabla \cdot (\sigma\nabla u) = 0 \quad \text{in } \Omega, \\ I = \displaystyle\int_{\mathcal{E}^+} \sigma\frac{\partial u}{\partial \mathbf{n}}\,\mathrm{d}s = -\int_{\mathcal{E}^-} \sigma\frac{\partial u}{\partial \mathbf{n}}\,\mathrm{d}s, \quad \nabla u \times \mathbf{n} = 0 \quad \text{on } \mathcal{E}^+ \cup \mathcal{E}^-, \\ \sigma\dfrac{\partial u}{\partial \mathbf{n}} = 0 \quad \text{on } \partial\Omega \setminus \overline{\mathcal{E}^+ \cup \mathcal{E}^-}, \end{cases} \tag{9.19}$$

where $\partial u/\partial \mathbf{n} = \nabla u \cdot \mathbf{n}$. The above non-standard boundary value problem (9.19) is well-posed and has a unique solution up to a constant. We omit the proof of the uniqueness (up to a constant) within the class $W^{1,2}(\Omega)$ since it follows from standard arguments in PDEs.

Let us briefly discuss the boundary conditions, which are essentially related to the size of the electrodes. The condition $\nabla u \times \mathbf{n}|_{\mathcal{E}^\pm} = 0$ ensures that each of $u|_{\mathcal{E}^+}$ and $u|_{\mathcal{E}^-}$ is a constant, since ∇u is normal to its level surface. The term

$$\pm I = \int_{\mathcal{E}^\pm} \sigma \frac{\partial u}{\partial \mathbf{n}}\, \mathrm{d}s$$

means that the total amount of injection current through the electrodes is I milliamps. Let us denote

$$g := -\sigma \frac{\partial u}{\partial \mathbf{n}}\bigg|_{\partial\Omega}.$$

In practice, it is difficult to specify the Neumann data g in a pointwise sense because only the total amount of injection current I is known. It should be noted that the boundary condition in (9.19) leads to $|g| = \infty$ on $\partial\mathcal{E}^\pm$, singularity along the boundary of the electrodes, and $g \notin L^2(\partial\Omega)$. But, fortunately, $g \in H^{-1/2}(\partial\Omega)$, which also can be proven by the standard regularity theory in PDEs.

The exact model (9.19) can be converted into the following standard problem of an elliptic equation with mixed boundary conditions.

Lemma 9.2.2 *Assume that $\tilde{u}$ solves*

$$\begin{cases} \nabla \cdot (\sigma \nabla \tilde{u}) = 0 \quad \text{in } \Omega, \\ \tilde{u}|_{\mathcal{E}^+} = 1, \quad \tilde{u}|_{\mathcal{E}^-} = 0, \\ -\sigma \dfrac{\partial \tilde{u}}{\partial \mathbf{n}} = 0 \quad \text{on } \partial\Omega \setminus (\mathcal{E}^+ \cup \mathcal{E}^-). \end{cases} \tag{9.20}$$

If u is a solution of the mixed boundary value problem (9.19), then

$$u = \frac{I}{\int_{\partial\mathcal{E}^+} \sigma (\partial \tilde{u}/\partial \mathbf{n})\, \mathrm{d}s}\, \tilde{u} \quad \text{in } \Omega \text{ (up to a constant)}. \tag{9.21}$$

Proof. The proof is elementary by looking at the energy of $w = u - c\tilde{u}$ for a constant c:

$$\begin{aligned} \int_\Omega \sigma |\nabla w|^2 \, \mathrm{d}\mathbf{r} &= \int_{\partial\Omega} \sigma \frac{\partial w}{\partial \mathbf{n}} w \, \mathrm{d}s \\ &= \int_{\mathcal{E}^+} \sigma \frac{\partial w}{\partial \mathbf{n}} \, \mathrm{d}s \, (u|_{\mathcal{E}^+} - c) + \left(\int_{\mathcal{E}^-} \sigma \frac{\partial w}{\partial \mathbf{n}} \, \mathrm{d}s \right) u|_{\mathcal{E}^-} \\ &= (u|_{\mathcal{E}^+} - u|_{\mathcal{E}^-} - c) \left(I - c \int_{\partial\mathcal{E}^+} \sigma \frac{\partial \tilde{u}}{\partial \mathbf{n}} \, \mathrm{d}s \right). \end{aligned}$$

Hence, for

$$c = \frac{I}{\int_{\partial\mathcal{E}^+} \sigma (\partial \tilde{u}/\partial \mathbf{n})\, \mathrm{d}s},$$

the above relation generates $|\nabla w| = 0$ in Ω, which means that w is a constant in Ω. □

Now, we explain the inverse problem for the MREIT model, in which we try to reconstruct σ. The presence of the internal current density $\mathbf{J} = -\sigma\nabla u$ generates a magnetic flux density $\mathbf{B} = (B_x, B_y, B_z)$ such that Ampère's law $\mathbf{J} = \nabla \times \mathbf{B}/\mu_0$ holds in Ω. With the z axis pointing in the direction of the main magnetic field of the MRI scanner, the relation between the measurable quantity B_z and the unknown σ is governed by the Biot–Savart law:

$$B_z(\mathbf{r}) = \frac{\mu_0}{4\pi}\int_\Omega \frac{\langle \mathbf{r}-\mathbf{r}', \sigma(\mathbf{r}')\mathbb{L}\nabla u(\mathbf{r}')\rangle}{|\mathbf{r}-\mathbf{r}'|^3}\,\mathrm{d}\mathbf{r}' \quad \text{for } \mathbf{r}\in\Omega, \tag{9.22}$$

where

$$\mathbb{L} = \begin{pmatrix} 0 & 1 & 0 \\ -1 & 0 & 0 \\ 0 & 0 & 0 \end{pmatrix}.$$

Here, we must read u as a nonlinear function of σ. The following lemma is crucial to understand why we need at least two injection currents with the requirement (9.26) in the sequel.

Lemma 9.2.3 *Suppose u is a solution of (9.19) and the pair (σ, u) satisfies (9.22). Then B_z in (9.22) can be expressed as*

$$B_z = \frac{\mu_0}{4\pi}\int_\Omega \frac{-1}{|\mathbf{r}-\mathbf{r}'|}\begin{vmatrix} \partial\sigma/\partial x & \partial\sigma/\partial y \\ \partial u/\partial x & \partial u/\partial y \end{vmatrix}\mathrm{d}\mathbf{r}' + \frac{\mu_0}{4\pi}\int_{\partial\Omega}\frac{1}{|\mathbf{r}-\mathbf{r}'|}\mathbf{n}\cdot(\sigma\mathbb{L}\nabla u)\,\mathrm{d}s. \tag{9.23}$$

Moreover, there exist infinitely many pairs $(\tilde\sigma, \tilde u)$ such that

$$\begin{vmatrix} \partial\sigma/\partial x & \partial\sigma/\partial y \\ \partial u/\partial x & \partial u/\partial y \end{vmatrix} = \begin{vmatrix} \partial\tilde\sigma/\partial x & \partial\tilde\sigma/\partial y \\ \partial\tilde u/\partial x & \partial\tilde u/\partial y \end{vmatrix}$$

in Ω and $\mathbf{n}\cdot(\sigma\mathbb{L}\nabla u) = \mathbf{n}\cdot(\tilde\sigma\mathbb{L}\nabla\tilde u)$ on $\partial\Omega$.

Proof. From (9.22), we have

$$\begin{aligned} B_z &= \frac{\mu_0}{4\pi}\int_\Omega \nabla_{\mathbf{r}'}\frac{1}{|\mathbf{r}-\mathbf{r}'|}\cdot\big(\sigma(\mathbf{r}')\mathbb{L}\nabla u(\mathbf{r}')\big)\,\mathrm{d}\mathbf{r}' \\ &= \frac{\mu_0}{4\pi}\int_\Omega \frac{-1}{|\mathbf{r}-\mathbf{r}'|}\nabla\cdot(\sigma\mathbb{L}\nabla u)\,\mathrm{d}\mathbf{r}' + \frac{\mu_0}{4\pi}\int_{\partial\Omega}\frac{1}{|\mathbf{r}-\mathbf{r}'|}\mathbf{n}\cdot(\sigma\mathbb{L}\nabla u)\,\mathrm{d}s. \end{aligned}$$

Then (9.23) follows from

$$\nabla\cdot(\sigma\mathbb{L}\nabla u) = \mathbf{e}_z\cdot[\nabla\sigma\times\nabla u] = \begin{vmatrix} \partial\sigma/\partial x & \partial\sigma/\partial y \\ \partial u/\partial x & \partial u/\partial y \end{vmatrix},$$

where $\mathbf{e}_z = (0, 0, 1)$.

Now, we will show that there are infinitely many pairs $(\tilde\sigma, \tilde u)$ such that $\mathbf{e}_z\cdot[\nabla\sigma\times\nabla u] = \mathbf{e}_z\cdot[\nabla\tilde\sigma\times\nabla\tilde u]$ and $\tilde u$ is a solution of (9.19) with σ replaced by $\tilde\sigma$. Indeed, we can construct infinitely many pairs $(\tilde\sigma, \tilde u)$ satisfying the much stronger condition $\sigma\nabla u = \tilde\sigma\nabla\tilde u$. From the maximum–minimum principle for the elliptic equation, $u|_{\mathcal{E}^+}$ and $u|_{\mathcal{E}^-}$ are the maximum and minimum values of u in $\bar\Omega$, respectively. Choose a and b such that

$\inf_\Omega u = u|_{\mathcal{E}^-} < a < b < u|_{\mathcal{E}^+} = \sup_\Omega u$. For any increasing function $\phi \in C^2([a,b])$ satisfying

$$\phi'(a) = \phi'(b) = 1, \quad \phi''(a) = \phi''(b) = 0, \quad \phi(a) = a, \quad \phi(b) = b, \tag{9.24}$$

we define

$$\tilde{u}(\mathbf{r}) = \begin{cases} \phi(u(\mathbf{r})) & \text{if } \mathbf{r} \in \hat{\Omega}, \\ u(\mathbf{r}) & \text{if } \mathbf{r} \in \Omega \setminus \hat{\Omega}, \end{cases} \qquad \tilde{\sigma}(\mathbf{r}) = \begin{cases} \sigma(\mathbf{r})/\phi'(u(\mathbf{r})) & \text{if } \mathbf{r} \in \hat{\Omega}, \\ \sigma(\mathbf{r}) & \text{if } \mathbf{r} \in \Omega \setminus \hat{\Omega}, \end{cases}$$

where $\hat{\Omega} := \{\mathbf{r} \in \Omega : a \le u(\mathbf{r}) \le b\}$. The conditions on ϕ guarantee that $\tilde{\sigma} \in C^1(\Omega)$ and $\tilde{\sigma} > 0$ in Ω. Since $\nabla\tilde{u} = \phi'(u)\nabla u$, we have

$$\tilde{\sigma}\nabla\tilde{u} = \frac{\sigma}{\phi'(u)}\nabla\hat{u} = \sigma\nabla u.$$

So it is clear that

$$\begin{vmatrix} \partial\sigma/\partial x & \partial\sigma/\partial y \\ \partial u/\partial x & \partial u/\partial y \end{vmatrix} = \begin{vmatrix} \partial\hat{\sigma}/\partial x & \partial\hat{\sigma}/\partial y \\ \partial\hat{u}/\partial x & \partial\hat{u}/\partial y \end{vmatrix}$$

and $\mathbf{n} \cdot (\sigma\mathbb{L}\nabla u) = \mathbf{n} \cdot (\tilde{\sigma}\mathbb{L}\nabla\tilde{u})$ on $\partial\Omega$. Since $\tilde{u} = u$ near the electrodes $\mathcal{E}^+$ and $\mathcal{E}^-$, $\tilde{u}$ has the same boundary condition on the electrodes as u. Therefore, $\tilde{u}$ is a solution of (9.19) with σ replaced by $\tilde{\sigma}$. This completes the proof since ϕ can be chosen arbitrarily under the constraints (9.24). □

9.3 Inverse Problem Formulation using B or J

When we measure **B**, including all its three components, we can compute **J** using (9.7). For the case of $\sigma = \sigma^*$, which is the true conductivity, our problem is reduced to the following nonlinear boundary value problem:

$$\begin{aligned} \nabla \cdot \left(\frac{J^*}{|\nabla u_{\sigma^*}|} \nabla u_{\sigma^*} \right) &= 0 \quad \text{in } \Omega, \\ \frac{J^*}{|\nabla u_{\sigma^*}|} \frac{\partial u_{\sigma^*}}{\partial n} &= g \quad \text{on } \partial\Omega, \end{aligned} \tag{9.25}$$

where $J^* = |\mathbf{J}^*| = \sigma^*|\nabla u_{\sigma^*}|$. Imaging the conductivity σ^* means finding a constructive map $\{I, J^*\} \to \sigma^*$ from the above highly nonlinear equation. Owing to this intricate relation, it is almost impossible to find an explicit expression for σ^* in terms of J^* and I. So, we may adopt an iterative scheme to search for the true solution σ^*.

Before developing an algorithm, it is necessary to check if the data pair $\{I, J^*\}$ has sufficient information to determine $\sigma*$. Unfortunately, as shown in Figure 9.7, it is possible that two different conductivity distributions may correspond to the same pair of $\{I, B_z\}$ or $\{I, J^*\}$.

Let us consider a conducting material consisting of two regions with different conductivity values. It is well known that the normal component of **J** is continuous across the interface while its tangential component changes. Thus the magnitude of the current density will change if the tangential components of **J** at the interface is non-zero. Hence, **J** plays an important role in reconstructing a conductivity image due to its change when current crosses the interface non-orthogonally between two regions. However, **J** may not provide any information for imaging a portion of the interface where current flows orthogonally.

To deal with this uniqueness issue, we may use four electrodes at four sides (east, south, west, north) on the boundary so that we can apply two different current flows using two pairs of electrodes. Let I^1 and I^2 be the two currents from the two pairs of electrodes. Two sets of current density data, $J^1 = |\mathbf{J}^1|$ and $J^2 = |\mathbf{J}^2|$, induced by I^1 and I^2, respectively, can now be used to image the conductivity distribution. Owing to the positions of the electrodes, the two vectors $\mathbf{J}^1$ and $\mathbf{J}^2$ are not in the same (or opposite) direction(s) and thus satisfy

$$\left| \mathbf{J}^1 \times \mathbf{J}^2 \right| \neq 0. \tag{9.26}$$

Therefore, at least one of J^1 and J^2 changes abruptly at the interface of two regions where any change of conductivity occurs.

9.4 Inverse Problem Formulation using B_z

Based on the observations in previous sections, the harmonic B_z algorithm was developed, which will be explained later. It provides a scaled conductivity image of each transverse slice $\Omega_{z_0} = \Omega \cap \{z = z_0\}$. According to the identity (9.16) and the non-uniqueness result, we should produce at least two linearly independent currents. With two data $B_{z,1}$ and $B_{z,2}$ corresponding to two current densities $\mathbf{J}_1$ and $\mathbf{J}_2$, respectively, satisfying $(\mathbf{J}_1 \times \mathbf{J}_2) \cdot \mathbf{e}_z \neq 0$ in Ω_{z_0}, we can perceive a transverse change of σ on the slice Ω_{z_0} using the approximation (9.17). This is the main reason why we usually use two pairs of surface electrodes $\mathcal{E}_1^{\pm}$ and $\mathcal{E}_2^{\pm}$ as shown in Figures 9.4 and 9.6.

We inject two linearly independent currents I_1 and I_2 into an imaging object using two pairs of electrodes. In general, one may inject N different currents using N pairs of electrodes with $N \geq 2$ at the expense of an increased data acquisition time. In order to simplify the electrode attachment procedure, it is desirable to attach four surface electrodes so that, in the imaging region, the area of the parallelogram made by the two vectors $\mathbf{J}_1 \times \mathbf{e}_z$ and $\mathbf{J}_2 \times \mathbf{e}_z$ is as large as possible. We may then spend a given fixed data acquisition time to collect $B_{z,1}$ and $B_{z,2}$ data with a sufficient number of data averaging for a better signal-to-noise ratio (SNR).

9.4.1 Model with Two Linearly Independent Currents

Throughout this section, we assume that we inject two linearly independent currents through two pairs of surface electrodes $\mathcal{E}_1^{\pm}$ and $\mathcal{E}_2^{\pm}$. For a given $\sigma \in C_+^1(\bar{\Omega}) := \{\sigma \in C^1(\bar{\Omega}) : 0 < \sigma < \infty\}$, we denote by $u_j[\sigma]$ the induced voltage corresponding to the injection current I_j with $j = 1, 2$, that is, $u_j[\sigma]$ is a solution of the following boundary value problem:

$$\begin{cases} \nabla \cdot \big(\sigma \nabla u_j[\sigma]\big) = 0 \quad \text{in } \Omega, \\ I_j = \displaystyle\int_{\mathcal{E}_j^+} \sigma \frac{\partial u_j[\sigma]}{\partial \mathbf{n}} \, \mathrm{d}s = -\int_{\mathcal{E}_j^-} \sigma \frac{\partial u_j[\sigma]}{\partial \mathbf{n}} \, \mathrm{d}s, \\ \nabla u_j[\sigma] \times \mathbf{n} = 0 \quad \text{on } \mathcal{E}_j^- \cup \mathcal{E}_j^+, \\ \sigma \dfrac{\partial u_j[\sigma]}{\partial \mathbf{n}} = 0 \quad \text{on } \partial\Omega \setminus \overline{\mathcal{E}_j^+ \cup \mathcal{E}_j^-}. \end{cases} \tag{9.27}$$

We define a map $\Lambda : C_+^1(\bar{\Omega}) \to H^1(\Omega) \times H^1(\Omega) \times \mathbb{R}$ by

$$\Lambda[\sigma](\mathbf{r}) = \begin{pmatrix} \dfrac{\mu_0}{4\pi}\displaystyle\int_\Omega \frac{\langle \mathbf{r}-\mathbf{r}', \sigma\nabla u_1[\sigma](\mathbf{r}')\times \mathbf{e}_z\rangle}{|\mathbf{r}-\mathbf{r}'|^3}\,d\mathbf{r}' \\ \dfrac{\mu_0}{4\pi}\displaystyle\int_\Omega \frac{\langle \mathbf{r}-\mathbf{r}', \sigma\nabla u_2[\sigma](\mathbf{r}')\times \mathbf{e}_z\rangle}{|\mathbf{r}-\mathbf{r}'|^3}\,d\mathbf{r}' \\ u_1[\sigma]\big|_{\mathcal{E}_2^+} - u_1[\sigma]\big|_{\mathcal{E}_2^-} \end{pmatrix}, \quad \mathbf{r}\in\Omega. \tag{9.28}$$

We should note that, according to (9.13), we have

$$\Lambda[\sigma] = (B_{z,1} - \mathcal{H}_1,\ B_{z,2} - \mathcal{H}_2,\ V_{12}^{\pm}), \tag{9.29}$$

where $B_{z,j}$ is the z component of the magnetic flux density corresponding to the current density $\mathbf{J}_j = -\sigma\nabla u_j[\sigma]$ and $V_{12}^{\pm}$ is the voltage difference $u_1[\sigma]$ between the electrodes $\mathcal{E}_2^+$ and $\mathcal{E}_2^-$, that is, $V_{12}^{\pm} = u_1[\sigma]\big|_{\mathcal{E}_2^+} - u_1[\sigma]\big|_{\mathcal{E}_2^-}$. Here, $\mathcal{H}_1$ and $\mathcal{H}_2$ are the lead wire effects from the pairs $\mathcal{E}_1^{\pm}$ and $\mathcal{E}_2^{\pm}$, respectively. Since we know $\Delta\mathcal{H}_j = 0$ in Ω, the first two components of $\Lambda[\sigma]$ are available up to harmonic factors.

The inverse problem of MREIT is to identify σ from knowledge of $\Lambda[\sigma]$ up to harmonic factors. In practice, for given data $B_{z,1}$, $B_{z,2}$ and $V_{12}^{\pm}$, we should develop a robust image reconstruction algorithm to find σ within the admissible class $C_+^1(\bar{\Omega})$ so that such σ minimizes

$$\Phi(\sigma) = \sum_{j=1}^{2} \|\Delta(\Lambda_j[\sigma] - B_{z,j})\|_{L^2(\Omega)}^2 + \alpha|\Lambda_3[\sigma] - V_{12}^{\pm}|^2, \tag{9.30}$$

where $\Lambda[\sigma] = (\Lambda_1[\sigma], \Lambda_2[\sigma], \Lambda_3[\sigma])$ and α is a positive constant.

Regarding the smoothness constraint of $\sigma \in C_+^1(\bar{\Omega})$, we would like to emphasize again that it is not an important issue in practice since practically available B_z data are always a blurred version of the true B_z (see Remark 9.2.1).

9.4.2 Uniqueness

For uniqueness, we need to prove that $\Lambda[\sigma] = \Lambda[\tilde{\sigma}]$ implies that $\sigma = \tilde{\sigma}$. The following condition is essential for uniqueness:

$$\left|\left(\nabla u_1[\sigma](\mathbf{r}) \times \nabla u_2[\sigma](\mathbf{r})\right)\cdot \mathbf{e}_z\right| > 0 \quad \text{for } \mathbf{r}\in\Omega. \tag{9.31}$$

However, we still do not have a rigorous theory for the issue related to (9.31) in a three-dimensional domain though there are some two-dimensional results based on the geometric index theory (Alessandrini and Magnanini 1992; Bauman *et al.* 2000; Seo 1996).

We explain the two-dimensional uniqueness. Assume that $\sigma, \tilde{\sigma}, u_j[\sigma], u_j[\tilde{\sigma}]$ in a cylindrical domain Ω do not change along the z direction and $\Lambda[\sigma] = \Lambda[\tilde{\sigma}]$. This two-dimensional problem has some practical meaning because many parts of the human body are locally cylindrical in their shape.

Theorem 9.4.1 *Let Ω be a smooth domain in $\mathbb{R}^2$ with a connected boundary. If $\sigma, \tilde{\sigma} \in C^1(\bar{\Omega})$ such that $\Lambda_j[\sigma] = \Lambda_j[\tilde{\sigma}]$, $j = 1, 2$, then*

$$\sigma = \tilde{\sigma} \quad in\ \Omega.$$

Proof. By taking the Laplacian of $\Lambda_j[\sigma] = \Lambda_j[\tilde{\sigma}]$, $j = 1, 2$, we have

$$\mu_0 \nabla \cdot [\sigma \nabla u_j \times \mathbf{e}_z] = \Delta \Lambda_j[\sigma] = \Delta \Lambda_j[\tilde{\sigma}] = \mu_0 \nabla \cdot [\tilde{\sigma} \nabla \tilde{u}_j \times \mathbf{e}_z] \quad \text{in } \Omega,$$

where $u_j = u_j[\sigma]$ and $\tilde{u}_j = u_j[\tilde{\sigma}]$. The above identity leads to $\nabla \cdot [\sigma \nabla u_j \times \mathbf{e}_z - \tilde{\sigma} \nabla \tilde{u}_j \times \mathbf{e}_z] = 0$, which can be rewritten as

$$0 = \nabla_{xy} \times \left(\sigma \frac{\partial u_j}{\partial x} - \tilde{\sigma} \frac{\partial \tilde{u}_j}{\partial x}, \sigma \frac{\partial u_j}{\partial y} - \tilde{\sigma} \frac{\partial \tilde{u}_j}{\partial y} \right),$$

where $\nabla_{xy} = (\partial/\partial x, \partial/\partial y)$ is the two-dimensional gradient. Hence, there exists a scalar function $\phi_j(\mathbf{r})$ such that

$$\nabla_{xy}\phi_j := \left(\sigma \frac{\partial u_j}{\partial x} - \tilde{\sigma} \frac{\partial \tilde{u}_j}{\partial x}, \sigma \frac{\partial u_j}{\partial y} - \tilde{\sigma} \frac{\partial \tilde{u}_j}{\partial y} \right) \quad \text{in } \Omega. \tag{9.32}$$

Then ϕ_j satisfies the two-dimensional Laplace equation $\Delta_{xy}\phi_j = 0$ in Ω with zero Neumann data, and hence ϕ_j is a constant function. Using $\sigma \nabla_{xy} u_j - \tilde{\sigma} \nabla_{xy}\ \tilde{u}_j = \nabla_{xy}\phi^j = 0$ and (9.16), we can derive

$$\begin{bmatrix} \sigma \dfrac{\partial u_1}{\partial x} & -\sigma \dfrac{\partial u_1}{\partial y} \\ \sigma \dfrac{\partial u_2}{\partial x} & -\sigma \dfrac{\partial u_2}{\partial y} \end{bmatrix} \begin{bmatrix} \dfrac{\partial}{\partial y} \ln \dfrac{\sigma}{\tilde{\sigma}} \\ \dfrac{\partial}{\partial x} \ln \dfrac{\sigma}{\tilde{\sigma}} \end{bmatrix} = \begin{bmatrix} 0 \\ 0 \end{bmatrix} \quad \text{in } \Omega.$$

Based on the result of the geometric index theory in Alessandrini and Magnanini (1992) and Kwon *et al.* (2006), we can show that the matrix

$$\begin{bmatrix} \sigma \dfrac{\partial u_1}{\partial x} & -\sigma \dfrac{\partial u_1}{\partial y} \\ \sigma \dfrac{\partial u_2}{\partial x} & -\sigma \dfrac{\partial u_2}{\partial y} \end{bmatrix}$$

is invertible for all points in Ω. This shows that $\ln(\sigma/\tilde{\sigma})$ is constant or $\sigma = c\tilde{\sigma}$ for a scaling constant c. Owing to the fact that $u_1|_{\mathcal{E}_2^+} - u_1|_{\mathcal{E}_2^-} = \Lambda_3[\sigma] = \Lambda_3[\tilde{\sigma}] = \tilde{u}_1|_{\mathcal{E}_2^+} - \tilde{u}_1|_{\mathcal{E}_2^-}$, we have $c = 1$, which leads to $\sigma = \tilde{\sigma}$.

Now we wish to prove

$$\nabla u_1(\mathbf{r}) \times \nabla u_2(\mathbf{r}) = \begin{vmatrix} \dfrac{\partial u_1}{\partial x}(\mathbf{r}) & \dfrac{\partial u_1}{\partial y}(\mathbf{r}) \\ \dfrac{\partial u_2}{\partial x}(\mathbf{r}) & \dfrac{\partial u_2}{\partial y}(\mathbf{r}) \end{vmatrix} \neq 0$$

for all $\mathbf{r} \in \Omega$.

To derive a contradiction, we assume that there is a point $\mathbf{r}_0 \in \Omega$ such that $\nabla u_1(\mathbf{r}_0) \times \nabla u_2(\mathbf{r}_0) = 0$. Then there exist two constants t_1 and t_2, which are not both zero, satisfying $t_1 \nabla u_1(\mathbf{r}_0) + t_2 \nabla u_2(\mathbf{r}_0) = 0$. Let $\partial\Omega^+ = \{\mathbf{r} \in \partial\Omega \mid t_1 g^1(\mathbf{r}) + t_2 g^2(\mathbf{r}) \geq 0\}$ and $\partial\Omega^- = \{\mathbf{r} \in \partial\Omega \mid t_1 g^1(\mathbf{r}) + t_2 g^2(\mathbf{r}) \leq 0\}$. For $\mathcal{E}_+^1 \cap \mathcal{E}_+^2 = \emptyset$ and $\mathcal{E}_-^1 = \mathcal{E}_-^2$, it can be easily checked that at least one of them is connected. According to the theory of the index of isolated critical points (Alessandrini and Magnanini 1992; Seo 1996), with Neumann data $g = t_1 g^1 + t_2 g^2$, the corresponding solution u satisfies

$$\nabla u(\mathbf{r}) = t_1 \nabla u_1(\mathbf{r}) + t_2 \nabla u_2(\mathbf{r}) \neq 0$$

for all $\mathbf{r} \in \Omega$, especially at $\mathbf{r}_0$. This is a contradiction. Hence we can conclude that

$$\nabla u_1(\mathbf{r}) \times \nabla u_2(\mathbf{r}) \neq 0$$

for all $\mathbf{r} \in \Omega$. This linear independence of ∇u_1 and ∇u_2 implies that $\nabla \log(\sigma/\tilde{\sigma}) = 0$ or $\log(\sigma/\tilde{\sigma})$ is a constant. From this, $\sigma(\mathbf{r}) = \beta\tilde{\sigma}(\mathbf{r})$ for some constant β. Owing to the fact that $|u_j(\mathbf{r}_1) - u_j(\mathbf{r}_2)| = |\tilde{u}_j(\mathbf{r}_1) - \tilde{u}_j(\mathbf{r}_2)| = \alpha^j$, we conclude that $\beta = 1$ or $\sigma = \tilde{\sigma}$ and immediately $u_j = \tilde{u}_j$. □

Although uniqueness in three dimensions is still an open problem, we can expect three-dimensional uniqueness by looking at the roles of the three components $\Lambda_1[\sigma]$, $\Lambda_2[\sigma]$ and $\Lambda_3[\sigma]$ with appropriate attachments of electrodes. Typical experimental and simulated B_z data sets are shown in Figure 9.8 and 9.9, respectively.

- Comparing Figure 9.8(a) and 9.8(c), we can see that the first component $\Lambda_1[\sigma]$ probes the vertical change of $\ln\sigma$ where the current density vector field $\mathbf{J}_1$ flows mostly in the horizontal direction. Figure 9.9(b) shows the simulated $\Lambda_1[\sigma]$ data with a horizontally oriented current. It is more clear that the B_z data subject to the horizontal current flow distinguishes the conductivity contrast along the vertical direction.
- Comparing Figure 9.8(b) and 9.8(d), the second component $\Lambda_2[\sigma]$ probes the horizontal change of $\ln\sigma$ where $\mathbf{J}_2$ flows mostly in the vertical direction. Figure 9.9(c) shows the simulated $\Lambda_2[\sigma]$ data with a vertically oriented current. It is clear that the B_z data subject to the vertical current flow distinguishes the conductivity contrast along the horizontal direction.
- The third component $\Lambda_3[\sigma]$ is used to fix the scaling uncertainty mentioned earlier in section 2.2.

In general, if we could produce two currents such that $\mathbf{J}_1(\mathbf{r}) \times \mathbf{e}_z$ and $\mathbf{J}_2(\mathbf{r}) \times \mathbf{e}_z$ are linearly independent for all $\mathbf{r} \in \Omega$, we can expect uniqueness roughly by observing the roles of $\Lambda[\sigma]$. Taking account of the uniqueness and stability, we carefully attach two pairs of surface electrodes (which determine the two different Neumann data) as shown in Figure 9.10 so that the area of the parallelogram $\left|(\mathbf{J}_1 \times \mathbf{J}_2) \cdot \mathbf{e}_z\right|$ is as large as possible in the truncated cylindrical region. However, the proof of $\left|(\mathbf{J}_1(\mathbf{r}) \times \mathbf{J}_2(\mathbf{r})) \cdot \mathbf{e}_z\right| > 0$ for $\mathbf{r} \in \Omega$ would be difficult.

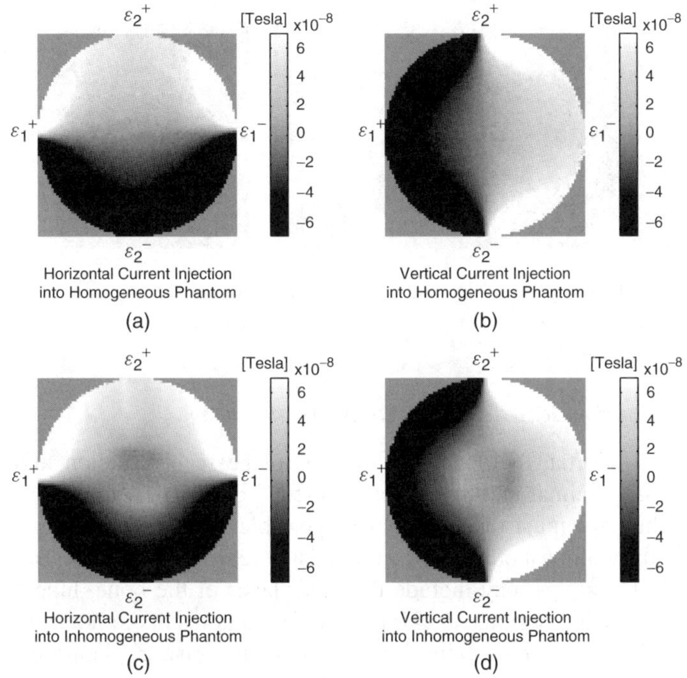

Figure 9.8 (a) and (b) Measured B_z data from a cylindrical homogeneous saline phantom subject to current injections along the horizontal and vertical directions, respectively. (c) and (d) Measured B_z data from the same phantom containing an agar anomaly with a different conductivity value from the background saline

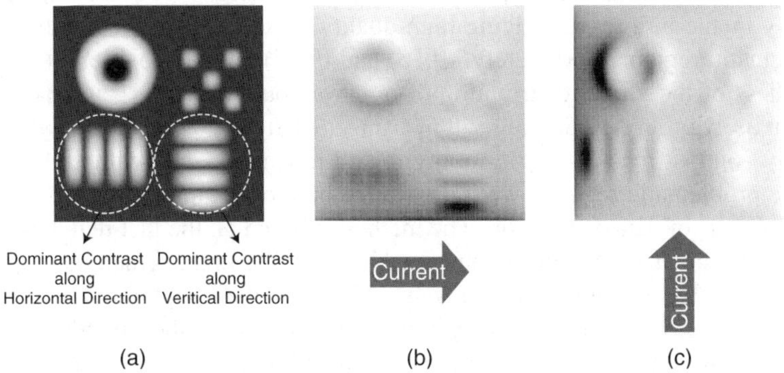

Figure 9.9 (a) Conductivity distribution of a model. Electrodes are attached along four sides of the model. (b) and (c) Simulated B_z data subject to current injections along the horizontal and vertical directions, respectively

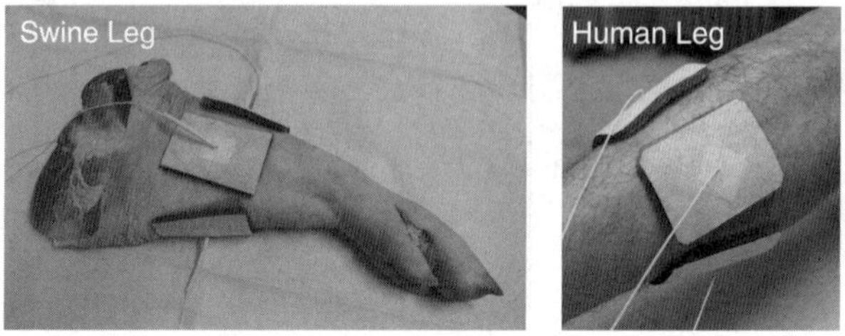

Figure 9.10 Typical examples of electrode attachment to maximize the area of the parallelogram $|(\mathbf{J}_1 \times \mathbf{J}_2) \cdot \mathbf{e}_z|$

9.4.3 *Defected B_z Data in a Local Region*

In MREIT, it is important to develop a robust image reconstruction algorithm that is applicable to *in vivo* animal and human experiments. Before developing an image reconstruction algorithm, we must take account of possible fundamental defects of measured B_z data. Inside the human body, there may exist a region where MR magnitude image values are small. Examples may include the outer layer of the bone, lungs and gas-filled internal organs. In such a region, $M \approx 0$ in (9.1) and this results in noise amplification. If the MR magnitude image M contains Gaussian random noise $\mathcal{Z}$, then the noise standard deviation in measured B_z data, denoted by $\mathrm{sd}(B_z)$, can be expressed in the following way (Sadleir 2005; Scott *et al.* 1992):

$$\mathrm{sd}(B_z) = \frac{1}{\sqrt{2}\gamma T_c} \frac{\mathrm{sd}(\mathcal{Z})}{M}. \tag{9.33}$$

From the above formula, the data B_z are not reliable inside an internal region where the MR magnitude image value M is small. It would be desirable to provide a high-resolution conductivity image in a region having high-quality B_z data regardless of the presence of such problematic regions. Fortunately, (9.16) and (9.17) would provide a local change of $\ln \sigma$ regardless of the global distribution of σ if we could predict $\mathbf{J}_1$ and $\mathbf{J}_2$ in that local region. This is why an MREIT algorithm using (9.16) and (9.17) can provide a robust conductivity contrast reconstruction in any region having B_z data with enough SNR.

For those problematic regions, we may use the harmonic inpainting method (Lee *et al.* 2006) as a process of data restoration. The method is based on the fact that $\Delta B_z = 0$ inside any local region having a homogeneous conductivity. We first segment each problematic region where the MR magnitude image value M is near zero. Defining a boundary of the region, we solve $\Delta B_z = 0$ using the measured B_z data along the boundary where noise is small. Then, we replace the original noisy B_z data inside the problematic region by the computed synthetic data. We must be careful in using this harmonic inpainting method since the problematic region will appear as a local homogeneous region in a reconstructed conductivity image. When there are multiple small local regions with large amounts of noise, we may consider using a harmonic decomposition denoising method (Lee *et al.* 2005) or other proper denoising techniques instead of the harmonic inpainting.

9.5 Image Reconstruction Algorithm

9.5.1 *J*-substitution Algorithm

Kwon *et al.* (2002b) solved the inverse problem of finding σ in (9.25) by developing a novel conductivity image reconstruction method called the J-substitution algorithm. Related to the boundary value problem in (9.25), we can introduce the cost functional

$$\Psi(\sigma) := \int_{\Omega} \left|J^*(\mathbf{r}) - \sigma(\mathbf{r})E_\sigma(\mathbf{r})\right|^2 \, d\mathbf{r}, \tag{9.34}$$

where $J^*(\mathbf{r})$ is the magnitude of the observed interior current density and $E_\sigma(\mathbf{r}) := |\nabla u_\sigma(\mathbf{r})|$ is the magnitude of the calculated electric field intensity obtained by solving (9.25) for a given σ. After discretization of the model $\bar{\Omega} = \bigcup_{k=0}^{N-1} \bar{\Omega}_k$ with the same area for all Ω_k, we get the squared residual sum

$$R(\sigma_0, \ldots, \sigma_{N-1}) := \sum_{k=0}^{N-1} \int_{\Omega_k} \left|J^*(\mathbf{r}) - \sigma_k E_\sigma(\mathbf{r})\right|^2 \, d\mathbf{r}, \tag{9.35}$$

where Ω_k is the kth element of the model and σ_k is the conductivity in Ω_k, which is assumed to be a constant on each element. Note that, in this case, the conductivity distribution is expressed by $\sigma(\mathbf{r}) = \sum_{k=0}^{N-1} \sigma_k \chi_{\Omega_k}(\mathbf{r})$, where $\chi_{\Omega_k}(\mathbf{r})$ denotes the indicator function of Ω_k, hence the electric field intensity $E_\sigma(\mathbf{r})$ in (9.35) is also a function of $(\sigma_0, \ldots, \sigma_{N-1})$. To update the conductivity from the zero gradient argument for the minimization of the squared residual sum, we differentiate (9.35) with respect to σ_m for $m = 0, \ldots, N-1$ to get

$$\begin{aligned} 0 = \frac{\partial R}{\partial \sigma_m} &= 2\int_{\Omega_m} E_\sigma(\mathbf{r})[\sigma_m E_\sigma(\mathbf{r}) - J^*(\mathbf{r})] \, d\mathbf{r} \\ &+ 2\sum_{k=0}^{N-1} \int_{\Omega_k} \sigma_k \frac{\partial E_\sigma(\mathbf{r})}{\partial \sigma_m} [\sigma_k E_\sigma(\mathbf{r}) - J^*(\mathbf{r})] \, d\mathbf{r}. \end{aligned} \tag{9.36}$$

This leads to the following approximate identity:

$$\begin{aligned} 0 &\approx E_\sigma(\mathbf{r}_m)[\sigma_m E_\sigma(\mathbf{r}_m) - J^*(\mathbf{r}_m)] \\ &+ \sum_{k=0}^{N-1} \sigma_k \frac{\partial E_\sigma(\mathbf{r}_k)}{\partial \sigma_m} [\sigma_k E_\sigma(\mathbf{r}_k) - J^*(\mathbf{r}_k)] \end{aligned} \tag{9.37}$$

for $m = 0, \ldots, N-1$, where $\mathbf{r}_k$ is the center point of the element Ω_k and we have used the simplest quadrature rule. Hence we obtain the following updating strategy to minimize the residual sum in (9.35):

$$\bar{\sigma}_m := \frac{J^*(\mathbf{r}_m)}{E_\sigma(\mathbf{r}_m)} \quad \text{for } m = 0, \ldots, N-1, \tag{9.38}$$

where $\bar{\sigma}_m$ is a new conductivity value on Ω_m and $E_\sigma(\mathbf{r}_m)$ is the calculated electric field intensity at the center point of Ω_m from an old conductivity distribution $\sigma(\mathbf{r}) = \sum_{k=0}^{N-1} \sigma_k \chi_{\Omega_k}(\mathbf{r})$.

Our inverse problem is to determine σ^* from two pairs of data $(I^q, J^q), q = 1, 2$ and the goal is to develop a reconstruction algorithm for σ^*. The reconstruction algorithm called the J-substitution algorithm for this nonlinear problem in (9.25) is as follows:

1. *Initial guess*. For the initial guess, we may choose a homogeneous conductivity σ^0, for example, $\sigma^0 := 1$.
2. *Forward solver*. For given conductivity σ^{2p+q} ($q = 1, 2$ and $p = 0, 1, 2, \ldots$), we solve the forward problem given by

$$\begin{gathered} \nabla \cdot (\sigma^{2p+q} \nabla u_p^q) = 0 \quad \text{in } \Omega, \\ \sigma^{2p+q} \frac{\partial u_p^q}{\partial n} = j_{I^q} \quad \text{on } \tilde{}\partial\Omega \quad \text{and} \quad \int_{\partial\Omega} u_p^q \, ds = 0. \end{gathered} \tag{9.39}$$

   The finite element method or finite difference method is commonly used to solve the forward problem in (9.39).
3. Update σ^{2p+q+1} as follows:

$$\sigma^{2p+q+1} := \frac{J^q}{|\nabla u_p^q|}. \tag{9.40}$$

   Now let us consider two conductivity distributions σ and $\sigma_\alpha := \alpha\sigma$ with a positive constant α for the same object Ω. If we inject the same current, the induced internal current density J corresponding to σ is identical to J_α corresponding to σ_α. However, the corresponding voltages satisfy $u_\alpha = (1/\alpha)u$. Therefore, in order to reconstruct the absolute conductivity distribution, the updating strategy in (9.40) should be modified as

$$\sigma^{2p+q+1} := \frac{J^q}{|\nabla u_p^q|} \frac{f^q_{\sigma^{2p+q}}}{f^q_{\sigma^*}}, \tag{9.41}$$

   where $f^q_{\sigma^*}$ is the measured voltage difference between two current injection electrodes for the injection current I^q and $f^q_{\sigma^{2p+q}}$ is the corresponding voltage difference when the conductivity distribution is given by σ^{2p+q}.
4. If $\left|\sigma^{2p+q+1} - \sigma^{2p+q}\right| < \epsilon$ for some measurement precision ϵ, stop; otherwise, go back to step 2 with $q = q + 1$ when $q = 1$ or $p = p + 1$ and $q = 1$ when $q = 2$.

The J-substitution algorithm can effectively and stably recover the conductivity distribution as long as the measured data sets of J_1^* and J_2^* are available. For cases where J_z is negligible, one may estimate J_1^* and J_2^* from measured B_z^1 and B_z^2, respectively, without rotating the imaging object. Carefully designing an experimental protocol including electrode size and configuration, the J-substitution method could be a suitable practical method to solve the inverse problem in MREIT. For other nonlinear inverse problems where internal data are available, one may also consider applying the J-substitution algorithm or its variations.

9.5.2 Harmonic B_z Algorithm

We consider the case where only B_z data are available for conductivity image reconstructions. The harmonic B_z algorithm is based on the following identity:

$$\mathbb{A}[\sigma](\mathbf{r})\begin{bmatrix}\dfrac{\partial \ln\sigma}{\partial x}(\mathbf{r})\\ \dfrac{\partial \ln\sigma}{\partial y}(\mathbf{r})\end{bmatrix} = \begin{bmatrix}\Delta\Lambda_1[\sigma](\mathbf{r})\\ \Delta\Lambda_2[\sigma](\mathbf{r})\end{bmatrix}, \quad \mathbf{r}\in\Omega, \tag{9.42}$$

where

$$\mathbb{A}[\sigma](\mathbf{r}) = \mu_0\begin{bmatrix}\sigma\dfrac{\partial u_1[\sigma]}{\partial y}(\mathbf{r}) & -\sigma\dfrac{\partial u_1[\sigma]}{\partial x}(\mathbf{r})\\ \sigma\dfrac{\partial u_2[\sigma]}{\partial y}(\mathbf{r}) & -\sigma\dfrac{\partial u_2[\sigma]}{\partial x}(\mathbf{r})\end{bmatrix}, \quad \mathbf{r}\in\Omega.$$

Noting that $\Delta\Lambda_j[\sigma] = \Delta B_{z,j}$ for $j = 1, 2$ from (9.13), we have

$$\begin{bmatrix}\dfrac{\partial \ln\sigma}{\partial x}(\mathbf{r})\\ \dfrac{\partial \ln\sigma}{\partial y}(\mathbf{r})\end{bmatrix} = (\mathbb{A}[\sigma](\mathbf{r}))^{-1}\begin{bmatrix}\Delta B_{z,1}(\mathbf{r})\\ \Delta B_{z,2}(\mathbf{r})\end{bmatrix}, \quad \mathbf{r}\in\Omega, \tag{9.43}$$

provided that $\mathbb{A}[\sigma]$ is invertible. The above identity (9.43) leads to an implicit representation formula for σ on each slice $\Omega_{z_0} := \Omega\cap\{z = z_0\}$ in terms of the measured data set $(B_{z,1}, B_{z,2}, V_{12}^{\pm})$. Denoting $\mathbf{x} = (x, y)$ and $\mathbf{x}' = (x', y')$, we have

$$\mathcal{L}_{z_0}\ln\sigma(\mathbf{x}) = \Phi_{\Omega_{z_0}}[\sigma](\mathbf{x}) \quad \text{for all } (\mathbf{x}, z_0)\in\Omega_{z_0}, \tag{9.44}$$

where

$$\Phi_{\Omega_{z_0}}[\sigma](\mathbf{x}) = \frac{1}{2\pi}\int_{\Omega_{z_0}}\frac{\mathbf{x}-\mathbf{x}'}{|\mathbf{x}-\mathbf{x}'|^2}\cdot\left((\mathbb{A}[\sigma](\mathbf{x}', z_0))^{-1}\begin{bmatrix}\Delta B_{z,1}(\mathbf{x}', z_0)\\ \Delta B_{z,2}(\mathbf{x}', z_0)\end{bmatrix}\right)\mathrm{d}s_{\mathbf{x}'} \tag{9.45}$$

and

$$\mathcal{L}_{z_0}\ln\sigma(\mathbf{x}) = \ln\sigma(\mathbf{x}, z_0) + \frac{1}{2\pi}\int_{\partial\Omega_{z_0}}\frac{(\mathbf{x}-\mathbf{x}')\cdot\nu(\mathbf{x}')}{|\mathbf{x}-\mathbf{x}'|^2}\ln\sigma(\mathbf{x}', z_0)\,\mathrm{d}\ell_{\mathbf{x}'}. \tag{9.46}$$

Here, ν is the unit outward normal vector to the curve $\partial\Omega_{z_0}$ and $\mathrm{d}\ell$ is the line element. From the trace formula for the double-layer potential in (9.46), the identity (9.44) on the boundary $\partial\Omega_{z_0}$ can be expressed as

$$\mathcal{T}_{z_0}\ln\sigma(\mathbf{x}) = \Phi_{\Omega_{z_0}}[\sigma](\mathbf{x}) \quad \text{for all } (\mathbf{x}, z_0)\in\partial\Omega_{z_0}, \tag{9.47}$$

where

$$\mathcal{T}_{z_0}\ln\sigma(\mathbf{x}) = \frac{\ln\sigma(\mathbf{x}, z_0)}{2} + \frac{1}{2\pi}\int_{\partial\Omega_{z_0}}\frac{(\mathbf{x}-\mathbf{x}')\cdot\nu(\mathbf{x}')}{|\mathbf{x}-\mathbf{x}'|^2}\ln\sigma(\mathbf{x}', z_0)\,\mathrm{d}\ell_{\mathbf{x}'}.$$

Note that the operator $\mathcal{T}_{z_0}$ is invertible on $L_0^2(\partial\Omega_{z_0}) = \{\phi \in L^2(\partial\Omega_{z_0}) : \int_{\partial\Omega_{z_0}} \phi \, d\ell = 0\}$ from the well-known potential theory (Folland 1976).

Lemma 9.5.1 *The operator $\mathcal{L}_z : H_\diamond^{1/2}(\Omega_z) \to H_\diamond^{1/2}(\Omega_z)$ is invertible where $H_\diamond^{1/2}(\Omega_z) := \{\eta \in H^{1/2}(\Omega_z) : \int_{\partial\Omega_z} \eta = 0\}$.*

Proof. The invertibility of $\mathcal{L}$ can be proven by the standard layer potential theory (Folland 1976; Verchota *et al.* 1984). For $w \in H_\diamond^{1/2}(\Omega_z)$, we need to find $v \in H_*^{1/2}(\Omega_z)$ such that $\mathcal{L}_z v = w$. Note that $w|_{\partial\Omega_z} \in L_\diamond^2(\partial\Omega_z) := \{\phi \in L^2(\Omega_z) : \int_{\partial\Omega_z} \phi = 0\}$. It is well known that there exists a unique $\psi \in L_\diamond^2(\partial\Omega_z)$ such that $\frac{1}{2}\psi - \mathcal{K}\psi = w|_{\partial\Omega_z}$ on $\partial\Omega_z$, where

$$\mathcal{K}\psi(\mathbf{x}) = \frac{1}{2\pi}\int_{\partial\Omega_{z_0}} \frac{(\mathbf{x}-\mathbf{x}')\cdot\nu(\mathbf{x}')}{|\mathbf{x}-\mathbf{x}'|^2}\psi(\mathbf{x}')\, d\ell_{\mathbf{x}'}, \quad \mathbf{x} \in \partial\Omega_z.$$

Now, we define

$$v(\mathbf{x}) = w(x,y) + \frac{1}{2\pi}\int_{\partial\Omega_{z_0}} \frac{(\mathbf{x}-\mathbf{x}')\cdot\nu(\mathbf{x}')}{|\mathbf{x}-\mathbf{x}'|^2}\psi(\mathbf{x}')\, d\ell_{\mathbf{x}'}, \quad \mathbf{x} \in \Omega_z. \tag{9.48}$$

Owing to the trace formula of the double-layer potential, $v = \psi$ on $\partial\Omega_z$. By replacing ψ in (9.48) with v, we have $w = \mathcal{L}_z v$ and this completes the proof. □

Because of the invertibility of the operators $\mathcal{L}_z : H_\diamond^{1/2}(\Omega_z) \to H_\diamond^{1/2}(\Omega_z)$ and $\mathcal{T}_{z_0}$, we can expect that the following iterative algorithm based on the identities (9.44) and (9.47) can determine σ up to a scaling factor:

$$\begin{cases} \nabla_{xy}\sigma^{n+1}(\mathbf{x}, z_0) = \mathbb{A}[\sigma^n]^{-1}\begin{bmatrix} \Delta B_{z,1} \\ \Delta B_{z,2} \end{bmatrix} & \text{for } (\mathbf{x}, z_0) \in \Omega_{z_0}, \\ \mathcal{L}_{z_0} \ln\sigma^{n+1}(\mathbf{x}) = \Phi_{\Omega_{z_0}}[\sigma^{n+1}](\mathbf{x}) & \text{for } (\mathbf{x}, z_0) \in \Omega_{z_0}. \end{cases} \tag{9.49}$$

From the first step in (9.49), we can update $\nabla_{xy}\sigma^{n+1}$ for all imaging slices of interest within the object as long as the measured data B_z are available for the slices. Next, we obtain $\sigma^{n+1}|_{\partial\Omega}$ by solving the integral equation (9.47) for the given right-hand side of the second step in (9.49). Since $\sigma^{n+1}|_{\partial\Omega_{z_0}}$ is known, so is the value σ^{n+1} inside Ω_{z_0} by simple substitution of $\sigma^{n+1}|_{\partial\Omega_{z_0}}$ and $\nabla_{xy}\sigma^{n+1}$ into the corresponding integrals. This harmonic B_z algorithm has shown a remarkable performance in various numerical simulations (Oh *et al.* 2003; Seo *et al.* 2003b) and imaging experiments.

Early MREIT methods have used all three components of the magnetic flux density $\mathbf{B} = (B_x, B_y, B_z)$, and they require impracticable rotations of the imaging object inside the MRI scanner. The invention of the harmonic B_z algorithm using only B_z instead of $\mathbf{B}$ (Seo *et al.* 2003b) changed the problem of impracticable rotations into a mathematical problem (9.28) with achievable data through applications of two linearly independent Neumann data. This harmonic B_z algorithm has been widely used in experimental studies, including *in vivo* animal and human imaging experiments (Kim *et al.* 2007, 2008a,b, 2009, 2011).

We now briefly mention the convergence behavior of (9.49). When σ has a low contrast in Ω, the direction of the vector field $\sigma\nabla u_j[\sigma]$ is mostly dictated by the geometry of the boundary $\partial\Omega$ and the electrode positions $\mathcal{E}_j^{\pm}$ (or Neumann boundary data) instead of the distribution of σ. This ill-posedness was the fundamental drawback of the corresponding inverse problem of EIT. But, in MREIT, we take advantage of this insensitivity of EIT. This means that the direction of the vector field $\sigma\nabla u_j[\sigma]$ is similar to that of $\sigma_0\nabla u_j[\sigma_0]$ with $\sigma_0 = 1$, and therefore the data $B_{z,1}$ and $B_{z,2}$ hold the major information of the conductivity contrast. Various numerical simulations show that only one iteration of (9.49) may provide a conductivity image σ^1 that is quite similar to the true conductivity σ. Rigorous mathematical theories regarding its convergence behavior have not been proven yet. There are some convergence results on (9.49) under *a priori* assumptions on the target conductivity (Liu *et al.* 2007).

9.5.3 Gradient B_z Decomposition and Variational B_z Algorithm

It would be better to minimize the amplitude of the injection current. However, the amplitude of the signal B_z is proportional to the amplitude of the injection current. For a given noise level of an MREIT system, this means that we have to deal with B_z data sets with a low SNR. Numerical implementation methods of an image reconstruction algorithm affect the quality of a reconstructed conductivity image since noise in B_z data is transformed into noise in the conductivity image. Depending on the chosen method, noise could be amplified or weakened.

Since double differentiation of B_z data tends to amplify its noise, the performance of the harmonic B_z algorithm could deteriorate when the SNR in the measured B_z data is low. To deal with this noise amplification problem, algorithms to reduce the number of differentiations have been developed. They include the gradient B_z decomposition algorithm (Park *et al.* 2004a) and the variational gradient B_z algorithm (Park *et al.* 2004b), which need to differentiate B_z only once. They show a better performance in some numerical simulations and we discuss only one of them for pedagogical purposes.

We briefly explain the gradient B_z decomposition algorithm in a special cylindrical domain $\Omega = \{\mathbf{r} = (x, y, z) \mid (x, y) \in D,\ -\delta < z < \delta\}$, where D is a two-dimensional, smooth and simply connected domain. Suppose that u is a solution of $\nabla\cdot(\sigma\nabla u) = 0$ in Ω with Neumann data g. We parameterize ∂D as $\partial D{:} = \{(x(t), y(t)) : 0 \le t \le 1\}$ and define

$$\tilde{g}(x(t), y(t), z) := \int_0^t g((x(t), y(t), z))\sqrt{|x'(t)|^2 + |y'(t)|^2}\,\mathrm{d}t$$

for $(x, y, z) \in \partial\Omega \setminus \{z = \pm\delta\}$. The gradient B_z decomposition algorithm is based on the following implicit reconstruction formula:

$$\sigma = \frac{|-(\partial\Upsilon/\partial y + \Theta_x[u])\,\partial u/\partial x + (\partial\Upsilon/\partial x + \Theta_y[u])\,\partial u/\partial y|}{(\partial u/\partial x)^2 + (\partial u/\partial y)^2} \quad \text{in } \Omega, \tag{9.50}$$

where

$$\Theta_x[u] := \frac{\partial\psi}{\partial y} - \frac{\partial W_z}{\partial x} + \frac{\partial W_x}{\partial z}, \quad \Theta_y[u] := \frac{\partial\psi}{\partial x} + \frac{\partial W_z}{\partial y} - \frac{\partial W_y}{\partial z} \quad \text{in } \Omega$$

and

$$\Upsilon = \phi + \frac{1}{\mu_0}B_z, \quad W(\mathbf{r}) := \int_{\Omega_\delta} \frac{1}{4\pi|\mathbf{r}-\mathbf{r}'|}\frac{\partial(\sigma\nabla u(\mathbf{r}'))}{\partial z}\,\mathrm{d}\mathbf{r}'.$$

Here, ϕ is a solution of

$$\begin{cases} \nabla^2\phi = 0 \quad \text{in } \Omega, \\ \phi = \tilde{g} - \dfrac{1}{\mu_0} B_z \quad \text{on } \partial\Omega \setminus \{z = \pm\delta\}, \\ \dfrac{\partial \phi}{\partial z} = -\dfrac{1}{\mu_0}\dfrac{\partial B_z}{\partial z} \quad \text{on } \partial\Omega \cap \{z = \pm\delta\} \end{cases} \tag{9.51}$$

and ψ is a solution of

$$\begin{cases} \nabla^2\psi = 0 \quad \text{in } \Omega, \\ \nabla\psi \cdot \tau = \nabla \times W \cdot \tau \quad \text{on } \partial\Omega \setminus \{z = \pm\delta\}, \\ \dfrac{\partial \psi}{\partial z} = -\nabla \times W \cdot \mathbf{e}_z \quad \text{on } \partial\Omega \cap \{z = \pm\delta\}, \end{cases} \tag{9.52}$$

where $\tau := (-\nu_y, \nu_x, 0)$ is the tangent vector on the lateral boundary $\partial\Omega \setminus \{z = \pm\delta\}$.

We may use an iterative reconstruction scheme with multiple Neumann data g_j, $j = 1, \ldots, N$, to find σ. Denoting by u_j^m a solution of $\nabla \cdot (\sigma^m \nabla u) = 0$ in Ω with Neumann data g_j, the reconstructed σ is the limit of a sequence σ^m that is obtained by the following formula:

$$\sigma^{m+1} = \frac{\sum_{i=1}^{N} |-(\partial\Upsilon_i/\partial y + \Theta_x[u_i^m])\, \partial u_i^m/\partial x + (\partial\Upsilon_i/\partial x + \Theta_y[u_i^m])\, \partial u_i^m/\partial y|}{\sum_{i=1}^{N} [(\partial u_i^m/\partial x)^2 + (\partial u_i^m/\partial y)^2]}. \tag{9.53}$$

This method needs to differentiate B_z only once, in contrast to the harmonic B_z algorithm, where the numerical computation of $\nabla^2 B_z$ is required. It has an advantage of much improved noise tolerance, and numerical simulations with added random noise of a realistic amount showed its feasibility and robustness against measurement noise. However, in practical environments, it shows poor performance compared with the harmonic B_z algorithm and may produce some artifacts.

The major reason is that the updated conductivity σ^{m+1} by the iteration process (9.53) is influenced by the global distribution of σ^m. We should note that there always exist some local regions having defective B_z data in human or animal experiments, and we always deal with a truncated region of the imaging object, which causes geometric errors. Hence, it would be very difficult to reconstruct the conductivity distribution in the entire region of the human or animal subject with reasonable accuracy, and it would be best to achieve robust reconstruction of σ in local regions where measured B_z data are reliable. In order to achieve a stable local reconstruction of conductivity contrast with moderate accuracy, poor conductivity reconstruction at one local region should not adversely influence conductivity reconstructions in other regions. This means that a conductivity image reconstruction algorithm should not depend too much on the global distribution of B_z, the global structure of σ and the geometry $\partial\Omega$.

9.5.4 *Local Harmonic B_z Algorithm*

Noting that there inevitably exist defective regions inside the human body where measured B_z data are not reliable, Seo *et al.* (2008) and Jeon (2010) proposed a modified version of the harmonic B_z algorithm called the local harmonic B_z algorithm to improve its practical applicability. Assume that we sequentially inject two currents I_1 and I_2 through two pairs of surface electrodes $\mathcal{E}_1^{\pm}$ and $\mathcal{E}_2^{\pm}$, respectively. For $j = 1, 2$, we let $u_j[\sigma]$ be a solution of the following boundary value problem:

$$\begin{cases} \nabla \cdot (\sigma \nabla u_j[\sigma]) = 0 \quad \text{in } \Omega, \\ I_j = \int_{\mathcal{E}_j^+} \sigma \dfrac{\partial u_j[\sigma]}{\partial \mathbf{n}} \, ds = -\int_{\mathcal{E}_j^-} \sigma \dfrac{\partial u_j[\sigma]}{\partial \mathbf{n}} \, ds, \\ \nabla u_j[\sigma] \times \mathbf{n}|_{\mathcal{E}_j^+ \cup \mathcal{E}_j^-} = 0, \\ \sigma \dfrac{\partial u_j[\sigma]}{\partial \mathbf{n}} = 0 \quad \text{on } \partial\Omega \setminus \overline{\mathcal{E}_j^+ \cup \mathcal{E}_j^-}. \end{cases} \tag{9.54}$$

The z component of the curl of Ampère's law $\nabla \times \mathbf{J} = (1/\mu_0)\nabla \times \nabla \times \mathbf{B}$ is

$$\langle \nabla\sigma, \mathbb{L}\nabla u_j[\sigma] \rangle = \frac{1}{\mu_0} \nabla^2 B_{z,j} \quad \text{where} \quad \mathbb{L} = \begin{pmatrix} 0 & 1 & 0 \\ -1 & 0 & 0 \\ 0 & 0 & 0 \end{pmatrix} \tag{9.55}$$

and $B_{z,j}$ is the z component of the induced magnetic flux density subject to the injected current I_j. This identity indicates that $(1/\mu_0)\nabla^2 B_{z,j}$ conveys information on any local change of σ along the direction $\mathbb{L}\nabla u_j[\sigma]$, which is a nonlinear function of σ.

We rewrite (9.55) as

$$\langle \nabla \ln \sigma, \mathbb{L}\sigma\nabla u_j[\sigma] \rangle = \frac{1}{\mu_0} \nabla^2 B_{z,j}. \tag{9.56}$$

Combining (9.56) for $j = 1, 2$, we get

$$\begin{bmatrix} \dfrac{\partial \ln \sigma}{\partial x}(\mathbf{r}) \\ \dfrac{\partial \ln \sigma}{\partial y}(\mathbf{r}) \end{bmatrix} = \frac{1}{\mu_0} (\mathbb{A}[\sigma](\mathbf{r}))^{-1} \begin{bmatrix} \nabla^2 B_{z,1}(\mathbf{r}) \\ \nabla^2 B_{z,2}(\mathbf{r}) \end{bmatrix}, \quad \mathbf{r} \in \Omega, \tag{9.57}$$

where

$$\mathbb{A}[\sigma](\mathbf{r}) = \begin{bmatrix} \sigma(\mathbf{r}) \dfrac{\partial u_1[\sigma]}{\partial y}(\mathbf{r}) & -\sigma(\mathbf{r}) \dfrac{\partial u_1[\sigma]}{\partial x}(\mathbf{r}) \\ \sigma(\mathbf{r}) \dfrac{\partial u_2[\sigma]}{\partial y}(\mathbf{r}) & -\sigma(\mathbf{r}) \dfrac{\partial u_2[\sigma]}{\partial x}(\mathbf{r}) \end{bmatrix}.$$

We should choose an electrode configuration including their size and positions in such a way that the condition number of $\mathbb{A}$ is presumably minimized. Two equally spaced pairs of large and flexible electrodes are advantageous in reducing the condition number.

We convert (9.57) into the following second-order differential equation by taking the transverse divergence:

$$\left(\frac{\partial^2}{\partial x^2} + \frac{\partial^2}{\partial y^2} \right) \ln \sigma(\mathbf{r}) = \frac{\partial}{\partial x} \Theta_x[\sigma](\mathbf{r}) + \frac{\partial}{\partial y} \Theta_y[\sigma](\mathbf{r}), \quad \mathbf{r} \in \Omega, \tag{9.58}$$

where

$$\Theta[\sigma](\mathbf{r}) = \begin{bmatrix} \Theta_x[\sigma](\mathbf{r}) \\ \Theta_y[\sigma](\mathbf{r}) \end{bmatrix} := \frac{1}{\mu_0}(\mathbb{A}[\sigma](\mathbf{r}))^{-1} \begin{bmatrix} \nabla^2 B_{z,1}(\mathbf{r}) \\ \nabla^2 B_{z,2}(\mathbf{r}) \end{bmatrix}.$$

One may apply (9.58) to the harmonic B_z algorithm as well to reconstruct conductivity images of an entire imaging domain. In this case, we may need to adopt an iteration scheme to reconstruct σ in each slice Ω_{z_0} since (9.58) is a nonlinear equation of σ.

We now assume that the imaging object Ω contains a local region R with small conductivity values, that is, $\sigma \approx 0$ in R. Examples may include the outer layers of bones, lungs and gas-filled tubular organs. Note that they usually coincide with the defective regions of MR signal void discussed early. We let D be a two-dimensional smooth subdomain in Ω_{z_0} excluding all problematic regions. Let σ_0 be an initial guess of σ. We change the nonlinear equation (9.58) into the following Poisson equation with a Neumann boundary condition in D:

$$\begin{cases} \left(\dfrac{\partial^2}{\partial x^2} + \dfrac{\partial^2}{\partial y^2}\right) \ln\sigma(x,y,z_0) = \nabla_{x,y} \cdot \Theta[\sigma_0](x,y,z_0) & \text{for } (x,y,z_0) \in D, \\ \nu \cdot \nabla_{x,y} \ln\sigma(x,y,z_0) = \nu \cdot \Theta[\sigma_0](x,y,z_0) & \text{for } (x,y,z_0) \in \partial D, \end{cases} \tag{9.59}$$

where $\nabla_{x,y} = (\partial/\partial x, \partial/\partial y)$ and ν is the outward unit normal vector to the two-dimensional boundary ∂D. Note that (9.59) has a unique solution up to a constant and does not require any information other than measured data $B_{z,j}$ in D. In this local harmonic B_z algorithm, the conductivity image quality is determined by the quality of the measured $B_{z,j}$ data only, without requiring any assumption on the conductivity or voltage values on the boundary $\partial\Omega$.

Provided that $|\nabla\sigma|$ is small, the vector field $\sigma\nabla u_j[\sigma]$ is dictated mainly by the injected current I_j and the global geometry of the boundary $\partial\Omega$ instead of the local distribution of σ. In such a case, we may approximate $\sigma\nabla u_j[\sigma] \approx \sigma_0 \nabla u_j[\sigma_0]$ with $\sigma_0 = 1$. This means that we can perceive a local change of $\ln\sigma$ in the direction $\sigma\mathbb{L}\nabla u_j[\sigma]$, which is approximately estimated by $\mathbb{L}\nabla u_j[1]$. In the single-step local harmonic B_z algorithm, we can perform a conductivity image reconstruction using $\sigma_0 = 1$ without iteration and produce a scaled conductivity image in D or any chosen region of interest.

9.5.5 Sensitivity Matrix-Based Algorithm

Using a sensitivity matrix $\mathbf{S}$ derived from (9.13) with the assumption of $\mathcal{H} = 0$, we may linearize the relationship between B_z and σ as follows (Birgul and Ider 1995; Birgul and Ider 1996):

$$\Delta B_z = \mathbf{S}\Delta\sigma, \tag{9.60}$$

where ΔB_z is the difference in B_z from the imaging object with homogeneous and perturbed conductivity distributions, σ_0 and $\sigma_0 + \Delta\sigma$, respectively. Inverting the sensitivity matrix, one can reconstruct a conductivity image from measured B_z data. This approach is similar to those used in time-difference EIT imaging.

Birgul *et al.* (2003) elaborated this method and presented experimental results using a two-dimensional saline phantom with 20 electrodes. Muftuler *et al.* (2004) and Birgul *et al.* (2006) studied the sensitivity-based method in terms of image resolution and contrast. Hamamura *et al.* (2006) demonstrated that this sensitivity-based method can image time changes of ion diffusion in agar phantoms. Muftuler *et al.* (2006) performed animal experiments on rats and imaged tumors using an iterative version of the sensitivity-based method. They showed that the conductivity values of tumor areas are increased in reconstructed conductivity images. This method cannot deal with the unknown term $\mathcal{H}$, which is not zero unless lead wires are perfectly parallel to the z axis.

9.5.6 Anisotropic Conductivity Reconstruction Algorithm

Some biological tissues are known to have anisotropic conductivity values, and the anisotropy ratio depends on the type of tissue. For example, human skeletal muscle shows an anisotropy of up to 1–10 between the longitudinal and transverse directions.

Seo *et al.* (2004) applied the MREIT technique to anisotropic conductivity image reconstructions. Investigating how an anisotropic conductivity

$$\boldsymbol{\sigma} = \begin{pmatrix} \sigma_{11} & \sigma_{12} & \sigma_{13} \\ \sigma_{12} & \sigma_{22} & \sigma_{23} \\ \sigma_{13} & \sigma_{23} & \sigma_{33} \end{pmatrix}$$

affects the internal current density and thereby the magnetic flux density, they understood that at least seven different injection currents are necessary for the anisotropic conductivity image reconstruction algorithm. The algorithm is based on the following two identities:

$$\mathbf{Us} = \mathbf{b} \quad \text{and} \quad \nabla \cdot \left[\begin{pmatrix} \sigma_{11} & \sigma_{12} & \sigma_{13} \\ \sigma_{12} & \sigma_{22} & \sigma_{23} \\ \sigma_{13} & \sigma_{23} & \sigma_{33} \end{pmatrix} \nabla u^j \right] = 0, \tag{9.61}$$

where

$$\mathbf{b} = \frac{1}{\mu_0} \begin{bmatrix} \nabla^2 B_{1,z} \\ \vdots \\ \nabla^2 B_{N,z} \end{bmatrix}, \quad \mathbf{s} = \begin{bmatrix} -\partial_y \sigma_{11} + \partial_x \sigma_{12} \\ -\partial_y \sigma_{12} + \partial_x \sigma_{22} \\ -\partial_y \sigma_{13} + \partial_x \sigma_{23} \\ \sigma_{12} \\ -\sigma_{11} + \sigma_{22} \\ \sigma_{23} \\ \sigma_{13} \end{bmatrix}$$

and

$$\mathbf{U} = \begin{bmatrix} u_x^1 & u_y^1 & u_z^1 & u_{xx}^1 - u_{yy}^1 & u_{xy}^1 & u_{xz}^1 & -u_{yz}^1 \\ \vdots & \vdots & \vdots & \vdots & \vdots & \vdots & \vdots \\ u_x^N & u_y^N & u_z^N & u_{xx}^N - u_{yy}^N & u_{xy}^N & u_{xz}^N & -u_{yz}^N \end{bmatrix}.$$

Here, u^j is the voltage corresponding to the jth injection current, $u_x^j = \partial u^j/\partial x$ and $\boldsymbol{\sigma}$ is assumed to be a symmetric positive definite matrix. As in the harmonic B_z algorithm, we may use an iterative procedure to compute $\mathbf{s}$ in (9.61). Assuming that we have computed all seven terms of $\mathbf{s}$, we can immediately determine $\sigma_{12}(\mathbf{r}) = s_4(\mathbf{r}), \sigma_{13}(\mathbf{r}) = s_7(\mathbf{r})$ and $\sigma_{23}(\mathbf{r}) = s_6(\mathbf{r})$. To determine σ_{11} and σ_{22} from $\mathbf{s}$, we use the relation between $\mathbf{s}$ and $\boldsymbol{\sigma}$:

$$\frac{\partial \sigma_{11}}{\partial x} = s_2 - \frac{\partial s_5}{\partial x} + \frac{\partial s_4}{\partial y} \quad \text{and} \quad \frac{\partial \sigma_{11}}{\partial y} = -s_1 + \frac{\partial s_4}{\partial x}. \tag{9.62}$$

The last component σ_{33} can be obtained by using the physical law $\nabla \cdot \mathbf{J} = 0$.

Numerical simulation results using a relatively simple two-dimensional model shown in Figure 9.11 demonstrated that the algorithm can successfully reconstruct images of an anisotropic conductivity tensor distribution provided that the B_z data have a high SNR. Unfortunately, this algorithm is not successful in practical environments since it is very sensitive to the noise and the matrix **U** is ill-conditioned in the interior region.

9.5.7 Other Algorithms

The algebraic reconstruction method (Ider and Onart 2004) may be considered as a variation of the harmonic B_z algorithm. They discussed numerous issues, including uniqueness, region-of-interest reconstruction and noise effects. Assuming that B_z data subject to an injection current into the head are available, Gao *et al.* (2005) developed a method to determine the conductivity values of the brain, skull and scalp layers using the radial basis function and simplex method. This kind of parametric approach may find useful

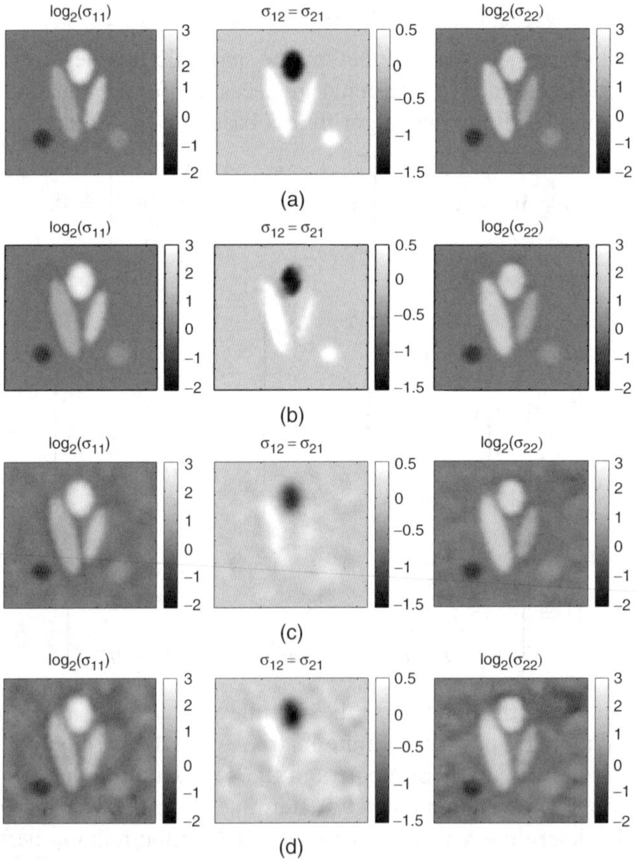

Figure 9.11 Numerical simulation of the anisotropic conductivity image reconstruction. (a) The target conductivity tensor image. (b)–(d) Reconstructed images when SNR is ∞, 300 and 150, respectively. Here, SNR means the SNR of the corresponding MR magnitude image. From Seo *et al.* (2004)

applications in EEG/MEG source imaging problems. Gao *et al.* (2006) also suggested the so-called RSM-MREIT algorithm, where the total error between measured and calculated magnetic flux densities is minimized as a function of a model conductivity distribution by using the response surface methodology algorithm.

9.6 Validation and Interpretation

9.6.1 Image Reconstruction Procedure using Harmonic B_z Algorithm

Based on the harmonic B_z algorithm, the Impedance Imaging Research Center (IIRC) in Korea developed MREIT software to offer various computational tools, from preprocessing to reconstruction of conductivity and current density images. Figure 9.12 shows a screen capture of the MREIT software, CoReHA (conductivity reconstructor using harmonic algorithms) (Jeon *et al.* 2009a,b). It includes three major tasks of preprocessing, model construction and data recovery, and conductivity image reconstruction.

- *Preprocessing*. We obtain magnetic flux density images $B_{z,1}$ and $B_{z,2}$ corresponding to two injection currents I_1 and I_2, respectively, from the k-space data after applying proper phase unwrapping and unit conversion. Since the magnetic flux density images could be quite noisy in practice, due to many factors, we may use a PDE-based denoising method such as harmonic decomposition.

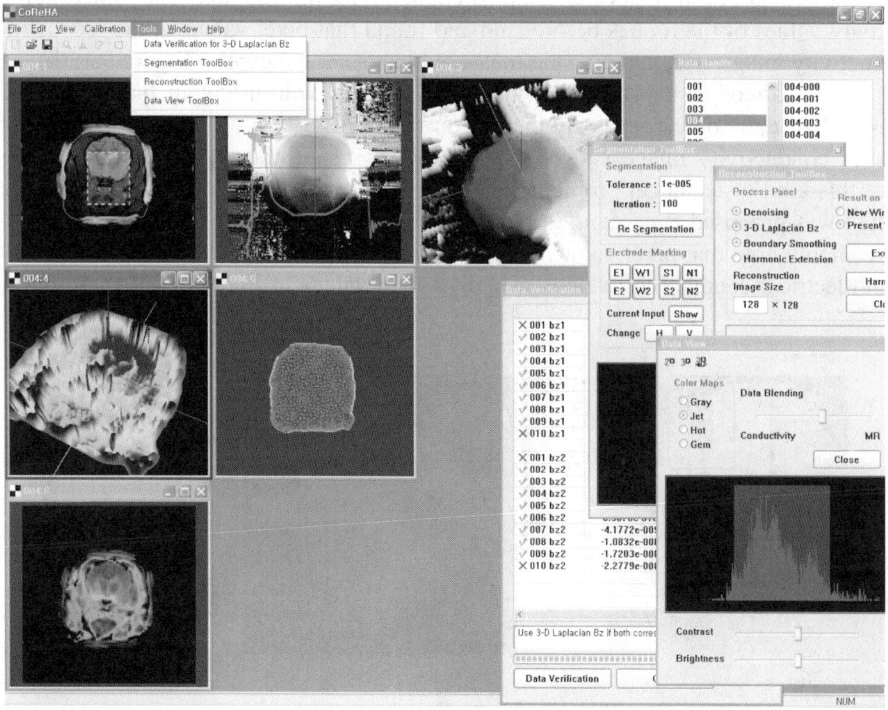

Figure 9.12 Screen capture of CoReHA. It provides main menus for image viewing, calibration or coordinate setting and data processing, including data verification, segmentation, meshing and image reconstruction

- *Model construction and data recovery*. In the geometrical modeling of the conducting domain, identifications of the outermost boundary and electrode locations are critical to impose boundary conditions. We may use a semi-automatic tool employing a level-set-based segmentation method. There could be an internal region where an MR signal void occurred. In such a problematic region, measured B_z data are defective. We may use the harmonic inpainting method to recover B_z data, assuming that the local region is homogeneous in terms of the conductivity.
- *Conductivity image reconstruction*. We can use the harmonic B_z algorithm as the default algorithm for three-dimensional conductivity image reconstructions. We may apply the local harmonic B_z algorithm (Seo *et al.* 2008) for conductivity image reconstructions in chosen regions of interest.

9.6.2 Conductivity Phantom Imaging

Since Woo and Seo (2008) have summarized most of the published results of conductivity phantom imaging experiments (Oh *et al.* 2003, 2004, 2005), we introduce only one of them. Figure 9.13(a) shows a tissue phantom including chunks of three different biological tissues in a background of agar gel. Its MR magnitude and reconstructed conductivity images are shown in Figure 9.13(b) and 9.13(c) (Oh *et al.* 2005). Compared with the MR magnitude image in Figure 9.13(b), the reconstructed conductivity image in (c) shows excellent structural information as well as conductivity information. They measured conductivity values of the tissues beforehand and found that pixel values in the reconstructed conductivity image were close to the measured values. As shown in Figure 9.13(b), an air bubble was formed inside the phantom. The MR signal void in the air bubble caused the measured B_z data to be very noisy there. From Figure 9.13(c), we can see that the reconstructed conductivity image shows spurious spikes inside the region of the air bubble. Since this kind of technical problem can occur in a living body, the harmonic inpainting method was proposed (Lee *et al.* 2006).

We should note that pixel values in Figure 9.13(c) provide totally different information about electrical conductivity values, whereas pixel values in Figure 9.13(b) are basically

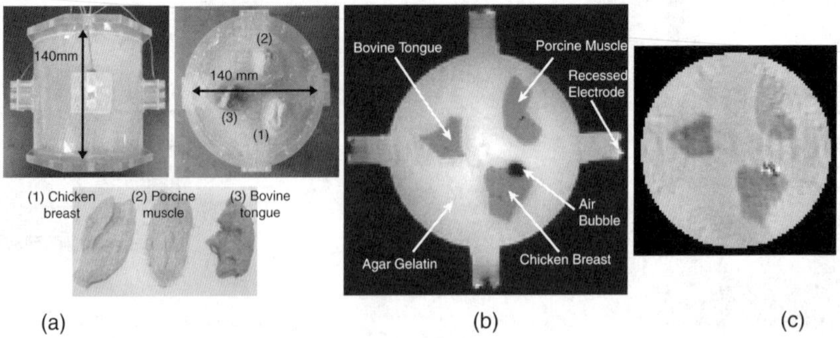

Figure 9.13 Biological tissue phantom imaging using a 3 T MRI scanner. (a) The phantom, (b) its MR magnitude image and (c) reconstructed conductivity image using the harmonic B_z algorithm. From Oh *et al.* (2005)

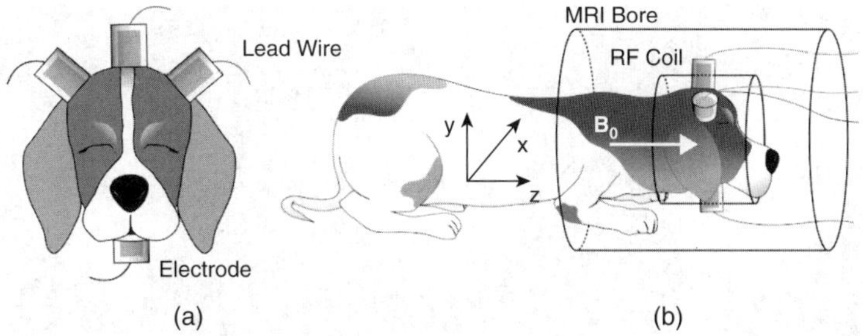

Figure 9.14 (a) Attachment of electrodes around a chosen imaging region and (b) placement of an imaging object inside an MRI scanner. $\mathbf{B}_0$ is the main magnetic field of the MRI scanner. From Kim *et al.* (2007)

related to proton densities. There are enough examples showing that a conductivity image clearly distinguishes two objects whereas they are indistinguishable in the corresponding conventional MR image. This happens, for example, when two objects have almost the same proton densities but significantly different amounts of mobile ions.

9.6.3 Animal Imaging

Figure 9.14 shows an experimental set-up for *post mortem* canine brain imaging experiments. Figure 9.15 shows reconstructed multi-slice conductivity images of a *post mortem* canine brain (Kim *et al.* 2007). These high-resolution conductivity images with a pixel size of 1.4 mm were obtained by using a 3 T MRI scanner and 40 mA injection currents. Restricting the conductivity image reconstruction only within the brain region to avoid technical difficulties related with the skull, these conductivity images of the intact canine brain clearly distinguish white and gray matter. Since the harmonic B_z algorithm cannot handle the tissue anisotropy, the concept of the equivalent isotropic conductivity should be adopted to interpret the reconstructed conductivity images. Figure 9.16 compares (a) an MR magnitude image, (b) a conductivity image of the brain region only and (c) a conductivity image of the entire head obtained from a *post mortem* canine head.

The image quality can be improved by using flexible electrodes with a larger contact area. Minhas *et al.* (2008) proposed a thin and flexible carbon-hydrogel electrode for MREIT imaging experiments. Using a pair of carbon-hydrogel electrodes with a large contact area, the amplitude of the injection current can be increased primarily due to a reduced average current density underneath the electrodes. Using two pairs of such electrodes, they reconstructed equivalent isotropic conductivity images of a swine leg, as shown in Figure 9.17, demonstrating the good contrast among different muscles and bones. From the reconstructed images, we can observe spurious spikes in the outer layers of bones, primarily due to the MR signal void there.

Figure 9.18(a) and 9.18(b) are MR magnitude and reconstructed conductivity images of a *post mortem* canine abdomen (Jeon *et al.* 2009b). Since the abdomen includes a complicated mixture of different organs, interpretation of the reconstructed conductivity image needs further investigation. They found that conductivity image contrast in the canine kidney is quite different from that of the MR magnitude image, clearly distinguishing the cortex, internal medulla, renal pelvis and urethra.

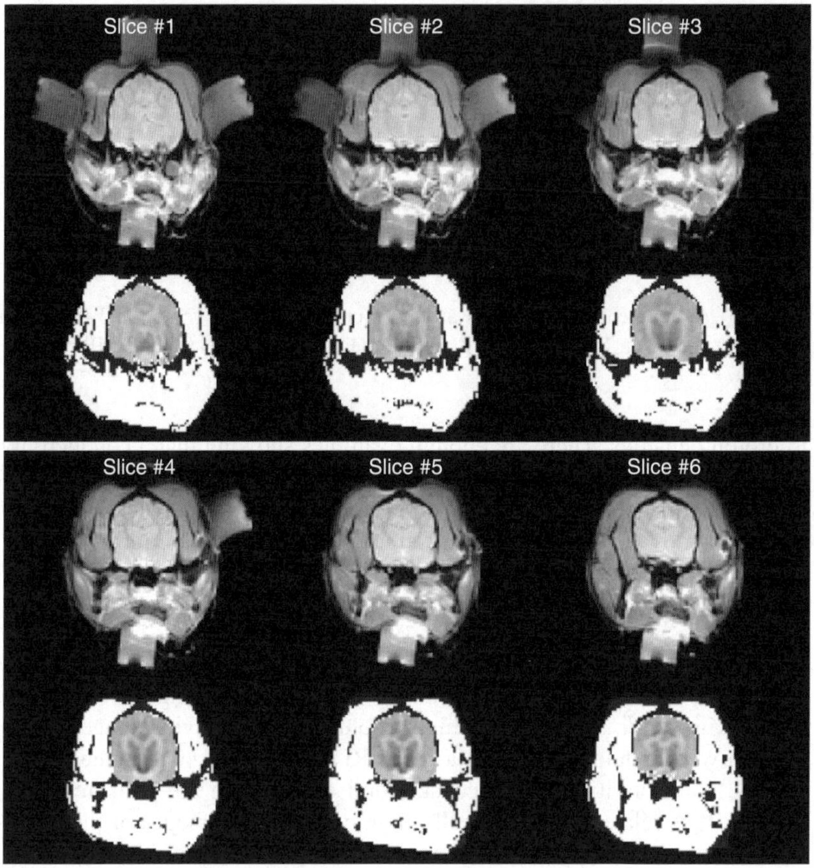

Figure 9.15 *Post mortem* animal imaging of a canine head using a 3 T MRI scanner. Multi-slice MR magnitude images of a canine head are shown in the upper panel, and reconstructed equivalent isotropic conductivity images of its brain are shown in the lower panel. From Kim *et al.* (2007)

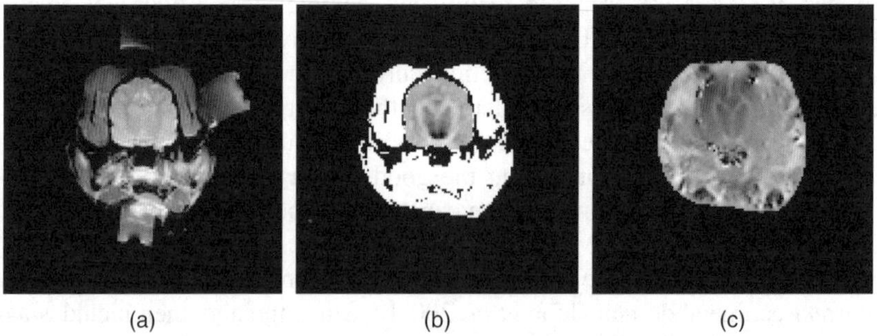

Figure 9.16 Comparison of (a) MR magnitude image, (b) conductivity image of the brain only, and (c) conductivity image of the entire head from a *post mortem* canine head

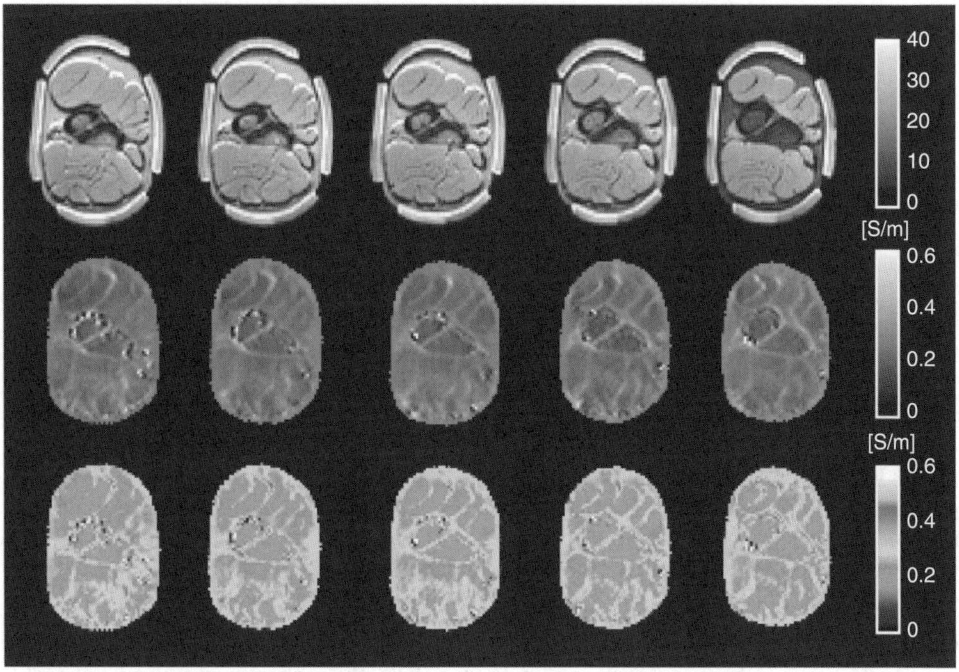

Figure 9.17 *Post mortem* animal imaging of a swine leg using a 3 T MRI scanner: multi-slice MR magnitude (top row), conductivity (middle row) and color-coded conductivity (bottom row). From Minhas *et al.* (2008)

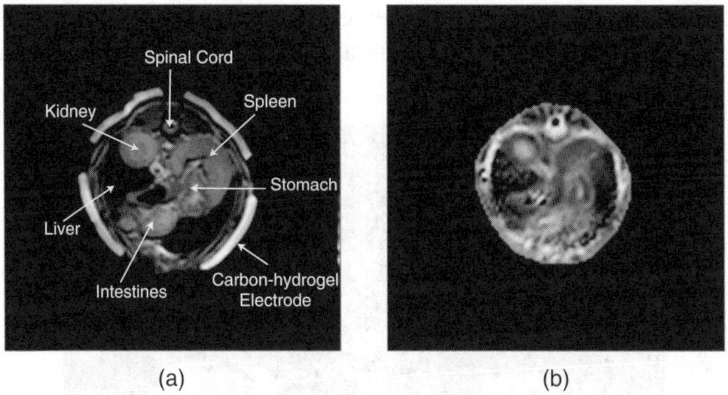

Figure 9.18 (a) MR magnitude image and (b) reconstructed conductivity image from a *post mortem* canine abdomen. The conductivity image in (b) shows a significantly different image contrast compared with the MR magnitude image in (a). From Jeon *et al.* (2009b)

Figure 9.19 compares *in vivo* and *post mortem* conductivity images of the same canine brain (Kim *et al.* 2008a). Though the *in vivo* conductivity image is noisier than the *post mortem* image, primarily due to the reduced amplitude of injection currents, the *in vivo* image shows a good contrast among white matter, gray matter and other brain

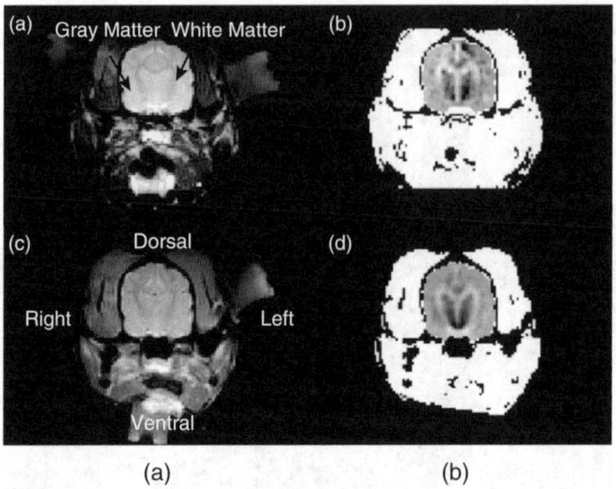

Figure 9.19 (a) *In vivo* and (c) *post mortem* MR magnitude images of a canine head. (b) *In vivo* and (d) *post mortem* equivalent isotropic conductivity images of the brain. The same animal was used for both *in vivo* and *post mortem* experiments. The image in (b) was obtained by using 5 mA injection currents, whereas 40 mA was used in (d). From Kim *et al.* (2008a)

tissues. Figure 9.20 shows *in vivo* imaging experiments of canine brains without and with a regional brain ischemia. As shown in Figure 9.20, the ischemia produced noticeable conductivity changes in reconstructed images.

9.6.4 Human Imaging

For an *in vivo* human imaging experiment, Kim *et al.* (2008b, 2009) chose the lower extremity as the imaging region. After a review by the Institutional Review Board, they performed an MREIT experiment of a human subject using a 3 T MRI scanner. They

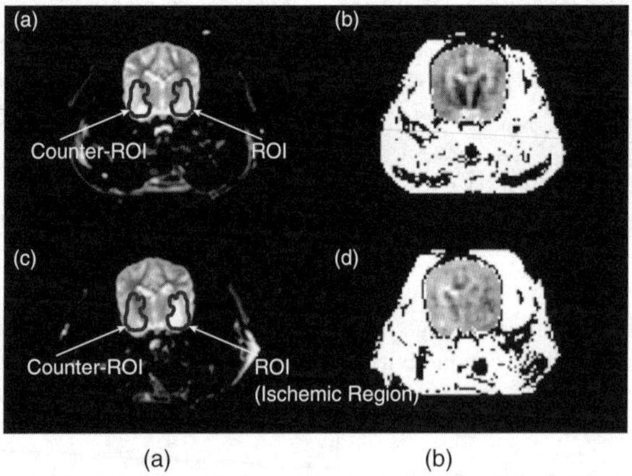

Figure 9.20 T_2-weighted MR images of a canine head (a) before and (c) after the embolization. (b) and (d) The corresponding equivalent isotropic conductivity images. The region of interest (ROI) defines the ischemic region, and counter-ROI defines the symmetrical region in the other side of the brain. From Kim *et al.* (2008a)

adopted thin and flexible carbon-hydrogel electrodes with conductive adhesive for current injections (Minhas *et al.* 2008). Owing to their large surface area of $80 \times 60\,\text{mm}^2$ and good contact with the skin, they could inject pulse-type currents with an amplitude as high as 9 mA into the lower extremity without producing a painful sensation. Sequential injections of two currents in orthogonal directions were used to produce the cross-sectional equivalent isotropic conductivity images in Figure 9.21 with 1.7 mm pixel size and 4 mm slice gap. The conductivity images distinguished well between different parts of muscles and bones. The outermost fatty layer was also clearly shown in each conductivity image. We could observe excessive noise in the outer layers of two bones due to the MR signal void phenomenon there.

9.7 Applications

MREIT provides conductivity images of an electrically conducting object with a pixel size of about 1 mm. It achieves such a high spatial resolution by adopting an MRI scanner to measure internal magnetic flux density distributions induced by externally injected imaging currents. Theoretical and experimental studies in MREIT demonstrate that it is expected to be a new clinically useful bio-imaging modality. Its capability to distinguish the conductivity values of different biological tissues in their living wetted states is unique.

Following the *in vivo* imaging experiment of the canine brain (Kim *et al.* 2008b), numerous *in vivo* animal imaging experiments are being conducted for imaging regions of extremities, abdomen, pelvis, neck, thorax and head. Animal models of various diseases are also being tried. To reach the stage of clinical applications, *in vivo* human imaging experiments are also in progress (Kim *et al.* 2009). These trials are expected to

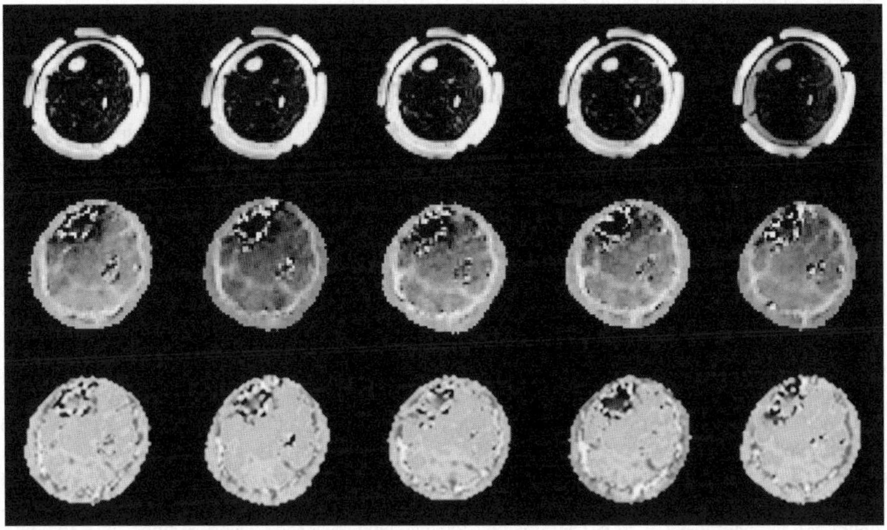

Figure 9.21 *In vivo* MREIT imaging experiment of a human leg using a 3 T MRI scanner. Multi-slice MR magnitude images, reconstructed equivalent isotropic conductivity images, and color-coded conductivity images of a human leg are shown in the top, middle and bottom rows, respectively. From Kim *et al.* (2009). Reproduced with permission from IEEE

accumulate new diagnostic information based on *in vivo* conductivity values of numerous biological tissues.

MREIT has been attempted to overcome the ill-posed nature of the inverse problem in EIT and to provide high-resolution conductivity images. Even though current EIT images have a relatively poor spatial resolution, the high temporal resolution and portability in EIT could be advantageous in several biomedical application areas (Holder 2005). Instead of competing in a certain application area, MREIT and EIT will be supplementary to each other. Taking advantage of the high spatial resolution in MREIT, Woo and Seo (2008) discussed numerous application areas of MREIT in biomedicine, biology, chemistry and material science. We should note that it is possible to produce a current density image for any electrode configuration once the conductivity distribution is obtained.

Future studies should overcome a few technical barriers to advance the method to the stage of routine clinical use. The biggest hurdle at present is the amount of injection current, which may stimulate muscle and nerve. Reducing it down to a level that does not produce undesirable side effects is the key to the success of this new bio-imaging modality. This demands innovative data processing methods based on rigorous mathematical analysis as well as improved measurement techniques to maximize SNRs for a given data collection time.

References

Alessandrini G and Magnanini R 1992 The index of isolated critical points and solutions of elliptic equations in the plane. *Ann. Scuola Norm. Sup. Pisa Cl. Sci. (4)* **19**, 567–589.

Alessandrini G, Isakov V and Powell J 1995 Local uniqueness in the inverse problem with one measurement. *Trans. Am. Math. Soc*. **347**, 3031–3041.

Bauman P, Marini A and Nesi V 2000 Univalent solutions of an elliptic system of partial differential equations arising in homogenization. *Indiana Univ. Math. J*. **128**, 53–64.

Bernstein MA, King KF and Zhou XJ 2004 *Handbook of MRI Pulse Sequences*. Elsevier, Burlington, MA.

Birgul O and Ider YZ 1995 Use of the magnetic field generated by the internal distribution of injected currents for electrical impedance tomography. In *Proc. 9th Int. Conf. on Electrical Bio-Impedance*, Heidelberg, Germany, pp. 418–419.

Birgul O and Ider YZ 1996 Electrical impedance tomography using the magnetic field generated by injected currents. In *Proc. 18th Annu. Int. Conf. IEEE Engineering in Medicine and Biology Society*, Amsterdam, The Netherlands, pp. 784–785. IEEE, New York.

Birgul O, Eyuboglu BM and Ider YZ 2003 Experimental results for 2D magnetic resonance electrical impedance tomography (MREIT) using magnetic flux density in one direction. *Phys. Med. Biol*. **48**, 3485–3504.

Birgul O, Hamamura MJ, Muftuler T and Nalcioglu O 2006 Contrast and spatial resolution in MREIT using low amplitude current. *Phys. Med. Biol*. **51**, 5035–5049.

Folland G 1976 *Introduction to Partial Differential Equations*. Princeton University Press, Princeton, NJ.

Gao G, Zhu SA and He B 2005 Estimation of electrical conductivity distribution within the human head from magnetic flux density measurement. *Phys. Med. Biol*. **50**, 2675–2687.

Gao N, Zhu SA and He BA 2006 New magnetic resonance electrical impedance tomography (MREIT) algorithm: the RSM-mREIT algorithm with applications to estimation of human head conductivity. *Phys. Med. Biol*. **51**, 3067–3083.

Haacke EM, Brown RW, Thompson MR and Venkatesan R 1999 *Magnetic Resonance Imaging: Physical Principles and Sequence Design*. John Wiley & Sons, Inc., New York.

Hamamura MJ and Muftuler LT 2008 Fast imaging for magnetic resonance electrical impedance tomography. *Magn. Reson. Imag*. **26**, 739–745.

Hamamura MJ, Muftuler LT, Birgul O and Nalcioglu O 2006 Measurement of ion diffusion using magnetic resonance electrical impedance tomography. *Phys. Med. Biol*. **51**, 2753–2762.

Hasanov KF, Ma AW, Nachman AI and Joy MLG 2008 Current density impedance imaging. *IEEE Trans. Med. Imag*. **27**, 1301–1309.

Hedeen RA and Edelstein WA 1997 Characterization and prediction of gradient acoustic noise in MR imagers. *Magn. Reson. Med*. **37**, 7–10.

Holder D (ed.) 2005 *Electrical Impedance Tomography: Methods, History and Applications*. IOP Publishing, Bristol.

Ider YZ and Onart S 2004 Algebraic reconstruction for 3D MREIT using one component of magnetic flux density. *Physiol. Meas*. **25**, 281–294.

Ider YZ, Onart S and Lionheart WRB 2003 Uniqueness and reconstruction in magnetic resonance electrical impedance tomography (MREIT). *Physiol. Meas*. **24**, 591–604.

Jeon K, Lee CO, Kim HJ, Woo EJ and Seo JK 2009a CoReHA: conductivity reconstructor using harmonic algorithms for magnetic resonance electrical impedance tomography (MREIT). *J. Biomed. Eng. Res*. **30**, 279–287.

Jeon K, Minhas AS, Kim YT, Jeong WC, Kim HJ, Kang BT, Park HM, Lee CO, Seo JK and Woo EJ 2009b MREIT conductivity imaging of the postmortem canine abdomen using CoReHA. *Physiol. Meas*. **30**, 957–966.

Jeong WC, Kim YT, Minhas AS, Kim HJ, Woo EJ and Seo JK 2008 Design of carbon-hydrogel electrode for MREIT. In *Proc. 9th Int. Conf. on Electrical Impedance Tomography*, Dartmouth, NH.

Joy MLG, Scott GC and Henkelman RM 1989 In vivo detection of applied electric currents by magnetic resonance imaging. *Magn. Reson. Imag*. **7**, 89–94.

Khang HS, Lee BI, Oh SH, Woo EJ, Lee SY, Cho MH, Kwon O, Yoon JR and Seo JK 2002 J-substitution algorithm in magnetic resonance electrical impedance tomography (MREIT): phantom experiments for static resistivity images. *IEEE Trans. Med. Imag*. **21**, 695–702.

Kim HJ, Lee BI, Cho Y, Kim YT, Kang BT, Park HM, Lee SY, Seo JK and Woo EJ 2007 Conductivity imaging of canine brain using a 3T MREIT system: postmortem experiments. *Physiol. Meas*. **28**, 1341–1353.

Kim HJ, Oh TI, Kim YT, Lee BI, Woo EJ, Seo JK, Lee SY, Kwon O, Park C, Kang BT and Park HM 2008a In vivo electrical conductivity imaging of a canine brain using a 3T MREIT system. *Physiol. Meas*. **29**, 1145–1155.

Kim HJ, Kim YT, Jeong WC, Minhas AS, Woo EJ, Kwon OJ and Seo JK 2008b In vivo conductivity imaging of a human leg using a 3T MREIT system. In *Proc. 9th Int. Conf. on Electrical Impedance Tomography*, Dartmouth, NH.

Kim HJ, Kim YT, Minhas AS, Jeong WC, Woo EJ, Seo JK and Kwon OJ 2009 In vivo high-resolution conductivity imaging of the human leg using MREIT: the first human experiment. *IEEE Trans. Med. Imag*. **28**, 1681–1611.

Kim YT, Yoo PJ, Oh TI and Woo EJ 2011 Magnetic flux density measurement in magnetic resonance electrical impedance tomography using a low-noise current source. *Meas. Sci. Technol*. **22**, 105803.

Kwon O, Lee JY and Yoon JR 2002a Equipotential line method for magnetic resonance electrical impedance tomography (MREIT). *Inv. Prob*. **18**, 1089–1100.

Kwon O, Woo EJ, Yoon JR and Seo JK 2002b Magnetic resonance electrical impedance tomography (MREIT): simulation study of J-substitution algorithm. *IEEE Trans. Biomed. Eng*. **48**, 160–167.

Kwon O, Park C, Park EJ, Seo JK and Woo EJ 2005 Electrical conductivity imaging using a variational method in B_z-based MREIT. *Inv. Prob*. **21**, 969–980.

Kwon O, Pyo H, Seo JK and Woo EJ 2006 Mathematical framework for B_z-based MREIT model in electrical impedance imaging. *Int. J. Comput. Math. Appl*. **51**, 817–828.

Kwon OI, Lee BI, Nam HS and Park C 2007 Noise analysis and MR pulse sequence optimization in MREIT using an injected current nonlinear encoding (ICNE) method. *Physiol. Meas*. **28**, 1391–1404.

Lee BI, Oh SH, Woo EJ, Lee SY, Cho MH, Kwon O, Seo JK and Baek WS 2003a Static resistivity image of a cubic saline phantom in magnetic resonance electrical impedance tomography (MREIT). *Physiol. Meas*. **24**, 579–589.

Lee BI, Lee SH, Kim TS, Kwon O, Woo EJ and Seo JK 2005 Harmonic decomposition in PDE-based denoising technique for magnetic resonance electrical impedance tomography. *IEEE Trans. Biomed. Eng*. **52**, 1912–1920.

Lee BI, Park C, Pyo HC, Kwon O and Woo EJ 2006 Optimization of current injection pulse width in MREIT. *Physiol. Meas*. **28**, N1–7.

Liu JJ, Seo JK, Sini M and Woo EJ 2007 On the convergence of the harmonic B_z algorithm in magnetic resonance electrical impedance tomography. *SIAM J. Appl. Math*. **67**, 1259–1282.

Minhas AS, Kim HJ, Kim YT, Jeong WC, Woo EJ and Seo JK 2008 Conductivity imaging of postmortem swine leg using MREIT. In *Proc. 9th Int Conf. on Electrical Impedance Tomography*, Dartmouth, NH.

Minhas AS, Woo EJ and Lee SY 2009 Magnetic flux density measurement with balanced steady state free precession pulse sequence for MREIT: a simulation study. In *Conf. Proc. IEEE Engineering in Medicine and Biology Society*, Minneapolis, MN, pp. 2276–2278.

Muftuler LT, Hamamura MJ, Birgul O and Nalcioglu O 2004 Resolution and contrast in magnetic resonance electrical impedance tomography (MREIT) and its application to cancer imaging. *Technol. Cancer Res. Treat.* **3**, 599–609.

Muftuler LT, Hamamura MJ, Birgul O and Nalcioglu O 2006 In vivo MRI electrical impedance tomography (MREIT) of tumors. *Technol. Cancer Res. Treat.* **5**, 381–387.

Nachman A, Tamasan A and Timonov A 2007 Conductivity imaging with a single measurement of boundary and interior data. *Inv. Prob.* **23**, 2551–2563.

Nachman A, Tamasan A and Timonov A 2009 Recovering the conductivity from a single measurement of interior data. *Inv. Prob.* **25**, 035014.

Nam HS and Kwon OI 2010 Optimization of multiply acquired magnetic flux density $B(z)$ using ICNE-multiecho train in MREIT. *Phys. Med. Biol.* **55**, 2743–2759.

Oh SH, Lee BI, Woo EJ, Lee SY, Cho MH, Kwon O and Seo JK 2003 Conductivity and current density image reconstruction using harmonic B_z algorithm in magnetic resonance electrical impedance tomography. *Phys. Med. Biol.* **48**, 3101–3116.

Oh SH, Lee BI, Park TS, Lee SY, Woo EJ, Cho MH, Kwon O and Seo JK 2004 Magnetic resonance electrical impedance tomography at 3 Tesla field strength. *Magn. Reson. Med.* **51**, 1292–1296.

Oh SH, Lee BI, Woo EJ, Lee SY, Kim TS, Kwon O and Seo JK 2005 Electrical conductivity images of biological tissue phantoms in MREIT. *Physiol. Meas.* **26**, S279–288.

Oh TI, Cho Y, Hwang YK, Oh SH, Woo EJ and Lee SY 2006 Improved current source design to measure induced magnetic flux density distributions in MREIT. *J. Biomed. Eng. Res.* **27**, 30–37.

Park C, Kwon O, Woo EJ and Seo JK 2004a Electrical conductivity imaging using gradient B_z decomposition algorithm in magnetic resonance electrical impedance tomography (MREIT). *IEEE Trans. Med. Imag.* **23**, 388–394.

Park C, Park EJ, Woo EJ, Kwon O and Seo JK 2004b Static conductivity imaging using variational gradient B_z algorithm in magnetic resonance electrical impedance tomography. *Physiol. Meas.* **25**, 275–269.

Park C, Lee BI, Kwon O and Woo EJ 2006 Measurement of induced magnetic flux density using injection current nonlinear encoding (ICNE) in MREIT. *Physiol. Meas.* **28**, 117–127.

Reilly JP 1998 *Applied Bioelectricity: From Electrical Stimulation to Electropathology*. Springer, New York.

Sadleir R 2005 Noise analysis in MREIT at 3 and 11 Tesla field strength. *Physiol. Meas.* **26**, 875–884.

Scott GC, Joy MLG, Armstrong RL and Henkelman RM 1991 Measurement of nonuniform current density by magnetic resonance. *IEEE Trans. Med. Imag.* **10**, 362–374.

Scott GC, Joy MLG, Armstrong RL and Hankelman RM 1992 Sensitivity of magnetic resonance current density imaging. *J. Magn. Reson.* **97**, 235–254.

Seo JK 1996 A uniqueness result on inverse conductivity problem with two measurements. *J. Fourier Anal. Appl.* **2**, 515–524.

Seo JK and Woo EJ 2011 Magnetic resonance electrical impedance tomography (MREIT). *SIAM Rev.* **53**, 40–68.

Seo JK, Kwon O, Lee BI and Woo EJ 2003a Reconstruction of current density distributions in axially symmetric cylindrical sections using one component of magnetic flux density: computer simulation study. *Physiol. Meas.* **24**, 565–577.

Seo J K, Yoon J R, Woo E J and Kwon O 2003b Reconstruction of conductivity and current density images using only one component of magnetic field measurements. *IEEE Trans. Biomed. Eng.* **50**, 1121–1124.

Seo JK, Pyo HC, Park CJ, Kwon O and Woo EJ 2004 Image reconstruction of anisotropic conductivity tensor distribution in MREIT: computer simulation study. *Phys. Med. Biol.* **49**, 4371–4382.

Seo JK, Kim SW, Kim S, Liu J, Woo EJ, Jeon K and Lee CO 2008 Local harmonic B_z algorithm with domain decomposition in MREIT: computer simulation study. *IEEE Trans. Med. Imag.* **27**, 1754–1761.

Verchota G 1984 Layer potentials and boundary value problems for Laplace equation in Lipschitz domains. *J. Funct. Anal.* **59**, 572–611.

Woo EJ and Seo JK 2008 Magnetic resonance electrical impedance tomography (MREIT) for high-resolution conductivity imaging. *Physiol. Meas.* **29**, R1–26.

Woo EJ, Lee SY and Mun CW 1994 Impedance tomography using internal current density distribution measured by nuclear magnetic resonance. *Proc. SPIE* **2299**, 377–385.

Zhang N 1992 Electrical impedance tomography based on current density imaging. MS Thesis, University of Toronto, Canada.

10

Magnetic Resonance Elastography

Magnetic resonance elastography (MRE) is an imaging modality capable of visualizing the stiffness of biological tissues by measuring the propagating strain waves in an object of interest (Low *et al.* 2010; Muthupillai *et al.* 1995; Papazoglou *et al.* 2005; Sack *et al.* 2002; Sinkus *et al.* 2000). Since MRE provides non-invasive assessment of variations in tissue elasticity, it has been used for non-invasive diagnosis of liver disease or for detecting prostate cancer. Tissue elasticity can change with disease, and the shear modulus (or modulus of rigidity) varies over a wide range, differentiating various pathological states of tissues (Venkatesh *et al.* 2008; Yin *et al.* 2007). It also has potential applications in studying skeletal muscle biomechanics (Papazoglou *et al.* 2005; Uffmann *et al.* 2004).

Tissue elasticity refers to the ability of a tissue to deform its shape when a mechanical force is applied and to regain its original shape after the force is removed; tumor tissue is less compressible than normal tissue. For centuries, palpation using the surface of the finger or palm has been used to measure tissue stiffness; it can be viewed as an elasticity measurement technique to feel the degree of tissue distortion (strain) due to pressure (stress) on the tissue. As a visual palpation, elastography using ultrasound was developed in the late 1980s (Lerner and Parker 1987; Ophir *et al.* 1991). This ultrasound elastography influenced the early development of MRE.

MRE provides a quantitative assessment of tissue stiffness (shear modulus) with a non-invasive method. It is based on the fact that the speed of shear wave propagation is closely related to tissue stiffness; the stiffer the tissue, the faster the speed. For a linear elastic medium, the shear modulus is proportional to the square of the shear velocity. To measure the shear velocity, MRE uses magnetic resonance imaging (MRI) techniques to detect the propagation of transverse acoustic strain waves in the object of interest (Muthupillai *et al.* 1995). It visualizes the time-harmonic displacement in the tissue induced by a harmonically oscillating mechanical vibration. With a fixed frequency, the shear velocity is proportional to the wavelength, which can be viewed as the peak-to-peak distance of the time-harmonic displacement. Hence, we can evaluate local values of tissue elasticity from the time-harmonic displacement. We refer to the review articles of Doyley (2012) and Mariappan *et al.* (2010) for overviews of MRE techniques. In this chapter, we will introduce several MRE approaches, from basic physics to potential applications.

Nonlinear Inverse Problems in Imaging, First Edition. Jin Keun Seo and Eung Je Woo.

10.1 Representation of Physical Phenomena

When a force is applied to a solid body, the body deforms in shape and volume to some extent. If the deformed body returns to its original shape once the force is removed, it is said to have experienced elastic deformation. If the deformation is irreversible, it is said to be plastic deformation. Stress is a description of the average force per unit normal area of a surface within a body. The unit of stress is the pascal (Pa $=$ N m^2, newtons per square meter). Strain measures the extent of deformation in terms of a relative displacement, that is, the change of distance between two adjacent points in the deformed state with respect to that distance in the undeformed state. In this section, we will deal only with linear elastic materials, and we give a brief overview of some elementary concepts in linear elasticity regarding the relationship between stress and strain. We refer to the books of Landau and Lifshitz (1986) and Pujol (2002) for detailed explanations.

10.1.1 Overview of Hooke's Law

Hooke's law states that strain is proportional to stress. In a one-dimensional simple model of a rod of elastic material,

$$\underbrace{\text{axial stress}}_{\sigma} = E \times \underbrace{\text{axial strain}}_{\epsilon}, \tag{10.1}$$

where E is Young's modulus. Here, we regard the rod of elastic material as a spring; Young's modulus can be viewed as the spring constant. Young's modulus $E = \sigma/\epsilon$ can be used to predict compression or elongation as a result of axial stress. A compressive force causes the rod to get shorter, whereas a tensile force makes the rod longer.

A generalized Hooke's law for a three-dimensional elastic body can be derived by viewing the elastic body as a network of linear springs. For a three-dimensional inhomogeneous elastic body, the stress at any given point $\mathbf{r}$ within the body is defined by a 3×3 matrix called the Cauchy stress tensor:

$$\boldsymbol{\sigma}(\mathbf{r}) = \begin{pmatrix} \sigma_{11}(\mathbf{r}) & \sigma_{12}(\mathbf{r}) & \sigma_{13}(\mathbf{r}) \\ \sigma_{21}(\mathbf{r}) & \sigma_{22}(\mathbf{r}) & \sigma_{23}(\mathbf{r}) \\ \sigma_{31}(\mathbf{r}) & \sigma_{32}(\mathbf{r}) & \sigma_{33}(\mathbf{r}) \end{pmatrix}.$$

The first column of the stress tensor $\boldsymbol{\sigma}$ represents the force acting on a differential area $\mathrm{d}A$ normal to the x axis (see Figure 10.1). Similarly, the second and third columns of $\boldsymbol{\sigma}$ are the forces acting on the differential area $\mathrm{d}A$ normal to the y and z axes, respectively. There are three main types of stress: compression, tension and shear stress. The diagonal components σ_{11}, σ_{22} and σ_{33} are normal stresses, the forces perpendicular to the area $\mathrm{d}A$. The remaining off-diagonal components are shear stresses, the forces parallel to the area $\mathrm{d}A$.

Strain can be obtained by comparing deformed geometry with undeformed geometry. The strain at any given point $\mathbf{r}$ within the body is expressed by the 3×3 matrix

$$\boldsymbol{\epsilon}(\mathbf{r}) = \begin{pmatrix} \epsilon_{11}(\mathbf{r}) & \epsilon_{12}(\mathbf{r}) & \epsilon_{13}(\mathbf{r}) \\ \epsilon_{21}(\mathbf{r}) & \epsilon_{22}(\mathbf{r}) & \epsilon_{23}(\mathbf{r}) \\ \epsilon_{31}(\mathbf{r}) & \epsilon_{32}(\mathbf{r}) & \epsilon_{33}(\mathbf{r}) \end{pmatrix}.$$

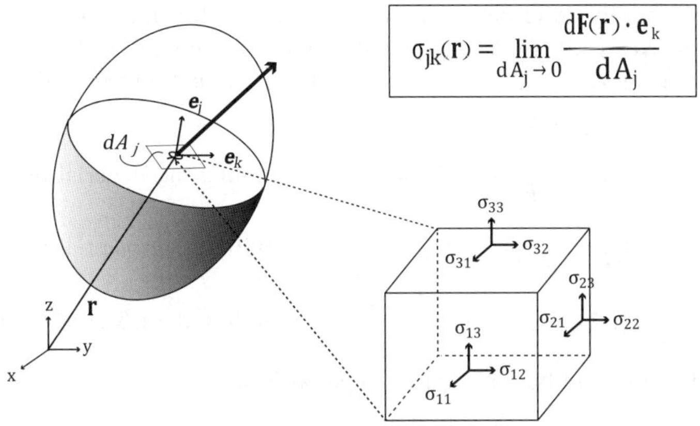

Figure 10.1 Stress tensor

Then, the generalized three-dimensional Hooke's law can be expressed as

$$\sigma_{ij} = \sum_{k=1}^{3} \sum_{\ell=1}^{3} C_{ijk\ell}\epsilon_{k\ell} \quad (i, j = 1, 2, 3), \tag{10.2}$$

where $C = (C_{ijk\ell})$ is the fourth-order elastic tensor, which is symmetrical: $C_{ijk\ell} = C_{ij\ell k} = C_{jik\ell}$. The tensor $(C_{ijk\ell})$ is called the stiffness tensor, and it links the stress tensor and the strain tensor.

Adopting the Voigt notation (e.g. $C_{1123} = C_{14}$), we can rewrite (10.2) in the following matrix representation:

$$\begin{pmatrix} \sigma_{11} \\ \sigma_{22} \\ \sigma_{33} \\ \sigma_{23} \\ \sigma_{31} \\ \sigma_{12} \end{pmatrix} = \begin{pmatrix} C_{11} & C_{12} & C_{13} & C_{14} & C_{15} & C_{16} \\ & C_{22} & C_{23} & C_{24} & C_{25} & C_{26} \\ & & C_{33} & C_{34} & C_{35} & C_{36} \\ & & & C_{44} & C_{45} & C_{46} \\ & \text{sym} & & & C_{55} & C_{56} \\ & & & & & C_{66} \end{pmatrix} \begin{pmatrix} \epsilon_{11} \\ \epsilon_{22} \\ \epsilon_{33} \\ \epsilon_{23} \\ \epsilon_{31} \\ \epsilon_{12} \end{pmatrix}. \tag{10.3}$$

In the case of isotropic materials, (10.3) can be simplified as

$$\begin{pmatrix} \sigma_{11} \\ \sigma_{22} \\ \sigma_{33} \\ \sigma_{23} \\ \sigma_{31} \\ \sigma_{12} \end{pmatrix} = \begin{pmatrix} \widehat{E}(1-\nu) & \widehat{E}\nu & \widehat{E}\nu & 0 & 0 & 0 \\ \widehat{E}\nu & \widehat{E}(1-\nu) & \widehat{E}\nu & 0 & 0 & 0 \\ \widehat{E}\nu & \widehat{E}\nu & \widehat{E}(1-\nu) & 0 & 0 & 0 \\ 0 & 0 & 0 & \mu & 0 & 0 \\ 0 & 0 & 0 & 0 & \mu & 0 \\ 0 & 0 & 0 & 0 & 0 & \mu \end{pmatrix} \begin{pmatrix} \epsilon_{11} \\ \epsilon_{22} \\ \epsilon_{33} \\ 2\epsilon_{23} \\ 2\epsilon_{31} \\ 2\epsilon_{12} \end{pmatrix}, \tag{10.4}$$

with

$$\widehat{E} = \frac{E}{(1+\nu)(1-2\nu)},$$

where E is Young's modulus, ν is Poisson's ratio and μ is the shear modulus.

Next, we will briefly describe the derivation of (10.4) by considering a cubic material aligned with the axes $\{x, y, z\}$. We assume that three "tension tests", labeled as $'$, $''$ and $'''$, are conducted along x, y and z, respectively. Then, normal strains will be produced as follows:

$$\begin{cases} \epsilon'_{11} = \frac{1}{E}\sigma_{11}, \quad \epsilon'_{22} = -\frac{\nu}{E}\sigma_{11}, \ \epsilon'_{33} = -\frac{\nu}{E}\sigma_{11}, \ \text{pulling the material by } \sigma_{11} \text{ along } x, \\ \epsilon''_{11} = -\frac{\nu}{E}\sigma_{22}, \ \epsilon''_{22} = \frac{1}{E}\sigma_{22}, \quad \epsilon''_{33} = -\frac{\nu}{E}\sigma_{22}, \ \text{pulling the material by } \sigma_{22} \text{ along } y, \\ \epsilon'''_{11} = -\frac{\nu}{E}\sigma_{33}, \ \epsilon'''_{22} = -\frac{\nu}{E}\sigma_{33}, \ \epsilon'''_{33} = \frac{1}{E}\sigma_{33}, \quad \text{pulling the material by } \sigma_{33} \text{ along } z. \end{cases}$$

The combined strains can be obtained by superposition:

$$\epsilon_{11} = \epsilon'_{11} + \epsilon''_{11} + \epsilon'''_{11} = \frac{1}{E}[(1+\nu)\sigma_{11} - \nu(\sigma_{11} + \sigma_{22} + \sigma_{33})],$$

$$\epsilon_{22} = \epsilon'_{22} + \epsilon''_{22} + \epsilon'''_{22} = \frac{1}{E}[(1+\nu)\sigma_{22} - \nu(\sigma_{11} + \sigma_{22} + \sigma_{33})],$$

$$\epsilon_{33} = \epsilon'_{33} + \epsilon''_{33} + \epsilon'''_{33} = \frac{1}{E}[(1+\nu)\sigma_{33} - \nu(\sigma_{11} + \sigma_{22} + \sigma_{33})].$$

This provides us with the three equations

$$\begin{aligned} \sigma_{11} &= \frac{E}{(1+\nu)(1-2\nu)}[(1-\nu)\epsilon_{11} + \nu\epsilon_{22} + \nu\epsilon_{33}], \\ \sigma_{22} &= \frac{E}{(1+\nu)(1-2\nu)}[(1-\nu)\epsilon_{22} + \nu\epsilon_{33} + \nu\epsilon_{11}], \\ \sigma_{33} &= \frac{E}{(1+\nu)(1-2\nu)}[(1-\nu)\epsilon_{33} + \nu\epsilon_{11} + \nu\epsilon_{22}]. \end{aligned} \tag{10.5}$$

The shear strains and stresses are connected by the shear modulus μ as

$$2\epsilon_{23} = \frac{\sigma_{23}}{\mu}, \quad 2\epsilon_{13} = \frac{\sigma_{13}}{\mu}, \quad 2\epsilon_{12} = \frac{\sigma_{12}}{\mu},$$

where

$$\mu = \frac{E}{2(1+\nu)},$$

which leads to

$$\sigma_{23} = \frac{E}{1+\nu}\epsilon_{23}, \quad \sigma_{13} = \frac{E}{1+\nu}\epsilon_{13}, \quad \sigma_{12} = \frac{E}{1+\nu}\epsilon_{12}. \tag{10.6}$$

The identity (10.4) follows from (10.5) and (10.6).

Exercise 10.1.1 *For transversely isotropic materials, explain why (10.3) can be expressed as*

$$\begin{pmatrix} \sigma_{11} \\ \sigma_{22} \\ \sigma_{33} \\ \sigma_{23} \\ \sigma_{31} \\ \sigma_{12} \end{pmatrix} = \begin{pmatrix} C_{11} & C_{12} & C_{13} & 0 & 0 & 0 \\ C_{12} & C_{11} & C_{13} & 0 & 0 & 0 \\ C_{13} & C_{13} & C_{33} & 0 & 0 & 0 \\ 0 & 0 & 0 & C_{44} & 0 & 0 \\ 0 & 0 & 0 & 0 & C_{44} & 0 \\ 0 & 0 & 0 & 0 & 0 & \frac{1}{2}(C_{11} - C_{12}) \end{pmatrix} \begin{pmatrix} \epsilon_{11} \\ \epsilon_{22} \\ \epsilon_{33} \\ 2\epsilon_{23} \\ 2\epsilon_{31} \\ 2\epsilon_{12} \end{pmatrix}, \tag{10.7}$$

where

$$C_{11} = E_1(1 - \nu_{13}\nu_{31})\Upsilon,$$
$$C_{33} = E_3(1 - \nu_{12}^2)\Upsilon,$$
$$C_{12} = E_1(\nu_{12} + \nu_{13}\nu_{31})\Upsilon,$$
$$C_{13} = E_1(\nu_{31} + \nu_{12}\nu_{31})\Upsilon = E_3(\nu_{13} + \nu_{12}\nu_{13})\Upsilon,$$
$$C_{44} = \mu_{13},$$
$$C_{66} = \mu_{12},$$
$$\Upsilon = \frac{1}{1 - \nu_{12}^2 - 2\nu_{13}\nu_{31} - 2\nu_{12}\nu_{13}\nu_{31}}.$$

Here, E_i is the Young's modulus along axis i, μ_{ij} is the shear modulus in direction j on the plane whose normal is in direction i, and ν_{ij} is the Poisson's ratio that corresponds to a contraction in direction j when an extension is applied in direction i.

10.1.2 Strain Tensor in Lagrangian Coordinates

We now provide basic descriptions of the strain tensor. Let $\Phi_t : \Omega \to \Omega_t$ be the mapping from the undeformed body Ω at time $t = 0$ to the deformed state $\Phi_t(\Omega) = \Omega_t$ at time t. Let $\mathbf{R} = (X, Y, Z)$ denote a position of a particle in the undeformed frame, and let $\mathbf{r} = \mathbf{r}(\mathbf{R}, t) = \Phi_t(\mathbf{R})$ indicate the position of the same particle in the deformed frame at time t (see Figure 10.2). Hence, the motion of the specified particle at $\mathbf{R}$ in the reference frame is described by $\mathbf{r}(\mathbf{R}, t) = \Phi_t(\mathbf{R})$. This approach is called the Lagrangian description. One may use the Eulerian description $\mathbf{R} = \mathbf{R}(\mathbf{r}, t)$ when we are interested in a particle that occupies a given point $\mathbf{r}$ at a given time.

The displacement vector (the total movement of a particle with respect to the undeformed frame) is given by

$$\mathbf{u}(\mathbf{R}, t) = \mathbf{r}(\mathbf{R}, t) - \mathbf{R} = \Phi_t(\mathbf{R}) - \mathbf{R}.$$

With the assumption of small deformations, we can approximate

$$\Phi_t(\mathbf{R} + \mathrm{d}\mathbf{R}) - \Phi_t(\mathbf{R}) \approx \nabla\Phi_t(\mathbf{R})\,\mathrm{d}\mathbf{R} = (\nabla\mathbf{u}(\mathbf{R}, t) + I)\,\mathrm{d}\mathbf{R},$$

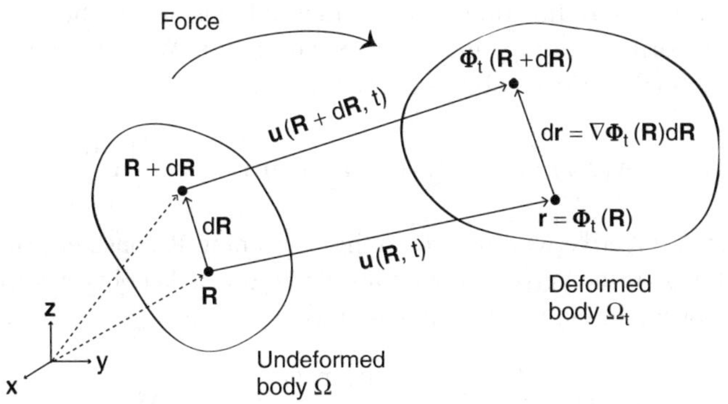

Figure 10.2 Relative particle movement in the continuum

where I is the identity matrix and $\nabla \mathbf{u}$ is the 3×3 matrix

$$\nabla \mathbf{u} = \begin{pmatrix} \partial u_1/\partial X & \partial u_1/\partial Y & \partial u_1/\partial Z \\ \partial u_2/\partial X & \partial u_2/\partial Y & \partial u_2/\partial Z \\ \partial u_3/\partial X & \partial u_3/\partial Y & \partial u_3/\partial Z \end{pmatrix}.$$

To analyze the change in length elements, we compare the distance $|\mathrm{d}\mathbf{r}|^2$ in the deformed frame with the distance $|\mathrm{d}\mathbf{R}|^2$ in the reference frame:

$$|\mathrm{d}\mathbf{r}|^2 - |\mathrm{d}\mathbf{R}|^2 = \mathrm{d}\mathbf{R}^{\mathrm{T}}(\nabla\Phi_t \nabla\Phi_t^{\mathrm{T}} - I)\,\mathrm{d}\mathbf{R} \approx \mathrm{d}\mathbf{R}^{\mathrm{T}}[(\nabla\mathbf{u} + I)(\nabla\mathbf{u} + I)^{\mathrm{T}} - I]\,\mathrm{d}\mathbf{R}.$$

From the assumption of small deformations, we can neglect the term $\nabla\mathbf{u}(\nabla\mathbf{u})^{\mathrm{T}} \approx 0$ to get the approximation

$$(\nabla\mathbf{u} + I)(\nabla\mathbf{u} + I)^{\mathrm{T}} - I \approx \nabla\mathbf{u} + \nabla\mathbf{u}^{\mathrm{T}}.$$

Introducing Cauchy's infinitesimal strain tensor

$$\epsilon = \tfrac{1}{2}(\nabla\mathbf{u} + \nabla\mathbf{u}^{\mathrm{T}}),$$

the difference $|\mathrm{d}\mathbf{r}|^2 - |\mathrm{d}\mathbf{R}|^2$ can be approximated by

$$|\mathrm{d}\mathbf{r}|^2 - |\mathrm{d}\mathbf{R}|^2 \approx 2\,\mathrm{d}\mathbf{R}^{\mathrm{T}}\,\underbrace{[\tfrac{1}{2}(\nabla\mathbf{u} + \nabla\mathbf{u}^{\mathrm{T}})]}_{\epsilon}\,\mathrm{d}\mathbf{R}.$$

Since $|\mathrm{d}\mathbf{r}|^2 - |\mathrm{d}\mathbf{R}|^2 \approx 2|\mathrm{d}\mathbf{R}|(|\mathrm{d}\mathbf{r}| - |\mathrm{d}\mathbf{R}|)$, the relative change in length in the $\mathrm{d}\mathbf{R}$ direction is

$$\frac{|\mathrm{d}\mathbf{r}| - |\mathrm{d}\mathbf{R}|}{|\mathrm{d}\mathbf{R}|} = \left\langle \epsilon \frac{\mathrm{d}\mathbf{R}}{|\mathrm{d}\mathbf{R}|}, \frac{\mathrm{d}\mathbf{R}}{|\mathrm{d}\mathbf{R}|} \right\rangle.$$

The above identity provides some geometric meaning of the strain tensor.

10.2 Forward Problem and Model

Newton's second law (force = mass × acceleration) provides a description of motion. We will use the Eulerian description of motion in order to focus on a particle at $\mathbf{r} = \mathbf{r}(\mathbf{R}, t)$. As t varies, different particles occupy the same spatial point $\mathbf{r}$. With the Eulerian description, the velocity $\mathbf{v} = \partial\mathbf{u}(\mathbf{R}, t)/\partial t$ can be expressed as

$$\mathbf{v}(\mathbf{r}, t) = \frac{\partial}{\partial t}\mathbf{u}(\mathbf{r}, t) + (\mathbf{u}\cdot\nabla)\mathbf{u}(\mathbf{r}, t) = \frac{\mathrm{D}}{\mathrm{D}t}\mathbf{u}.$$

Here, $\mathrm{D}/\mathrm{D}t$ is the derivative with respect to time t keeping $\mathbf{R}$ constant (called the material derivative), and $\partial/\partial t$ is the derivative with respect to time t keeping $\mathbf{r}$ constant. Similarly, the acceleration of a particle can be expressed as

$$\mathbf{a}(\mathbf{r}, t) = \frac{\mathrm{D}}{\mathrm{D}t}\mathbf{v}(\mathbf{r}, t) = \frac{\partial\mathbf{v}(\mathbf{r}, t)}{\partial t} + (\mathbf{v}\cdot\nabla)\mathbf{v}(\mathbf{r}, t).$$

Assuming that the density of the body, denoted by ρ, is locally constant, let us examine the equation of motion in a small cube Q with surface ∂Q. From the balance of linear momentum,

$$\underbrace{\frac{\mathrm{d}}{\mathrm{d}t}\int_Q \rho \mathbf{v}\,\mathrm{d}V}_{\text{momentum rate}} = \underbrace{\int_{\partial Q} \boldsymbol{\sigma}\cdot\mathbf{n}\,\mathrm{d}S}_{\text{surface force}} + \underbrace{\int_Q \rho \mathbf{f}\,\mathrm{d}V}_{\text{body force}}, \tag{10.8}$$

where $\mathbf{n}$ is the unit outward normal vector and $\mathbf{f}$ is the body force per unit volume. From the divergence theorem, (10.8) becomes

$$\int_Q \rho\frac{\mathrm{D}\mathbf{v}}{\mathrm{D}t}\,\mathrm{d}V = \int_Q [\nabla\cdot\boldsymbol{\sigma} + \rho\mathbf{f}]\,\mathrm{d}V. \tag{10.9}$$

Since the cube Q is arbitrarily small, (10.8) leads to

$$\rho\frac{\mathrm{D}\mathbf{v}}{\mathrm{D}t} = \nabla\cdot\boldsymbol{\sigma} + \rho\mathbf{f}. \tag{10.10}$$

From the assumption of small deformations, the acceleration $\mathrm{D}\mathbf{v}/\mathrm{D}t$ is approximated by

$$\frac{\mathrm{D}\mathbf{v}}{\mathrm{D}t} = \frac{\partial\mathbf{v}}{\partial t} + \underbrace{(\mathbf{v}\cdot\nabla)\mathbf{v}}_{\approx 0} \approx \frac{\partial\mathbf{v}}{\partial t} = \frac{\partial}{\partial t}\left(\frac{\partial\mathbf{u}}{\partial t} + \underbrace{(\mathbf{u}\cdot\nabla)\mathbf{u}}_{\approx 0}\right) \approx \frac{\partial^2\mathbf{u}}{\partial t^2}.$$

Assuming that the body is isotropic, the generalized Hooke's law (10.2) leads to

$$\boldsymbol{\sigma} = \lambda\nabla\cdot\mathbf{u}\,I + \mu[\nabla\mathbf{u} + (\nabla\mathbf{u})^{\mathrm{T}}], \tag{10.11}$$

where μ and λ are the Lamé coefficients given by

$$\mu = \frac{E}{2(1+\nu)}, \quad \lambda = \frac{\nu E}{(1+\nu)(1-2\nu)}$$

with Poisson's ratio ν. Hence, substituting (10.11) into (10.10) yields the equation of motion in terms of the displacement $\mathbf{u}$:

$$\rho\frac{\partial^2\mathbf{u}}{\partial t^2} = \nabla\cdot(\mu\nabla\mathbf{u}) + \nabla\left((\lambda+\mu)\nabla\cdot\mathbf{u}\right) + \rho\,\mathbf{f}. \tag{10.12}$$

In the case of a locally homogeneous isotropic medium, (10.12) can be simplified to

$$\rho\frac{\partial^2\mathbf{u}}{\partial t^2} = \mu\nabla^2\mathbf{u} + (\lambda+\mu)\nabla\nabla\cdot\mathbf{u} + \rho\,\mathbf{f}. \tag{10.13}$$

In the case of an anisotropic medium, $\mathbf{u} = (u_1, u_2, u_3)$ is dictated by the following elasticity system (Landau and Lifshitz 1986):

$$\rho\frac{\partial^2 u_i}{\partial t^2} = \sum_{j,k,\ell}\frac{\partial}{\partial x_j}(C_{ijk\ell}\epsilon_{k\ell}) + \rho f_i \quad (i = 1, 2, 3). \tag{10.14}$$

10.3 Inverse Problem in MRE

Imaging methods in MRE can be roughly divided into three steps (Mariappan *et al.* 2010):

- **Excitation**. Apply a sinusoidal vibration with an angular frequency of oscillation ω, $50\,\text{Hz} \leq \omega/\pi \leq 200\,\text{Hz}$ through the surface of the object. This induces tissue vibrations in an imaging region inside the human body.
- **Tissue response measurement**. Measure the induced tissue vibrations that are magnetically encoded by oscillating magnetic field gradients (Muthupillai *et al.* 1995).
- **Shear modulus reconstruction**. Visualize the shear modulus distribution using the measured tissue-displacement field.

The sinusoidal excitation at the angular frequency ω induces the internal displacement $\mathbf{u}(\mathbf{r}, t)$ within the body at the same angular frequency ω, and its time-harmonic displacement $\mathbf{u}(\mathbf{x})$ is given by

$$\mathbf{u}(\mathbf{r}, t) = \Re\{\mathbf{u}(\mathbf{r})\,\mathrm{e}^{\mathrm{i}\omega t}\}.$$

Here, for simplicity, we use the same notation for the time-harmonic displacement $\mathbf{u}(\mathbf{x})$ as the time-dependent displacement $\mathbf{u}(\mathbf{r}, t)$. Hopefully, this will not cause any confusion from the context. From now on, the notation $\mathbf{u}$ will be used for the time-harmonic displacement. Assume that $\rho\mathbf{f} \approx 0$ in (10.13) and (10.14).

For a linear isotropic body, substituting $\mathbf{u}(\mathbf{r}, t) \leftarrow \mathbf{u}(\mathbf{r})\,\mathrm{e}^{\mathrm{i}\omega t}$ into (10.13) leads to the governing equation of the induced internal time-harmonic displacement vector $\mathbf{u}$:

$$\nabla \cdot (\mu \nabla \mathbf{u}) + \nabla((\lambda + \mu)\nabla \cdot \mathbf{u}) + \omega^2 \rho \mathbf{u} = 0. \tag{10.15}$$

Here, taking the viscosity effect into account, μ and λ can be complex-valued. The real part $\Re(\mu)$ is the shear modulus and $\Re(\lambda)$ is the Lamé coefficient, with

$$\Re(\mu) = \frac{E}{2(1+\nu)} \quad \text{and} \quad \Re(\lambda) = \frac{\nu E}{(1-2\nu)(1+\nu)},$$

$\Im(\mu)$ is the shear viscosity accounting for attenuation within the medium and $\Im(\lambda)$ is the viscosity of the compressible wave.

Similarly, in the case of an anisotropic material, substituting $\mathbf{u}(\mathbf{r}, t) \leftarrow \mathbf{u}(\mathbf{r})\,\mathrm{e}^{\mathrm{i}\omega t}$ into (10.14) with $\boldsymbol{\epsilon} = \frac{1}{2}(\nabla \mathbf{u} + \nabla \mathbf{u}^{\mathrm{T}})$ leads to

$$-\omega^2 \rho u_i = \sum_{j,k,\ell} \frac{\partial}{\partial x_j}(C_{ijk\ell}\epsilon_{k\ell}) \quad (i = 1, 2, 3). \tag{10.16}$$

The corresponding inverse problem is to recover the distribution of tissue elasticity from the time-harmonic displacement $\mathbf{u}(\mathbf{r})$ inside the body.

10.4 Reconstruction Algorithms

Let Ω be the domain occupying the object to be imaged. Assume that soft tissues exhibit linear, isotropic mechanical properties. Most reconstruction methods for shear modulus imaging have used the following scalar equation:

$$\nabla \cdot (\mu(\mathbf{r})\nabla u(\mathbf{r})) + \omega^2 \rho u(\mathbf{r}) = 0 \quad (\mathbf{r} \in \Omega). \tag{10.17}$$

This model has the major advantage of requiring one component of $\mathbf{u}$, and it simplifies the underlying mathematical theory to the corresponding inverse problem; the inverse problem using the vector equation (10.15) can be reduced to the inverse problem using the scalar equation (10.17). We refer the interested reader to Manduca *et al.* (2001), McLaughlin and Renzi (2006), McLaughlin and Yoon (2004), Oliphant *et al.* (2001) and Sinkus *et al.* (2005b).

This model (10.17) uses the assumption that the longitudinal wave can be filtered out, since the longitudinal wave varies slowly compared with the shear wave (McLaughlin and Renzi 2006; Sinkus *et al.* 2005b). We should note that the scalar equation (10.17) for $\mathbf{u}$ is not accurate because $\lambda\nabla \cdot \mathbf{u}$ is not negligible; although $\nabla \cdot \mathbf{u}$ is very small, λ is very large. However, it seems that the scalar equation (10.17) for μ is reasonably accurate; for a given axial component of time-harmonic shear wave in (10.15), the shear modulus μ approximately satisfies (10.17).

The scalar equation (10.17) can be derived under the assumptions that $\nabla\mu$ and $\nabla \cdot \mathbf{u}$ are small (Lee *et al.* 2010). From the elasticity equation (10.15), $\mathbf{u}$ can be decomposed into

$$\omega^2 \rho \mathbf{u} = \nabla \times (\mu \nabla \times \mathbf{u}) - \nabla((\lambda + 2\mu)\nabla \cdot \mathbf{u}) - \Upsilon, \tag{10.18}$$

where

$$\Upsilon := 2\sum_{j=1}^{3} \frac{\partial \mu}{\partial x_j} \nabla u_j - 2\nabla\mu(\nabla \cdot \mathbf{u}).$$

Neglecting $\Upsilon \approx 0$, we get the following approximation:

$$\mathbf{u} \approx \frac{1}{\omega^2 \rho}\nabla \times (\mu\nabla \times \mathbf{u}) - \frac{1}{\omega^2 \rho}\nabla((\lambda + 2\mu)\nabla \cdot \mathbf{u}). \tag{10.19}$$

Exercise 10.4.1 *Prove the following four identities.*

1. $\nabla \cdot (\mu\nabla\mathbf{u}) = -\nabla \times (\mu\nabla \times \mathbf{u}) + \mu\nabla\nabla \cdot \mathbf{u} + \nabla\mu \times (\nabla \times \mathbf{u}) + (\nabla\mu \cdot \nabla)\mathbf{u}$
2. $\nabla\mu \times (\nabla \times \mathbf{u}) + (\nabla\mu \cdot \nabla)\mathbf{u} = \nabla \cdot (\mu\nabla\mathbf{u}^{\mathrm{T}}) - \mu\nabla\nabla \cdot \mathbf{u}$
3. $\nabla \cdot (\mu\nabla\mathbf{u}^{\mathrm{T}}) = \mu\nabla\nabla \cdot \mathbf{u} + \sum_{j=1}^{3} \frac{\partial \mu}{\partial x_j}\nabla u_j$
4. $-\nabla \times (\mu\nabla \times \mathbf{u}) = \nabla \cdot (\mu\nabla\mathbf{u}) - \mu\nabla\nabla \cdot \mathbf{u} - \nabla\mu \times (\nabla \times \mathbf{u}) - (\nabla\mu \cdot \nabla)\mathbf{u}$
$\quad = \nabla \cdot (\mu\nabla\mathbf{u}) - \mu\nabla\nabla \cdot \mathbf{u} - \nabla \cdot (\mu\nabla\mathbf{u}^{\mathrm{T}}) + \mu\nabla\nabla \cdot \mathbf{u}$

Here, $\nabla\mathbf{u}^{\mathrm{T}}$ *is the transpose of the matrix* $\nabla\mathbf{u}$.

Exercise 10.4.2 *Prove (10.18) using the above four identities and the elasticity equation (10.15).*

Writing

$$\mathbf{u}_T = \frac{1}{\omega^2 \rho}\nabla \times (\mu\nabla \times \mathbf{u}) \quad \text{and} \quad \mathbf{u}_L = -\frac{1}{\omega^2 \rho}\nabla\left((\lambda + 2\mu)\nabla \cdot \mathbf{u}\right),$$

the decomposition (10.19) is expressed as

$$\mathbf{u} \approx \mathbf{u}_T + \mathbf{u}_L. \tag{10.20}$$

Application of divergence to both sides of (10.19) leads to

$$\nabla \cdot \mathbf{u} - \nabla \cdot \mathbf{u}_L \approx 0. \tag{10.21}$$

Since $\nabla \cdot \mathbf{u} \approx 0$, the above approximation yields $\nabla \cdot \mathbf{u}_L \approx 0$. Therefore, $\mathbf{u}_L$ satisfies both $\nabla \times \mathbf{u}_L = 0$ and $\nabla \cdot \mathbf{u}_L \approx 0$. This means that each component of $\mathbf{u}_L$ satisfies the Laplace equation approximately

$$\nabla^2 \mathbf{u}_L = \frac{1}{\omega^2 \rho} \nabla^2 \left(\nabla \left((\lambda + 2\mu)\nabla \cdot \mathbf{u}\right)\right) \approx 0. \tag{10.22}$$

From the elasticity equation (10.15) with the assumptions $\nabla \cdot \mathbf{u} \approx 0$ and $\nabla \mu \approx 0$, we have

$$\nabla \cdot (\mu \nabla \mathbf{u}) + \omega^2 \rho \mathbf{u} = -\nabla \cdot (\mu \nabla \mathbf{u}^{\mathrm{T}}) - \nabla(\lambda \nabla \cdot \mathbf{u}) \approx -\nabla((\lambda + 2\mu)\nabla \cdot \mathbf{u}) = -\omega^2 \rho \mathbf{u}_L. \tag{10.23}$$

From (10.22), $\mathbf{u}_L$ is approximately harmonic in the region of interest, thereby varying very slowly in the internal region with negligible contribution to u. Hence, u_L can be treated as noise in the data (McLaughlin and Renzi 2006) and we may assume that the shear modulus μ and the displacement $\mathbf{u}$ satisfy the forward equation

$$\nabla \cdot (\mu \nabla \mathbf{u}) + \omega^2 \rho \mathbf{u} = 0. \tag{10.24}$$

For the above derivation of (10.24), we have used the assumption of $\nabla \mu \approx 0$ in two places.

10.4.1 Reconstruction of μ with the Assumption of Local Homogeneity

Assuming local homogeneity on μ (or $\nabla \mu \approx 0$), (10.24) can be expressed as

$$\mu(\mathbf{r}) \nabla^2 u(\mathbf{r}) + \omega^2 \rho u(\mathbf{r}) = 0 \quad (\mathbf{r} \in \Omega). \tag{10.25}$$

Then, μ can be directly recovered from

$$\mu(\mathbf{r}) = \omega^2 \rho \frac{u(\mathbf{r})}{\nabla^2 u(\mathbf{r})} \quad (\mathbf{r} \in \Omega). \tag{10.26}$$

This direct inversion method is the most commonly used MRE algorithm, which requires the local homogeneity assumption on μ (Kruse *et al.* 2000; Manduca *et al.* 2001; Manduca *et al.* 2002; Manduca *et al.* 2003; Oliphant *et al.* 2000a; Oliphant *et al.* 2000b; Oliphant *et al.* 2001; Sinkus *et al.* 2005b). Figure 10.3 shows the performance of the direct algebraic inversion method (Mariappan *et al.* 2010).

One drawback of the direct algebraic inversion method (10.26) is that double differentiation of the measured displacement data u can cause undesirable noise effects owing to the tendency of the operation to amplify noise. To alleviate the noise amplification from measured data, this method typically needs some filtering to reduce high-frequency noise (Manduca *et al.* 2003). Another drawback of this method is that the modeling error from the assumption of local homogeneity produces artifacts around regions of differing elastic properties even with noiseless data (Kwon *et al.* 2009).

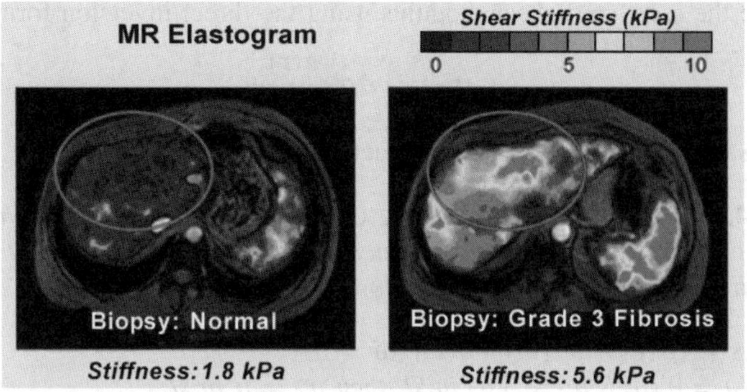

Figure 10.3 Typical MR elastograms. Reproduced from the GE Healthcare website at http://www.gehealthcare.com/euen/mri/products/MR-Touch/index.html

10.4.2 Reconstruction of μ without the Assumption of Local Homogeneity

Next, we explain a method for reconstructing shear modulus images without the assumption of local homogeneity (Kwon *et al.* 2009; Lee *et al.* 2010).

According to the Helmholtz–Hodge decomposition, the vector field $\mu\nabla u$ in (10.17) can be decomposed into a curl-free component and a divergence-free component:

$$\mu\nabla u(\mathbf{r}) = \nabla f(\mathbf{r}) + \nabla \times W(\mathbf{r}), \quad \mathbf{r} \in \Omega, \tag{10.27}$$

where the scalar potential f satisfies

$$\begin{cases} \nabla^2 f(\mathbf{r}) = \nabla \cdot (\mu\nabla u(\mathbf{r})) = -\rho\omega^2 u(\mathbf{r}), & \mathbf{r} \in \Omega, \\ \nabla f(\mathbf{r}) \cdot \mathbf{n} = \mu\nabla u(\mathbf{r}) \cdot \mathbf{n}, & \mathbf{r} \in \partial\Omega, \end{cases} \tag{10.28}$$

and the vector field W satisfies

$$\begin{cases} \nabla^2 W(\mathbf{r}) = -\nabla\mu \times \nabla u(\mathbf{r}), & \mathbf{r} \in \Omega, \\ W(\mathbf{r}) = 0, & \mathbf{r} \in \partial\Omega. \end{cases} \tag{10.29}$$

Taking the inner product of $\nabla\bar{u}$, the complex conjugate of ∇u, on both sides of the identity (10.27) leads to the following identity:

$$|\nabla u(\mathbf{r})|^2 \mu(\mathbf{r}) = \nabla f(\mathbf{r}) \cdot \nabla\bar{u}(\mathbf{r}) + \nabla \times W(\mathbf{r}) \cdot \nabla\bar{u}(\mathbf{r}). \tag{10.30}$$

Assuming $|\nabla u(\mathbf{r})| \neq 0$, μ can be decomposed into

$$\mu(\mathbf{r}) = \mu^*(\mathbf{r}) + \mu^{**}(\mathbf{r}), \tag{10.31}$$

where

$$\mu^*(\mathbf{r}) := \frac{\nabla f(\mathbf{r}) \cdot \nabla\bar{u}(\mathbf{r})}{|\nabla u(\mathbf{r})|^2}, \quad \mu^{**}(\mathbf{r}) := \frac{\nabla \times W(\mathbf{r}) \cdot \nabla\bar{u}(\mathbf{r})}{|\nabla u(\mathbf{r})|^2}. \tag{10.32}$$

Let μ_d denote the recovered shear modulus using the direct inversion formula (10.26):

$$\mu_d(\mathbf{r}) := \omega^2 \rho \frac{u(\mathbf{r})}{\nabla^2 u(\mathbf{r})}.$$

The following theorem provides some characteristics of μ^* and μ_d.

Theorem 10.4.3 Lee *et al.* (2010) *Let $\mu \in C^1(\Omega)$ be the true shear modulus satisfying (10.17) and let u be the non-vanishing displacement in $H^2(\Omega)$. Assume that $\mu^* = \mu_d = \mu$ on the boundary $\partial\Omega$. Then, we have the following:*

(a) μ^* satisfies $\int_\Omega (\mu(\mathbf{r}) - \mu^*(\mathbf{r}))|\nabla u(\mathbf{r})|^2 \, d\mathbf{r} = 0$;
(b) if $\nabla \mu_d \cdot \nabla u = 0$ (or $\nabla \mu_d \perp \nabla u$) in Ω, then $\mu_d = \mu$ in Ω;
(c) if $\nabla u \times \nabla f = 0$ in Ω, then $\mu^* = \mu$ in Ω.

Proof.

(a) From (10.29) and integrating by parts, we have

$$\int_\Omega \nabla \times W(\mathbf{r}) \cdot \nabla \bar{u}(\mathbf{r}) \, d\mathbf{r} = \int_{\partial\Omega} \mathbf{n} \times W \cdot \nabla \bar{u} \, dS = 0$$

where dS is the surface element. Hence, it follows from (10.27) and (10.31) that

$$\begin{aligned} &\int_\Omega [\mu(\mathbf{r}) - \mu^*(\mathbf{r})]|\nabla u(\mathbf{r})|^2 \, d\mathbf{r} \\ &= \int_\Omega \left(\nabla f + \nabla \times W - \frac{\nabla f(\mathbf{r}) \cdot \nabla \bar{u}(\mathbf{r})}{|\nabla u(\mathbf{r})|^2} \nabla u \right) \cdot \nabla \bar{u}(\mathbf{r}) \, d\mathbf{r} = 0. \end{aligned}$$

(b) The reconstructed μ_d satisfies

$$\nabla \cdot (\mu_d(\mathbf{r}) \nabla u(\mathbf{r})) = \mu_d(\mathbf{r}) \nabla^2 u(\mathbf{r}) + \nabla \mu_d(\mathbf{r}) \cdot \nabla u(\mathbf{r}) = \mu_d(\mathbf{r}) \nabla^2 u(\mathbf{r}) = -\rho\omega^2 u(\mathbf{r}).$$

Since the true shear modulus $\mu(\mathbf{r})$ also satisfies $\nabla \cdot (\mu(\mathbf{r}) \nabla u(\mathbf{r})) = -\rho\omega^2 u(\mathbf{r})$, we have

$$\begin{cases} \nabla \cdot ((\mu_d(\mathbf{r}) - \mu(\mathbf{r})) \nabla u(\mathbf{r})) = 0, & \mathbf{r} \in \Omega, \\ (\mu_d(\mathbf{r}) - \mu(\mathbf{r})) \nabla u(\mathbf{r}) \cdot \mathbf{n} = 0, & \mathbf{r} \in \partial\Omega. \end{cases} \tag{10.33}$$

Now, we will prove $\mu = \mu_d$ in Ω based on the results in McLaughlin and Yoon (2004) and Richter (1981). From (10.33), we have

$$\int_{\Omega_{\mathrm{re}}^{\pm}} \Re\{\mu_d(\mathbf{r}) - \mu(\mathbf{r})\} |\nabla u(\mathbf{r})|^2 \, d\mathbf{r} = 0$$

and

$$\int_{\Omega_{\mathrm{im}}^{\pm}} \Im\{\mu_d(\mathbf{r}) - \mu(\mathbf{r})\} |\nabla u(\mathbf{r})|^2 \, d\mathbf{r} = 0,$$

where $\Omega_{\mathrm{re}}^{+} = \{\mathbf{r} \in \Omega : \Re\{\mu_d(\mathbf{r}) - \mu(\mathbf{r})\} > 0)\}$, $\Omega_{\mathrm{re}}^{-} = \{\mathbf{r} \in \Omega : \Re\{\mu_d(\mathbf{r}) - \mu(\mathbf{r})\} < 0)\}$, $\Omega_{\mathrm{im}}^{+} = \{\mathbf{r} \in \Omega : \Im\{\mu_d(\mathbf{r}) - \mu(\mathbf{r})\} > 0)\}$ and $\Omega_{\mathrm{im}}^{-} = \{\mathbf{r} \in \Omega : \Im\{\mu_d(\mathbf{r}) - \mu(\mathbf{r})\} < 0)\}$. Here, $\Re\{v\}$ and $\Im\{v\}$ denote the real and imaginary parts of v, respectively. Since $\nabla u \neq 0$ in any open subset of Ω, $\Omega_{\mathrm{re}}^{\pm} = \emptyset$ and $\Omega_{\mathrm{im}}^{\pm} = \emptyset$. This completes the proof of (b).

(c) By the assumption $\nabla u \times \nabla f = 0$ in Ω, the stress vector $\mu \nabla u$ is parallel to ∇f. From the decomposition $\mu(\mathbf{r})\nabla u(\mathbf{r}) = \nabla f(\mathbf{r}) + \nabla \times W(\mathbf{r})$, there exists a scalar $\beta(\mathbf{r})$ such that $\nabla \times W(\mathbf{r}) = \beta(\mathbf{r})\nabla u(\mathbf{r})$. Thus we have

$$\mu(\mathbf{r})\nabla u(\mathbf{r}) = \nabla f(\mathbf{r}) + \nabla \times W(\mathbf{r}) = \nabla f(\mathbf{r}) + \beta(\mathbf{r})\nabla u(\mathbf{r}).$$

Hence, the divergence-free term $\nabla \times W$ satisfies

$$\nabla \cdot (\nabla \times W(\mathbf{r})) = \nabla \cdot (\beta(\mathbf{r})\nabla u(\mathbf{r})) = \nabla \cdot (\mu(\mathbf{r})\nabla u(\mathbf{r})) - \nabla^2 f(\mathbf{r}) = 0.$$

Therefore, β satisfies

$$\begin{cases} \nabla \cdot (\beta(\mathbf{r})\nabla u(\mathbf{r})) = 0, & \mathbf{r} \in \Omega, \\ \beta(\mathbf{r})\nabla u(\mathbf{r}) \cdot \mathbf{n} = 0, & \mathbf{r} \in \partial\Omega. \end{cases} \tag{10.34}$$

This leads to $\beta = 0$ using the same argument as in (b). This means that $\mu(\mathbf{r})\nabla u(\mathbf{r}) = \nabla f(\mathbf{r})$ in Ω. By taking an inner product of $\nabla \bar{u}$, the shear modulus μ satisfies

$$\mu(\mathbf{r}) = \frac{\nabla f(\mathbf{r}) \cdot \nabla \bar{u}(\mathbf{r})}{|\nabla u(\mathbf{r})|^2} = \mu^*(\mathbf{r}).$$

This completes the proof of (c). □

For the image reconstruction, the first quantity μ^* in (10.31) can be computed explicitly by solving the problem (10.28) from knowledge of $\rho\omega^2 u(\mathbf{r})$ and the Neumann boundary condition. However, we cannot compute μ^{**} directly since the equation contains the unknown quantity $\nabla\mu \times \nabla u$. Hence, we need an iterative procedure to get μ^{**}. We summarize the reconstruction procedure with the iterative method.

1. Get u, one component of the displacement vector, that can be measured by either an MRI or ultrasound device.
2. Given data u, compute the potential f from the PDE in (10.28).
3. Compute the principal component μ^* using (10.32).
4. Compute the residual part μ^{**} using the following iterative procedure.

   - Initial guess $\mu_0^{**} = 0$.
   - For each $n = 1, 2, \ldots$, compute the vector potential W^n that is the solution of the elliptic equation

$$\begin{cases} \nabla^2 W^n = \nabla(\mu^* + \mu_{n-1}^{**}) \times \nabla u, & \text{in } \Omega, \\ \nabla W^n \times \mathbf{n} = 0, & \text{on } \partial\Omega. \end{cases} \tag{10.35}$$

   - Update

$$\mu_n^{**}(\mathbf{r}) = \frac{\nabla \times W^n(\mathbf{r}) \cdot \nabla \bar{u}(\mathbf{r})}{|\nabla u(\mathbf{r})|^2}. \tag{10.36}$$

5. Display $\mu = \mu^* + \mu^{**}$.

Remark 10.4.4 *From Theorem 10.4.3, the reconstructed shear modulus μ_d and μ^* may play complementary non-overlapping roles since they have different characteristics with respect to the direction of wave propagation: μ^* is related to $\nabla\mu \cdot \nabla u$, while μ_d reflects the structure of $\nabla\mu \times \nabla u$. Since evaluation of μ^* is relatively robust against measured noise in the displacement u, μ^* can be used to capture the main feature of μ using a relatively noise-insensitive inversion method. The other μ^{**} is used to correct the residual caused by the quantity $\nabla\mu \times \nabla u$.*

We can test the iterative method via numerical simulations using a rectangular two-dimensional model of $10 \times 10\,\text{cm}^2$ with the origin at its bottom-left corner. Figure 10.4(a) shows the image of the simulated target shear modulus. The target shear modulus included vertical and lateral thin bars with shear storage modulus $\Re(\mu) = 2\text{–}3\,\text{kPa}$ with 0.2 kPa viscosity and background shear storage modulus $\Re(\mu_0) = 1\,\text{kPa}$ with 0.1 kPa viscosity. To get the displacement data, we solved the wave equation in (10.25) with the boundary conditions given by $u(x, y) = 1$ for $y = 10, 0 \le x \le 10$ and $\nabla u \cdot \mathbf{n} = 0$ otherwise. Figure 10.4(b) and (c) show the real and imaginary parts of the simulated displacement image, respectively. Figure 10.5(a) and (b) show the intensity of $|\nabla\mu \cdot \nabla u|$ and $|\nabla\mu \times \nabla u|$, respectively. Since the wave propagates from the top to bottom, Figure 10.5(a) highlights $\partial\mu/\partial y$ and Figure 10.5(b) emphasizes $\partial\mu/\partial x$.

Figure 10.6 shows the reconstructed image of the shear storage moduli μ_d, μ^* and $\mu^5 = \mu^* + \mu_5^{**}$. The quantity μ^* itself is a good approximation of the true μ, as shown

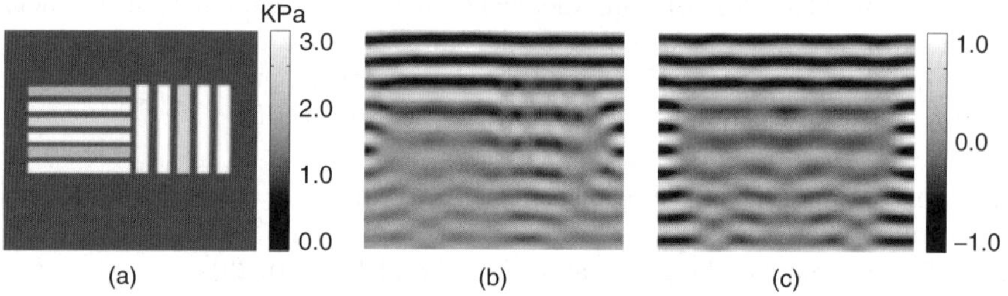

Figure 10.4 Simulation set-up. (a) Target shear modulus image. (b) and (c) Simulated real and imaginary displacement images, respectively. From Lee *et al.* (2010)

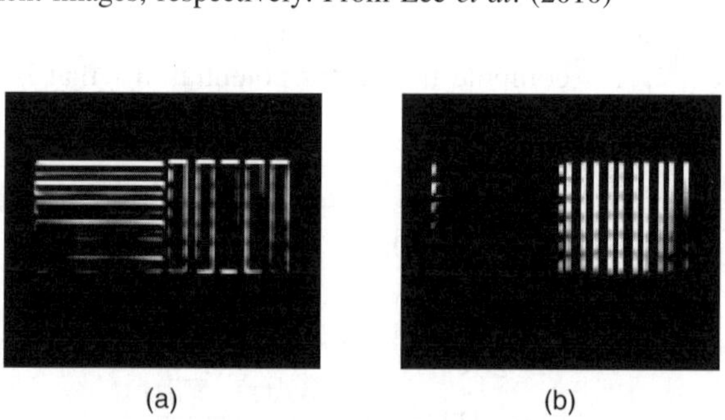

Figure 10.5 (a) Simulated image of $|\nabla\mu \cdot \nabla u|$ and (b) simulated image of $|\nabla\mu \times \nabla u|$. From Lee *et al.* (2010)

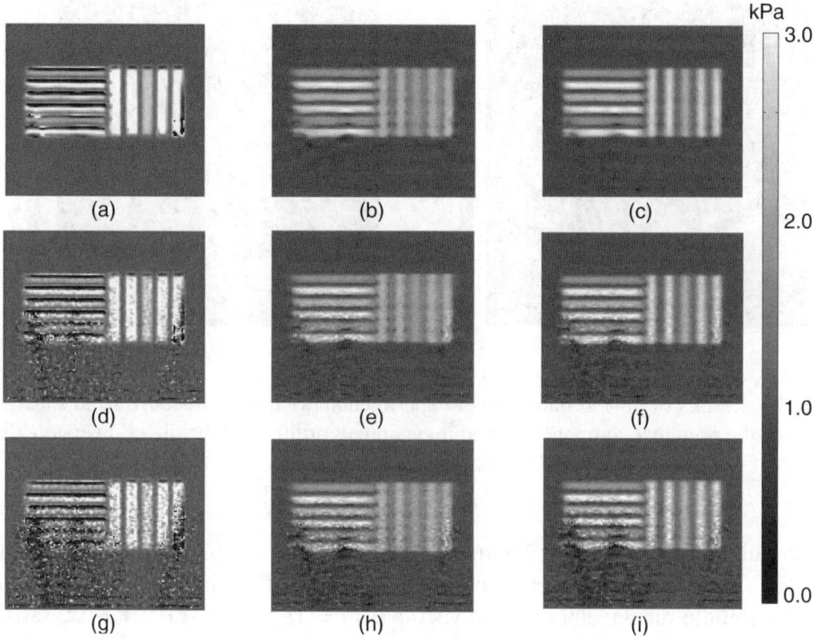

Figure 10.6 Simulation results. The first column (a), (d) and (g) are the real parts of reconstructed shear modulus using the direct inversion method, with added random noise of 0%, 3% and 6% , respectively. The second and third columns are the real parts of the principal and the fifth-updated shear storage modulus images, respectively, using the shear modulus decomposition algorithm corresponding to the first column. From Lee *et al.* (2010)

in Figure 10.6. The reconstructed μ^* was close to the target μ in the left region where $\nabla\mu \times \nabla u \approx 0$ and the updated μ^5 recovered the missed information of μ^* in the region where the intensity of $|\nabla\mu \times \nabla u|$ was high in Figure 10.5(b). The recovered $\Re(\mu_d)$ in Figure 10.6(a), even without added noise, shows some artifacts by neglecting the term $\nabla\mu \cdot \nabla u$ in the left region where the intensity of $\nabla\mu \cdot \nabla u$ was high in Figure 10.4(a). Figure 10.7 shows the results of *in vivo* human liver experiments (Kwon *et al.* 2009).

10.4.3 Anisotropic Elastic Moduli Reconstruction

Most research in MRE has made the assumption that soft tissues exhibit linear and isotropic mechanical properties. However, various biological tissues such as skeletal muscle are known to have anisotropic elastic properties (Chaudhry *et al.* 2008; Dresner *et al.* 2001; Gao *et al.* 1996; Gennisson *et al.* 2003; Heers *et al.* 2003; Humphrey 2003; Kruse *et al.* 2000; McLaughlin *et al.* 2007). Techniques using the scalar equation (10.17) definitely cannot be applied to such anisotropic cases.

The simplest model of the anisotropy would be a transversely isotropic model. Papazoglou *et al.* (2005) adopted the model with transverse isotropy in which the principal axis of symmetry $x_3 = z$ is aligned to be parallel to the muscle fibers. In this model, the full elasticity system (10.16) is reduced to the two-dimensional elasticity system with

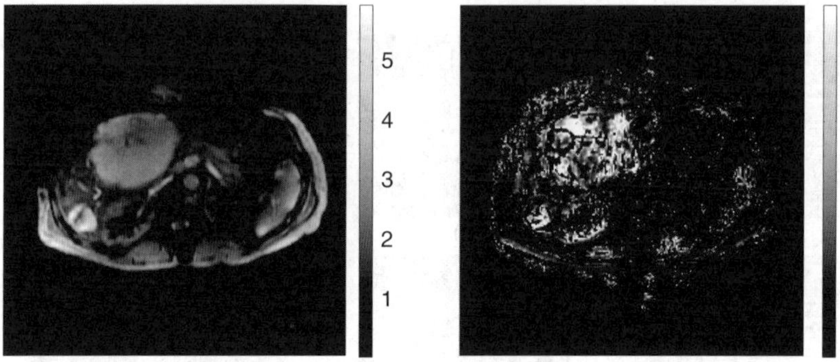

(a) (b)

Figure 10.7 Results of *in vivo* human liver experiments. (a) and (b) Reconstructed shear modulus images using the principal component-based inversion algorithm and the direct inversion algorithm, respectively. From Kwon *et al.* (2009). Reproduced with permission from IEEE

elastic moduli (shear modulus, Young's modulus and Poisson's ratio) depending only on $\mathbf{r} := (x, z)$. To be precise, let Ω be the xz plane of the elastic object to be imaged. The time-harmonic displacement wave vector $\mathbf{u}(\mathbf{r}) = (u_1(\mathbf{r}), u_3(\mathbf{r}))^{\mathrm{T}}$, $\mathbf{r} \in \Omega$, satisfies the following simplified elasticity system:

$$\nabla_{xz} \cdot \begin{bmatrix} 4\mu_{12}\dfrac{\partial u_1}{\partial x} + 2\mu_{12}\dfrac{\partial u_3}{\partial z} & \mu_{13}\dfrac{\partial u_1}{\partial z} + \mu_{13}\dfrac{\partial u_3}{\partial x} \\ \mu_{13}\dfrac{\partial u_1}{\partial z} + \mu_{13}\dfrac{\partial u_3}{\partial x} & 2\mu_{12}\dfrac{\partial u_1}{\partial x} + \mu_\beta\dfrac{\partial u_3}{\partial z} \end{bmatrix} = -\rho\omega^2 \begin{bmatrix} u_1 \\ u_3 \end{bmatrix}, \tag{10.37}$$

where

$$\nabla_{xz} = \left(\frac{\partial}{\partial x}, \frac{\partial}{\partial z}\right),$$

ρ is the tissue density and $\mu_\beta := 4\mu_{12}E_3/E_1$. Here, E_3 and E_1 are Young's modulus with respect to the x_3 and x_1 axes, respectively.

This transversely isotropic model for the elasticity has a potential application of determining the mechanical properties of human skeletal muscle, including the elastic properties of bundles of parallel fibers aligned in one direction. Papazoglou *et al.* (2005) and Sinkus *et al.* (2005a) studied this transversely isotropic model with the assumption of local homogeneity. Papazoglou *et al.* (2005) made a step toward a better description of two-dimensional shear wave patterns to reveal the anisotropy of the muscle fibers. Sinkus *et al.* (2005a) developed a direct reconstruction scheme for imaging the anisotropic property of breast tissues.

10.5 Technical Issues in MRE

MRE is a non-invasive imaging technique for recovering the mechanical properties of tissues. The image reconstruction algorithm in MRE is not perfect yet, owing to difficulties in handling the general elasticity equations. There are many challenging issues in the inverse problem of MRE. For soft tissues, $\nabla \cdot \mathbf{u}$ is very small, whereas the Lamé parameter λ is very large. Thus, even though $\nabla \cdot \mathbf{u} \approx 0$, the total effect of $(\lambda + \mu)\nabla\nabla \cdot \mathbf{u}$ may not be negligible and may produce errors or artifacts in reconstructed images. Hence, it would be desirable to investigate a stable algorithm to reconstruct the Lamé parameters simultaneously without the incompressibility assumption.

There have been increasing demands for anisotropic models, but it is very difficult to achieve a robust reconstruction. If MRE alone is insufficient to provide a robust reconstruction of high-resolution images, one may try combining it with other techniques to get some complementary information.

References

Chaudhry H, Bukiet B and Findley T 2008 Mathematical analysis of applied loads on skeletal muscles during manual therapy. *J. Am. Osteopath Assoc*. **108**, 680–688.

Doyley MM 2012 Topical review: Model-based elastography: a survey of approaches to the inverse elasticity problem. *Phys. Med. Biol*. **57**, R35.

Dresner MA, Rose GH, Rossman PJ, Muthupillai R, Manduca A and Ehman RL 2001 Magnetic resonance elastography of skeletal muscle. *J. Magn. Reson. Imag*. **13**, 269–276.

Gao L, Parker KJ, Lerner RM and Levinson SF 1996 Imaging of the elastic properties of tissue – a review. *Ultrasound Med. Biol*. **22**, 959–977.

Gennisson JL, Catheline S, Chaffai S and Fink M 2003 Transient elastography in anisotropic medium: application to the measurement of slow and fast shear wave speeds in muscles. *J. Acoust. Soc. Am*. **114**, 536–541.

Heers G, Jenkyn T, Dresner MA, Klein MO, Basford JR, Kaufman KR, Ehman RL and An KN 2003 Measurement of muscle activity with magnetic resonance elastography. *Clin. Biomech*. **18**, 537–542.

Humphrey JD 2003 Continuum biomechanics of soft biological tissues. *Proc. R. Soc. Lond. A* **459**, 3–46.

Kruse SA, Smith JA, Lawrence AJ, Dresner MA, Manduca A, Greenleaf JF and Ehman RL 2000 Tissue characterization using magnetic resonance elastography: preliminary results. *Phys. Med. Biol*. **45**, 1579–1590.

Kwon OI, Park C, Nam HS, Woo EJ, Seo JK, Glaser KJ, Manduca A and Ehman RL 2009 Shear modulus decomposition algorithm in magnetic resonance elastography. *IEEE Trans. Med. Imag*. **28**, 1526–1533.

Landau LD and Lifshitz EM 1986 *Theory of Elasticity*, 3rd edn. Pergamon Press, Oxford.

Lee TH, Ahn CY, Kwon OI and Seo JK 2010 A hybrid one-step inversion method for shear modulus imaging using time-harmonic vibrations. *Inv. Prob*. **26**, 085014.

Lerner RM and Parker KJ 1987 Sono-elasticity in ultrasonic tissue characterization and echographic imaging. In *Proc. 7th European Communities Workshop*, Nijmegen, The Netherlands, ed. J Thijssen. European Communities, Luxembourg.

Low RN, Bonekamp S, Motosugi U, Lee JM, Reeder S, Bensamoun SF and Charleux F 2010 *MRE in Clinical Practice: Case Review Compendium*. A GE Healthcare MR publication.

Manduca A, Oliphant TE, Dresner MA, Mahowald JL, Kruse SA, Amromin E, Felmlee JP, Greenleaf JF and Ehman RL 2001 Magnetic resonance elastography: non-invasive mapping of tissue elasticity. *Med. Image Anal*. **5**, 237–254.

Manduca A, Oliphant TE, Lake DS, Dresner MA and Ehman RL 2002 Characterization and evaluation of inversion algorithms for MR elastography. *Proc. SPIE* **4684**, 1180–1185.

Manduca A, Lake DS and Ehman RL 2003 Spatio-temporal directional filtering for improved inversion of MR elastography images. *Med. Image Anal*. **7**, 465–473.

Mariappan YK, Glaser KJ and Ehman RL 2010 Magnetic resonance elastography: a review. *Clin. Anat*., **23**, 497–511.

McLaughlin JR and Renzi D 2006 Shear wave speed recovery in transient elastography and supersonic imaging using propagating fronts. *Inv. Prob*. **22**, 681–706.

McLaughlin JR and Yoon JR 2004 Unique identifiability of elastic parameters from time dependent interior displacement measurement. *Inv. Prob*. **20**, 25–45.

McLaughlin JR, Renzi D and Yoon JR 2007 Anisotropy reconstruction from wave fronts in transversely isotropic acoustic media. *SIAM J. Appl. Math* **68**, 24–42.

Muthupillai R, Lomas DJ, Rossman PJ, Greenleaf JF, Manduca A and Ehman RL 1995 Magnetic resonance elastography by direct visualization of propagating acoustic strain waves. *Science* **269**, 1854–1857.

Oliphant TE, Manduca A, Greenleaf JF and Ehman RL 2000a Direct, fast estimation of complex-valued stiffness for magnetic resonance elastography. International Society for Magnetic Resonance in Medicine, Berkeley, CA.

Oliphant TE, Kinnick RR, Manduca A, Ehman RL and Greenleaf JF 2000b An error analysis of Helmholtz inversion for incompressible shear, vibration elastography with application to filter design for tissue characterization. In *Proc. IEEE Ultrasonics Symp.*, vol. 2, pp. 1795–1798. IEEE, New York.

Oliphant TE, Manduca A, Ehman RL and Greenleaf JF 2001 Complex-valued stiffness reconstruction for magnetic resonance elastography by algebraic inversion of the differential equation. *Magn. Reson. Med.* **45**, 299–310.

Ophir J, Cespedes I, Ponnekanti H, Yazdi Y and Li X 1991 Elastography: a quantitative method for imaging the elasticity of biological tissues. *Ultrason. Imag.* **13**, 111–134.

Papazoglou S, Braun J, Hamhaber U and Sack I 2005 Two-dimensional waveform analysis in MR elastography of skeletal muscles. *Phys. Med. Biol.* **50**, 1313–1325.

Pujol J 2002 *Elastic Wave Propagation and Generation in Seismology*. Cambridge University Press, Cambridge.

Richter GR 1981 An inverse problem for the steady state diffusion equation. *SIAM J. Appl. Math.* **41**, 210–221.

Sack I, Bernarding J and Braun J 2002 Analysis of wave patterns in MR elastography of skeletal muscle using coupled harmonic oscillator simulations. *Magn. Reson. Imag.* **20**, 95–104.

Sinkus R, Lorenzen J, Schrader D, Lorenzen M, Dargatz M and Holz D 2000 High-resolution tensor MR elastography for breast tumour detection. *Phys. Med. Biol.* **45**, 1649–1664.

Sinkus R, Tanter M, Catheline S, Lorenzen J, Kuhl C, Sondermann E and Fink M 2005a Imaging anisotropic and viscous properties of breast tissue by magnetic resonance elastography. *Magn. Reson. Med.* **53**, 372–387.

Sinkus R, Tanter M, Xydeas T, Catheline S, Bercoff J and Fink M 2005b Viscoelastic shear properties of in vivo breast lesions measured by MR elastography. *J. Magn. Reson. Imag.* **23**, 159–165.

Venkatesh SK, Yin M, Glockner JF, Takahashi N, Araoz PA, Talwalkar JA and Ehman RL 2008 MR elastography of liver tumors: preliminary results. *Am. J. Roent.* **190**, 1534–1540.

Uffmann K, Maderwald S, Ajaj W, Galban CG, Mateiescu S, Quick HH and Ladd ME 2004 In vivo elasticity measurements of extremity skeletal muscle with MR elastography. *NMR Biomed.*, **17**, 181–190.

Yin M, Talwalkar JA, Glaser KJ, Manduca A, Grimm RC, Rossman PJ, Fidler JL and Ehman RL 2007 Assessment of hepatic fibrosis with magnetic resonance elastography. *Clin. Gastroenterol. Hepatol.* **5**, 1207–1213.

Further Reading

Braun J, Buntkowsky G, Bernarding J, Tolxdorff T and Sack I 2001 Simulation and analysis of magnetic resonance elastography wave images using coupled harmonic oscillators and Gaussian local frequency estimation. *Magn. Reson. Imag.* **19**, 703–713.

Kallel F and Cespedes I 1995 Determination of elasticity distribution in tissue from spatio-temporal changes in ultrasound signals. *Acoust. Imag.* **22**, 433–443.

Kallel F and Bertrand M 1996 Tissue elasticity reconstruction using linear perturbation method. *IEEE Trans. Med. Imag.* **15**, 299–313.

Lai WM, Rubin D and Krempl E 2010 *Introduction to Continuum Mechanics*, 4th edn. Butterworth-Heinemann, Burlington, MA.

Lerner RM, Huang SR and Parker KJ 1990 Sonoelasticity images derived from ultrasound signals in mechanically vibrated tissues. *Ultrasound Med. Biol.* **16**, 231–239.

Levinson SF, Shinagawa M and Sato T 1995 Sonoelastic determination of human skeletal muscle elasticity. *J. Biomech.* **28**, 1145–1154.

Lin K and McLaughlin J 2009 An error estimate on the direct inversion model in shear stiffness imaging. *Inv. Prob.* **25**, 075003.

Lin K, McLaughlin J and Zhang N 2009 Log-elastographic and non-marching full inversion schemes for shear modulus recovery from single frequency elastographic data. *Inv. Prob.* **25**, 075004.

Nightingale K, Nightingale R, Stutz D and Trahey G 2002 Acoustic radiation force impulse imaging of in vivo vastus medialis muscle under varying isometric load. *Ultrason. Imag.* **24**, 100–108.

Papazoglou S, Rump J, Braun J and Sack I 2006 Shear wave group velocity inversion in MR elastography of human skeletal muscle. *Magn. Reson. Med.* **56**, 489–497.

Parker KJ, Huang SR, Musulin RA and Lerner RM 1990 Tissue response to mechanical vibrations for sonoelasticity imaging. *Ultrasound Med. Biol*. **16**, 241–246.

Qin EC, Sinkus R, Rae C and Bilston LE 2011 Investigating anisotropic elasticity using MR-elastography combined with diffusion tensor imaging: validation using anisotropic and viscoelastic phantoms. In *Proc. Int. Soc. for Magnetic Resonance in Medicine*, Montreal, Canada.

Ringleb SI, Bensamoun SF, Chen Q, Manduca A, An KN and Ehman RL 2007 Applications of magnetic resonance elastography to healthy and pathologic skeletal muscle. *J. Magn. Reson. Imag*. **25**, 301–309.

Sack I, Samani A, Plewes D and Braun J 2003 Simulation of in vivo MR elastography wave patterns of skeletal muscles using a transverse isotropic elasticity model. *Proc. 11th Int. Soc. for Magnetic Resonance in Medicine*.

Salo J and Salomaa MM 2003 Nondiffracting waves in anisotropic media. *Phys. Rev. E* **67**, 056609.

Sinkus R, Daire JL, Beers BEV and Vilgrain V 2010 Elasticity reconstruction: beyond the assumption of local homogeneity. *C. R. Mécanique* **338**, 474–479.

Taylor B, Maris HJ and Elbaum C 1969 Phonon focusing in solids. *Phys. Rev. Lett*. **23**, 416–419.

Yamakoshi Y, Sato J and Sato T 1990 Ultrasonic imaging of internal vibration of soft tissue under forced vibration. *IEEE Trans. Ultrason. Ferroelectr. Freq. Control* **37**, 45–53.

Index

Nonlinear Inverse Problems in Imaging, First Edition. Jin Keun Seo and Eung Je Woo.